Carlos Bravo-Diaz
Editor

Lipid Oxidation in Food and Biological Systems

A Physical Chemistry Perspective

Springer

Editor
Carlos Bravo-Diaz
Facultad de Química, Dpto Química Física
Universidad de Vigo
Vigo, Spain

ISBN 978-3-030-87221-2 ISBN 978-3-030-87222-9 (eBook)
https://doi.org/10.1007/978-3-030-87222-9

This Springer imprint is published by the registered company Springer Nature Switzerland AG.
The registered company address is: Gewerbestrasse 11, 6330 Cham, Switzerland

Lipid Oxidation in Food and Biological Systems

Preface

Lipid peroxidation is a complex radical process that takes place wherever oxygen and unsaturated lipids coexist, including plants, animals, and their derivative products such as foods. It is currently considered one of the main molecular mechanisms involved in the oxidative damage of cell structures and in the deterioration of alimentary oils and fats, involving the uptake of oxygen, the formation and propagation of lipid radicals, the rearrangement of the double bonds on unsaturated lipids, and the eventual production of a variety of breakdown products. Contributions of lipid oxidation to neurodegenerative diseases and cancer are currently being explored because low levels of reactive oxygen and nitrogen species are necessary for signaling to regulate cell functions, but too much leads to lipid oxidation and pathology. Beyond biochemical systems, lipid oxidation is also relevant in the cosmetic and pharmaceutical industries, and it even reaches faraway fields such as art, where it is responsible, at least in part, for the yellowing of paints and varnishes.

Even though the environments and materials where lipid oxidation takes place may be quite different, the components of the starting oils contain a relatively limited range of unsaturated fatty acids, so their degradation should share some commonalities, including kinetic laws and catalytic and solvent effects. Nevertheless, because environments vary considerably, we should expect important differences in the mechanisms of the reactions, undergoing distinctive oxidation and inhibition pathways which may lead to a plethora of different reaction products.

A comprehensive picture of the singularities of the reactions in different systems, and under different conditions, may be assembled by applying holistic molecular approaches. To achieve this, we need more detailed elucidation of the chemical interactions between antioxidants and lipid radicals and how the kinetics of the reactions changes in different systems. Such information can only be obtained by controlling all aspects of the reactions—intrinsic physicochemical properties (H-atom donating ability, polarity, partitioning, etc.), formation of oxidized lipids, and the physical form of the system (solvents, interfaces, solid molecular matrices)—and then tracking the various fates of the antioxidants, oxidizing lipids, and co-oxidation targets, particularly proteins. Information of particular importance will be the chemical fate of antioxidants under processing stress without lipid challenges, the kinetics and mechanisms of the reactions involved, the rates and pathways of reactions of lipid radicals with antioxidants, the nature of the competition of other molecules with oxidizing lipids—antioxidant interactions, the role of interfaces in reactions in bulk oils and in emulsions or in protein-rich food matrices, the role of solvents, etc. All these issues must be carefully addressed.

During the last few decades, a huge effort has been focused on unraveling the mechanism of lipid oxidation, developing analytical methods to detect lipid oxidation products, and minimizing the harmful effects through the addition of exogenous antioxidants. Lipid peroxides and their scission products are commonly measured to evaluate the progress of lipid oxidation in food systems, and several reviews and books on lipid oxidation and on antioxidants are available in the recent literature. Unfortunately, too much attention has been given to qualitative or semi-qualitative interpretations of the experimental observations without careful considerations of the physical-chemistry phenomena underlying the various reactions and processes involved. In this sense, the works carried out, among others, by Ned Porter, Keith Ingold, Etsuo Nikki, and their coworkers should serve as inspiration in rationalizing the complex chemistry embodied in lipid peroxidation and inhibition reactions.

In attempting to cover the existing gaps in the literature, the aim of this book is to collect current insights into the mechanisms of lipid oxidation with a special focus on physical-chemistry approaches, including the role of different reactive oxygen species, mode of actions of synthetic and natural antioxidants, and any interaction between oxidized lipids, antioxidants, and oxidizing systems. The book also aims to provide the latest research on strategies to monitor and control lipid oxidation in complex systems, and some aspects of the role of antioxidants in biology, nutrition, and health are covered. Chemical reactivity in biological matrices has been less investigated, from the physical-chemistry perspective, than that in simpler bulk oils and/or lipid-based emulsions, but they are equally important to get a full picture of the complexity and importance of these reactions in nature.

Bearing this in mind, we were fortunate to have an international group of top-level leading researchers share their knowledge, expertise, and latest research findings. For the sake of convenience, the book has been divided into three sections. The first one describes recent methods to investigate lipid oxidation and its inhibition by antioxidants. The second one is mainly focused on the mechanisms of lipid oxidation and inhibition, and the third one covers distinctive health, biological, and nutritional aspects.

The methodological section starts with the contribution of Marie Hennebele, who focuses on recent analytical advancements in terms of the identification and quantification of multiple key oxidation products and of spatiotemporal resolution of the reactions. In chapter "Determining Antioxidant Distributions in Intact Emulsions by Kinetic Methods: Application of Pseudophase Models", Carlos Bravo-Díaz describes the main features of a unique kinetic method, grounded in thermodynamics, which allows for determining the distributions of antioxidants in intact emulsions without disturbing the existing equilibria. Grzegorz Litwinienko reports, in chapter "Fluorescent Probes for Monitoring Oxidation of Lipids and Assessment of

Antioxidant Activity", on the use of fluorescent probes for monitoring the oxidation of lipids and assessment of antioxidant activity in liposomes and biological organisms. To close this methodological section, Mario Estévez describes in chapter "Lipid Oxidation in Meat Systems: Updated Means of Detection and Innovative Antioxidant Strategies" some of the current problems to detect lipid oxidation status of meats, suggesting some innovative solutions to the problem.

The kinetics and mechanistic aspects of the lipid oxidation inhibition reactions start with chapter "The Underrecognized Role of the Hydroperoxyl ($HOO^{\bullet}$) Radical in Chain Propagation of Lipids and its Implication in Antioxidant Activity" by Ricardo Amoratti, who provides a general but complete overview of the role of hydroperoxyl radicals, discussing on their antioxidant character. In chapter "Location, Orientation and Buoyance Effects of Radical Probes as Studied by EPR", Carolina Aliaga describes how EPR can provide insights into the location, distribution, and orientation effects of radical probes in micellar and emulsified systems. The role of lipid antioxidants is examined by Karen Schaich in chapter "Lipid Antioxidants: More than Just Lipid Radical Quenchers", opening the mind to the readers regarding the complexity of reactions occurring during inhibition of peroxidation by phenols. Thio compounds have long been known as peroxide destroyers, but can also lead to accelerated lipid spoilage, and Olga Kasaikina reports in chapter "Effect of Lipophilic and Hydrophilic Thiols on the Lipid Oxidation" on the role of thiols during lipid oxidation. In chapter "Control of Lipid Oxidation in Oil-in Water Emulsions: Effects of Antioxidant Partitioning and Surfactant Concentration", emulsions are introduced as reaction media, and Sonia Losada-Barreiro describes the effects of the interfacial concentrations of antioxidants and of the surfactant concentrations on antioxidant efficiencies. To complement these studies, Marlene Costa reports in chapter "Effects of Emulsion Droplet Size on the Distribution and Efficiency of Antioxidants" on a long-time unsolved problem in food science, related to the effects of droplet sizes on antioxidant efficiencies. The potential uses of nanoemulsions in the food industry for encapsulating, protecting, and delivering lipid compounds are discussed in chapter "Nanoemulsions as Encapsulation System to Prevent Lipid Oxidation" by Carla Arancibia. Reactivity in bulk oils is covered in chapter "The Effects of Association Colloids on Lipid Autoxidation in Bulk Oils", where Krzysztof Dwiecki describes the formation of association colloids in bulk oils and discusses on the role of water in micelle formation and on their effects on the lipid autoxidation process. Antioxidants are not limited to phenolic compounds, and Grzegorz Litwinienko reports, in chapter "Antioxidant Activity of Edible Isothiocyanates", on several aspects of the chemical and biological activity of isothiocyanates, with an emphasis on the effects of their chemical structures and on the role of side chains. The section ends with the bridging chapter by Mauricio Baptista (Chapter "Photosensitized Lipid Oxidation: Mechanisms and Consequences to Health Sciences") on photosensitized membrane lipid oxidation, illustrating both basic mechanistic principles and advances in important pathologic and therapeutic aspects including UV-induced skin lesions and anticancer and antimicrobial photodynamic therapies.

The section focused on the health, biological, and nutritional aspects starts with chapter "Antioxidants in Dentistry: Oxidative Stress and Periodontal Diseases", where Sofía Fernández-Bravo covers health aspects related to the effects of lipid oxidation on periodontal diseases. Plant sterols can promote cancer cell apoptosis and inhibit their growth and invasiveness through reducing angiogenesis, and Magdalena Rudzinska provides, in chapter "Phytosterols as Functional Compounds and Their Oxidized Derivatives", an overview of studies investigating plant sterols in food products, their degradation during thermo-oxidation, and the formation of oxyphytosterols. chapter "Amino Acids, Amino Acid Derivatives and Peptides as Antioxidants" by Luis Monteiro focuses on the use of structurally modified peptides with antioxidant properties, with improved ability to cross the cell membranes, in the development of new therapeutic strategies to minimize mitochondrial dysfunction. Some biological aspects of the lipid oxidation and inhibition reactions are covered by Pascual García-Pérez in chapter "Plant Phenolics as Dietary Antioxidants: Insights on Their Biosynthesis, Sources, Health-Promoting Effects, Sustainable Production, and Effects on Lipid Oxidation", exploring the use of plant antioxidants as natural antioxidants, describing their biosynthesis, most relevant dietary sources, and approaches to improve their extraction and purification, together with a brief description of their health-enhancing properties. The nutritional aspects of olive oil antioxidants are covered by Fátima Paiva-Martins (Chapter "Olive Oil Phenolic Compounds as Antioxidants in Functional Foods: Description, Sources and Stability") and Santiago Aubourg (Chapter "Nutritional and Preservative Properties of Polyphenol-Rich Olive Oil: Effect on Seafood Processing and Storage"). Fátima Paiva-Martins describes the search for olive oil phenolic antioxidants and alternative sources of these compounds, as well as the understanding of their activity and stability in food matrices. Santiago Aubourg reviews published information related to the use of olive oil in seafood preparation, with an emphasis on thermal processing such as frying and canning, describing the composition and nutritional characteristics of some olive oils and their effects on human health benefits.

We certainly thank all the authors who have contributed to this book and invested their valuable time in preparing their chapters during the long, hard, period caused by the COVID-19 pandemic. We also thank those colleagues who were willing to contribute but who finally, for different reasons, were not able to do it. We hope that these up-to-date, expert reviews in the field of lipid peroxidation and antioxidants will be useful and inspiring to a broad audience, not only scientists, but also to people with a general background in biological and medical sciences.

Vigo, Spain Carlos Bravo-Diaz

Acknowledgments

This book was prepared during the sabbatical leave of Carlos Bravo-Díaz supported by Universidad de Vigo.

Contents

Part I
Methods to Investigate Lipid Oxidation and Its Inhibition

Lipid Oxidation in Food Emulsions: Analytical Challenges and Recent Developments

Sten ten Klooster, Vincent Boerkamp, Eleni Lazaridi, Suyeon Yang, Machi Takeuchi, Claire Berton-Carabin, Karin Schroën, Hans-Gerd Janssen, Heiner Friedrich, Johannes Hohlbein, John van Duynhoven, and Marie Hennebelle

1 Introduction

Lipid oxidation is one of the main causes of quality deterioration in food emulsions, leading to the generation of off-flavors and a concomitant reduction in shelf-life. Over the past decades, the classical radical chain reaction pathway has been questioned with the description of alternative chemical reactions (Schaich 2020) and the added complexity in emulsion systems in which the interfacial composition

Sten ten Klooster, Vincent Boerkamp, Eleni Lazaridi, Suyeon Yang and Machi Takeuchi contributed equally with all other contributors.

S. ten Klooster · K. Schroën
Laboratory of Food Process Engineering, Wageningen University and Research, WG, Wageningen, The Netherlands

V. Boerkamp · E. Lazaridi · M. Hennebelle (✉)
Laboratory of Food Chemistry, Wageningen University and Research, WG, Wageningen, The Netherlands
e-mail: Marie.Hennebelle@wur.nl

S. Yang
Laboratory of Biophysics, Wageningen University and Research, WE, Wageningen, The Netherlands

M. Takeuchi
Laboratory of Physical Chemistry, Department of Chemical Engineering and Chemistry, Eindhoven University of Technology, MB, Eindhoven, The Netherlands

C. Berton-Carabin
Laboratory of Food Process Engineering, Wageningen University and Research, WG, Wageningen, The Netherlands

INRAE, UR BIA, Nantes, France

C. Bravo-Diaz (ed.), *Lipid Oxidation in Food and Biological Systems*,
https://doi.org/10.1007/978-3-030-87222-9_1

and colloidal transport mechanisms may play a key role (Laguerre et al. 2020). The mainstream methods currently used to assess lipid oxidation in food products (e.g. peroxide value, *para*-anisidine, TBARs, hexanal) focus on one (type of) oxidation product and fail to grasp the chemical and multiscale complexity of lipid oxidation reactions. This stands in contrast to the methodological advances that have been made in the biomedical field to assess lipid oxidation. There, the introduction of (oxy)lipidomics and advanced imaging techniques enabled breakthroughs in understanding the role of lipid oxidation in mediating biological processes. The multiscale complexity of food emulsions have also raised the need to get more insights into the spatio-temporal resolution of the reactions (Laguerre et al. 2020) and the role of interfacial composition, including surface-active lipid compounds, which have been overlooked so far.

In this chapter, we will discuss how the introduction of new methodological approaches has opened avenues to identify and quantify multiple key products generated at different stages of lipid oxidation. After shortly introducing the challenges related to lipid extraction prior to analysis, we will focus on the use of magnetic resonance and chromatography to elucidate the reaction pathways in oil-in-water (O/W) emulsions. Finally, we will discuss recent developments in electron and fluorescence microscopy to monitor the structural changes associated with lipid oxidation and the localization of the reaction sites in emulsions.

H.-G. Janssen
Laboratory of Organic Chemistry, Wageningen University and Research, WE, Wageningen, The Netherlands

Unilever Food Innovation Center, WH, Wageningen, The Netherlands

H. Friedrich
Laboratory of Physical Chemistry, Department of Chemical Engineering and Chemistry, Eindhoven University of Technology, MB, Eindhoven, The Netherlands

Center for Multiscale Electron Microscopy, Department of Chemical Engineering and Chemistry, Eindhoven University of Technology, MB, Eindhoven, The Netherlands

J. Hohlbein
Laboratory of Biophysics, Wageningen University and Research, WE, Wageningen, The Netherlands

Microspectroscopy Research Facility, Wageningen University and Research, WE, Wageningen, The Netherlands

J. van Duynhoven
Laboratory of Biophysics, Wageningen University and Research, WE, Wageningen, The Netherlands

Unilever Food Innovation Center, WH, Wageningen, The Netherlands

MAGNetic resonance research FacilitY (MAGNEFY), Wageningen University and Research, WE, Wageningen, The Netherlands

2 Lipid Extraction and Sample Preparation

Before lipid oxidation can be measured, an extraction step is often required to isolate the lipids from the emulsion. The effect of this extraction step on the measured results is not always considered, yet may be of high importance. If the targeted molecules are not fully extracted, their concentration will be underestimated, and some components may even be totally overlooked. In this section, the most frequently used methods for lipid extraction from emulsions are described and potential pros and cons are discussed.

2.1 Common Lipid Extraction Methods

Many different methods may be used to extract lipids (Christie 2012). The most common ones in the lipid oxidation field are the so-called 'Folch' (Folch et al. 1957) and the hexane-isopropanol (HIP) or isooctane-isopropanol (IOIP) methods. The Folch extraction method uses a mixture of chloroform/methanol/water in a ratio of 8: 4:3 (v/v/v) (Folch et al. 1957). First used to isolate brain lipids from animals, it was later adapted to isolate lipids from emulsions (Donnelly et al. 1998). The main disadvantage of this method is that chloroform and methanol are both toxic and chloroform is even categorized as an undesired solvent (Joshi and Adhikari 2019). Moreover, proteins might end up in the bottom layer as well (Radin 1981). HIP or IOIP mixtures in a ratio of 3:2 (v/v) or 3:1 (v/v) have also been frequently used to extract lipids from biological samples (Hara and Radin 1978) and emulsions (Silvestre et al. 2000), using an emulsion-to-solvent ratio of 1:5 (v/v) (Silvestre et al. 2000). Lipid oxidation products can be quantified in the upper phase with colorimetric methods such as the *para*-anisidine value and the iron-based hydroperoxide value. Due to safety considerations, the HIP- and IOIP-based methods are nowadays often preferred over the Folch method (Joshi and Adhikari 2019).

Besides the aforementioned methods, many other solvent extraction methods have been suggested, such as the Bligh & Dyer method, using a chloroform: methanol:water ratio of 2:2:1.8 (v/v/v) (Bligh and Dyer 1959). Furthermore, slight adjustments have been suggested, for example, the use of acetone instead of isopropanol in the IOIP method (Chaiyasit et al. 2000), or modification of the ratios of the solvents used in the Folch, HIP and IOIP methods. It is also possible to extract lipids without the use of solvents. A procedure involving a freeze-thaw step to break the emulsion, followed by a centrifugation step, can be used to collect the lipid phase from mayonnaise (Merkx et al. 2018). This procedure can also be applied to other high internal phase emulsions, including the creamed phase of more diluted emulsions, as obtained after centrifugation. No significant differences in the hydroperoxide content were found between lipid fractions obtained by the freeze-thaw method, or by Folch extraction (whether that also holds for aldehydes was not verified) (Merkx et al. 2018).

Samples obtained from (spray-)drying emulsions, such as milk powders, raise an additional challenge. In these systems, lipids are either entrapped as droplets in a glassy matrix, or present as free fat on the surface (Vignolles et al. 2007). From a final product's perspective, high amounts of free (surface) fat are often undesired, as they can decrease the product quality (e.g., reconstitution properties and stickiness), and it has been argued that free fat oxidizes faster than embedded fat (Vignolles et al. 2007). Therefore, it may be important to extract these lipid fractions separately. This can be achieved by first washing the dried product (e.g., milk powder) with a nonpolar solvent e.g., benzene, petroleum ether, n-hexane, isooctane, toluene, tert-butyl methyl ether, carbon disulphide hexane or ethyl acetate to remove the free fat (Vignolles et al. 2007; Buma 1971) and then by extracting the remaining lipids using one of the previously mentioned solvent-based lipid extraction methods on the rehydrated/redispersed powder (Velasco et al. 2006).

It is clear that there is no consensus about the method that should be used, and reaching a underpinned choice is difficult because actual lipid extraction yields are rarely measured or reported. One should also keep in mind that the differences in polarity of the solvents used (e.g., hexane and isooctane compared to chloroform) may alter the extraction yields, and thus make comparisons between methods not that straightforward.

2.2 *Limitations of Current Lipid Extraction Methods*

In general, solvent-based extractions are labor-intensive, require chemicals that may be toxic or harmful to the environment and it remains questionable whether all lipids (i.e. classes of lipids, lipid oxidation products) are extracted either totally, or in a representative ratio. We ourselves found that lipids can be readily extracted from surfactant-stabilized emulsions, but when proteins are present, extraction yield can be rather low (unpublished data), a fact that is hardly recognized in literature.

It is challenging to achieve a complete extraction of lipids by any method (Christie 2012). Results obtained in our lab (unpublished) showed that the lipid extraction yield for a model O/W emulsion (made with dairy proteins in phosphate buffer 10 mM, pH 7.0 and of 10 wt.% stripped vegetable oil) can be below 50%, in particular when using HIP or IOIP in a 3:1 ratio. Such low extraction yields decrease the 'measured' concentration of lipid oxidation products greatly, leading to underestimation of lipid oxidation. Moreover, extraction yields might be different for the different classes of oxidized species (see below). A phenomenon that seems associated to low extraction yield is the formation of a white, opaque layer between the upper and lower layer when carrying out the extraction, as also observed by Hara and co-workers (Radin 1981; Hara and Radin 1978). The formation of this organogel-like phase is probably caused by incomplete precipitation of the proteins. This effect seems to be enhanced when oxidized proteins are present, and thus affects the lipid extraction yield in an increasing manner during oxidation. To overcome this, protein precipitation can be enhanced, e.g. by adding salt (NaCl).

Obtaining representative samples is a major issue in lipid extractions. Nonpolar lipids, such as triacylglycerols (TAGs), are especially soluble in hydrocarbons such as n-hexane and isooctane, but also in somewhat more polar solvents such as chloroform (Christie 2012). Other lipid components, such as lipid oxidation products and minor components naturally present in oils (e.g., phospholipids, mono- and diglycerides) are expected to be more prominently present at the oil-water interface, where lipid oxidation is thought to be initiated (Laguerre et al. 2020; Berton-Carabin et al. 2014). These components have thus a slightly lower solubility in hydrocarbon solvents (Christie 2012), and some lipid oxidation products, such as 4-hydroxyhex-2-enal, are even appreciably soluble in water (64.9 g/L; logP ~0.5) (Vandemoortele et al. 2020). When extracting these components, it is important to realise that they may partly partition into the polar phase during extraction, especially when using a hydrocarbon as extraction solvent. Therefore, the lipid extraction method has to be carefully chosen according to the system studied and research question that needs to be addressed. The extraction method can be validated by the use of internal standards, and if needed, a correction factor can be applied.

2.3 Considerations Regarding Lipid Extraction from Emulsions

When a lipid extraction step is required prior to lipid oxidation measurement in an O/W emulsion, it should be good practice to optimize the extraction methods and to check the extraction yield (also as a function of time) to prevent possible underestimation of concentrations. Furthermore, the extraction solvents have to be chosen based on the polarity of the targeted molecule(s) to avoid class-selective extraction. It is recommended to measure the partition coefficient of the targeted molecule (s) between the two phases used for extraction in order to reach a rational decision on which solvents to use, and also which correction to apply if needed.

3 Unravelling Lipid Oxidation Pathways by Magnetic Resonance Spectroscopy

Common conventional methods for unravelling lipid oxidation pathways, such as the peroxide value, *para*-anisidine value, or hexanal quantification, only focus on a single type of oxidation products and do not provide the comprehensive overview of lipid oxidation species required for gaining mechanistical insights (Schaich 2020). In this regards, magnetic resonance spectroscopy is a promising candidate to probe both the early stage of lipid oxidation using electron spin resonance (ESR) and the various lipid oxidation species using nuclear magnetic resonance (NMR).

3.1 Early Assessment of Lipid Oxidation Using ESR Spectroscopy

An early assessment of lipid oxidation can be attained by investigation of the radical intermediates formed during the initial stage of oxidation. ESR spectroscopy can specifically and with a high sensitivity detect the unpaired electrons in radical species (Eaton et al. 2010).

Direct ESR measurement of lipid radicals generated during lipid oxidation is difficult due to their short lifetime and subsequent low steady state concentrations. Peroxy-radicals are predominantly detected by ESR as their half-lifetime (~7 s) is much longer as compared to those of other reactive oxygen species ($LO^{\bullet}$, 10^{-7} s; $L^{\bullet}$, 10^{-9} s; $O_2^{\bullet-}$, 10^{-6} s; $^1O_2$, 10^{-7} s; $HO^{\bullet}$, 10^{-10} s) (Pryor 1984). Lipid radicals are therefore commonly detected by adding spin traps, which react with free radicals to form more stable radical adducts. The spin trap N-tert-butyl-α-phenylnitrone (PBN) is liposoluble and is therefore frequently applied, for instance to monitor autooxidation in vegetable oils (Velasco et al. 2005; Jn et al. 2004), and to assess emulsion stability and antioxidant efficacy at an early stage (hours to days) (Merkx et al. 2021a).

A drawback of spin traps is that they need to be homogeneously dispersed in the sample. For wet emulsions, gentle stirring can be used to disperse lipid soluble spin traps (Merkx et al. 2021a). For dried emulsions, more cumbersome procedures need to be followed (Velasco et al. 2021). The moment of spin trap addition also needs to be considered. Several studies added PBN before oxidation (Velasco et al. 2005; Jn et al. 2004; Velasco et al. 2021) to measure the accumulation of lipid radicals over time. While this method allows the discrimination between different sample formulations, it also retards lipid oxidation by scavenging radicals and therefore perturbing oxidation reactions (Merkx et al. 2021a). This issue can be mitigated by adding the spin trap after oxidation and just before ESR measurements (Cui et al. 2017). However, as enough radicals should already be present before PBN addition to obtain a quantifiable signal, this approach is not applicable at a very early stage of oxidation, i.e. during the lag phase (Cui et al. 2017).

An important consideration when using PBN trapping in the ESR approach is the rapid decomposition of PBN-OOL-adducts into alkoxy-radicals ($LO^{\bullet}$), benzaldehyde and 2-methyl-2-nitrosopropane (MNP) (Merkx et al. 2021a; Silvagni et al. 2010) (Fig. 1). The trapping of lipid radicals by PBN and its degradation product, MNP, inhibits lipid hydroperoxide (LOOH) generation, while the generation of $LO^{\bullet}$ promotes the formation of aldehydes.

Overall, the use of PBN-trapping in combination with ESR allows the early assessment of antioxidant strategies in O/W emulsions, but its use for mechanistic studies needs careful consideration (Merkx et al. 2021a). In future studies, the use of amphiphilic spin traps which partition at the interface (Berton-Carabin et al. 2013) could aid in localizing radical reactions in O/W emulsions (Laguerre et al. 2020).

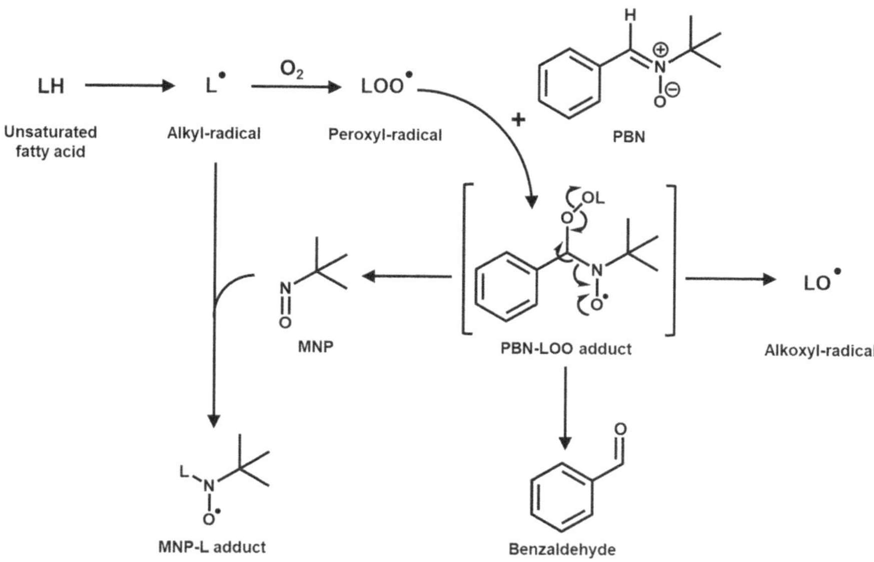

Fig. 1 Proposed mechanism of N-tert-butyl-α-phenylnitrone (PBN) adduct formation and degradation in the presence of oxygen (adapted from (Merkx et al. 2021a))

3.2 Mechanistic Insight on Lipid Oxidation Using NMR Spectroscopy

As previously mentioned, mechanistic insights into lipid oxidation can only be obtained by monitoring a diverse set of lipid oxidation species rather than single species, e.g. peroxide value, *para*-anisidine value, or hexanal (Schaich 2020). Nuclear magnetic resonance (NMR) spectroscopy is well suited to provide such insights as it can detect various lipid oxidation species simultaneously. It is reproducible, easy to automate, high-throughput, does not require derivatization and allows for quantification without the need for authentic standards.

In the past decades, many lipid oxidation species have already been analyzed using ^{1}H NMR (see Martínez-Yusta, Goicoechea (Martínez-Yusta et al. 2014) for a comprehensive review). Table 1 provides an overview of the frequently detected lipid oxidation species and conventional methods to which the results were correlated. Due to susceptibility mismatches between water/oil interfaces, the ^{1}H NMR signals of intact O/W emulsions are significantly broadened. Therefore, a carefully chosen oil extraction is recommended/required prior analysis (see also Sect. 2).

Unfortunately, NMR has an inherently low sensitivity that, when combined with the low analyte concentration, can make accurate quantification of lipid oxidation products difficult (Alexandri et al. 2017). A strategy to increase sensitivity is to use band-selective excitation pulses, which resulted in a limit of quantification (LOQ) of 0.03 mmol/kg oil for both LOOH and aldehydes (Merkx et al. 2018). This LOQ is comparable to the one obtained with more conventional methods, such as iron-based

Table 1 Frequently detected lipid oxidation species using ^{1}H NMR and the corresponding conventional methods. The detected protons are colored in red

Oxidation products		Chemical shift (ppm)	Conventional method	References
Aldehydes	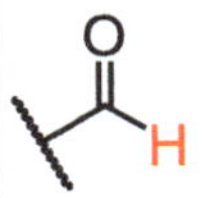	9.8–9.4[a]	GC (hexanal); *p*-anisidine value	Merkx et al. (2018), Skiera et al. (2012a) Goicoechea and Guillen (2010)
Carboxylic acids	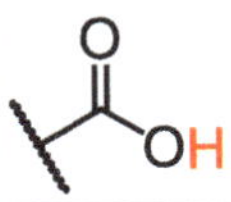	11.4[a]	Acid value; Saponification value	Skiera et al. (2014)
Epoxides	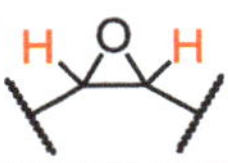	3.2–2.9[b]	HBr-acetic acid titration; Peroxide value	Xia et al. (2015)
Hydroperoxides	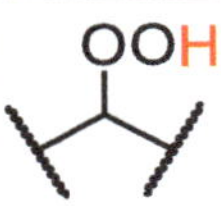	11.2-10[a]	Peroxide value	Merkx et al. (2018), Skiera et al. (2012b)
Hydroxides	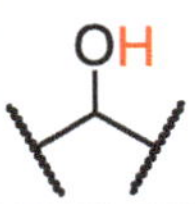	3.6[b]	Polar compounds	Martínez-Yusta and Guillén (2014)

[a]Denotes chemical shifts in $CDCl_3$:DMSO-d_6 (5:1)
[b]In $CDCl_3$

spectrophotometric method (Shantha and Decker 1994; Bou et al. 2008) or peroxide value by titration (AOCS 1998–2008). This approach was recently applied to develop a predictive model of the onset of aldehyde generation based on the early LOOH formation (Merkx et al. 2021b). This strategy is only applicable when the analytes of interest are fully resolved, i.e. they do not overlap with other major lipid signals.

When the analytes of interest are not resolved, 2D-NMR experiments, such as correlation spectroscopy (COSY) or heteronuclear single quantum coherence (HSQC), could be considered to separate and quantify the signals (Alexandri et al. 2017). An example are the epoxides, which partially overlap with bisallylic protons in ^{1}H NMR spectra. While these epoxides have been quantified in epoxidized oil where the bisallylic hydrogen signal is greatly reduced (Xia et al. 2016), it is impossible to accurately quantify and assess their role at the early stage of lipid oxidation using ^{1}H NMR.

NMR spectroscopy is an ever-evolving field. The development and establishment of routine NMR experiments for the simultaneous quantification of multiple lipid oxidation species will help resolving the complexity of lipid oxidation reactions in food emulsions. Moreover, the application of benchtop ^{1}H NMR (Hwang 2015) or ultrafast 2D NMR (Giraudeau and Frydman 2014) for lipid oxidation can be a promising screening technique.

4 Contribution of Minor Lipid Species by Liquid Chromatography Mass Spectrometry

In O/W emulsions, the key role played by the interface in lipid oxidation processes (Laguerre et al. 2020; Berton-Carabin et al. 2014) can be partly explained by the close proximity of compounds of different polarities, chemistries and solubilities. Triacylglycerols (TAGs) are mainly located inside the lipid droplets, but their oxidation might yield more polar species that would have greater affinity for the interface. On the other hand, minor surface-active lipids such as phospholipids (PLs), mono- (MAGs) and di-acylglycerols (DAGs) are initially more abundant at the interface where lipid oxidation is likely to start. To elucidate lipid oxidation pathways in emulsified foods, it is therefore important to gain more insights into the concentration and partitioning of the various lipid substrates and their oxidation products.

4.1 Non-volatile Oxidation Products in Oil-in-Water Emulsions

Although other analytical techniques, including NMR (see Sect. 3.2), have emerged, high performance liquid chromatography (HPLC) remains the most versatile technique for the study of non-volatile lipid oxidation products due to the variety of separation mechanisms available, e.g. based on size, polarity or hydrophobicity. Amongst the various separation modes available, normal-phase HPLC (NPLC) and reversed-phase HPLC (RPLC) have been of particular interest for the analysis of oxidized lipids. NPLC separates lipids based on their polarity resulting from hydroxyl groups and double bonds or other functional groups, and neglects mostly the non-polar lipid chain. RPLC, and more specifically non-aqueous RPLC (NARP-HPLC), is used for the separation of TAGs according to their non-polar moiety.

Besides the variety in separation mechanisms, HPLC can be hyphenated with a wide variety of detectors, such as ultraviolet, fluorescence, refractive index, evaporative light scattering (ELSD) and mass spectrometer (MS). Amongst them, mass spectrometry is the most powerful tool for lipid analysis, as it provides the high sensitivity and resolution required for the study of complex samples. Several MS techniques have been coupled with HPLC with the most widely used for the identification of oxidized TAGs being electrospray ionization (ESI), atmospheric pressure chemical ionization (APCI), and less frequently atmospheric pressure photoionization (APPI). The use of HPLC-MS approaches to set up (oxy)lipidomics platforms has been widely developed in the biomedical field to better understand the role of (enzymatic) lipid oxidation in human health. More recently, similar approaches have been applied to determine the profile of free and esterified oxidized fatty acids in various oils (Richardson et al. 2017; Emami et al. 2020) and human milk (Gan et al. 2020). While these methods require a hydrolysis step prior to analysis, other studies have focused on intact lipid species (Table 2).

NPLC has been applied in combination with post-column addition of sodium iodide and ESI-MS for the identification and quantitation of oxidized TAG in rapeseed and linseed oils (Steenhorst-Slikkerveer et al. 2000) and in combination with APPI-MS for the detection of hydroperoxides, oxo-2½ glycerides, and epoxides during different stages of lipid oxidation in complex food emulsions (Hollebrands and Janssen 2017). NPLC was also applied with prior derivatization of the aldehyde functionality to identify and quantify a variety of aldehydes, and epoxides in fresh and aged mayonnaise samples either by the use of ESI-MS/MS or fluorescence detection (Hollebrands et al. 2018).

NARP-HPLC-APCI-MS has been implemented in the analysis of several oxidized TAGs produced by autoxidation of TAG standards, including hydroperoxides, epoxides (Neff and Byrdwell 1998) and a ketone (Byrdwell and Neff 1999). The protonated molecular TAG ions and their DAG fragments were identified using the APCI technique. NARP has also been combined with ESI- and APCI-MS in one method (dual parallel MS) with addition of ammonium formate for the complimentary identification of oxidized TAGs in canola oil including hydroxy and ketone functional groups (Byrdwell and Neff 2002). Different mobile phase additives have been applied to increase the ionization efficiency of the rather non-polar (oxidized) lipids in ESI, to facilitate the identification of oxidized TAGs. Hydroperoxides and epoxides including their isomers originating from TAG standards were characterized by NARP-HPLC–ESI-MS/MS with post-column addition of ammonium formate (Giuffrida et al. 2004). Their identification was done using the characteristic MS/MS fragmentation pattern of dual neutral losses of hydrogen peroxide and water. Addition of acetic acid into the mobile phase has also been applied for the study of oxidized TAGs, including epoxides, epoxy-epidioxides, epoxy-hydroperoxides, hydroxy-epidioxides, and hydroxy-hydroperoxides derived from oxidized TAG standards (Zeb and Murkovic 2010), and autoxidized camellia oil (Zeb 2012) and autoxidized corn oil (Zeb and Murkovic 2013). Kato et al. used NARP-HPLC-ESI-MS/MS with addition of sodium acetate to identify different hydroperoxides and investigate the oxidation mechanisms (i.e. autoxidation and photo-oxidation) in canola oil (Kato et al. 2018).

Although several HPLC methods have been developed to identify non-volatile oxidation products, most studies have focused so far on oxidized TAGs (Table 2), neglecting the presence of other oxidized minor lipid compounds in dispersed systems.

Table 2 Overview of the liquid chromatography approaches developed to monitor non-volatile lipid oxidation products with their different separation modes, detection methods, and mobile phase additives

Ox-TAGs	Separation	Detection	Additive	Adducts	Samples	References
Epoxy, oxo, hydroperoxy, hydroxy (+oxo-2½ glycerides)	NPLC	ESI-MS	NaI (post-column)	$[M + Na]^+$	Standards, vegetable oils	Steenhorst-Slikkerveer et al. (2000)
Epoxy, oxo, hydroperoxy (+oxo-2½ glycerides)	NPLC	APPI-MS			Mayonnaise	Hollebrands and Janssen (2017)
Epoxy (+oxo-2½ glycerides)	NPLC	ESI-MS/MS, fluorescence		$[M + H]^+$	Mayonnaise	Hollebrands et al. (2018)
Epoxy, oxo, hydroperoxy	NARP	ACPI-MS		$[M + H]^+$	Standards	Neff and Byrdwell (1998), Byrdwell and Neff (1999)
Oxo, hydroperoxy	NARP	Dual-MS	NH_4HCO_2		Canola oil	Byrdwell and Neff (2002)
Epoxy, hydroperoxy	NARP	ESI-MS/MS	NH_4HCO_2 (post-column)	$[M + NH_4]^+$	Standards	Giuffrida et al. (2004)
Epoxy, epoxy-epodioxy, epoxy-hydroperoxy, hydroxy-hydroperoxy, hydroperoxy, epidioxy-hydroperoxy	NARP	ESI-MS/MS	CH_3COOH	$[M + H]^+$, $[M + Na]^+$, $[M + NH_4]^+$	Standards	Zeb and Murkovic (2010)
					Camelia oil	Zeb (2012)
					Corn oil	Zeb and Murkovic (2013)
Hydroperoxy	NARP	ESI-MS/MS	$C_2H_3NaO_2$		Canola oil	Kato et al. (2018)

4.2 *Study of Minor Lipid Species: Missing Gap and Perspectives*

Although most studies have focused so far on the analysis of oxidized TAGs, the key role played by the interface in such O/W emulsions encourages to have a closer look at its composition. The analysis and identification of oxidation products generated from minor surface-active lipid components (e.g. phospholipids, DAG or MAG) will be of importance in order to fully elucidate the lipid oxidation mechanism in emulsified foods. This has been hindered so far by the lack of sensitive methods for detecting and quantifying lipid oxidation products generated from various lipid classes, as well as the limited available methods that are able to monitor multiple oxidation products in the same sample.

One of the main challenges in developing such a method is that non-oxidized DAGs and MAGs interfere with the detection of a variety of oxidized TAG products of similar polarity. Moreover, lipid oxidation products and especially the secondary non-volatile ones are very low in concentration in comparison to non-oxidized TAGs. Multidimensional chromatography set-ups are a promising tool in order to overcome these challenges. By using a combination of different chromatographic techniques and separation modes, they can provide a higher resolution power and peak capacity than conventional one-dimensional chromatography. Recently, Lazaridi et al. (2021) developed a novel on-line comprehensive LC × LC-ELSD method that combines size exclusion chromatography (SEC) and NPLC, as the first and second dimension respectively, for the separation of lipid classes (e.g. polymerized TAG, TAG, DAG and MAG) and their oxidation products (Lazaridi et al. 2021). This analytical approach opens new avenues for the study of lipid oxidation in complex systems such as emulsions and, combined with MS detection, could allow for identification of a wide range of non-volatile oxidation products generated from various lipid species.

5 Assessment of Volatile Lipid Oxidation Products by Headspace Gas Chromatography

In lipid-containing food products, the generation of off-flavors following lipid oxidation is negatively perceived by the consumer and therefore strongly determines the shelf-life of a product. In this section, the assessment of volatile oxidation products by headspace gas chromatography (HS-GC), along with the associated challenges, will be discussed.

5.1 Profiling of Volatile Organic Compounds to Assess Lipid-Derived off-Odors

The rancid off-odor generated by lipid oxidation is caused by volatile organic compounds (VOCs). The most described class of VOCs are aldehydes, of which hexanal is most frequently analyzed. However, other secondary oxidation products also contribute to the headspace profile, including alkanes, alkenes, alcohols, acids, ketones, and epoxides. Small quantities of secondary oxidation products can significantly influence the odor of a food product, especially if they have a low odor detection threshold (Kochhar 1996). Moreover, the interplay of the volatiles with each other and the matrix (e.g. masking) influences the perceived odor (Böttcher et al. 2015). Therefore, to get a complete understanding of the off-odor related to lipid oxidation, an overview of all VOCs is required.

VOCs in (oxidized) oils are often analyzed in the headspace of a food sample. The low boiling point of these VOCs make them suitable for gas chromatography analysis. Because many of the off-odor species have very low odor thresholds, the use of enrichment methods is required to accurately detect and quantify the VOCs generated during lipid oxidation by gas chromatography. Over the years, several enrichment techniques have been developed and applied, including static headspace (SHS), dynamic headspace (DHS), headspace sorptive extraction (HSSE), and solid phase microextraction (SPME). An overview of methods for VOCs analysis of oils has been previously published by Sghaier, Vial (Sghaier et al. 2016).

Amongst the three main methods (SHS, DHS, and SPME), SHS is the simplest, where a sample is drawn directly from the headspace after establishing an equilibrium between the oil and the headspace. The advantage of SHS is its instrumental simplicity and non-selectivity, while its sensitivity is usually low, in the order of mg/kg. In DHS, a continuous stream of inert gas strips the sample from its VOCs that are then concentrated on an absorbent trap before injection in the GC. This increases the sensitivity dramatically to μg/kg, but requires complex instrumentation and adds analyte selectivity due to the trap. Moreover, in aqueous emulsions, the water vapor can cause problems. As a hybrid method, SPME inserts an adsorbent fiber into the headspace to concentrate its VOCs, leading to an intermediate sensitivity and selectivity, but uses simpler instrumentation than DHS.

5.2 Analytical Challenges for the Absolute Quantification of VOCs

The major challenge for headspace analysis is the identification and quantification of VOCs. Relative quantification or semi-quantification can be performed by assuming equal response factors for all VOCs (Multari et al. 2019; Gorji et al. 2019). For absolute quantification, authentic and appropriate standards are required, which should be homogenously dispersed in the sample before analysis without disturbing its structural integrity. The partitioning of VOCs between the sample and headspace is influenced by the matrix (Sghaier et al. 2016). Thus, for external calibration curves, the standards need to be dissolved homogeneously in a comparable matrix as the sample, e.g. an oil stripped from volatiles by deodorization or multiple headspace extraction for the study of bulk oils (Kolb and Ettre 2006). External calibration curves have been applied successfully for liquid emulsions, such as dressings (Let et al. 2007) and model O/W emulsions (Berton et al. 2011; Berton et al. 2012). However, establishing an appropriate matrix and reproducible incorporation of standards is not trivial for the analysis of more complex samples, such as highly viscous emulsions (e.g. mayonnaise) or dried emulsions, for which lipid extraction is often applied prior to analysis (Gorji et al. 2019; Hartvigsen et al. 2000).

During oxidation, structural and chemical changes of the matrix can occur which can affect the partition coefficients of different VOCs and therefore their quantification (Azarbad and Jeleń 2015). The application of an internal standard could mitigate these matrix effects. As the partition coefficients vary amongst VOCs, the reference compounds need to be representative of the VOCs of interest (e.g. similar structure, partitioning coefficient, and volatility), making isotope-labelled standards the ideal choice (Gómez-Cortés et al. 2015). Because of the large range of compounds that need to be quantified, the use of a single internal standard is unfortunately often not sufficient.

By providing an overview of all VOCs generated during lipid oxidation, HS-GC-MS is of interest to correlate the sensory perception to the reactions at play. However, the challenges of quantification should be tackled or at least considered in the interpretation of the results.

6 Probing Lipid Oxidation by Multi-Scale Imaging Techniques (from nm to μm)

The microstructure of food emulsions plays a major role in mediating lipid oxidation. One aspect is the biomolecular arrangement at droplet interfaces (Berton-Carabin et al. 2014). Recently, the role of the continuous phase in mediating transport of reactive intermediates was highlighted (Laguerre et al. 2017). Co-oxidation of proteins and the presence of reactive species in the continuous phase can cause structural changes in food emulsions. To probe these changes with multi-scale imaging, different types of electron and fluorescence microscopy have been implemented (Fig. 2). Whereas transmission electron microscopy (TEM) and scanning electron microscopy (SEM) can cover the nm to μm range, fluorescence microscopy and specifically confocal laser scanning microscopy (CLSM) provides access to the μm to mm range allowing the direct observation of lipid oxidation as well as studying structural changes upon oxidation.

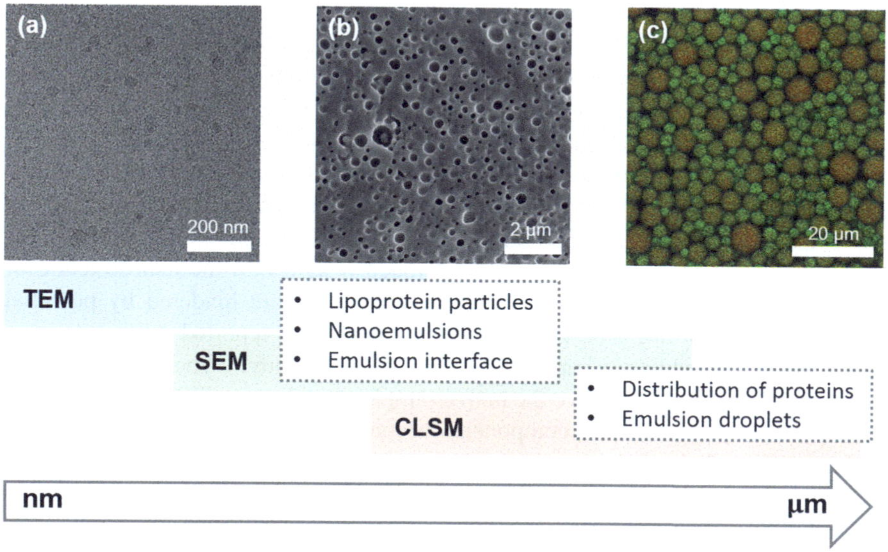

Fig. 2 Summary of microscopy techniques and key features found in food emulsions. (**a**) Cryo-TEM image of low-density lipoproteins (LDLs) from egg yolks. (**b**) SEM image of a dried emulsion. (**c**) CLSM image of lipid droplets in mayonnaise with BODIPY 665/676. Green and orange color shows oxidized and non-oxidized lipid droplets, respectively (Yang et al. 2020)

6.1 Electron Microscopy (EM) Studies

EM refers to a set of techniques that use electrons for imaging of samples with (sub)nanometer resolution. EM can be roughly divided in two general approaches, TEM which generates projection images (Williams and Carter 2009) and SEM which is a surface imaging technique (Goldstein et al. 2018). In conventional dry EM, the sample is placed in a vacuum which requires solid samples that are commonly obtained by drying of dispersions/emulsions on a sample support or by complicated sample preparation procedures.

6.1.1 Dry and Cryo-TEM

Dry TEM is mainly utilized to investigate nanoscale features in food emulsions such as absorbed proteins and interfacial layers. Localization of emulsifier proteins is of importance for understanding physicochemical properties of food emulsions. For example, the structure of proteins like low-density lipoproteins (LDLs) from egg yolk (Anton et al. 2003) and liposome dispersions prepared from a milk fat globule membrane and soy phospholipids have been examined using dry TEM (Thompson et al. 2006). The study of the oil-water interface in emulsions is essential for developing effective barriers that can prevent lipid oxidation (Livney et al. 2017).

These studies, however, require significant sample preparation efforts such as fixation, embedding, sectioning and potentially staining (Ayache et al. 2010). This is on account that the typical size of droplets in food emulsions, roughly 0.1 μm to 10 μm in diameter depending on product type, are too thick to be transmitted by an electron beam. Only when the size of emulsion droplets is in the order of tens to hundreds of nm, size and morphological changes of these droplets can be investigated by dry TEM with negative staining for understanding emulsion stability (Ke et al. 2020) and flocculation during lipid oxidation (Guerra-Rosas et al. 2016).

While dry TEM studies have provided much insight into the nanostructure of food emulsions and their respective components, they are hindered by potential sample preparation artifacts. For instance, Livney et al. reported artifacts arising from agarose embedding (Livney et al. 2017). More recently, cryogenic approaches have been utilized that preserve the native sample state. In cryo-TEM, the sample is rapidly cooled so that liquid components become almost instantaneously vitrified (Schultz 1988). This vitrification can be carried out either by high pressure freezing followed by cryo-(ultra)microtomy for samples up to 100 μm thick to obtain an electron transparent thin section, or directly by plunge vitrification for samples that can be accommodated in liquid layers thinner than 1 μm (Schultz 1988; Michler 2008). Food emulsions such as mayonnaise consist of relatively large droplets (>1 μm in diameter), thus, high pressure freezing and (cryo)microtomy are required. In previous works, plunge vitrification has been mainly employed for studying the native nanostructure of model food emulsions and oil separated food emulsions (Horn et al. 2012; Waninge et al. 2004). The oil fractions of these model food emulsions are typically low ($\leq$10%) due to limitations on sample thickness for plunge vitrification. Mean distances of emulsion droplets were analyzed from cryo-TEM images to interpret interactions of droplets stabilized by two different milk proteins (Horn et al. 2012). It was reported that the longer distances between droplets could potentially indicate repulsive forces between particles, resulting in a better oxidative stability.

6.1.2 Dry and Cryo-SEM

Dry SEM has been mainly used for analyzing the surface microstructure of spray-dried emulsions. Characterization of the porosity of the dried emulsion has been a main interest to evaluate the encapsulation efficiency for oxidation-stable dairy emulsions (Hernandez Sanchez et al. 2016; Azizi et al. 2018). The observed hollow structure of the powder was associated with the spray drying process and the initial composition of the emulsion (Hernandez Sanchez et al. 2016). It was suggested that the holes act as an oxygen reservoir and hence promote oxidation of the encapsulated oil during storage. However, it should be noted that oxygen permeability depends on the type of solid matrix and when high, can limit the impact of internal oxygen on lipid oxidation (Linke et al. 2020). Furthermore, SEM has been used to study the thermal/mechanical stability of liquid emulsions during the spray-drying process. It was found that there were fewer pores visible on the surface of the spray dried emulsions when lipid nanoparticles are incorporated into the emulsions, which indicates enhanced stability of emulsion droplets (Azizi et al. 2018).

Cryo-preservation of food emulsions can also be employed for SEM to observe the native state of liquid emulsions. Here, freeze fracture, which is a technique to crack open a rapidly frozen sample by a cooled knife (Verkleij and Ververgaert 1978), has been utilized to obtain the internal morphology of food emulsions (Miguel et al. 2019; Jiang et al. 2020; Horn et al. 2011; Xiao et al. 2016). The fracture plane usually appears along weak points (Stillwell 2016), which are often emulsion interfaces. Incorporation of zein capsules in an aqueous phase of mayonnaise was observed by cryo-SEM (Miguel et al. 2019). It was shown that the mayonnaise containing the zein capsules had lower amount of primary oxidation products during storage than mayonnaise containing glucose syrup capsules. The interfacial microstructure of the emulsion droplets with added zein-pectin nanocomposites was studied by cryo-SEM (Jiang et al. 2020). A gel-like network of nanoparticles was found between the surfaces of the emulsion droplets, which resulted in a higher thermal stability of the emulsion.

6.2 *Fluorescence Microscopy Studies*

Fluorescence microscopy has been applied in many different research areas allowing for multi-color detection and measurements under ambient conditions. Confocal laser scanning microscopy (CLSM) is a widely used variant reaching a lateral resolution of ~200 nm and featuring effective optical sectioning capabilities by removing the out of focus light with a pinhole placed in the first conjugated image plane. In food research, CLSM is an established technique allowing to study the stability of emulsions and to observe dynamic changes in food structures (Blonk and van Aalst 1993). Furthermore, fluorescence microscopy enables the characterization of oxidation processes in food emulsions (Cardona et al. 2013). In addition to monitoring structural changes of, for example, lipid droplets, fluorophores offer additional readouts allowing to report on polarity, pH, and sensitivity to oxygen and free radicals within the target sample. Fluorophores are either specifically attached to biomolecules, such as fluorescently labelled secondary antibodies targeting primary antibodies attached to proteins of interest, or the fluorophores are directly monitored whilst freely diffusing in the lipid or aqueous phase of food emulsions. Due to the presence of oxidizable proteins in foods, detecting auto-fluorescence provides an additional read-out that can be used to co-localize lipids and proteins (Yang et al. 2020).

In the following section, we will provide a short overview of fluorophores suitable to monitor the structural impact of lipid oxidation processes and to localize oxidation events.

6.2.1 Assessment of Microstructural Impact of Oxidation Using Fluorophores

The fluorophore Nile red is frequently used for staining intracellular lipid droplets and neutral lipids. Its fluorescence emission spectrum ranges from yellow to deep red depending on the polarity of the solvent. In water, Nile red is non-fluorescent, but turns fluorescent when entering lipid-rich or hydrophobic environments. Therefore, Nile Red is suitable for studying the microstructural stability of emulsions that contain lipids. Zafeiri et al. used Nile red with CLSM for studying the potential of solid lipid particles as Pickering stabilizers in O/W emulsions (Zafeiri et al. 2017).

The fluorophore Nile blue, which is structurally similar to Nile red, is also suitable for staining lipid structures. Nile blue was used to elucidate the oil volume fraction impact on the formation and properties of cold-set soy protein isolate (SPI)-stabilized emulsion gels (Yang et al. 2013).

BODIPY 493/503 (4,4-Difluoro-1,3,5,7,8-Pentamethyl-4-Bora-3a,4a-Diaza-s-Indacene) is a stain for neutral lipids and emits a peak at 515 nm. Barden et al. studied the relationship between physical structure and oxidative stability in crackers using CLSM (Barden et al. 2015). For evaluating the distribution of natural antioxidants in poultry meal and extruded kibbles, the BODIPY dye was applied in the work of Ye et al. (Liyun Ye et al. 2019).

Rhodamine B can be added to the water-phase to detect the proteins. In a number of studies, Rhodamine B was used in combination with lipid-staining fluorophores (e.g. Nile Red, Nile Blue, BODIPY) to simultaneously study lipid and protein structures (Zafeiri et al. 2017; Yang et al. 2013; Barden et al. 2015; Liyun Ye et al. 2019).

6.2.2 Localization of Lipid Oxidation Events

BODIPY 665/676 is a peroxidation sensor that exhibits a spectral change in fluorescence after interaction with peroxyl radicals. In a non-oxidized environment, the fluorophore shows fluorescence emission in the red spectral region (650–720 nm) but shifts the emission to green (580–650 nm) in an oxidized environment. Raudsepp et al. used CLSM and BODIPY 665/676 to detect radicals at the single droplet level in O/W emulsions. Due to the long excitation wavelength, BODIPY 665/676 minimizes the potential spectral overlap with the auto-fluorescence of oxidized proteins making it ideal for detecting lipid oxidation in food system comprising both lipids and proteins (Raudsepp et al. 2014). Monitoring the green and red spectral regions of BODIPY 665/676 emission together with auto fluorescence in the blue detection channel from oxidation-dependent proteins, Yang et al. quantitatively mapped lipid and protein oxidation in mayonnaise (Yang et al. 2020). From the spectral changes of BODIPY 665/676, different lipid oxidation rates depending on the presence of antioxidants were analyzed while protein autofluorescence was used to analyze the degree of protein oxidation. BODIPY

665/676 is resistant to photobleaching, does not diffuse into the aqueous phase or between lipid droplets and can be combined with additional fluorescent probes (Raudsepp et al. 2014).

BODIPY 581/591 C-11 is another free radical sensor that changes fluorescence intensity upon oxidation. This fluorophore shifts its peak emission and excitation wavelength from 581/591 nm to 500/510 nm upon oxidation. BODIPY 581/591 C-11 is, however, less sensitive toward peroxyl radicals than BODIPY 665/676, due to its lower number of conjugated dienes (Naguib 2000; Laguerre et al. 2007). Banerjee et al. showed lipid oxidation reactions at oil-water interfaces featuring polyunsaturated fatty acids from cod liver cell with BODIPY 581/591 C-11 and high-resolution, structured-illumination microscopy (SIM) (Banerjee et al. 2017).

DPPP (diphenyl-1-Pyrenylphosphine) allows for direct fluorometric detection of hydroperoxides in lipids. In its unoxidized state, DPPP is non-fluorescent but turns fluorescent after oxidation by peroxides having a peak fluorescence at 380 nm with excitation at 352 nm. Mosca et al. investigated the effect of the emulsifiers' layer structure in olive O/W emulsions by comparing the changes in fluorescence intensity of DPPP upon oxidation (Monica Mosca et al. 2013).

Table 3 provides an overview of the chemical structures of the fluorophores discussed in Sect. 6.2, that can be used to monitor structural changes and localize oxidation events.

6.3 *Overview of Food Microscopy Methods*

Figure 3 shows a summary of the currently applied microscopy techniques with a selection of discussed examples and includes an outlook to potentially relevant techniques.

Besides chemical/cryofixation of samples, imaging of the sample in liquid using environmental SEM (ESEM) (Mathews and Donald 2002) and liquid-phase (LP) (S)-TEM (Wu et al. 2020) are now being developed. With these techniques, the dynamics of a system can be observed, which enable correlative analysis of imaging and scattering. The spatial resolution that can be achieved is limited due to the perturbing gas atmosphere in ESEM and beam damage from the high energy electrons in both ESEM and LP-TEM.

Furthermore, structural analysis EM provides possibilities for measuring the elemental or molecular composition using electron energy loss spectroscopy (EELS), energy-filtered TEM (EFTEM), energy-dispersive X-ray spectroscopy (EDX) (Leijten et al. 2020), and Raman spectroscopy (Lopez-Sanchez et al. 2011). These techniques usually require high electron doses and may not be easily accessible. They are, however, important to consider for future work as they allow to identify component distribution, such as the partitioning of pro-oxidant metals in food emulsions.

Apart from two-dimensional (2D) EM imaging, (cryo) electron tomography (cryo-ET) is an emerging technique that can provide three-dimensional structure of

Table 3 Chemical structures of fluorophores with excitation/emission information. Emission peak can be slightly different in different solvents

Target		*Fluorophores* (excitation/ emission) nm	Chemical structure
Structure	Lipid	*Nile red* (Zafeiri et al. 2017) (λ_{ex} 532/λ_{em} 585)	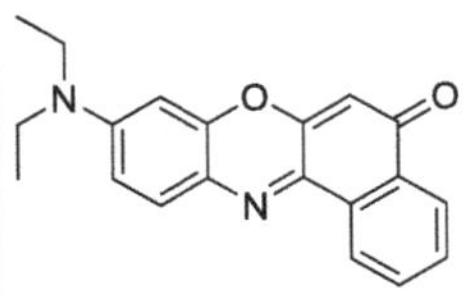
		Nile blue (Yang et al. 2013) (λ_{ex} 633/λ_{em} 660)	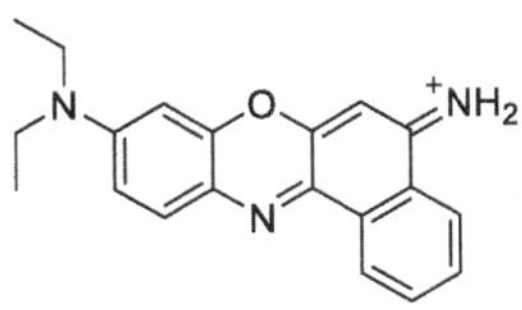
		BODIPY 493/503 (Barden et al. 2015) (λ_{ex} 488/λ_{em} 515)	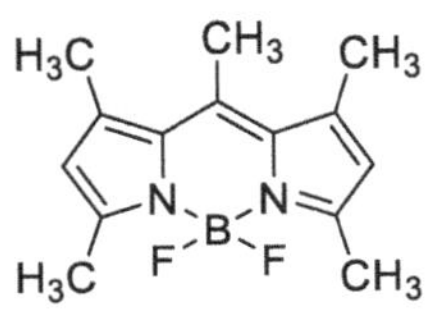
	Protein	*Rhodamine B* (λ_{ex} 553/λ_{em} 627)	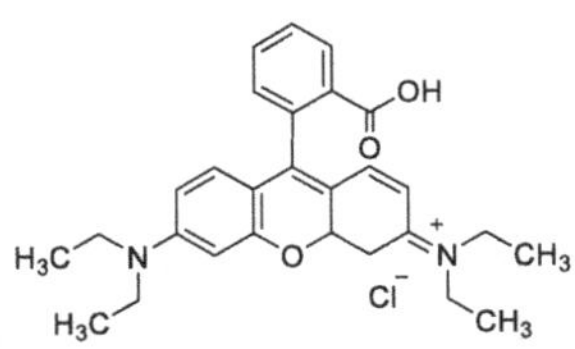
Oxidation mapping		*BODIPY 665/676* (Raudsepp et al. 2014) (λ_{ex} 580 & 675/ λ_{em} 605/635 & 685)	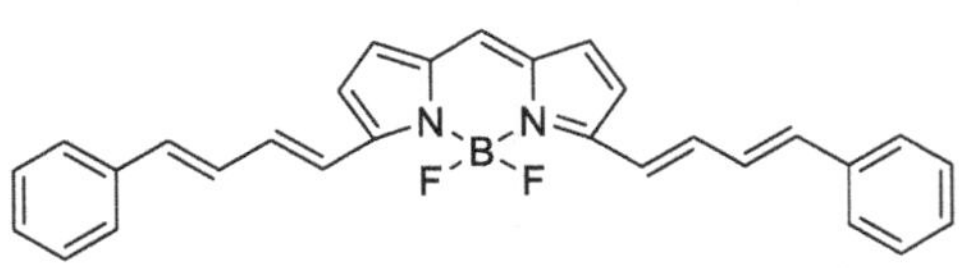
		BODIPY 581/591 C-11 (Banerjee et al. 2017) (λ_{ex} 488 & 561/ λ_{em} 500–545 & 570–640)	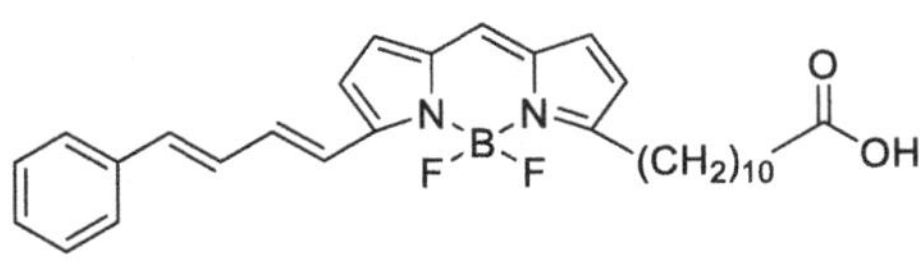
		DPPP (Monica Mosca et al. 2013) (λ_{ex} 352/λ_{em} 380)	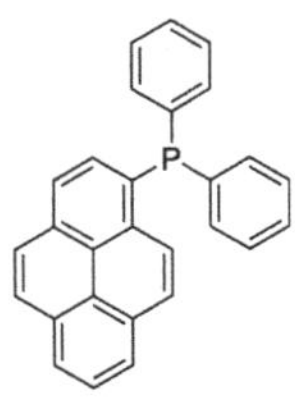

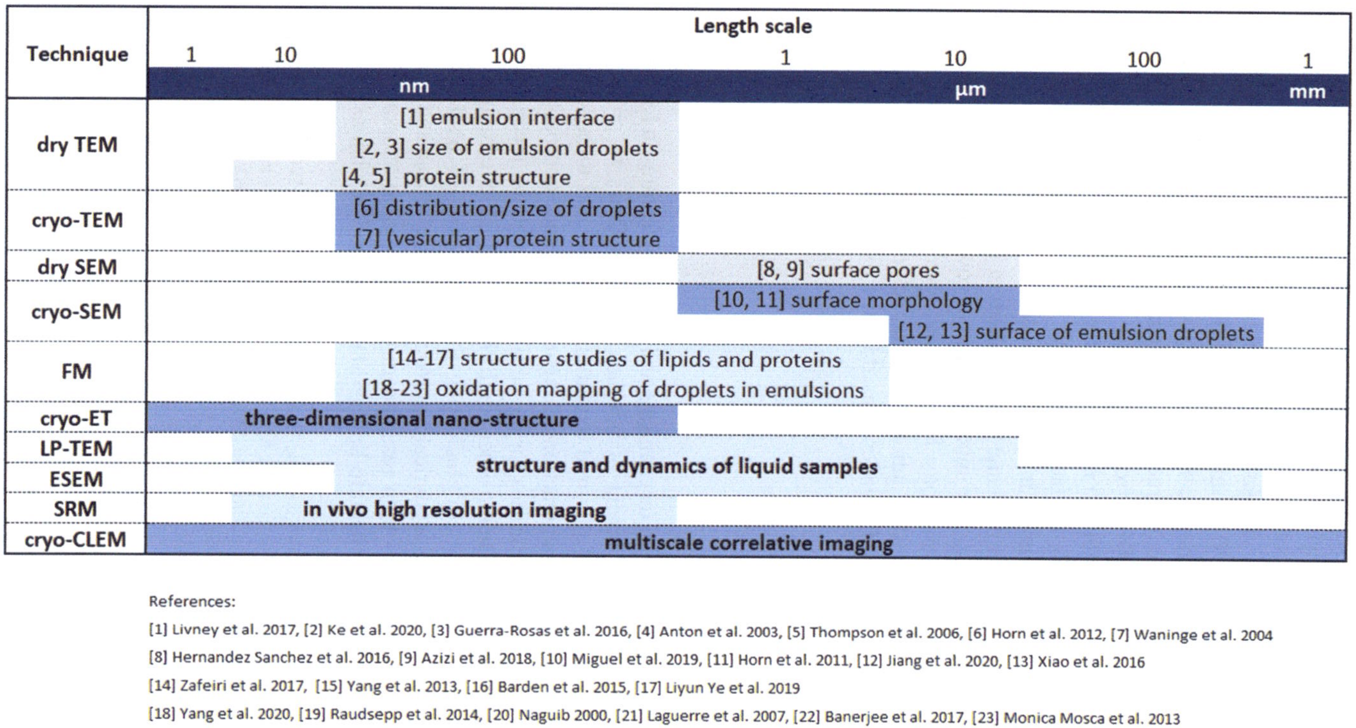

Fig. 3 Overview of microscope techniques, their accessible length scales and discussed examples. Grey, dark blue, and light blue indicate the dry, cryo-, and liquid state measurements, respectively

biological materials by reconstructing 2D images that are taken at multiple tilt angles (Koning et al. 2018). Cryo-ET can be applied to food emulsions, for instance to analyze distribution of emulsifiers at the emulsion interface.

Super-resolution microscopy (SRM) techniques allow for improving the resolution of FL microscopy for assessment of lipid oxidation under ambient room conditions (Banerjee et al. 2017) and further allow for simultaneous localization of lipids and proteins. The higher spatial resolution of SRM will enable to obtain local oxidation information on nanostructures which are smaller than 100 nm such as low-density lipoproteins (LDLs) from egg yolk.

Cryogenic correlative light and electron microscopy (cryo-CLEM), which is the combination of two complementary techniques such as FM/SRM and EM, has currently gained much attention in the life sciences, as it can provide functional and structural information at the same time (de Boer et al. 2015; Wolff et al. 2016). Cryo-CLEM is therefore also a promising technique towards a better understanding of lipid oxidation in food emulsions. For instance, localization of lipid/protein free radicals, which initiates lipid oxidation, could be achieved by SRM and its direct correlation with the structural changes caused by the oxidation can then be studied by EM.

7 Conclusion

The complexity of the reaction mechanisms and structural features in food emulsions has encouraged the development of new analytical methods to provide a more comprehensive overview of the lipid oxidation pathways and to better understand the spatial resolution of these reactions. Magnetic resonance approaches (ESR and NMR) have shown good promises for the early assessment and mechanistic underpinning of lipid oxidation. On the other hand, chromatographic techniques and mass spectrometry methods can help elucidating the complexity of oxidation product mixtures. Combining these analytical approaches with electron and fluorescence microscopy are expected to help researchers with the spatio-temporally resolved characterization of events related to lipid oxidation in food emulsions.

Acknowledgements The authors thank the Dutch Research Council (NWO)in the framework of the Innovation Fund for Chemistry and the Ministry of Economic Affairs in the framework of the TKI/PPS-Toeslagregeling for funding the PhD projects of S. ten Klooster, V. Boerkamp, E. Lazaridi, S. Yang, and M. Takeuchi (Grant number 731.017.301).

References

Alexandri E, Ahmed R, Siddiqui H, Choudhary MI, Tsiafoulis CG, Gerothanassis IP (2017) High resolution NMR spectroscopy as a structural and analytical tool for unsaturated lipids in solution. Molecules 22(10):1663

Anton M, Martinet V, Dalgalarrondo M, Beaumal V, David-Briand E, Rabesona H (2003) Chemical and structural characterisation of low-density lipoproteins purified from hen egg yolk. Food Chem 83(2):175–183

AOCS (1998–2008) American Oil Chemist's Association Official Method Cd 8–53, Cd 8b-90 and Ja 8–87. In: AOCS Press C, IL (ed) Official methods and recommended practices of the AOCS, 5th edn. AOCS, Champaign

Ayache J, Beaunier L, Boumendil J, Ehret G, Laub D (2010) Sample preparation handbook for transmission Electron microscopy, vol XXV, 1st edn. Springer-Verlag, New York, p 250

Azarbad MH, Jeleń H (2015) Determination of hexanal—an indicator of lipid oxidation by static headspace gas chromatography (SHS-GC) in fat-rich food matrices. Food Anal Methods 8(7): 1727–1733

Azizi M, Kierulf A, Connie Lee M, Abbaspourrad A (2018) Improvement of physicochemical properties of encapsulated echium oil using nanostructured lipid carriers. Food Chem 246:448–456

Banerjee C, Westberg M, Breitenbach T, Bregnhoj M, Ogilby PR (2017) Monitoring interfacial lipid oxidation in oil-in-water emulsions using spatially resolved optical techniques. Anal Chem 89(11):6239–6247

Barden L, Vollmer D, Johnson D, Decker E (2015) Impact of iron, chelators, and free fatty acids on lipid oxidation in low-moisture crackers. J Agric Food Chem 63(6):1812–1818

Berton C, Ropers MH, Viau M, Genot C (2011) Contribution of the interfacial layer to the protection of emulsified lipids against oxidation. J Agric Food Chem 59(9):5052–5061

Berton C, Ropers M-H, Bertrand D, Viau M, Genot C (2012) Oxidative stability of oil-in-water emulsions stabilised with protein or surfactant emulsifiers in various oxidation conditions. Food Chem 131(4):1360–1369

Berton-Carabin CC, Coupland JN, Elias RJ (2013) Effect of the lipophilicity of model ingredients on their location and reactivity in emulsions and solid lipid nanoparticles. Colloids Surf A Physicochem Eng Asp 431:9–17

Berton-Carabin CC, Ropers MH, Genot C (2014) Lipid oxidation in oil-in-water emulsions: involvement of the interfacial layer. Compr Rev Food Sci Food Saf 13(5):945–977

Bligh EG, Dyer WJ (1959) A rapid method of total lipid extraction and purification. Can J Biochem Physiol 37(8):911–917

Blonk JCG, van Aalst H (1993) Confocal scanning light microscopy in food research. Appl Technol 26(4):297–311

Böttcher S, Steinhäuser U, Drusch S (2015) Off-flavour masking of secondary lipid oxidation products by pea dextrin. Food Chem 169:492–498

Bou R, Codony R, Tres A, Decker EA, Guardiola F (2008) Determination of hydroperoxides in foods and biological samples by the ferrous oxidation-xylenol orange method: a review of the factors that influence the method's performance. Anal Biochem 377(1):1–15

Buma TJ (1971) Free fat and physical structure of spray-dried whole milk 2. An evaluation of methods for the determination of free-fat content. Neth Milk Dairy 25:42–52

Byrdwell WC, Neff WE (1999) Non-volatile products of triolein produced at frying temperatures characterized using liquid chromatography with online mass spectrometric detection. J Chromatogr A 852(2):417–432

Byrdwell WC, Neff WE (2002) Dual parallel electrospray ionization and atmospheric pressure chemical ionization mass spectrometry (MS), MS/MS and MS/MS/MS for the analysis of triacylglycerols and triacylglycerol oxidation products. Rapid Commun Mass Spectrom 16(4): 300–319

Cardona JAR, Iriart CH, Herrera ML (2013) Applications of confocal laser scanning microscopy (CLSM) in foods. Confocal laser microscopy-principles and applications in medicine, biology, and the food sciences. InTechOpen:203–234

Chaiyasit W, Silvestre M, McClements DJ, Decker EA (2000) Ability of surfactant hydrophobic tail group size to alter lipid oxidation in oil-in-water emulsions. J Agric Food Chem 48(8): 3077–3080

Christie WW, Han X (2012) Chapter 3 – Lipid extraction, storage and sample handling. In: Christie WW, Han X (eds) Lipid analysis, 4th edn. Woodhead Publishing, pp 55–66
Cui L, Lahti PM, Decker EA (2017) Evaluating electron paramagnetic resonance (EPR) to measure lipid oxidation lag phase for shelf-life determination of oils. J Am Oil Chem Soc 94(1):89–97
de Boer P, Hoogenboom JP, Giepmans BN (2015) Correlated light and electron microscopy: ultrastructure lights up! Nat Methods 12(6):503–513
Donnelly JL, Decker EA, McClements DJ (1998) Iron-catalyzed oxidation of menhaden oil as affected by emulsifiers. J Food Sci 63(6):997–1000
Eaton GR, Eaton SS, Barr DP, Weber RT (2010) Quantitative Epr. Springer Science & Business Media, Berlin
Emami S, Zhang Z, Taha AY (2020) Quantitation of Oxylipins in fish and algae oil supplements using optimized hydrolysis procedures and ultra-high performance liquid chromatography coupled to tandem mass-spectrometry. J Agric Food Chem 68(35):9329–9344
Folch J, Lees M, Stanley GHS (1957) A simple method for the isolation and purification of total lipides from animal tissues. J Biol Chem 226(1):497–509
Gan J, Zhang Z, Kurudimov K, German JB, Taha AY (2020) Distribution of free and esterified Oxylipins in cream, cell, and skim fractions of human Milk. Lipids 55(6):661–670
Giraudeau P, Frydman L (2014) Ultrafast 2D NMR: an emerging tool in analytical spectroscopy. Annu Rev Anal Chem 7:129–161
Giuffrida F, Destaillats F, Skibsted LH, Dionisi F (2004) Structural analysis of hydroperoxy- and epoxy-triacylglycerols by liquid chromatography mass spectrometry. Chem Phys Lipids 131(1): 41–49
Goicoechea E, Guillen MD (2010) Analysis of hydroperoxides, aldehydes and epoxides by 1H nuclear magnetic resonance in sunflower oil oxidized at 70 and 100 C. J Agric Food Chem 58(10):6234–6245
Goldstein JI, Newbury DE, Michael JR, Ritchie NWM, Scott JHJ, Joy DC (2018) Scanning Electron microscopy and X-ray microanalysis, 4th edn. Springer-Verlag, New York, p 550
Gómez-Cortés P, Sacks GL, Brenna JT (2015) Quantitative analysis of volatiles in edible oils following accelerated oxidation using broad spectrum isotope standards. Food Chem 174:310–318
Gorji SG, Calingacion M, Smyth HE, Fitzgerald M (2019) Comprehensive profiling of lipid oxidation volatile compounds during storage of mayonnaise. J Food Sci Technol 56(9): 4076–4090
Guerra-Rosas MI, Morales-Castro J, Ochoa-Martínez LA, Salvia-Trujillo L, Martín-Belloso O (2016) Long-term stability of food-grade nanoemulsions from high methoxyl pectin containing essential oils. Food Hydrocoll 52:438–446
Hara A, Radin NS (1978) Lipid extraction of tissues with a low-toxicity solvent. Anal Biochem 90 (1):420–426
Hartvigsen K, Lund P, Hansen LF, Hølmer G (2000) Dynamic headspace gas chromatography/mass spectrometry characterization of volatiles produced in fish oil enriched mayonnaise during storage. J Agric Food Chem 48(10):4858–4867
Hernandez Sanchez MDR, Cuvelier ME, Turchiuli C (2016) Effect of alpha-tocopherol on oxidative stability of oil during spray drying and storage of dried emulsions. Food Res Int 88 (Pt A):32–41
Hollebrands B, Janssen HG (2017) Liquid chromatography–atmospheric pressure photo ionization-mass spectrometry analysis of the nonvolatile precursors of rancid smell in mayonnaise. LC GC Europe 30(9):470–483
Hollebrands B, Varvaki E, Kaal S, Janssen H-G (2018) Selective labeling for the identification and semi-quantification of lipid aldehydes in food products. Anal Bioanal Chem 410(22): 5421–5429
Horn AF, Nielsen NS, Andersen U, Søgaard LH, Horsewell A, Jacobsen C (2011) Oxidative stability of 70% fish oil-in-water emulsions: impact of emulsifiers and pH. Eur J Lipid Sci Technol 113(10):1243–1257

Horn AF, Nielsen NS, Jensen LS, Horsewell A, Jacobsen C (2012) The choice of homogenisation equipment affects lipid oxidation in emulsions. Food Chem 134(2):803–810
Hwang HS (2015) NMR spectroscopy for assessing lipid oxidation. Lipid Technol 27(8):187–189
Jiang Y, Zhu Y, Li F, Du J, Huang Q, Sun-Waterhouse D et al (2020) Antioxidative pectin from hawthorn wine pomace stabilizes and protects Pickering emulsions via forming zein-pectin gel-like shell structure. Int J Biol Macromol 151:193–203
Jn V, Andersen ML, Skibsted LH (2004) Evaluation of oxidative stability of vegetable oils by monitoring the tendency to radical formation. A comparison of electron spin resonance spectroscopy with the Rancimat method and differential scanning calorimetry. Food Chem 85(4): 623–632
Joshi DR, Adhikari N (2019) An overview on common organic solvents and their toxicity. J Pharm Res Int 28(3):1–18
Kato S, Shimizu N, Hanzawa Y, Otoki Y, Ito J, Kimura F et al (2018) Determination of triacylglycerol oxidation mechanisms in canola oil using liquid chromatography–tandem mass spectrometry. NPJ Sci Food 2(1):1
Ke L, Xu Y, Gao G, Wang H, Yu Z, Zhou J et al (2020) Catalase to demulsify oil-in-water fish oil-polysorbate emulsion and affect lipid oxidation. Food Res Int 133:109169
Kochhar SP (1996) Oxidative pathways to the formation of off-flavours. In: Saxby MJ (ed) Food taints and off-Flavours. Springer US, Boston, MA, pp 168–225
Kolb B, Ettre LS (2006) Static headspace-gas chromatography: theory and practice. John Wiley & Sons, Hoboken
Koning RI, Koster AJ, Sharp TH (2018) Advances in cryo-electron tomography for biology and medicine. Ann Anat 217:82–96
Laguerre M, Lecomte J, Villeneuve P (2007) Evaluation of the ability of antioxidants to counteract lipid oxidation: existing methods, new trends and challenges. Prog Lipid Res 46(5):244–282
Laguerre M, Bily A, Roller M, Birtic S (2017) Mass transport phenomena in lipid oxidation and Antioxidation. Annu Rev Food Sci Technol 8:391–411
Laguerre M, Tenon M, Bily A, Birtić S (2020) Toward a spatiotemporal model of oxidation in lipid dispersions: a hypothesis-driven review. Eur J Lipid Sci Technol 122(3):1900209
Lazaridi E, Janssen H-G, Vincken J-P, Pirok B, Hennebelle M (2021) A comprehensive two-dimensional liquid chromatography method for the simultaneous separation of lipid species and their oxidation products. J Chromatogr A 1644:462106
Leijten Z, Wirix MJM, Strauss M, Plitzko JM, de With G, Friedrich H (2020) Low-dose (S)TEM elemental analysis of water and oxygen uptake in beam sensitive materials. Ultramicroscopy 208:112855
Let MB, Jacobsen C, Meyer AS (2007) Lipid oxidation in milk, yoghurt, and salad dressing enriched with neat fish oil or pre-emulsified fish oil. J Agric Food Chem 55(19):7802–7809
Linke A, Linke T, Kohlus R (2020) Contribution of the internal and external oxygen to the oxidation of microencapsulated fish oil. Eur J Lipid Sci Technol 122(8):1900381
Livney YD, Ruimy E, Ye AM, Zhu X, Singh H (2017) A milkfat globule membrane-inspired approach for encapsulation of emulsion oil droplets. Food Hydrocoll 65:121–129
Liyun Ye AP-M, Li J, Joseph P, Nahas R, Michel-Salaun F (2019) Using confocal microscopy to estimate the distribution of natural antioxidants in poultry meal and extruded kibbles. Eur J Lipid Sci Technol 121(9):1800374. https://doi.org/10.1002/ejlt.201800374
Lopez-Sanchez P, Schumm S, Pudney PD, Hazekamp J (2011) Carotene location in processed food samples measured by cryo in-SEM Raman. Analyst 136(18):3694–3697
Martínez-Yusta A, Guillén MD (2014) Deep-frying food in extra virgin olive oil: a study by 1H nuclear magnetic resonance of the influence of food nature on the evolving composition of the frying medium. Food Chem 150:429–437
Martínez-Yusta A, Goicoechea E, Guillén MD (2014) A review of thermo-oxidative degradation of food lipids studied by 1H NMR spectroscopy: influence of degradative conditions and food lipid nature. Compr Rev Food Sci Food Saf 13(5):838–859

Mathews RG, Donald AM (2002) Conditions for imaging emulsions in the environmental scanning electron microscope. Scanning 24(2):75–85

Merkx DW, Hong GS, Ermacora A, Van Duynhoven JP (2018) Rapid quantitative profiling of lipid oxidation products in a food emulsion by 1H NMR. Anal Chem 90(7):4863–4870

Merkx DWH, Plankensteiner L, Yu Y, Wierenga PA, Hennebelle M, Van Duynhoven JPM (2021a) Evaluation of PBN spin-trapped radicals as early markers of lipid oxidation in mayonnaise. Food Chem 334:127578

Merkx DW, Swager A, van Velzen EJ, van Duynhoven JP, Hennebelle M (2021b) Quantitative and predictive modelling of lipid oxidation in mayonnaise. Antioxidants 10(2):287

Michler GH (2008) Preparation of thin sections: (Cryo)ultramicrotomy and (Cryo)microtomy. In: Electron microscopy of polymers, 1st edn. Springer Laboratory, Springer, Berlin, Heidelberg

Miguel GA, Jacobsen C, Prieto C, Kempen PJ, Lagaron JM, Chronakis IS et al (2019) Oxidative stability and physical properties of mayonnaise fortified with zein electrosprayed capsules loaded with fish oil. J Food Eng 263:348–358

Monica Mosca FC, Lopez F, Ceglie A (2013) Role of emulsifier layer, antioxidants and radical initiators in the oxidation of olive oil-in-water emulsions. Food Res Int 50:377–383

Multari S, Marsol-Vall A, Heponiemi P, Suomela J-P, Yang B (2019) Changes in the volatile profile, fatty acid composition and other markers of lipid oxidation of six different vegetable oils during short-term deep-frying. Food Res Int 122:318–329

Naguib YM (2000) Antioxidant activities of astaxanthin and related carotenoids. J Agric Food Chem 48(4):1150–1154

Neff WE, Byrdwell WC (1998) Characterization of model triacylglycerol (triolein, trilinolein and trilinolenin) autoxidation products via high-performance liquid chromatography coupled with atmospheric pressure chemical ionization mass spectrometry. J Chromatogr A 818(2):169–186

Pryor WA (1984) Free radicals in autoxidation and in aging. Free radicals in molecular biology, aging and disease, pp 13–41

Radin NS (1981) Extraction of tissue lipids with a solvent of low toxicity. Methods Enzymol 72.: Elsevier:5–7

Raudsepp P, Bruggemann DA, Andersen ML (2014) Detection of radicals in single droplets of oil-in-water emulsions with the lipophilic fluorescent probe BODIPY(665/676) and confocal laser scanning microscopy. Free Radic Biol Med 70:233–240

Richardson CE, Hennebelle M, Otoki Y, Zamora D, Yang J, Hammock BD et al (2017) Lipidomic analysis of oxidized fatty acids in plant and algae oils. J Agric Food Chem 65(9):1941–1951

Schaich KM (2020) Lipid oxidation: new perspectives on an old reaction. In: Shahidi F (ed) Bailey's industrial oil and fat products, 7th edn. John Wiley & Sons, Hoboken

Schultz P (1988) Cryo-electron microscopy of vitrified specimens. Q Rev Biophys 21(2):129–228

Sghaier L, Vial J, Sassiat P, Thiebaut D, Watiez M, Breton S et al (2016) An overview of recent developments in volatile compounds analysis from edible oils: technique-oriented perspectives. Eur J Lipid Sci Technol 118(12):1853–1879

Shantha NC, Decker EA (1994) Rapid, sensitive, iron-based spectrophotometric methods for determination of peroxide values of food lipids. J AOAC Int 77(2):421–424

Silvagni A, Franco L, Bagno A, Rastrelli F (2010) Thermoinduced lipid oxidation of a culinary oil: a kinetic study of the oxidation products by magnetic resonance spectroscopies. J Phys Chem A 114(37):10059–10065

Silvestre MPC, Chaiyasit W, Brannan RG, McClements DJ, Decker EA (2000) Ability of surfactant headgroup size to alter lipid and antioxidant oxidation in oil-in-water emulsions. J Agric Food Chem 48(6):2057–2061

Skiera C, Steliopoulos P, Kuballa T, Holzgrabe U, Diehl B (2012a) 1 H NMR approach as an alternative to the classical p-anisidine value method. Eur Food Res Technol 235(6):1101–1105

Skiera C, Steliopoulos P, Kuballa T, Holzgrabe U, Diehl B (2012b) 1 H-NMR spectroscopy as a new tool in the assessment of the oxidative state in edible oils. J Am Oil Chem Soc 89(8): 1383–1391

Skiera C, Steliopoulos P, Kuballa T, Diehl B, Holzgrabe U (2014) Determination of free fatty acids in pharmaceutical lipids by 1H NMR and comparison with the classical acid value. J Pharm Biomed Anal 93:43–50
Steenhorst-Slikkerveer L, Louter A, Janssen H-G, Bauer-Plank C (2000) Analysis of nonvolatile lipid oxidation products in vegetable oils by normal-phase high-performance liquid chromatography with mass spectrometric detection. J Am Oil Chem Soc 77(8):837
Stillwell W (2016) Long-range membrane properties. In: An introduction to biological membranes. Elsevier, Amsterdam, pp 221–245
Thompson AK, Haisman D, Singh H (2006) Physical stability of liposomes prepared from milk fat globule membrane and soya phospholipids. J Agric Food Chem 54(17):6390–6397
Vandemoortele A, Simon M, Claes A, De Meulenaer B (2020) Behavior of Hexanal,(E)-Hex-2-enal, 4-Hydroxyhex-2-enal, and 4-Hydroxynon-2-enal in oil-in-water emulsions. J Agric Food Chem 68(41):11568–11577
Velasco J, Andersen ML, Skibsted LH (2005) Electron spin resonance spin trapping for analysis of lipid oxidation in oils: inhibiting effect of the spin trap α-phenyl-N-tert-butylnitrone on lipid oxidation. J Agric Food Chem 53(5):1328–1336
Velasco J, Marmesat S, Dobarganes C, Márquez-Ruiz G (2006) Heterogeneous aspects of lipid oxidation in dried microencapsulated oils. J Agric Food Chem 54(5):1722–1729
Velasco J, Andersen ML, Skibsted LH (2021) ESR spin trapping for in situ detection of radicals involved in the early stages of lipid oxidation of dried microencapsulated oils. Food Chem 341: 128227
Verkleij AJ, Ververgaert PHJT (1978) Freeze-fracture morphology of biological membranes. Biochim Biophys Acta Rev Biomembr 515(3):303–327
Vignolles M-L, Jeantet R, Lopez C, Schuck P (2007) Free fat, surface fat and dairy powders: interactions between process and product. A review. Le Lait 87(3):187–236
Waninge R, Kalda E, Paulsson M, Nylander T, Bergenståhl B (2004) Cryo-TEM of isolated milk fat globule membrane structures in cream. Phys Chem Chem Phys 6(7):1518–1523
Williams DB, Carter CB (2009) Transmission Electron microscopy, 2nd edn. Springer, Boston, MA, p 775
Wolff G, Hagen C, Grunewald K, Kaufmann R (2016) Towards correlative super-resolution fluorescence and electron cryo-microscopy. Biol Cell 108(9):245–258
Wu H, Friedrich H, Patterson JP, Sommerdijk N, de Jonge N (2020) Liquid-phase Electron microscopy for soft matter science and biology. Adv Mater 32(25):2001582
Xia W, Budge SM, Lumsden MD (2015) New 1H NMR-based technique to determine epoxide concentrations in oxidized oil. J Agric Food Chem 63(24):5780–5786
Xia W, Budge SM, Lumsden MD (2016) 1 H-NMR characterization of epoxides derived from polyunsaturated fatty acids. J Am Oil Chem Soc 93(4):467–478
Xiao J, Xa W, Perez Gonzalez AJ, Huang Q (2016) Kafirin nanoparticles-stabilized Pickering emulsions: Microstructure and rheological behavior. Food Hydrocolloids 54:30–39
Yang M, Liu F, Tang C-H (2013) Properties and microstructure of transglutaminase-set soy protein-stabilized emulsion gels. Food Res Int 52:409–418
Yang S, Verhoeff AA, Merkx DWH, van Duynhoven JPM, Hohlbein J (2020) Quantitative spatiotemporal mapping of lipid and protein oxidation in mayonnaise. Antioxidants (Basel) 9(12):1278
Zafeiri I, Smith P, Norton IT, Spyropoulos F (2017) Fabrication, characterisation and stability of oil-in-water emulsions stabilised by solid lipid particles: the role of particle characteristics and emulsion microstructure upon Pickering functionality. Food Funct 8(7):2583–2591
Zeb A (2012) Triacylglycerols composition, oxidation and oxidation compounds in camellia oil using liquid chromatography–mass spectrometry. Chem Phys Lipids 165(5):608–614
Zeb A, Murkovic M (2010) Characterization of the effects of β-carotene on the thermal oxidation of triacylglycerols using HPLC-ESI-MS. Eur J Lipid Sci Technol 112(11):1218–1228
Zeb A, Murkovic M (2013) Determination of thermal oxidation and oxidation products of β-carotene in corn oil triacylglycerols. Food Res Int 50(2):534–544

Determining Antioxidant Distributions in Intact Emulsions by Kinetic Methods: Application of Pseudophase Models

Laurence S. Romsted and Carlos Bravo-Díaz

1 Introduction

Lipophilicity is a fundamental physicochemical parameter or concept for basic science that has a powerful influence in economic areas such as medicine, pharmacy and food technology (Leo 2002; Korinth et al. 2012). The parameter is described quantitatively by the partition constant P_W^O (or log P_W^O), Eq. (1) (see below), which expresses the ratio of monomeric, neutral, solute concentrations in the organic and aqueous phases of a binary system under equilibrium conditions. P_W^O values are usually measured in, or calculated for, different model solvent systems depending on the particular scientific interest (Schulte et al. 1998; Amézqueta et al. 2020; Tetko et al. 2018; Leo et al. 1971).

P_W^O values enhance understanding the properties and behavior of a given compound between oil and aqueous regions in drug discovery, medicinal, environmental and food chemistry and in other chemical industries. For instance, the recovery of endogenous antioxidants that are lost when dissolved in the excess of water required to extract olive oil from olive fruits is of major economic importance in the oil industry (Rodis et al. 2002; de Leonardis 2014). Knowledge of their values is also important for basic science, because the partitioning of solutes between oil and water phases depends on their interactions with the components of the various phases, and provides useful information on the intermolecular forces involved in the transfer process. Thermodynamic information (ΔG, ΔH, ΔS) on the transfer of the

L. S. Romsted
Department of Chemistry and Chemical Biology, Rutgers, The State University of New Jersey, New Brunswick, NJ, USA
e-mail: lromsted@chem.rutgers.edu

C. Bravo-Díaz (✉)
Departamento de Química Física, Facultad de Química, Universidad de Vigo, Vigo, Spain
e-mail: cbravo@uvigo.es

C. Bravo-Diaz (ed.), *Lipid Oxidation in Food and Biological Systems*,
https://doi.org/10.1007/978-3-030-87222-9_2

solute can be obtained from the variation of partition constants with temperature. The distributions and the effective concentrations of the solute in each phase can be estimated once the $P_W{}^O$ values are determined (Meireles et al. 2020; Dar et al. 2017; Raimúndez-Rodríguez et al. 2019; Losada-Barreiro et al. 2015). Finally, $P_W{}^O$ values reflect the balance of all intermolecular interactions (electrostatic, dipole-dipole, and dispersion forces) between a solute and the solvent in the two phases in which it partitions (Leo 2002; Liu et al. 2011; Sangster 1997). Values determined in one solvent (e.g., octanol) cannot be, in general, extrapolated to other solvents and are best determined for each oil and solute of interest (Freiría-Gándara et al. 2018a).

For a variety of reasons, there is great interest in investigating antioxidant partitioning in foods (Frankel 2005; Shahidi 2015; Mozuraityte et al. 2016). For example, a growing number of lipid-based delivery systems (emulsions and nanoemulsions), composed of vegetable (olive, corn, soybean) and omega-3 rich fish oil formulations, are widely used in parenteral nutrition and in the delivery of bioactives (Mcclements 2018; Kumar et al. 2011). But, probably, the most important reason for investigating the partitioning of antioxidants in foods is because adding antioxidants is one of the most practical, and low-cost, strategies available to minimize peroxidation of lipid-based products, and the partitioning of the antioxidant partitioning in the system strongly affects their performance (Frankel 2005; Shahidi 2015; Shahidi and Zhong 2010; Losada-Barreiro and Bravo-Díaz 2017).

Solubilization of antioxidants in the different environments of lipid-based emulsions (oil, interfacial and aqueous regions) results in different physicochemical interactions when compared to homogeneous bulk or biphasic systems. This makes that a large number of physico-chemical effects can be analyzed because a number of parameters may modify the distribution of antioxidants affecting their effective concentration. Among them, the polarities of the various regions of the emulsions, the nature of the emulsion (oil and surfactant types, etc.), and environmental conditions (acidity, temperature, electrolyte concentration, etc.) are of interest in rationalizing the performance of antioxidants (Raimúndez-Rodríguez et al. 2019; Bravo-Díaz et al. 2015; Mitrus et al. 2019).

In a binary oil-water systems in which phases are in equilibrium, the transfer of an antioxidant transfer between oil (O) and aqueous (W) phases, Scheme 1, can be described, on thermodynamic grounds, by the change in the Gibbs free energy of transfer. $\Delta G^{O \rightarrow W}$, defines the ratio of chemical potentials μ_{AO} of antioxidant transfer, $\Delta G^{O \rightarrow W} = \mu_{AO}{}^O - \mu_{AO}{}^W$ (T, P are constants). At equilibrium, $\Delta G^{O \rightarrow W} = 0$ and Eq. (1), relating the Gibbs free energy with $P_W{}^O$, can be derived (Leo et al. 1971). Details on the assumptions and equations involved can be found elsewhere (Leo et al. 1971; Sangster 1997; Silvério et al. 2010).

$$\Delta G^{0,W \rightarrow O} = -RT \ln P_w^O = -RT \ln \frac{(AO_O)}{(AO_W)} \quad (1)$$

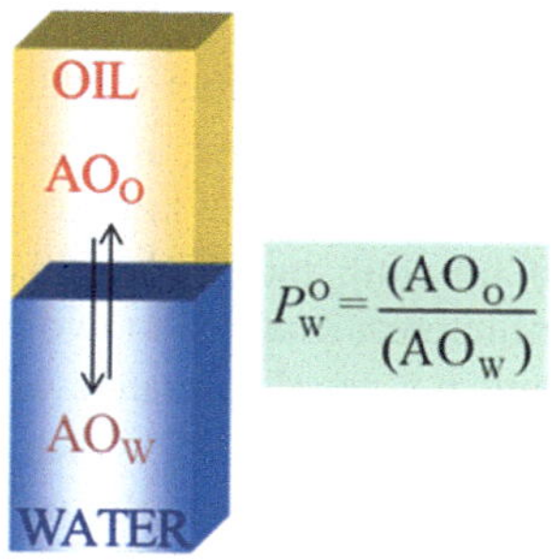

Scheme 1 Partitioning of AO between the oil and aqueous phases of a binary oil-water systems and definition of the corresponding partition constant $P_W{}^O$

The partition constant between n-octanol and water, $P_W{}^{OCT}$, is one of the most widely employed descriptors of the drug hydrophobicity (Leo 2002; Amézqueta et al. 2020; Sangster 1997). Octanol is often chosen as organic solvent because octanol-saturated aqueous buffered solutions possess physicochemical properties comparable to those of water pores of cell membranes, and because octanol saturated with buffer mimics the properties of a lipid layer (Korinth et al. 2012). The transport process from the aqueous (buffered) phase to the octanol phase is used to represent the partitioning of dissolved substance between hydrophilic and lipophilic media. Aqueous buffer/hexane mixtures are also employed to mimic the permeability of compounds through blood–brain barrier, a process that is largely associated with nonspecific interactions (Schulte et al. 1998).

2 Computational and Experimental Methods to Determine Partition Constants

Several software packages (ClogP, AlogPs, ACDLabs, Molinspiration, etc.) provide estimates of $P_O{}^W$ values, but the results are not always consistent because they employ different algorithms in the calculations (Amézqueta et al. 2020). Computational methods include a number of contributions that are presumed to be additive. Each contribution is described as the product of two properties, one characteristic of the solute and the second one stands for the influence of that property on the difference in the standard state chemical potentials in water and the organic solvent (Schulte et al. 1998; Meyer and Maurer 1995; Blokhina et al. 2013). Common properties include the polarity, polarizability, hydrogen bond acceptor and hydrogen bond donating capabilities and molar volume. Values for a variety of molecular structures are achieved by increasing the length and branching of aliphatic chains, changing the position and the number of substituents, as well as by introducing different atoms (O, N, P, S, X^-, etc.) into their chemical structures. All results require experimental confirmation of the computational calculations.

Experimental $P_w{}^O$ values are obtained by direct or indirect methods (Amézqueta et al. 2020). Direct methods include the classical shake-flask, which is illustrated commonly online, and potentiometric methods. More sophisticated and precise methods include high pressure liquid chromatography (HPLC) and ^{1}H NMR (Cumming and Rücker 2017). In the traditional shake-flask method, the compound of interest is added to two partitioning solvents (e.g., water-saturated with n-octanol and n-octanol-saturated with water) (Amézqueta et al. 2020, 1995; Andrés et al. 2015; OECD 1995). Once both solvents are placed in the same flask, the solute of interest is added to the mixture and the flask is shaken to speed equilibration. On standing, the phases separate and the amount of compound in each phase is measured in each phase by any suitable analytical method (often UV-vis spectroscopy) and log $P_O{}^W$ is calculated from their concentration ratio, Eq. (1). The shake-flask method is a deceptively simple but sometimes painful method because a number of parameters such as temperature, acidity, oil-to- water volume ratio, length of the shaking, etc., need to be controlled to minimize errors. Further details of the method can be found elsewhere (Amézqueta et al. 2020; Andrés et al. 2015; OECD 1995).

3 Why Are Partition Constants of Antioxidants in Emulsions Valuable? An Example, a Brief Kinetic Analysis of Lipid Oxidation and Its Inhibition by Antioxidants

Oxidation of lipids has deleterious effects on all lipid-based foods in contact with air (Frankel 2005; Shahidi 2015; Kancheva and Kasaikina 2012; Schaich 2005). Off-odors formed by air oxidation decreases food quality and sometimes produce potentially harmful products. A thorough description of the reaction is quite complex because of the many reactions involved, (Schaich 2005; Schaich et al. 2013) but the basic processes include initiation, propagation and termination steps, Scheme 2 (reactions 1–3). The most important step is the rate-limiting step, reaction 2 (Ross et al. 2003). Antioxidants are added to the system to minimize these radical reactions (Frankel 2005; Shahidi 2015; Ross et al. 2003). The main purpose of antioxidants is to react with the formed peroxyl radicals (reaction 4), donating a H-atom to regenerate the parent lipid (primary antioxidants). However, they may prevent the formation of radicals (secondary antioxidants) in the initiation step, reaction 1. (Shahidi 2015; Foti 2007; Litwinienko et al. 2005). Details of the reactions and on their mechanisms can be found elsewhere (Frankel 2005; Schaich 2005; Ross et al. 2003) as well as in other chapters of the present book.

Molecules (e.g. ArOH) added to inhibit lipid oxidation (reaction 4) must quench efficiently the production of peroxyl radicals to be effective. The rate of the inhibition reaction 4, $r_{inh} = k_{inh}(LOO^{\bullet})(ArOH)$, should be equal to, or higher than, the rate of radical production $r_p = k_p(LOO^{\bullet})(LH)$ (Mitrus et al. 2019; Freiría-Gándara et al. 2018b; Ferreira et al. 2018). k_{inh} values (not shown) depend on factors such as the

Initiation	$LH + R^{\bullet} \xrightarrow{k_i} L^{\bullet} + RH \overset{k_{ox}}{\rightarrow} LOO^{\bullet}$	(1)
Propagation	$LOO^{\bullet} + LH \overset{k_p}{\rightarrow} LOOH + L^{\bullet}$	(2)
Termination	$LOO^{\bullet} + LOO^{\bullet} \overset{k_t}{\rightarrow}$ non-radical products	(3)
Inhibition	$LOO^{\bullet} + ArOH \overset{k_{inh}}{\rightarrow} ArO^{\bullet} + LOOH$	(4)
	$LOO^{\bullet} + ArO^{\bullet} \overset{k_c}{\rightarrow}$ non-radical products	(5)

Scheme 2 Oversimplified mechanism for lipid oxidation and inhibition by antioxidants showing: initiation (1), propagation (2), termination, and inhibition (4, 5) reactions. LH is any unsaturated fatty acid, $R^{\bullet}$ is any radical initiator, $L^{\bullet}$ and $LOO^{\bullet}$ stand for the fatty acid radical and the peroxyl radical, respectively, LOOH stands for a lipid hydroperoxide and ArOH stands for any antioxidant

nature of the antioxidant and the polarity of the solvent and in, emulsions, the interfacial region (Bravo-Díaz et al. 2015; Mitrus et al. 2019; Freiría-Gándara et al. 2018b; Ferreira et al. 2018; Costa et al. 2016).

Substituents on the aromatic rings of antioxidants may have profound effects on the hydrogen atom donating properties of antioxidants (Ross et al. 2003). In general, electron-donating substituents at the *ortho* and/or *para* positions increase the reactivity of monohydroxy phenolic antioxidants. Catechols, 1,2 hydroxybenzene and derivatives, are remarkably reactive antioxidants because of the increased stabilization of the semiquinone radical formed from catechol (Frankel 2005; Ross et al. 2003). Solvents also have an effect on k_{inh} values, which depend on the how the solvation of the reactants and particular mechanism of the inhibition affect the reaction. Further discussion on substituents and solvents affect k_{inh} values can be found elsewhere (Schaich 2005; Ross et al. 2003).

In summary, (1) the *most reactive antioxidant may not be the most efficient antioxidant* because its effective concentration at the reaction site may not be very high, and (2) the rate of the inhibition reaction r_{inh} can be modulated by either modifying k_{inh} and/or optimizing its effective concentration in the interfacial region (Raimúndez-Rodríguez et al. 2019; Bravo-Díaz et al. 2015; Mitrus et al. 2019; Costa et al. 2020a). We note that once an antioxidant and emulsified system is chosen, its efficiency depends exclusively (other factors constant) on its effective concentration in the interfacial region.

Unfortunately, P_W^O values cannot be employed to predict the partitioning of antioxidants in emulsions because the formation of kinetically stable emulsions requires the addition of surfactants to oil-water binary systems, which alters interfacial compositions and properties and creates a new interfacial region between the oil and aqueous regions and new P_W^O values (Raimúndez-Rodríguez et al. 2019; Freiría-Gándara et al. 2018a; Bravo-Díaz et al. 2015). The distribution of neutral antioxidants is controlled by two partition coefficients, Scheme 3, those between the oil-interfacial, P_O^I, and aqueous-interfacial, P_W^I, regions, Eqs. (2–4) (Bravo-Díaz et al. 2015; Romsted and Bravo-Díaz 2013). Treatment of ionic emulsions with ionic

Scheme 3 Equilibria for the distribution of gallic acid (GA, R = H) and its hydrophobic derivatives (R = $(CH)_{n-1}CH_3$, n = 1–12) between the oil, interfacial and aqueous regions of the emulsion. Only GA may ionize significantly at the typical acidities (pH = 2–6) of emulsified foods and, depending on hydrophobicity, the antioxidants may be oil-insoluble ($P_O^I = 0$) or water-insoluble ($P_w^I = 0$). Figure in Scheme 3 is from (Galan et al. 2016), with permission

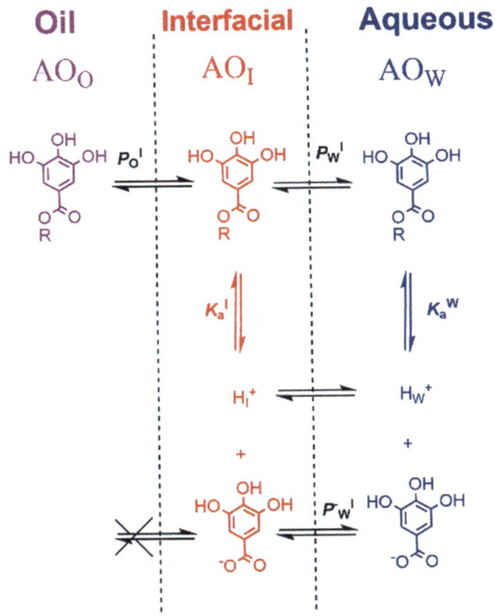

surfactants is a more complex situation and requires treatment of counterion or coion distributions. Details and mathematical treatment can be found elsewhere (Bravo-Díaz et al. 2015; Gao et al. 2013; Gu 2012).

In spite of this, knowledge of P_W^O values is certainly important and very helpful to determine the distribution of antioxidants in emulsions because the ratio P_O^I/P_w^I is numerically equal to P_W^O, Eq. (4), and this facilitates the experimental determination of P_O^I and P_w^I values (Dar et al. 2017; Bravo-Díaz et al. 2015; Romsted and Bravo-Díaz 2013; Gunaseelan et al. 2006).

$$P_W^I = \frac{(AO_I)}{(AO_W)} \tag{2}$$

$$P_O^I = \frac{(AO_I)}{(AO_O)} \tag{3}$$

$$\frac{P_W^I}{P_O^I} = \frac{(AO_I)/(AO_W)}{(AO_I)/(AO_O)} = \frac{(AO_O)}{(AO_W)} = P_W^O \tag{4}$$

3.1 Experimental Methods for Determining Partition Constants in Binary Oil-Water and Emulsified Systems

The experimental determination of the partition constants is much more complex in emulsions than in binary oil-water systems because of the physical impossibility of separating the interfacial region from the oil and aqueous regions without disrupting the existing equilibria. Thus, any attempt to determine the partition coefficients P_O^I and P_w^I needs to be done in the intact emulsions to ensure reproducibility of the results (Bravo-Díaz et al. 2015; Romsted and Bravo-Díaz 2013). Computational methods cannot be employed because the algorithms needed for calculations do not include the properties of food-grade oils and hence P_W^O values cannot be predicted by employing current software packages (Freiría-Gándara et al. 2018a).

In addition, common analytical methods (UV-vis, fluorescence, etc.) widely employed to determining P_W^O values in binary oil-water systems cannot be employed in emulsified systems because they are opaque. Magnetic spectroscopies (EPR, NMR) require the use of radical traps (EPR) and usually require large solute concentrations, making quantification difficult (Oehlke et al. 2008; Heins et al. 2007; Giraudeau and Frydman 2014).

Attempts were made to investigate the partitioning behavior of solutes in emulsions by employing diffusion-controlled electrochemical reactions (Russling 1994). Unfortunately the methods do not completely differentiate between oil-regions and aqueous surfactant environments by applying the solubilization capacities of oil and surfactant of the solutes because they disregarded the potential interactions of the two components within the emulsion. Such interactions are not included in proposed mathematical models (Oehlke et al. 2008; Stöckman and Schwarz 1999) despite significant differences in the solubilization capacities were reported (Stöckman and Schwarz 1999). Studies based on determining the solubilization capability of biphasic systems without modification for emulsions only roughly quantify the proportion of solutes in the different phases (Stöckman and Schwarz 1999).

Romsted and coworkers (Dar et al. 2017; Bravo-Díaz et al. 2015; Romsted and Bravo-Díaz 2013; Gunaseelan et al. 2006) developed, and successfully employed, a conceptually different approach grounded in thermodynamics and inspired by pseudophase kinetic models. The logic of the approach is, partially, based on the relationship embodied in the equation for P_W^O given in Scheme 1 and Eq. (4): if one can determine P_W^O values from the ratio of solute concentrations ($P_W^O = (AO_O)/(AO_W)$), then one should also be able to determine the local molarities of the antioxidant *IF* the partition constant is known. Thus, the method developed by Romsted et al. provides estimates of the partition constant values rather than of the concentrations in each region. For this purpose they used chemical kinetics in conjunction with pseudophase models.

4 Chemical Kinetic Method: A Unique Physical-Organic Approach for Determining Antioxidant Distributions and Interfacial Molarities in Intact Food Emulsions

Chemists have used a variety of association colloids and emulsified systems to understand their effects on chemical reactions for a long time (Bravo-Díaz et al. 2015; Pillai and Shah 1996; Bunton and Romsted 1999; Garti 2003; Lopez-Quintela et al. 2004; Romsted 2012). Kinetic models for interpreting chemical reactivity in colloidal systems were first introduced by Menger and Portnoy (Menger and Portnoy 1967) for bimolecular reactions of neutral reactants in micellar systems. Romsted expanded this approach to include ion exchange effects on reactions between uncharged reactants and reactive ions (Romsted 2012; Bunton et al. 1997; Romsted 1984). A main feature of these methods requires determination of reactant distributions in the system, which is needed to correctly model their chemical reactivity. Because the chemical reactions employed take place in intact systems without separating phases, problems caused by disrupting the existing equilibria are minimized (Bravo-Díaz et al. 2015; Romsted and Bravo-Díaz 2013).

4.1 Main Assumptions of the Pseudophase Model as Applied to Reactions in Emulsions (Dar et al. 2017; Bravo-Díaz et al. 2015; Romsted and Bravo-Díaz 2013)

Modeling chemical reactivity in emulsions does not require knowledge of the sizes of the droplets in the emulsion (discrete structure – separate continuous region duality), but the volumes of the reaction sites need to be known or assumed (Bravo-Díaz et al. 2015).

Bulk and Molecular Organization Kinetically stable emulsions, microemulsions or emulsions are generally formed after bulk mixing of oil, surfactant and water is complete. Except in reverse aggregates, the water region is continuous and the oil and interfacial regions are equal to the totality of the oil and interfacial regions in all droplets regardless of size or shape. The volumes of each region are set equal to the volumes of added oil (O), surfactant (I), and water (W), $V_T = V_O + V_I + V_W$, and are usually expressed as volume fractions, $\Phi = V_{region}/V_{emulsion}$. The interfacial region, which is a layer of surfactant between the oil and water, is composed of surfactant tails, some oil, and hydrated surfactant head groups and components such small amounts of reactants. The structure of the interfacial region is only partially understood, but full characterization is not required to estimate partition and rate constants of reactants.

Reactant Diffusivities and Dynamic Equilibrium Molecules and ions diffuse orders of magnitude faster than rates of most thermal reactions studied in association

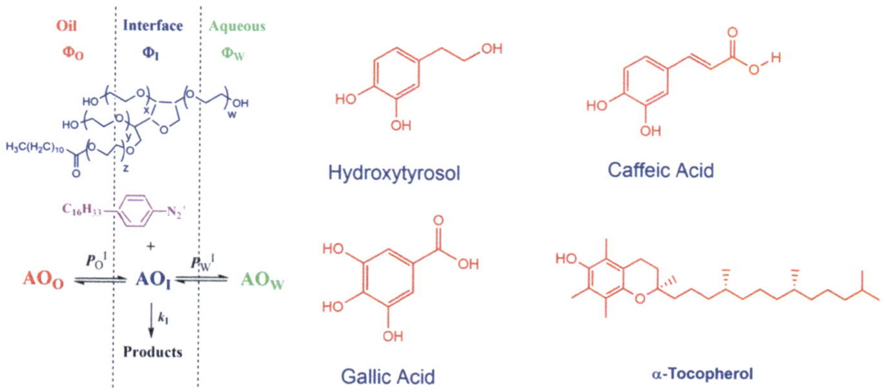

Scheme 4 Left. Two-dimensional section of the interfacial region of an emulsion showing the distribution of an antioxidant and the chemical structures and locations of the chemical probe 16-ArN_2^+ and nonionic surfactant. Right. Chemical structures of some of natural antioxidants whose distribution was determined by using the pseudophase kinetic model (Ferreira et al. 2018; Costa et al. 2020b; Lisete-Torres et al. 2012; Pastoriza-Gallego et al. 2009)

colloids and emulsions (Bravo-Díaz et al. 2017). Their diffusivities may be near their diffusion control limit within and between the emulsions droplets throughout the system. Reactant distributions throughout the total volume of a stirred emulsion are generally in dynamic equilibrium i.e., their concentrations in each region are constant after initial mixing is complete because of fast diffusion. The concentrations of reactants in each region are proportional to their relative solubilities in each region.

To determine antioxidant distributions, the reduction of a specifically designed chemical probe, 4-hexadecylbenzenediazonium, 16-ArN_2^+ (prepared as tetrafluoroborate salts), by antioxidants was employed. The chemical structure of 16-ArN_2^+ is shown in Scheme 4 (see schematic of interfacial region), and was selected for several reasons (Dar et al. 2017; Bravo-Díaz et al. 2015; Romsted and Bravo-Díaz 2013). The most outstanding are: (1) it reacts with virtually all antioxidants of interest and (2) that its reactive group, $-N_2^+$, is located within the interfacial region of the aggregates. Thus, the concentration of the probe, 16-ArN_2^+, in the water and oil regions is negligible and the reaction between the AO and 16-ArN_2^+ occurs only in the interfacial region, Scheme 4.

Emphasis must be done in that the reactive molecules (16-ArN_2^+ and the antioxidant) needs to encounter simultaneously at the same location within the interface, which constitutes a very different problem of that of a bulk phase adjacent to a reactive surface which requires the arrival of molecules to the surface, i.e., those reactions where the (irreversible) adsorption of a molecule onto a solid surface is required. Therefore, the distribution of the antioxidant, and the rate constant of the reaction between 16-ArN_2^+ and the antioxidant, do not depend on the shape or size of the droplets in the emulsion (Bravo-Díaz et al. 2015; Costa et al. 2020a; Costa et al. 2020b).

4.2 Relevant Equations Derived for the Application of the Pseudophase Model

The reaction between 16-ArN_2^+ and the antioxidant is usually carried out under pseudo-first order conditions, i.e., [16-ArN_2^+] < < [AO]. In a typical experiment, the stoichiometric concentration of 16-ArN_2^+ is ca. 10^{-4} M and that of the antioxidant ca. 10^{-3} M. The volume ratios of water and oil to AO are ca. 10^3. Thus, changes in the reactant concentrations with time do not significantly perturb the properties of the (stirred) emulsions. The values of the observed and interfacial bimolecular rate constants, k_{obs} and, k_I, respectively, are for reaction in the totality of the interfacial regions of the emulsions droplets. Because the chemical probe reacts exclusively in the interfacial region of the emulsion, Scheme 3, the rate of disappearance of 16-ArN_2^+ is given by Eq. (5).

$$\begin{aligned} -d[16\text{-ArN}_2{}^+{}_T]/dt &= k_{obs}[16\text{-ArN}_2{}^+{}_T] = k_2[16\text{-ArN}_2{}^+{}_T][AO_T] \\ &= k_I(16\text{-ArN}_2{}^+{}_T)(AO_I)\Phi_I \end{aligned} \tag{5}$$

Bearing in mind the definitions of the partition constants, Eqs. (2–3), the relationships of k_{obs} to both AO distribution ($P_O{}^I$, $P_W{}^I$) and medium (k_I) effects, can be derived (Bravo-Díaz et al. 2015), Eq. (6–8). Equations (7 and 8) are simplifications of the more general Eq. (6) and are employed for oil-insoluble and aqueous-insoluble AOs, respectively.

$$k_{obs} = \frac{[AO_T]k_I P_W^I P_O^I}{\Phi_O P_W^I + \Phi_I P_W^I P_O^I + \Phi_W P_O^I} \tag{6}$$

$$k_{obs} = \frac{[AO_T]k_I P_W^I}{\Phi_I P_W^I + \Phi_W} \tag{7}$$

$$k_{obs} = \frac{[AO_T]k_I P_O^I}{\Phi_I P_O^I + \Phi_O} \tag{8}$$

These equations predict that k_{obs} should decrease asymptotically when Φ_I increases at constant temperature, [AO_T], Φ_O and Φ_W. $P_O{}^I$. Equation (9) has the same form as the reciprocal of Eq. (6), where the parameters "*a*" and "*b*" are given by Eqs. (10 and 11), respectively. Parameters "*a*" and "*b*" are obtained from the fitting process and used to solve for the partition constants (Bravo-Díaz et al. 2015; Romsted and Bravo-Díaz 2013).

$$\frac{1}{k_{obs}} = \frac{1}{a} + \frac{b}{a}\Phi_I \tag{9}$$

$$a = \frac{[\mathrm{AO_T}]k_\mathrm{I}P^\mathrm{I}_\mathrm{W}P^\mathrm{I}_\mathrm{O}(1 + \Phi_\mathrm{W}/\Phi_\mathrm{O})}{P^\mathrm{I}_\mathrm{W} + \Phi_\mathrm{W}/\Phi_\mathrm{O}P^\mathrm{I}_\mathrm{O}} \tag{10}$$

$$b = \frac{P^I_W P^I_O(1 + \Phi_W/\Phi_O)}{P^I_W + \Phi_W/\Phi_O P^I_O} - 1 \tag{11}$$

$$\frac{1}{k_{obs}} = \frac{\Phi_\mathrm{W}}{k_\mathrm{I}[\mathrm{AO_T}]P^\mathrm{I}_\mathrm{W}} + \frac{1}{k_\mathrm{I}[\mathrm{AO_T}]}\Phi_\mathrm{I} \tag{12}$$

$$\frac{1}{k_{obs}} = \frac{\Phi_\mathrm{O}}{k_\mathrm{I}[\mathrm{AO}]_T P^\mathrm{I}_\mathrm{O}} + \frac{1}{k_\mathrm{I}[\mathrm{AO}]_\mathrm{T}}\Phi_\mathrm{I} \tag{13}$$

Equations (12 and 13) are the reciprocals of Eqs. (7 and 8), respectively, and they predict that plots of $1/k_{\mathrm{obs}}$ *vs* Φ_I should be linear with positive intercepts, and the values of the partition constants are obtained from the ratio between the slope and the intercept (Bravo-Díaz et al. 2015; Romsted and Bravo-Díaz 2013).

4.3 Determining $\mathbf{k}_{obs}$ *Values in Opaque Emulsions: Monitoring the Reaction Between 16-ArN*$_2^+$ *and AO by Employing Electrochemical and Azo Dye Derivatization Methods*

Because emulsions are opaque, spectroscopic methods commonly employed to monitor chemical reactions in solution cannot be employed. Two special protocols have been developed. One exploits the electrochemical properties of 16-ArN$_2^+$ by monitoring the variation in its concentration with time by using suitable electroanalytical techniques such as linear sweep voltammetry (LSV) (Bravo-Díaz et al. 2015; Romsted and Bravo-Díaz 2013; Gunaseelan et al. 2006). The second method exploits the rapid reaction of the chemical probe, 16-ArN$_2^+$ ions, with a suitable excess of the coupling agent such as N-(1-naphthyl)ethylenediamine dihydrochloride, NED, yielding an stable azo dye, Scheme 5. The final absorbance at the end of the reaction can be monitored spectrometrically at $\lambda = 572$ nm after dilution with a 50:50 (v:v) BuOH:EtOH mixture that makes an optically transparent, homogeneous solution. Typical $t_{1/2}$ values for the trapping reaction are less than 10 seconds, much faster than the reaction with the AOs, which have $t_{1/2}$ values in the range of 3–30 min.

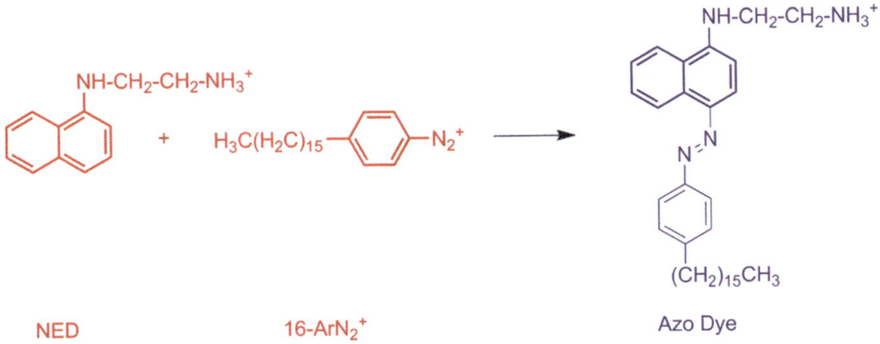

Scheme 5 Reaction between the trapping agent *N*-(1-Naphthyl)ethylenediamine, NED, with 16-ArN_2^+ to yield a stable purple azo dye

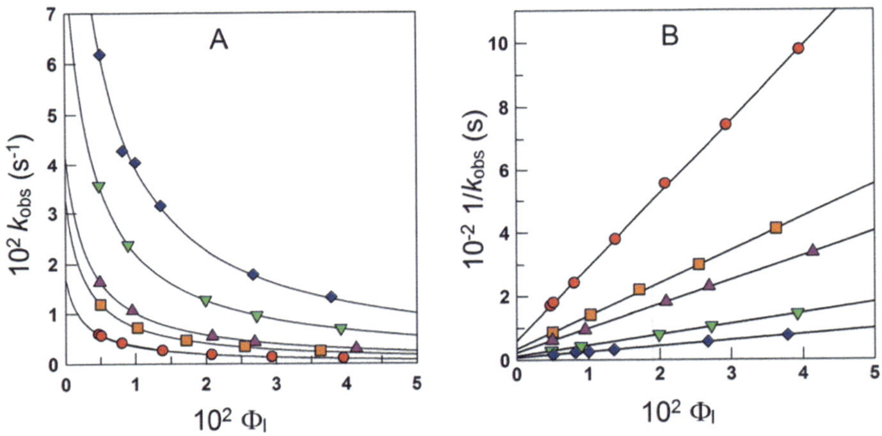

Fig. 1 Variation in k_{obs} and $1/k_{obs}$ values with Φ_I for the reaction of caffeic acid (A, B) with 16-ArN_2^+ in 1:9 (v:v) corn oil-in-water emulsions stabilized with Tween 20 at different temperatures. The solid lines were obtained by fitting the k_{obs} and $1/k_{obs}$ values to Eqs. (7 and 12), respectively. (red circles) T = 15, (Orange squares) T = 20, (pink triangles) T = 25, (green inverted triangles) T = 30, (blue diamond) T = 35. Measured pH = 3.66 (0.04 M citric/citrate buffer. [AO] ~ 4.3×10^{-3} M, [16-ArN_2^+] ≈ 2.9×10^{-4} M (Costa et al. 2015)

4.4 Determining Antioxidant Partition Constants in Intact Emulsions

The partition constants are obtained from the changes in the measured rate constant k_{obs} with the concentration of surfactant (often expressed as the surfactant volume fraction $\Phi_I = V_{surf/}V_{emulsion}$). Figure 1 shows typical plots of k_{obs} versus Φ_I and $1/k_{obs}$ versus Φ_I obtained for caffeic acid and α-Tocopherol as a function of temperature (Martinez-Aranda et al. 2014). The solid lines are the theoretical curves

Table 1 Chemical structures of some investigated AOs (R stands for the alkyl chain grafted to the parent AO) and values for their partition constants between the oil-interfacial, P_O^I, and aqueous-interfacial, P_W^I, regions of emulsions in olive oil O/W emulsions

Structure	–R	P_O^I	P_W^I
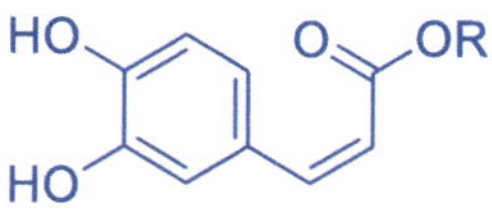	–H (CA)	–	349
	$-CH_3$ (C1)	312	720
	$-CH_2\ CH_3$ (C2)	405	3156
	$-(CH_2)_2\ CH_3$ (C3)	454	–
	$-(CH_2)_7\ CH_3$ (C8)	502	–
Caffeic acid	$-(CH_2)_{15}\ CH_3$ (C16)	376	–
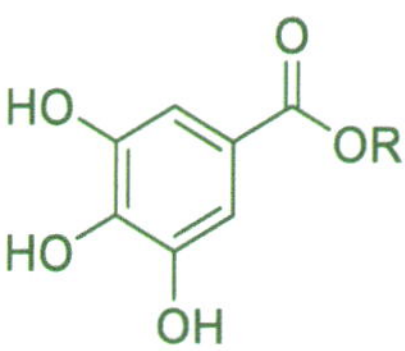	–H (GA)	–	101
	$-(CH_2)_2\ CH_3$ (C3)	449	328
	$-(CH_2)_7\ CH_3$ (C8)	27	–
Gallic acid	$-(CH_2)_{11}\ CH_3$ (C12)	17	–
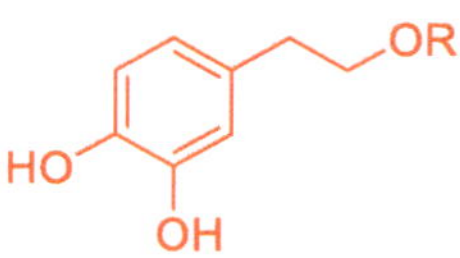	–H (HT)	–	53
	$-CO-CH_3$ (C2)	320	93
	$-CO-(CH_2)\ CH_3$ (C3)	197	123
	$-CO-(CH_2)_2\ CH_3$ (C4)	171	373
	$-CO-(CH_2)_5\ CH_3$ (C7)	184	–
Hydroxytyrosol	$-CO-(CH_2)_7\ CH_3$ (C9)	296	–
	$-CO-(CH_2)_9\ CH_3$ (C11)	125	–
	$-CO-(CH_2)_{11}\ CH_3$ (C13)	82	–
	$-CO-(CH_2)_{15}\ CH_3$ (C17)	52	–
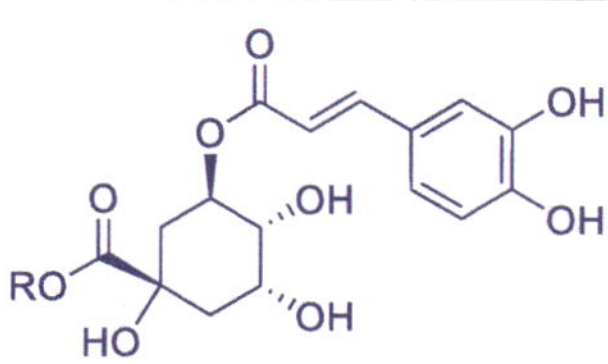	–H (CGA)	–	40
	$-CH_2\ CH_3$ (C2)	–	78
	$-(CH_2)_3\ CH_3$ (C4)	–	141
	$-(CH_2)_7\ CH_3$ (C8)	111	–
	$-(CH_2)_9\ CH_3$ (C10)	124	–
	$-(CH_2)_{11}\ CH_3$ (C12)	159	–
Chlorogenic	$-(CH_2)_{15}\ CH_3$ (C16)	89	–

Data extracted from refs. (Meireles et al. 2020; Costa et al. 2015; Almeida et al. 2016; Losada-Barreiro et al. 2013; Costa et al. 2017)

obtained by fitting the experimental data to Eqs. (7, 8, 12 and 13), from where the partition constants were obtained. Note the excellent linearity of the reciprocal plots.

Table 1 shows partition constants obtained by using the pseudophase kinetic model adjacent to the chemical structures of the antioxidants. The values of the partition constants for the transfer of the antioxidants from the aqueous and oil regions to the interfacial one are higher than one (Leo 2002), i.e., $P > 1$. They have a natural tendency to be located in the interfacial region and the Gibbs free energy of transfer ($\Delta G = -RT\ln P$) of antioxidants from the aqueous or oil phase to the interfacial region is negative, and depend on AO hydrophobicity. For the same series of antioxidants, P_W^I values increase upon increasing the hydrophobicity of the AO, in keeping with the hydrophobic effect. However, the variations of P_O^I are more complex since they increase upon increasing the hydrophobicity, with the least hydrophobic AOs having the smallest P_O^I values and the most hydrophobic the largest ones.

However, the reader must be aware that the percentage of AOs in the interfacial region is different because the antioxidants whose hydrophilic-lipophilic (HLB) closely matches that of the interfacial region (a melange of oil, polyoxyethylene chains and water) have the highest interfacial solubilities, such as esters of antioxidants bearing alkyl chains of 4–8 carbon atoms (Bravo-Díaz et al. 2015; Costa et al. 2015; Almeida et al. 2016; Lopez De Arbina et al. 2019; Laguerre et al. 2009).

5 Conclusions and Future Perspectives

The application of pseudophase kinetic models to emulsions is a unique method for estimating the partition constants of antioxidants between oil and water, water and interfacial and oil and interfacial regions in intact emulsions, i.e. without disrupting the existing equilibria. The values of these constants are needed to determining their local concentrations in the various domains of the emulsions and to compare their relative AO efficiencies (Meireles et al. 2020; Raimúndez-Rodríguez et al. 2019; Mitrus et al. 2019; Costa et al. 2020a; Costa et al. 2020b).

A number of studies demonstrate that the antioxidant efficiency correlates directly with fraction of the antioxidant in the interfacial region but not with that the aqueous or oil regions, providing physical evidence for the interfacial region as the main reaction site between antioxidants and peroxyl radicals (Lisete-Torres et al. 2012; Costa et al. 2015; Losada-Barreiro et al. 2013; Costa et al. 2021). In addition, knowledge of their values permits interpretation of the effects of AO hydrophobicity of several series of homologous antioxidants on their efficiency, which generally increases linearly with hydrophobicity (e.g., numbers of C atoms in their alkyl chains) up to a maxima at an intermediate (C4–C12,) chain length, Fig. 2 (Bravo-Díaz et al. 2015; Costa et al. 2015; Losada-Barreiro et al. 2013; Laguerre et al. 2012; Medina et al. 2009).

In addition, relevant thermodynamic parameters (ΔG, ΔH, ΔS) for the transfer of components between the regions of the emulsion can be obtained from the variations of the partition constants with temperature (Raimúndez-Rodríguez et al. 2019; Losada-Barreiro et al. 2015). Their knowledge is useful to (1) obtain insights into the transfer mechanism, i.e., on the enthalpic and entropic interactions with solvent, e.g., spontaneity of the transfer process, the hydrogen bond interactions between the component and the solvent, and (2) to predict the partitioning behavior of structurally similar compounds. The thermodynamic parameters $\Delta H_T^{0,W \rightarrow I}$ and $\Delta H_T^{0,O \rightarrow I}$ for transfer of a given antioxidant to the interfacial region are obtained from the

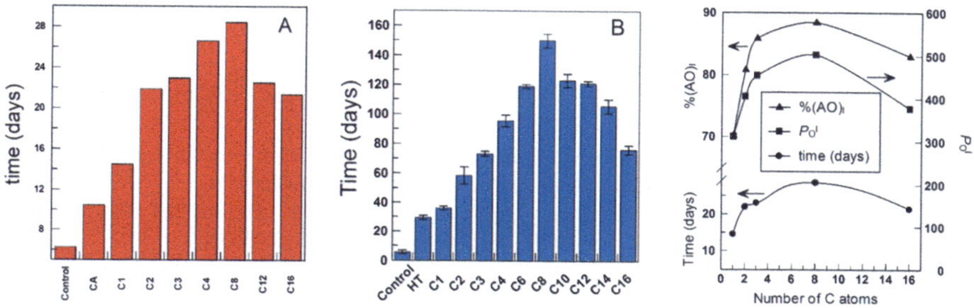

Fig. 2 Parabolic-like variations on the oxidative stability of oil-in-water emulsions as a function of the alkyl chain length of caffeic acid derivatives (**A**) and hydroxytyrosol derivatives (**B**). Relationships between the variations in the percentage of antioxidant in the interfacial region (%AO_I), in the partition constant of the antioxidants between the oil and interfacial region (P_O^I), and in the time necessary to increase the percentage of conjugated dienes 1%, an assessment of relative antioxidant efficiency as a function of the hydrophobicity of caffeic acid derivatives. Plots adapted from references (Costa et al. 2015) and (Almeida et al. 2016)

variation of $\ln P_W^I$ and $\ln P_O^I$ values with T (Raimúndez-Rodríguez et al. 2019; Losada-Barreiro et al. 2015).

The research is, however, far from being finished because any changes in oil, surfactant or AO structure will alter AO efficiency. Currently, polyphenolic antioxidants are primarily used in food systems. In addition to preventing lipid oxidation in foods, they are also used in the treatment of some human diseases. Research on a better understanding of the location and reaction mechanisms of polyphenols in emulsion-based delivery systems (for example, emulsions for parenteral nutrition), to control the release of incorporated bioactives in specific sites is crucial to boosting their efficiency in areas such as biology, medicine and pharmacy. In addition, a number of fundamental questions remain unanswered, for example, how to predict the best antioxidant or set of antioxidants for a given system, including the analyses of potential synergistic effects, (Kancheva and Kasaikina 2012; Kancheva and Angelova 2017) deserving futher investigations.

Acknowledgement This chapter was prepared during a sabbatical leave of C. B-D, supported by the University of Vigo. We would like to thank all coworkers who participated with enthusiasm for years in this project and made important contributions with their helpful discussions and support. Special thanks are given to Prof. Sonia Losada-Barreiro (Univ. Vigo), Dr. Marlene Costa (Univ. Porto) and Prof. Fátima Paiva-Martins (Univ. Porto). Finnancial support from Ministerio de Educaci6n y Ciencia (CTQ2006-13969-BQU), Xunta de Galicia (10TAL314003PR and REDUSO, Grant number ED431D 2017/18), and Universidad de Vigo is acknowledged.

Conflict of Interest The authors declare no conflict of interest.

References

Almeida J, Losada-Barreiro S, Costa M, Paiva-Martins F, Bravo-Díaz C, Romsted LS (2016) J Agric Food Chem 64:5274–5283
Amézqueta S, Subirats X, Fuguet E, Rosés M, Ràfols C (2020) Chapter 6 – Octanol-water partition constant. In: Poole CF (ed) Liquid-phase extraction. Elsevier, Amsterdam
Andrés A, Rosés M, Ràfols C, Bosch E, Espinosa S, Segarra V, Huerta JM (2015) Eur J Pharmaceutical Sci 76:181–191. https://doi.org/10.1016/j.ejps.2015.05.008
Blokhina SV, Ol'khovich MV, Sharapova AV, Proshin AN, Perlovich GL (2013) J Chem Thermodynamics 61:11–17. https://doi.org/10.1016/j.jct.2013.01.022
Bravo-Díaz C, Romsted LS, Liu C, Losada-Barreiro S, Pastoriza-Gallego MJ, Gao X, GU Q, Krishnan G, Sánchez-Paz V, Zhang Y, Ahmad-Dar A (2015) Langmuir 31:8961–8979. https://doi.org/10.1021/acs.langmuir.5b00112
Bravo-Díaz C, Romsted LS, Losada-Barreiro S, Paiva-Martins F (2017) Eur J Lipid Sci Technol 119:1600277. https://doi.org/10.1002/ejlt.201600277
Bunton CA, Romsted LS (1999) In: Kumar P, Mittal KL (eds) Handbook of microemulsion science and technology. Marcel-Dekker, New York
Bunton C, Yao J, Romsted LS (1997) Curr Opin Colloid Interf Sci 2:622–628
Costa M, Losada-Barreiro S, Paiva-Martins F, Bravo-Díaz C, Romsted LS (2015) Food Chem 175:233–242
Costa M, Losada-Barreiro S, Paiva-Martins F, Bravo-Diaz C (2016) RSC Advances 6:91483–91493
Costa M, Losada-Barreiro S, Paiva-Martins F, Bravo-Díaz C (2017) J Sci Food Agric 97:564–571. https://doi.org/10.1002/jsfa.7765
Costa M, Freiría-Gándara J, Losada-Barreiro S, Paiva-Martins F, Bravo-Díaz C (2020a) J Colloid Interf Sci 562:352–362. https://doi.org/10.1016/j.jcis.2019.12.011
Costa M, Losada-Barreiro S, Bravo-Díaz C, Vicente AA, Monteiro LS, Paiva-Martins F (2020b) Food Chem 310:125716. https://doi.org/10.1016/j.foodchem.2019.125716
Costa M, Losada-Barreiro S, Paiva-Martins F, Bravo-Díaz C (2021) Foods 10:539
Cumming H, Rücker C (2017) ACS Omega 2:6244–6249. https://doi.org/10.1021/acsomega.7b01102
Dar AA, Bravo-Diaz C, Nazir N, Romsted LS (2017) Current Opinion Colloid Interf Sci 32:84–93. https://doi.org/10.1016/j.cocis.2017.09.001
de Leonardis A (2014) Virgin olive oil: production, composition, uses and benefits for man. Nova Science Publishers, Incorporated
Ferreira I, Costa M, Losada-Barreiro S, Paiva-Martins F, Bravo-Díaz C (2018) Food Res Int 112:192–198. https://doi.org/10.1016/j.foodres.2018.06.007
Foti MC (2007) J Pharm Pharmacol 59:1673–1685. https://doi.org/10.1211/jpp.59.12.0010
Frankel EN (2005) Lipid oxidation. The Oily Press, PJ Barnes & Associates, Bridgwater
Freiría-Gándara J, Losada-Barreiro S, Paiva-Martins F, Bravo-Díaz C (2018a) J Chem Eng Data 63:2999–3007. https://doi.org/10.1021/acs.jced.8b00258
Freiría-Gándara J, Losada-Barreiro S, Paiva-Martins F, Bravo-Díaz C (2018b) Food & Function 9:4429–4442
Galan A, Losada-Barreiro S, Bravo-Díaz C (2016) Chem Phys Chem 17:296–304
Gao X, Bravo-Díaz C, Romsted LS (2013) Langmuir 29:4928–4933
Garti N (2003) Curr Opin Colloid Interf Sci 8:197–211
Giraudeau P, Frydman L (2014) Annu Rev Anal Chem 7:129–161. https://doi.org/10.1146/annurev-anchem-071213-020208
Gu Q, Bravo-Díaz C, Romsted LS (2012) J Colloid Interf Sci 400:41–48
Gunaseelan K, Romsted LS, Pastoriza-Gallego MJ, González-Romero E, Bravo-Díaz C (2006) Adv Colloid Interf Sci 123–126:303–311
Heins A, Sokolowski T, Stöckmann T, Schwarz K (2007) Lipids 42:561–572
1995. http://www.oecd.org/chemicalsafety/risk-assessment/1948169.pdf

Kancheva V, Angelova S (2017) Synergistic effects of antioxidant composition during inhibited lipid autooxidation. In: Catalá A (ed) Lipid peroxidation: inhibition, effects and mechanisms. Nova Science Publisher, Hauppauge

Kancheva VD, Kasaikina OT (2012) Lipid oxidation in homogeneous and micro-heterogeneous Media in presence of prooxidants, antioxidants and surfactants. In: Catala A (ed) Lipid peroxidation. InTech, Rijeka

Korinth G, Wellner T, Schaller KH, Drexler H (2012) Toxicol Letters 215:49–53. https://doi.org/10.1016/j.toxlet.2012.09.013

Kumar KS, Dhachinamoorthi D, Saravanan R, Gopal U, Shanmugam V (2011) Int J Pharm Sci Rev Res 10:37–45

Laguerre M, López-Giraldo LJ, Lecomte J, Figueroa-Espinoza MJ, Baréa B, Weiss J, Decker EA, Villeneuve P (2009) J Agric Food Chem 57:11335–11342

Laguerre M, Bayrasy C, Lecomte J, Chabi B, Decker EA, Wrutniak-Cabello C, Cabello G, Villeneuve P (2012) Biochimie 95:1–7

Leo AJ (2002) Octanol/water partition coefficients. Encyclopedia of computational chemistry. John Wiley & Sons, Ltd., Chichester

Leo A, Hansch C, Elkins D (1971) Chem Rev 71:525–616

Lisete-Torres P, Losada-Barreiro S, Albuquerque H, Sánchez-PAZ V, Paiva-Martins F, Bravo-Díaz C (2012) J. Agric. Food Chem 60:7318–7325

Litwinienko G, Kamal-Eldin A, Pokorny J (2005) Analysis of lipid oxidation. AOCS Press, Champaign, IL

Liu X, Testa B, Fahr A (2011) Pharm Res 28:962–977. https://doi.org/10.1007/s11095-010-0303-7

Lopez De Arbina A, Losada-Barreiro S, Rezende MC, Vidal M, Aliaga C (2019) Food Chem 279:288–293. https://doi.org/10.1016/j.foodchem.2018.12.020

Lopez-Quintela MA, Tojo C, Blanco MC, Garcia Rio L, Leis JR (2004) Curr Opin Colloid Interf Sci 9:264–278

Losada-Barreiro S, Bravo-Díaz C (2017) Eur J Med Chem 133:379–402

Losada-Barreiro S, Bravo Díaz C, Paiva Martins F, Romsted LS (2013) J Agric Food Chem 61:6533–6543

Losada-Barreiro S, Sánchez-Paz V, Bravo-Díaz C (2015) Org Biomol Chem 13:876–885

Martinez-Aranda N, Losada-Barreiro N, Bravo-Díaz C, Romsted LS (2014) Food Biophys 9:380–388. https://doi.org/10.1007/s11483-014-9332-9

Mcclements DJ (2018) Food Funct 9:22–41. https://doi.org/10.1039/C7FO01515A

Medina I, Lois S, Alcántara D, Lucas L, Morales JC (2009) J Agric Food Chem 57:9773–9779

Meireles M, Losada-Barreiro S, Costa M, Paiva-Martins F, Bravo-Díaz C, Monteiro LS (2020) J Sci Food Agric 99:3917–3925. https://doi.org/10.1002/jsfa.9615

Menger FM, Portnoy CE (1967) J Am Chem Soc 89:4698–4703. https://doi.org/10.1021/ja00994a023

Meyer P, Maurer G (1995) Ind Eng Chem Res 34:373–381. https://doi.org/10.1021/ie00040a041

Mitrus O, Żuraw M, Losada-Barreiro S, Bravo-Díaz C, Paiva-Martins F (2019) J Agric Food Chem 67:3266–3274. https://doi.org/10.1021/acs.jafc.8b06545

Mozuraityte R, Kristinova V, Rustad T (2016) Oxidation of food components. In: Encyclopedia of food and health. Academic, Oxford

OECD (1995) Test no. 107: partition coefficient (n-octanol/water): shake flask method. OECD, Paris

Oehlke K, Garamus V, Heins A, Stöckman H, Schwarz K (2008) J Colloid Interf Sci 322:294–303

Pastoriza-Gallego MJ, Sánchez-Paz V, Losada-Barreiro S, Bravo-Diaz C, Gunaseelan K, Romsted LS (2009) Langmuir 25:2646–2653

Pillai V, Shah DO (1996) Reaction kinetics in water-in-oil microemulsions. In: PILLAI V, SHAH DO (eds) Dynamic properties of interfaces and association structures. AOC Press, Illinois, USA

Raimúndez-Rodríguez EA, Losada-Barreiro S, Bravo-Díaz C (2019) J Colloid Interf Sci 555:224–233. https://doi.org/10.1016/j.jcis.2019.07.085

Rodis PS, Karathanos VT, Mantzavinou A (2002) J Agric Food Chem 50:596–601

Romsted LS (1984) Micellar effects on reaction rates and equilibria. In: Mittal KL, Lindman J (eds) Surfactants in solution. Plenum Press, New York
Romsted LS (2012) Introduction to surfactant self-assembly. In: Gale PA, Steed JW (eds) Supramolecular chemistry: from molecules to nanomaterials. J. Wiley & Sons Ltd., New York
Romsted LS, Bravo-Díaz C (2013) Curr Opin Colloid Interf Sci 18:3–14
Ross L, Barclay C, Vinqvist MR (2003) Phenols as antioxidants. In: Rappoport Z (ed) The chemistry of phenols. J. Wiley, West Sussex
Russling JF (1994) Electrochemistry in micelles, microemulsions and related microheterogeneous fluids. In: Electroanalytical chemistry, vol 18. Marcel Dekker, New York
Sangster J (1997) Octanol-water partition coefficients, fundamentals and Physcial chemistry. J. Wiley & Sons, Chichester
Schaich KM (2005) Lipid oxidation: theoretical aspects. In: Shahidi F (ed) Bailey's industrial oil and fat products. J. Wiley & Sons, New York
Schaich KM, Shahidi F, Zhong Y, Eskin NAM (2013) Chapter 11 – Lipid oxidation. In: Biochemistry of foods, 3rd edn. Academic, San Diego
Schulte J, Dürr J, Ritter S, Hauthal WH, Quitzsch K, Maurer G (1998) J Chem Eng Data 43:69–73. https://doi.org/10.1021/je970112e
Shahidi F (2015) Handbook of antioxidants for food preservation, 1st edn. Woodhead Publisher, Amsterdam
Shahidi F, Zhong Y (2010) Chem Soc Rev 39(11):4067–4079
Silvério SC, Rodríguez O, Teixeira JA, Macedo EA (2010) J Chem Thermodynamics 42:1063–1069. https://doi.org/10.1016/j.jct.2010.04.005
Stöckman H, Schwarz K (1999) Langmuir 15:6142–6149
Tetko IV, Yan A, Gasteiger J (2018) Prediction of physicochemical properties of compounds. Applied Chemoinformatics

Fluorescent Probes for Monitoring Oxidation of Lipids and Assessment of Antioxidant Activity

Jarosław Kusio and Grzegorz Litwinienko

1 Introduction

Oxidative stress is caused by an overwhelming production of a series of oxidants, including superoxide ($O_2^{\bullet-}$), hydrogen peroxide (H_2O_2), hydroxyl radical ($HO^{\bullet}$), and other related compounds collectively known as reactive oxygen species (ROS). Together with reactive nitrogen species (RNS), ROS might oxidize biomolecules, including proteins, lipids, and nucleic acids, leading to cell death, and, on the other hand, in non-living systems, ROS initiate and mediate the deterioration of food systems. Among biomolecules, lipids are especially sensitive to peroxidation. The understanding of the nature of free radicals and mechanisms of peroxidation has led to an enormous increase of research in biology and medicine and has been summarised in periodically re-edited excellent monography "Free Radicals in Biology and Medicine" written by Halliwell and Gutteridge (first edition in 1985, fifth edition in 2015) (Halliwell and Gutteridge 2015). A wide variety of methods (assays and protocols) have been developed over the years for studies of oxidation and antioxidant action in completely artificial model systems, in foods and in recomposed structures/systems formulated for pharmaceutical and cosmetic applications as well as in living cells. Recent advances in analytical chemistry and spectroscopy allow to elaborate the methods for detection of enormously small amounts of chemical species involved in peroxidation at any step. As far as ROS/RNS with an odd electron are considered, the first choice methodology should be Electron Paramagnetic Resonance (EPR) with various spin traps, usually nitroxides or nitrones, that can be conveniently applied to artificial systems and living organisms. Unfortunately, EPR is expensive, the methodology of experiments and their interpretation are rather difficult for non-specialists. Another method,

J. Kusio · G. Litwinienko (✉)
University of Warsaw, Faculty of Chemistry, Warsaw, Poland
e-mail: litwin@chem.uw.edu.pl

C. Bravo-Diaz (ed.), *Lipid Oxidation in Food and Biological Systems*,
https://doi.org/10.1007/978-3-030-87222-9_3

Magnetic Resonance Imaging (MRI), although broadly applied in diagnostics, is too expensive and time-consuming to be routinely used as a benchtop technique for studies of oxidative stress, antioxidant activity or fast, automatic or semi-automatic assessment of the quality of food products. Another line of spectroscopy, spectrofluorimetry, is broadly used as an analytical tool in pharmacy and biochemistry. However, more sophisticated applications of this technique flourished together with the intensive development of fluorescence microscopy, confocal microscopy, and other spectrofluorimetric techniques used in biochemistry, diagnostics, toxicology, genetic engineering, immunology and many other fields. This chapter provides an introduction to spectrofluorimetric methods employed for detecting oxidative stress, lipid oxidation/peroxidation, and antioxidant activity monitored with fluorescent probes. Their reaction with ROS should be recordable in real-time, giving the information about the progress of oxidation of the whole observed system. The same expectations are expressed for fluorescent techniques to be used for monitoring the oxidative deterioration of food, with all diversity of those systems, from simple bulk oils and fats to complex physical-chemical nature of the processes undergoing in lipid/water emulsions or in such multi-phase and multi-component systems as tissues or processed foods. For this reason, we decided not to describe the application of fluorescence spectroscopy in the analysis of food, food processing, oxidative deterioration of fats and edible oils. The excellent reviews are dedicated to those problems (Laguerre et al. 2007; Laguerre et al. 2020) and also some aspects of formation of fluorescent products of co-oxidation of lipids and proteins (Schaich 2008). In another book chapter, *Fluorescence Spectroscopy for the Monitoring of Food Processes,* the authors were focused on applications of fluorescence spectroscopy paired with chemometrics on various foods (Ahmad et al. 2017). As this chapter is focused on Reactive Oxygen Species, a reader who is interested in fluorescent probes dedicated to detection of nitrogen and sulphur species is directed to the recent publication by Wu and co-workers in *Acc. Chem. Res* (Wu et al. 2019a).

For anyone doing research in the field of antioxidants, the word "spectrofluorimetry" collocates with the ORAC method, but herein we describe the recent achievements in the design and synthesis of fluorescent probes for ROS sensing and imaging that have moved the methodology from simple spectrofluorimetric titrations to methods approaching the sensitivity on the level of single molecules observed in individual liposomes or living cells under ambient conditions in the presence of exogenous ROS (Godin et al. 2014).

2 Reactive Intermediates of Peroxidation

There are some common features that allow to describe the lipid peroxidation phenomena with the same mechanistic assumptions, regardless the process proceeds within (bio)membranes, in living cells, artificial biomimetic systems like liposomes, in emulsions or in a bulk lipid phase. This chain process is mediated by free radicals and is initiated by radicals or their non-radical precursors. The most popular term

Table 1 Non-exhaustive list of Reactive Oxygen Species and Reactive Nitrogen Species (Halliwell and Gutteridge 2015)

Reactive Oxygen Species (ROS)	Reactive Nitrogen Species (RNS)
Radicals:	*Radicals:*
$O_2^{\bullet-}$ (superoxide anion radical or superoxide)	$NO^{\bullet}$ (nitric oxide)
$HO^{\bullet}$ (hydroxyl)	$NO_2^{\bullet}$ (nitrogen dioxide)
$HOO^{\bullet}$ (hydroperoxyl)	$NO_3^{\bullet}$ (nitrate radical)
$ROO^{\bullet}$/$LOO^{\bullet}$ (alkylperoxyl/lipid peroxyl)	$^{\bullet}N_3$ (azide radical)
$RO^{\bullet}$ (alkoxyl)	
$CO_3^{\bullet-}$ (carbonate)	
$CO_2^{\bullet-}$ (carbon dioxide)	
Non-radicals:	*Non-radicals:*
HOOH (hydrogen peroxide)	HNO_2 (nitrous acid)
ROOH/LOOH (alkyl or lipid hydroperoxide)	NO^+ (nitrosyl cation)
$ONOOH \leftrightarrows ONOO^-, H^+$ (peroxynitrite)	NO^- (nitroxyl anion)
$O_2NOOH \leftrightarrows O_2NOO^-, H^+$ (peroxynitrate)	N_2O_4 (dinitrogen tetraoxide)
O_3 (ozone)	N_2O_3 (dinitrogen trioxide)
$^1O_2$ $^1\Delta g$ (singlet oxygen)	ROONO (alkyl peroxynitrite)
	RO_2ONO (alkyl peroxynitrate)

describing all those chemical individuals is Reactive Oxygen Species (ROS), also extended to Reactive Nitrogen Species (RNS). Examples of ROS and RNS are listed in Table 1, but the list of reactive species able to initiate peroxidation also covers the ones containing other elements like chlorine, bromine, and sulphur. Molecular oxygen and transition metal ions also match the exact definition of free radicals (molecules, atoms or ions with one or more unpaired electrons) but, compared to ROS, their ability to react with non-radical organic molecules and to initiate peroxidation depends on the presence of other ROS or RNS. ROS/RNS collected in Table 1 are representative for living/cellular systems, however, in non-living systems, such as foods, the oxidative processes might proceed with the participation of other reactive species, because biological defences and structures are either absent or destroyed, for example, a participation of intermediates generated from proteins being oxidized or lipid-protein co-oxidation (Schaich 2008).

Exogenous ROS are by-products of cellular metabolism and can be found in any part of organism. Some of them play an essential role in signal transduction, apoptosis, elimination of pathogens, healing wounds and other processes. In a healthy organism, even some mitochondrial leaking of ROS seems to be well controlled by antioxidant enzymes belonging to oxidoreductases like glutathione peroxidase, superoxide dismutase, and catalase. Besides, ROS/RNS might be converted into unreactive species by various small non-enzymatic molecules, acting via physical processes (quenchers) or chemical mechanisms (mainly reduction by antioxidants). The oxidation-reduction homeostasis might be dysregulated because of overproduction of ROS and dramatic decrease of the level of antioxidants or because of defects in expression of antioxidant genes, resulting in the oxidative stress and oxidative damage of cells and tissues. On the cellular level, oxidative damage involves all vital molecules, membrane lipids, proteins, sugars and nucleic acids. However, the most sensitive constituents of cells are phospholipids containing polyunsaturated fatty acids (PUFAs) and self-assembled into biomembranes. Although the lipid peroxyl radicals and lipid hydroperoxides generated during membrane oxidation trigger the cascade of signal molecules that respond to changed cell structure by inducing an appropriate gene response, they also transform cholesterol, phospholipids, plasmalogens, other polyunsaturated fatty acids, sugars, and proteins into deleterious products (Spiteller 2006).

Consequences of oxidative stress in organism/tissue or cell and lipid oxidation (in cells and in artificial systems) can be monitored, for instance, by detection of peroxidation products (conjugated dienes, hydroperoxides, hexanal, or adducts with thiobarbituric acid), however, those methods are time-consuming and do not allow to follow the course of oxidation in real-time mode. On the other hand, the intensity of oxidative stress (or initial rate of oxidation) can be monitored as a decrease of the amount of antioxidants present in the system and sacrificially consumed by ROS. Some of the methods employ the compounds that competitively react with ROS present/internally generated or deliberately introduced in the studied system. Monitoring the analytical signal produced by the compound reacting with ROS (called a probe) gives information about the amount of ROS, rate of their production or rate and efficacy of reduction of ROS by the antioxidants present in the sample. Below we describe the fluorescent probes used for the detection of ROS that can be applied for monitoring peroxidation course of spontaneous peroxidation of lipid / cellular systems as well as for monitoring the rate of inhibited peroxidation when fluorescent probe is used as a competitive scavenger of ROS.

3 Fluorescent Probes for Monitoring Peroxidation

3.1 Luminescence and Fluorescence

Luminescence, also called cold light or cold body radiation is a phenomenon of spontaneous light emission by a substance, not resulting from heating this substance to high temperature. Luminescence occurs from an electronic-vibrational excited molecule that is in thermal equilibrium with the surrounding. There are several kinds of luminescence because any form of energy delivered to a molecule can be converted into emission of photons: chemiluminescence is an effect of chemical reaction; electroluminescence is caused by electric current, field or electromagnetic wave passing through solid or gas, radioluminescence is caused by ionising radiation, and photoluminescence is a result of the absorption of photons by a substance. This last interaction between a photon and a molecule invokes the excitation of valence electron(s). The excited state is not stable (sometimes quasi-stable), and the system spontaneously returns to a more stable ground state. For most molecules, this relaxation process involves the energy transfer to a solvent or other molecules present in the solution. During the return to the ground state, the excess of energy might be emitted as light. The phenomenon of photoluminescence can be divided

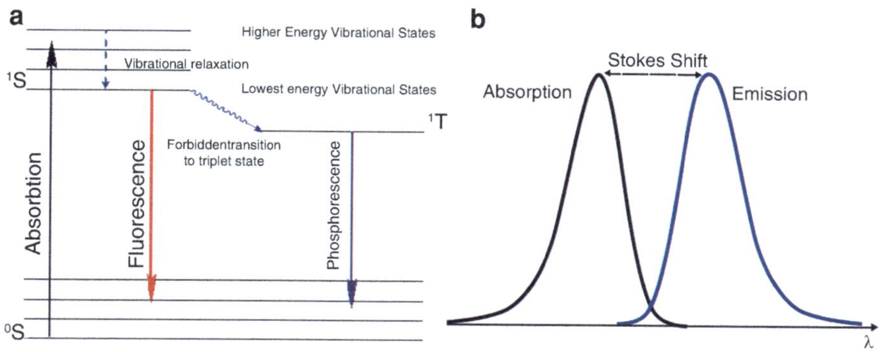

Fig. 1 (**a**) Jablonsky Energy Diagram for fluorescence and phosphorescence. (**b**) Emission and absorption spectra with highlighted Stokes shifts

into two main categories – fluorescence (singlet → singlet transitions, within nanoseconds) and phosphorescence (singlet → triplet transitions, that can last up to seconds or even hours) (Lakowicz 2006).

The mechanism of photoluminescence is presented in the Jablonsky Energy Diagram (Jablonski 1933) (Fig. 1a). If a molecule of fluorophore at singlet ground electronic state (bars labelled as ^{0}S) absorbs UV, visible, or infrared radiation, it will be promoted to a singlet first state (^{1}S). The molecule exists in several vibrational energy levels within the ground, first, and other energy levels. Fluorophore that has absorbed photon is usually excited to a higher vibrational energy level within the first singlet state (^{0}S → ^{1}S) and rapidly relaxes to the lowest vibrational energy level in the process of a non-emissive transition, referred to as vibrational relaxation, which occurs in less than a picosecond. Then, the photon emission occurs as the molecule returns to the ground singlet state (^{1}S → ^{0}S), see Fig. 1a. The fluorescence processes are usually four orders of magnitude slower than vibrational relaxation. Thus, the molecule has sufficient time to obtain a thermally equilibrated lowest energy ^{1}S state before the emission. Nevertheless, as the ^{1}S → ^{0}S relaxation lasts nanoseconds, the fluorescence can be usually observed and measured during the irradiation. In contrast, for phosphorescence, the electron undergoes intersystem crossing, with a spin conversion to a "forbidden" triplet state ^{1}T. After the transition to ^{1}T, the electron might be trapped. Transfers ^{1}T → ^{0}S are forbidden, kinetically unfavourable and very slow. Therefore the phosphorescence can be observed up to hours after the irradiation of fluorophore (Lakowicz 2006; Noomnarm and Clegg 2009).

3.2 *Fluorophore Properties*

A characteristic band of fluorescence emission (Fig. 1b) represents the waves of different energies (wavelength) even when the fluorophore was irradiated with monochromatic light. The reason for that is because during the relaxation (Fig. 1a; $^1S \rightarrow {}^0S$ bands) the molecule can return to many of the possible vibrational energy levels of the ground state. Light emitted by excited fluorophore has almost always longer wavelengths than the absorbed light (*Stokes shift;* Fig. 1b) (Lakowicz 2006; Banwell and EM 1994) and this is typical for fluorescent molecules in solution. A common cause of the Stokes shift is the vibrational relaxation to the lowest vibrational level of excited state ^1S. Moreover, fluorophores tend to decay to higher vibrational levels of ^0S (Fig. 1a), and further loss of excitation energy is realized by conversion of the excess vibrational energy into thermal energy. In addition to these effects, fluorophores can display more significant Stokes shifts due to solvent effects, excited-state reactions, complex formation, or energy transfer (Lakowicz 2006). As the exception, when the absorbing fluorophore is already excited, the relaxation can generate waves shorter than the absorbed light. Thus, the fluorescence spectrum contains the anti-Stokes lines. Such anti-Stokes behaviour can also be caused by thermal photons' dissipation in a crystal lattice, cooling the crystal in the process. Another anti-Stokes process is photon upconversion, which can occur when the sequential absorption of two or more photons leads to light emission with a wavelength than the excitation wavelength.

The fluorophores can be characterised by a number of parameters. One of the most important parameters needed for quantitative fluorescence is the excitation cross-section of the fluorophore as relevant to probability of photons to be absorbed and the molar attenuation coefficient (ε, expressed in $M^{-1} \times cm^{-1}$) is an efficiency at which a fluorophore absorbs excitation light at a given wavelength. The fluorophore's quantum yield ($\boldsymbol{\Phi}$, dimensionless quantity between 0 and 1) is a ratio of the number of emitted photons to those absorbed by the molecule. Energy gained by the molecule is spent on the emission and on non-radiative processes such as Förster resonance energy transfer, intersystem crossing or converted to heat during internal and external conversion (Lakowicz 2006). Some rhodamines have the highest $\boldsymbol{\Phi}$, close to 1, but even compounds with the quantum yield of 0.1 are still visibly fluorescent. The quantum yields are measured by comparison to well-known quantum yields standards such as fluorescein and quinine (Brouwer 2011; Würth et al. 2015). The final brightness of a fluorophore is then the result of its attenuation coefficient and quantum yield. Another important parameter is fluorescence lifetime that can be defined as the meantime after excitation that molecules stay in their excited state before relaxing to the ground state (Fig. 1a). Usually, this transition lasts approximately 10 nanoseconds (Lakowicz 2006). The above three parameters are intrinsic features of a particular fluorophore (the structure, functional groups, energy levels etc.) but the intensity and lifetime can also be dependent on some external conditions, as fluorophores can lose their emissive properties because of quenching and photobleaching processes. Quenching occurs due to short-range interactions of a fluorophore with the local molecular environment. For example, collisional quenching takes place when the excited-state fluorophore is returned to its ground state upon contact with a molecule of quencher. A wide variety of molecules can act as collisional quenchers. Examples include oxygen, halogens, amines, and some electron-deficient molecules like acrylamide. Sometimes quenching can be caused by formation of complexes with quenching molecules (static quenching) or by attenuation of the emitted light by the fluorophore itself (Lakowicz 2006). Both, static and collisional quenching, are often dependent on temperature, pressure and concentration of fluorophore, however, fluorophores are not altered chemically. On the other hand, photobleaching is a destruction of the fluorophore by physical or chemical factors, making it permanently non-fluorescent. Photobleaching can be, for example, caused by the prolonged exposure to high-intensity excitation light or can be caused by chemical destruction of the structurally important functional group of the fluorophore.

3.3 Fluorescent Probes: Development, Classification, Mechanisms

The success and rapid development of immunofluorescence triggered the search for applications of fluorophores in many other fields, and nowadays a term "fluorescent probe" (FP) is used for fluorophores (as whole molecules or as a part of molecular systems containing fluorophore moiety), used as sensors not only for chemical species being markers of biological processes but also to detect structural (microenvironmental) changes like membrane fusion, lipid domain formation, lipid transport mechanisms, lipid-protein interactions, nucleic acid dynamics, and protein conformation and conformational changes (Drummen 2012). Fluorescent lipophilic dyes (membrane dyes, membrane labels) provide important information on the cell structure and its dynamics, membrane heterogeneity and other local physical properties. Simple apolar probes like pyrene and diphenyl hexatriene (DPH) were used in the '80s. In the same decade, FPs were applied for monitoring Ca^{2+} – dependent signals transduction in cells (Grynkiewicz et al. 1985; Minta et al. 1989) and transport processes across the membrane. Since that time research upon FP has been vigorously continued and expanded, resulting, among others, in the development of a great variety of fluorescent dyes applied in fluorescence microscopy, fluorescent probes designed for sensing metal cations (K^+, Zn^{2+}, Cu^+, Cu^{2+}, Cd^{2+}, Hg^{2+}) (Fahrni and O'Halloran 1999; Baruah et al. 2005; Zeng et al. 2006; Peng et al. 2007; Yuan et al. 2007), markers of physiological processes, and fluorescent probes for reactive chemical compounds such as ROS and RNS (Wu et al. 2019a; Soh 2006; Gomes et al. 2005; Gomes et al. 2006; Chen et al. 2016; Krumova and Cosa 2013).

FPs can be broadly divided into two groups: intrinsic and extrinsic. Intrinsic fluorophores are naturally biomolecules that already exist or are generated in the system, and this group involves for example the aromatic amino acids, NADH, flavins, chlorophyll, parinaric acid and some of the amino acids such as tryptophan or tyrosine. Intrinsic probes are part of specific systems and their utilization as markers oxidative stress is limited, for example, they can be applied for monitoring the general oxidation status of complicated, natural (bio)materials. Another group, extrinsic fluorophores have to be introduced to the studied system. Extrinsic fluorophores include, for example, fluorescein, rhodamine, and numerous other substances described in this chapter, mostly focused on the development of extrinsic fluorescent markers that can be tailored to specific needs (localisation, sensitivity, selectivity etc.). An important problem is the interaction of the extrinsic probe with the studied system. The problem is very general as any probe (not only FP dedicated to ROS) might physically and chemically affect the studied system in terms of interactions with lipid membrane (adsorption of FP at the interface or inside the lipid bilayer can perturb the membrane fluidity, labelling procedures for living cells might induce cellular stress and therefore perturb the experimental results, (Drummen 2012) side reactions with other reactive components in the sample (reaction with proteins, enzymes, etc.) and finally, the probe should not be cytotoxic (if living systems are monitored). Apart from those general aspects, some specific interactions of FP with the studied system have to be considered. For example, a possible interferences of non-antioxidant components (proteins, tyrosine and tryptophane) with ROS resulting in an overestimation of the antioxidant power (Laguerre et al. 2007).

The fluorescent probe should be selective, i.e., designed in such a way to produce a signal only in the presence of the monitored analyte. Signalling should be manifested as either decrease or increase of the fluorescence. Because of that, there are two main classes of these compounds – fluorogenic off/on probes and quenchable on/off probes. The first type depends on enabling or increasing their emission properties during the reaction with the analyte, while the on/off probes diminish or completely lose their fluorescence through the detection process. The on/off switching might be caused by several reasons. The most common reason for ROS probes is the oxidation of fluorophore resulting in photobleaching which converts the fluorescent probe into a non-emissive molecule. Furthermore, quenching of the fluorescence might occur due to changes in the environment of the fluorophore. Thus, the on/off probes often lack specificity as there are many processes besides reaction with ROS that might decrease their emission.

An example of an on/off fluorescent probe is 1,3-diphenylisobenzofuran (DPBF), yellow, crystalline solid soluble in non-polar solvents, with maximum absorption at 420 nm. This compound can be used to detect singlet oxygen ($^1O_2$) in a biological system. After reaction with this ROS, the endoperoxide breaks the extended π-electron system and emissive properties of DPBF are lost. It then decomposes to colourless 1,2-dibenzoylbenzene (Fig. 2) (Entradas et al. 2020).

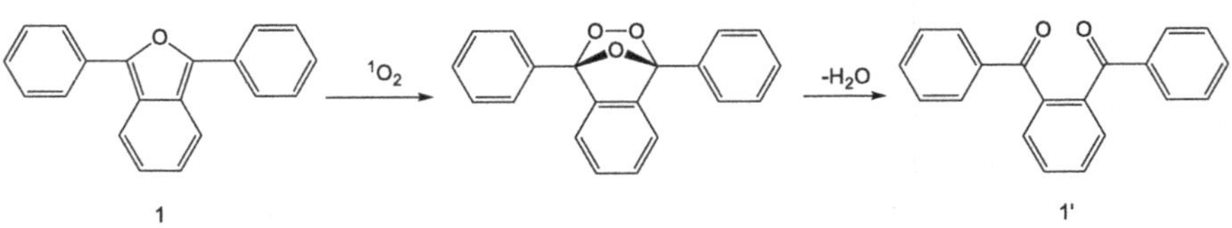

Fig. 2 DPBF reaction with singlet oxygen leading to loss of its fluorescence (Entradas et al. 2020; Singh et al. 1978)

Ohyashiki et al. (Ohyashiki et al. 1999) proposed DPBF for monitoring superoxide radical ($O_2^{\bullet -}$) in phospholipid liposomes. They reported a decrease of fluorescence when DPBF was incorporated into the liposomes and exposed to xanthine/xanthine oxidase system (the source of $O_2^{\bullet -}$), but the intensity of emission did not change when superoxide dismutase was also present in the system. The authors concluded that lipophilic DPBF can be used as FP for detecting hydrophilic $O_2^{\bullet -}$ inside the membrane lipid layer (lipid dynamics is responsible fur such penetration). Another example of a quenchable on/off probe for ROS detection is *cis*-parinaric acid (cis-PnA), a fluorescent polyunsaturated fatty acid with 4 conjugated π-bonds. Its fluorescence is irreversibly lost after reaction with peroxyl radicals, therefore, this fluorescent probe might be used as a marker of lipid peroxidation and for evaluation of the antioxidant activity of lipophilic compounds. Although researchers using cis-PnA for monitoring phase transitions of lipid bilayers or interactions with proteins consider its sensitivity towards oxygen (peroxidation) as a serious drawback (during preparation, storage and manipulating with this FP), however, the drawback turns into advantage when peroxidation processes are studied because the role of cis-PnA is to mimic other polyunsaturated fatty acids (Kuypers et al. 1987; Naguib 1998). However, this FP is photolabile and absorbs in the UV region at 320 nm (i.e. et the same region as unsaturated lipids). Table 2 presents an overview of commonly used quenchable (on/off) FPs used to detect ROS.

Among fluorescent on/off probes listed in Table 2, two compounds, fluorescein or b-phycoerythrin, found extensive application for assay named Oxygen Radical Absorbance Capacity (ORAC) initially used for assessment of human serum antioxidant activity expressed as the ability to scavenge $ROO^{\bullet}$ radicals. The method, developed in 1993 (Cao et al. 1993), employed b-phycoerythrin, a water-soluble, red protein-pigment complex isolated from red algae, which exhibits high fluorescence yield quenched by peroxyl radicals produced from thermal decomposition of water-soluble azoinitiator (azobisamidine propane, abbreviated as ABAP or AAPH). Because b-phycoerythrin has many disadvantages such as inconsistent reactivity towards ROS, photosensitivity, which may lead to loss of fluorescence in side-processes, and undesired reaction with phenolic antioxidants, other fluorescent indicators, such as fluorescein and 6-carboxyfluorescein, were used (Naguib 2000a; Huang et al. 2002; Ou et al. 2001). Both 6-carboxyfluorescein and fluorescein have high quantum yield, molar absorptivity, thermo- and photo-stability. The oxidation of the fluorescent probe by $ROO^{\bullet}$ in the presence of other compounds reacting with peroxyl radicals gives information on their radical scavenging ability (frequently described as antioxidant capacity or antioxidant activity). In the ORAC protocol, the fluorescence is recorded as a fluorescence intensity plot (AUC) versus time, and the antioxidant capacity is calculated as the integrated area under the curve. Trolox™ is used in this protocol as the calibration standard antioxidant and the results are expressed as ORAC value (Eq. 1) (Ou et al. 2013).

Table 2 Overview of some commonly used quenchable **[on/off]** fluorescent probes designed for detection of various ROS

No.	Compound	Excitation/ emission wavelengths (nm)	Detected ROS	Applications
1	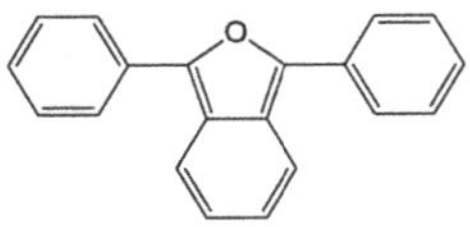 **1,3-Diphenylisobenzofuran (DPBF)**	410/455	$O_2^{\bullet-}$, $^1O_2$	A fluorescent indicator either for monitoring $O_2^{\bullet-}$ or $^1O_2$ in phospholipid liposomes (Ohyashiki et al. 1999).
2	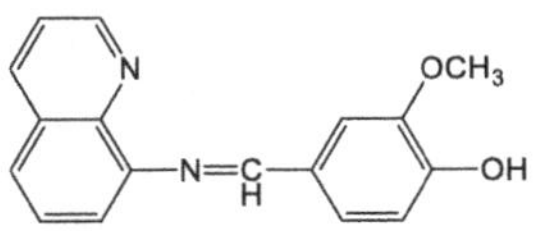 **Vanillin-8-aminoquinoline (VAQ)**	226/307	$O_2^{\bullet-}$	Determination of the rate of super-oxide generation in *red sage* (Tang et al. 2004a).
3	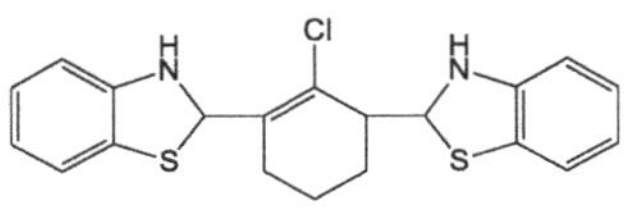 **2-chloro-1,3-dibenzothiazolinecyclohexene**	485/559	$O_2^{\bullet-}$	Imaging $O_2^{\bullet-}$ in living cells (Gao et al. 2007).
4	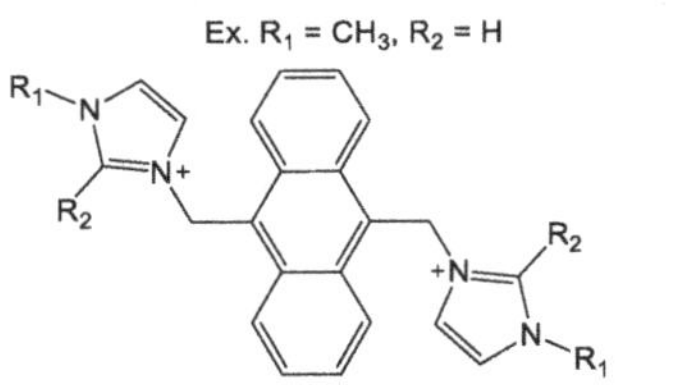 **Imidazole fluorescent ionic liquids with anthracene groups**	377/422	$O_2^{\bullet-}$	Imidazole fluores-cent ionic liquids with anthracene groups were developed for the determination of $O_2^{\bullet-}$ in aqueous systems. They were also used for determination of SOD activity in garlic and papaya extracts (Liu et al. 2013).
5	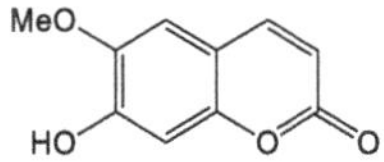 **Scopoletin**	360/460	H_2O_2 in presence of HRP[a]	Widely used as a probe for H_2O_2 monitoring, either in isolated mito-chondria, or in stimulated neutro-phils and eosino-phils (Bartoli et al. 1977; Mohanty et al. 1997; Zhou et al. 1997; Staniek and Nohl

(continued)

Table 2 (continued)

No.	Compound	Excitation/ emission wavelengths (nm)	Detected ROS	Applications
				1999; Barja 2002).
6	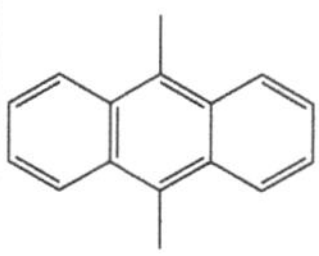 **9,10-Dimethylanthracene (DMA)**	375/436	$^1O_2$	Selective detection of singlet oxygen in many organic solvents and water (Corey and Taylor 1964; Usui 1973; Wilkinson and Brummer 1981).
7	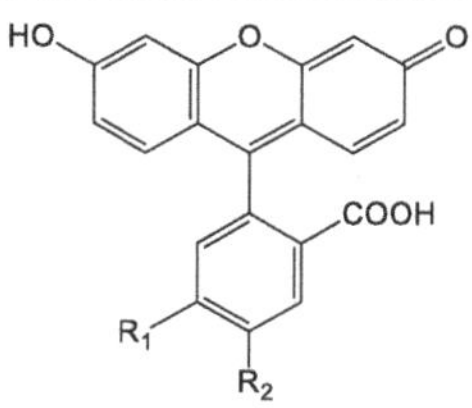 **Fluorescein ($R_1 = R_2 = H$)/ 6-Carboxyfluorescein ($R1 = CO_2H$, $R2 = H$)**	495/515	$ROO^•$	Evaluation of antioxidant capacity in commonly used ORAC[b] technique (see the text).
7a	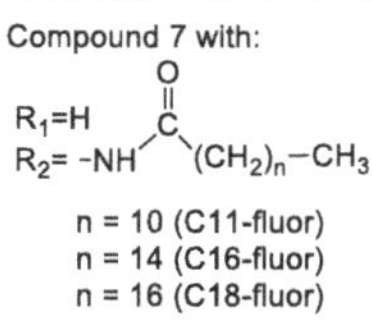 **5-(N-alkanoyl)aminofluorescein (C11, C16, and C18-fluor)**	495/515	$ROO^•$	Lipophilic derivative of comp. 7. Determination of lipid peroxidation associated to cellular membrane (Laguerre et al. 2007; Makrigiorgos et al. 1997).
7b	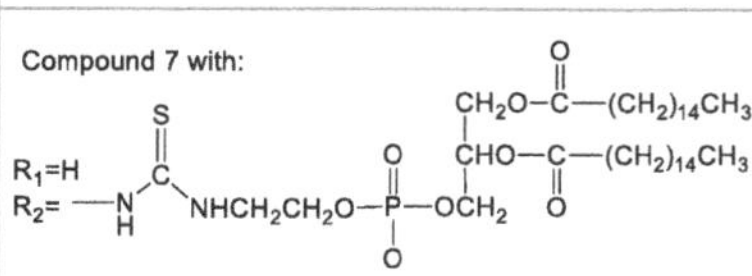 **Fluorescein with phospholipide (fluor-DHPE)**	488/520	$ROO^•$	Determination of lipid peroxidation associated to cellular membrane (Laguerre et al. 2007; Maulik et al. 1998).
8	**b-Phycoerythrin**	520/580	$ROO^•$	Evaluation of antioxidant capacity in commonly used ORAC technique (see the text).

(continued)

Table 2 (continued)

No.	Compound	Excitation/ emission wavelengths (nm)	Detected ROS	Applications
9	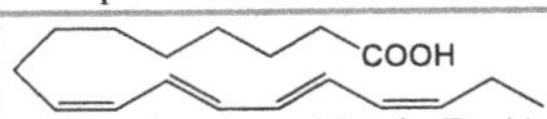 ***cis*-Parinaric acid (cis-PnA)**	320/432	$ROO^{\bullet}$	*cis*-PnA can be used as membrane probe for detecting the initial stages of lipid peroxidation in living cells (Drummen et al. 1999).
10	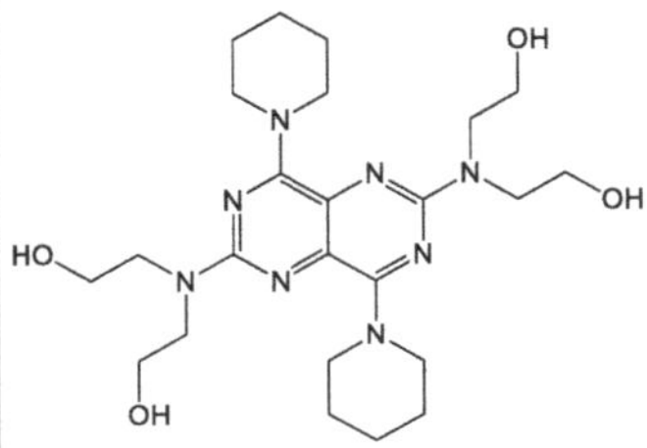 **Dipyridamole**	415/480	$ROO^{\bullet}$, $HO^{\bullet}$, $O_2^{\bullet-}$	This is an example of a well-known pharmaceutical drug used as a coronary vasodilator and anti-platelet agent which was also applied as a fluorescent probe (on/off type) for ROS in biological systems (Iuliano et al. 2000).

[a] *HRP* Horseradish peroxidase
[b] *ORAC* Oxygen Radical Absorbance Capacity

$$\text{ORAC value} = \left[\left(\text{AUC}_{\text{Sample}} - \text{AUC}_{\text{Blank}}\right)/\text{AUC}_{\text{Trolox}} - \text{AUC}_{\text{Blank}}\right] \times (\text{molarity of Trolox}/\text{molarity of sample}) \quad (1)$$

A linear relation of the ORAC value and the antioxidant concentration was confirmed for ORAC protocol applied in human serum samples and samples containing antioxidants (Cao et al. 1993), therefore, the ORAC assay was successfully used in the studies of lipophilic antioxidants such as α-tocopherol, β-carotene or bilirubin. On the other hand, it was suggested that this method gives unreliable results for lipophilic antioxidants because test is performed in aqueous solutions (Huang et al. 2002; Prior et al. 2003). To improve the assay, randomly methylated cyclodextrins were used to promote the solubility of lipophilic antioxidants in aqueous systems (Huang et al. 2002; Ou et al. 2013). The method is standardised and straightforward, however, the ORAC value can be affected by the secondary reactions (fluorescein might be recovered by the studied antioxidants changing the stoichiometry of the reaction) (Dorta et al. 2018).

The second type of fluorescent probes are the off/on probes, also called fluorogenic probes. They are non- or weakly emissive in their reduced form, but their fluorescence is enhanced as a result of reaction with ROS. These probes have an apparent advantage over the on/off probes because the processes that may lead to a false increase of the fluorescence are not frequent. However, such advantages can be apparent, as the generated fluorescence can be quenched in several ways, producing the underestimated results. Many mechanisms might affect the fluorogenic effect of the developed off/on probes. Some of those processes include changes in the fluorophore's photochemical properties due to redox reactions that result in shifts in the absorption and emission spectra. The overview of some commonly used off/on probes based on this processes is presented in Table 3. For example, 2,7-dichlorodihydrofluorescein (DCFH) can be mentioned as one of the oldest and commonly used fluorogenic sensor for reactive oxygen species. A reduced form of DCFH is non-fluorescent, but after the oxidation it exhibits a strong fluorescence (Fig. 3).

Initially, DCFH was believed as a specific indicator for hydrogen peroxide (Bystryak et al. 1995), but later the reactions with a variety of reactive oxygen and nitrogen species ($HO^•$, $ROO^•$, $NO^•$, $ONOO^-$) were reported as generating fluorescence of DCFH (Crow 1997; Wang and Joseph 1999). Its diacetate form, DCFH-DA, can diffuse across the cellular membrane and in cytosol it is hydrolysed by esterases to DFCH, see Fig. 3. In this way DCFH-DA/DCFH probe is a good tool for detection of ROS within the cells (Halliwell and Whiteman 2004). Probes based on DCFH derivatives are not selective towards ROS. Moreover, a significant background fluorescence is observed in some lipid-containing systems that make them not so useful for detecting specific ROS in the cellular system.

High selectivity of fluorescent probes can be achieved when the reaction is combined with enzymatic catalysis. For example, N-acetyl-3,7-dihydroxyphenoxazine (Amplex Red) is an example of fluorogenic probe that is highly specific towards hydrogen peroxide because in presence of horseradish peroxidase (HRP) Amplex Red is quantitatively oxidized by H_2O_2 to fluorescent resorufin, see Fig. 4 (Mohanty et al. 1997). Moreover, the same probe can be also used for detection and quantitative determination of $O_2^{•-}$ when another enzyme is added to the system, superoxide dismutase, responsible for the conversion of superoxide radical into hydrogen peroxide (Pastor et al. 2004).

In 2004 Maeda et al. (Maeda et al. 2004) proposed a probe for selective detection of hydrogen peroxide which, unlike DCFH and Amplex Red, was fluorescent after reaction with H_2O_2 and hydrolysis (perhydrolysis) of protected derivative, see Fig. 5. This different mechanism (release of the fluorescent product) allowed them to detect endogenous cellular hydrogen peroxide with higher selectivity than DCFH and its analogues. They designed and presented a series of pentafluorobenzenesulfonyl fluoresceins as selective fluorescent probes for H_2O_2 (Fig. 5).

In their later studies Maeda et al. (Maeda et al. 2005) also presented a series of fluorescent probes for specific detection of $O_2^{•-}$, bis(2,4-dinitrobenzenesulfonyl) fluoresceins, based on this same mode of action. Some of them are presented in Table 4.

Electron transfer and in general, charge transfer processes such as Intramolecular Charge Transfer (ICT), Twisted Intramolecular Charge Transfer (TICT), and

Table 3 Overview of some commonly used fluorogenic **[off/on]** fluorescent probes for ROS detection based on general oxidation of fluorophore that leads to emission increase and their applications

No.	Compound	Ex/em wavelengths (nm)	Detected ROS	Typical applications
11	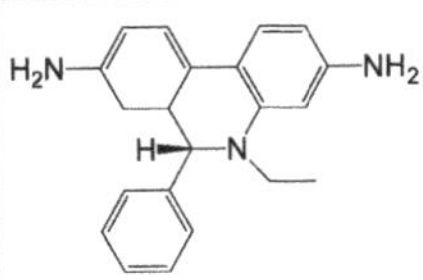 **Hydroethidine (HE)**	520/610	$O_2^{\bullet -}$	Widely used as probe for monitoring oxidative burst in cells (Bindokas et al. 1996).
12	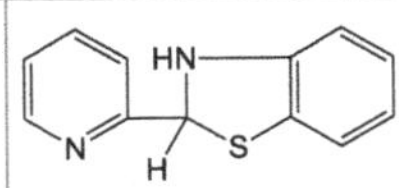 **2-(2-Pyridil)-benzothiazoline**	377/528	$O_2^{\bullet -}$	Probe was used to determine SOD activity in garlic, scallion, and onion. Research included the application of the probe to flow injection spectrofluorimetry for a rapid and automatic analysis (Tang et al. 2004b).
13	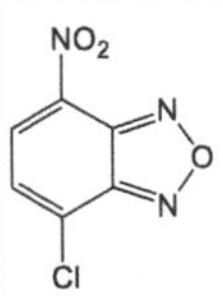 **4-chloro-7-nitrobenzo-2-oxa-1,3-diazole (NBD-Cl)**	470/550	$O_2^{\bullet -}$	For detection of $O_2^{\bullet -}$ derived from KO_2/DMSO solution as well as $O_2^{\bullet -}$ generated from the xanthine–xanthine oxidase reaction or by the addition of NADH to skeletal muscle sarcoplasmic reticulum (SR) vesicles (Olojo et al. 2005).
14	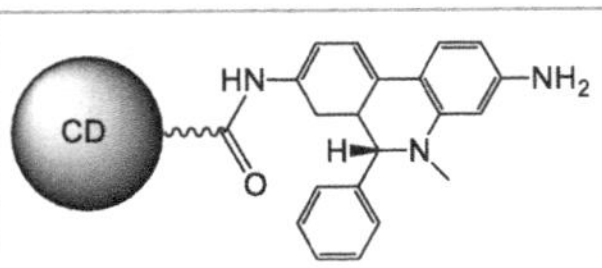 **Carbon dots and hydroethidine hybrid CD-HE**	488/525, 610	$O_2^{\bullet -}$	The probe has good cell-permeability, low cytotoxicity, and can be used for imaging superoxide in living cells (Gao et al. 2014).
15	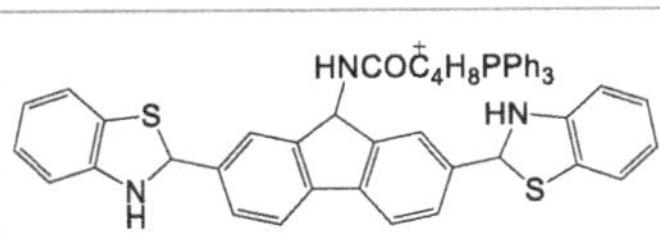 **9-butyltriphenylphosphoniu-macylamino-2,7-**	483/512 (OP)	$O_2^{\bullet -}$	First probe sensitive towards superoxide radicals used in one-photon (OP) and two-photon (TP) microscopy

(continued)

Table 3 (continued)

No.	Compound	Ex/em wavelengths (nm)	Detected ROS	Typical applications
	dibenzothiazolineflurene (MF-DBZH)			for imaging in mitochondria (Li et al. 2013).
16	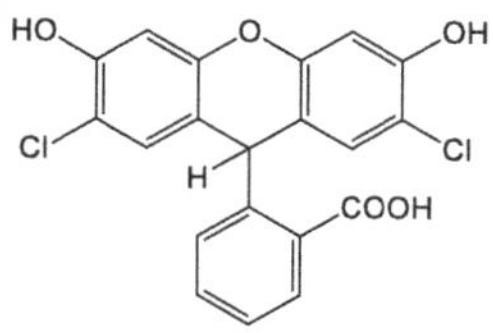 **2,7-Dichlorodihydrofluorescein (DCFH)**	498/522	H_2O_2, $HO^{\bullet}$, $ROO^{\bullet}$	See Fig. 3, diacetate form, DCFH-DA diffuses through the cellular membrane. After hydrolysis by esterases free DCFH becomes fluorescent probe useful as a marker of the cellular oxidative stress (Halliwell and Whiteman 2004). DCFH has been also used as an oxidative burst indicator in macrophages and neutrophils (Keston and Brandt 1965; Caldefie-Chézet et al. 2002).
17	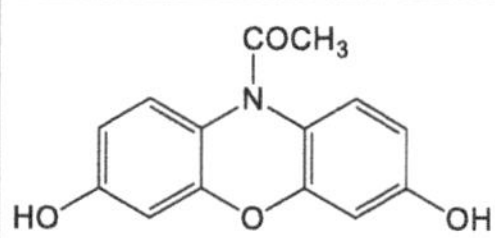 **N-Acetyl-3,7-dihydroxyphenoxazine (Amplex Red)**	563/587	H_2O_2 in presence of HRP O2	Amplex Red is a probe specific for the detection of H_2O_2, and can be used not only in activated phagocytic cells but also in other types of cells or even in non-cellular systems (Mohanty et al. 1997; Zhou et al. 1997).
18	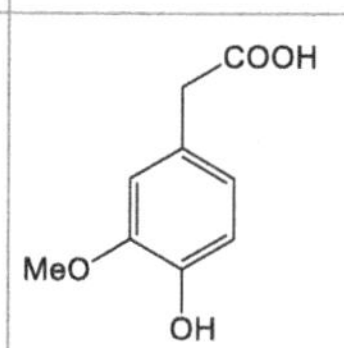 **Homovanillic acid**	312/420	H_2O_2 in presence of HRP; $O_2^{\bullet-}$ in the presence of HRP and SOD	Quantification of the fast and basal production rates of ROS in biological systems (Ruch et al. 1983). Fluorescent probe has been used for detecting the production of H_2O_2 in isolated

(continued)

Table 3 (continued)

No.	Compound	Ex/em wavelengths (nm)	Detected ROS	Typical applications
				mitochondria of various tissues, healthy, or otherwise affected by degenerative diseases (Barja 2002). HVA can be successfully used in the measurement of enzyme activity, when H_2O_2 is one of the final products as it has been shown in the measurement of for example lysyl oxidase (Palamakumbura and Trackman 2002) or acyl-CoA oxidases (Lageweg et al. 1991).
19	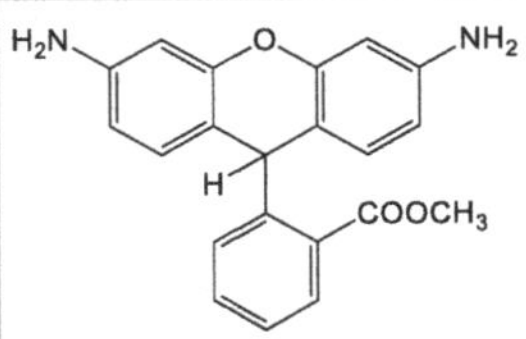 **Dihydrorhodamine 123 (DHR)**	505/529	H_2O_2 in presence of HRP	H_2O_2 but also HOCl, and ONOO detection in cells and also evaluation of scavenging activity (Crow 1997; Henderson and Chappell 1993; Kooy et al. 1994). Detection of singlet oxygen and determination of scavenging activity of $^1O_2$ by various agents (Costa et al. 2007).
20	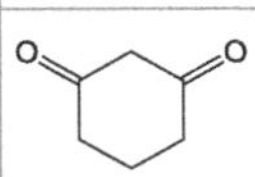 **1,3-Cyclohexanedione (CHD)**	400/452	$HO^•$	$HO^•$ and DMSO produces formaldehyde which at pH 4.5 reacts with ammonia and CHD to generate a product that is fluorescent upon heating (Tai et al. 2002).

(continued)

Table 3 (continued)

No.	Compound	Ex/em wavelengths (nm)	Detected ROS	Typical applications
21	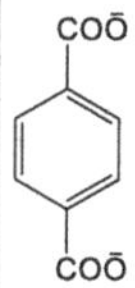 **Sodium terephthalate**	310/430	$HO^{\bullet}$	This compound was used as an agent trapping $HO^{\bullet}$ for verification of the scavenging effect of thiourea and mannitol (Tang et al. 2005). Because excitation and emission wavelengths of fluorescent 2-hydroxy-terephtalic acid ($\lambda_{ex} = 326$ nm, $\lambda_{em} = 432$ nm) are close to UV the probe is not suitable for the in vivo imaging of $HO^{\bullet}$ in living cells, but was used in vitro in flow systems (Qu et al. 2000; Yan et al. 2005).
22	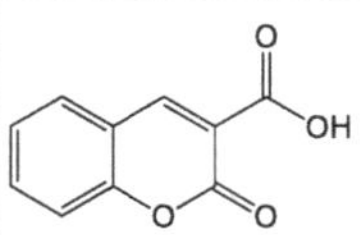 **Coumarin-3-carboxylic acid** (3-CCA)	350/450	$HO^{\bullet}$	Broadly used for the screening of scavenging activity (Crosby and Berthold 1962; Sherman and Robins 1968; Xiao and Parkin 2002) as well as $HO^{\bullet}$ generation activity (Lindqvist and Nordström 2001) of several compounds.
23	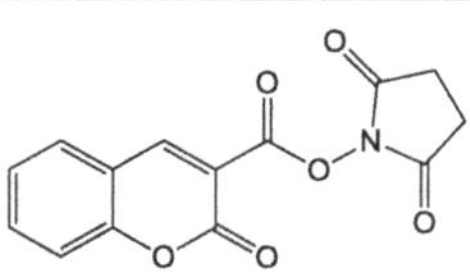 **N-succinimidyl ester of coumarin-3-carboxylic acid (SECCA)**	395/450	$HO^{\bullet}$	Similar to CCA. The amine-reactive SECCA can be coupled with various biomolecules such as albumin, avidin, histon, polylysine, and oligonucleotide, and such SECCA–biomolecule

(continued)

Table 3 (continued)

No.	Compound	Ex/em wavelengths (nm)	Detected ROS	Typical applications
				conjugates also react with HO$^•$. On the other hand, the excitation wavelengths of these probes are in short range (<400 nm) so biological applications of the probes for the in vivo measurements are limited (Makrigiorgos et al. 1993).
24	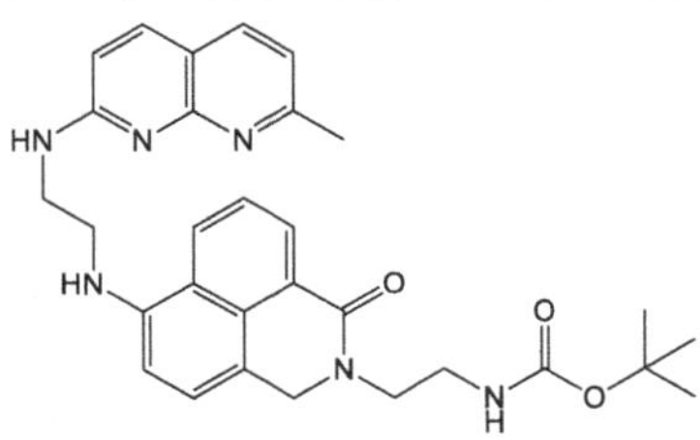 **Naphthalimide–naphthyridine derivative**	371/418, 552	HO$^•$	The probe does not respond to other ROS and RNS and has low cytotoxicity and high biocompatibility that enabled it to be used for detection of intracellular HO$^•$ (Meng et al. 2014).
25	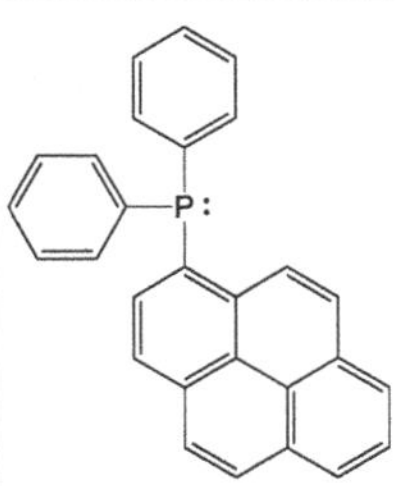 **Diphenyl-1-pyrenylphosphine (DPPP)**	351/380	ROO$^•$, ROOH	Also, triphenylphosphine (TPP). Non-polar DPPP and TPP react quantitatively with lipid hydroperoxides (and H_2O_2) giving stable phosphine oxides that can be separated by HPLC with UV or fluorescence detectors (Akasaka et al. 1987). Can be used for the measurement of the extent of oxidation in solution and in low-density lipoproteins (Noguchi et al. 1998) as well as markers for membrane peroxidation in living

(continued)

Table 3 (continued)

No.	Compound	Ex/em wavelengths (nm)	Detected ROS	Typical applications
				cells (Okimoto et al. 2000). DPPP was also used in hydroperoxides measurements in dietary oils and fats (Akasaka and Ohrui 2000; Tikekar et al. 2011; Uluata et al. 2021)

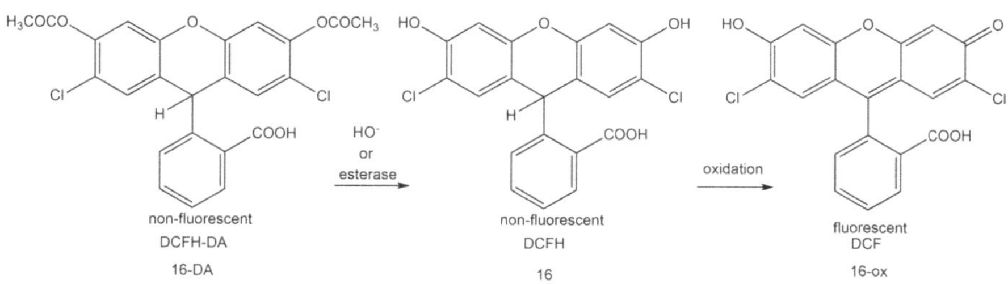

Fig. 3 Scheme of non-fluorescent 2,7-dichlorodihydrofluorescein diacetate hydrolysis and further oxidation to fluorescent 2,7-dichlorofluorescein (Crow 1997)

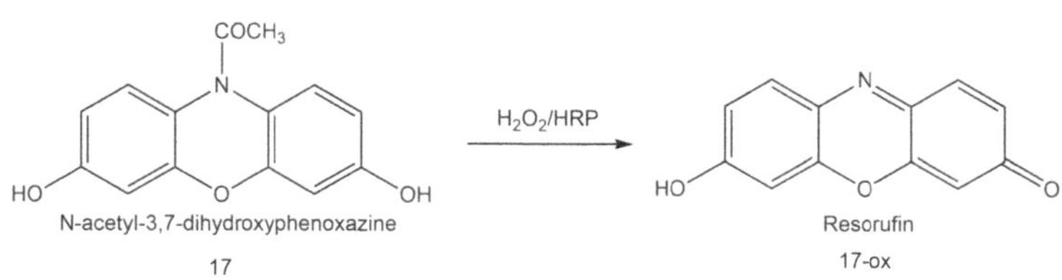

Fig. 4 Oxidation scheme of non-fluorescent N-acetyl-3,7-dihydroxyphenoxazine to fluorescent resorufin by hydrogen peroxide in presence of HRP

Photoinduced electron Transfer (PeT) are widely exploited in development of fluorogenic probes. Also, it has been presented that paramagnetic nitroxides are efficient quenchers of excited singlet state of aromatic hydrocarbons (Green and Singer 1974; Atik and Singer 1978). For example, 4-(9-anthroyloxy)-2,2,6,6-tetramethylpiperidine-1-oxyl was presented by Yang et al. (Fig. 6) as fluorogenic probe for hydroxyl radicals (Yang and Guo 2001). In this probe fluorescence is internally quenched by electron exchange between excited state of fluorophore (anthracene moiety) and nitroxide at its ground state. However, the reaction of this hybrid molecule with carbon-centred radical produces a diamagnetic product in which the intramolecular quenching is arrested, thus, a great enhancement of

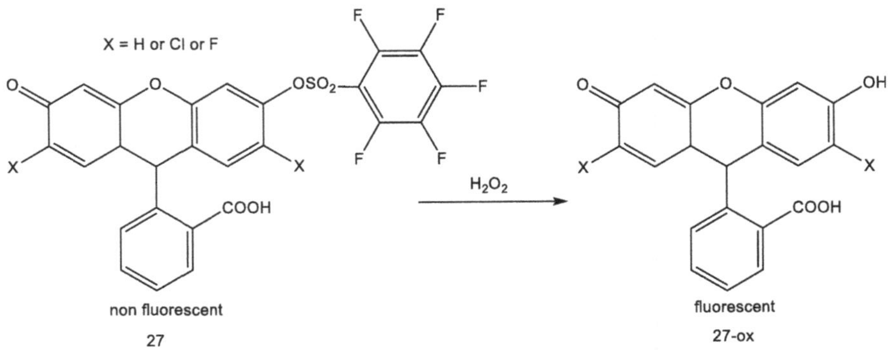

Fig. 5 Fluorescent probes for hydrogen peroxides developed by Maeda et al. and deprotection reaction that produces their fluorescence

fluorescence is observed. The probe can indirectly detect $HO^{\bullet}$ in the presence of DMSO with quantitative evolution of methyl radical that reacts with the probe (Yang and Guo 2001). DMSO is a good trapping agent for hydroxyl radicals and is relatively well tolerated by living systems (up to 1 M concentration), the proposed probe could be applied for biological studies, however, the interaction of nitroxyl moiety with endogenous reductants (ascorbate, glutathione) can false the results. The authors did not discuss the effect of diffusion controlled reaction of methyl radical with molecular oxygen present in biological samples.

Recently, Wu and co-workers designed and presented a simple fluorescent probe HMBT-LW (see Tables 5, 43) based on excited state intramolecular proton transfer (ESIPT) phenomenon (Wu et al. 2019b). Fluorophores that exhibit ESIPT usually exist in the ground state in their enolic (E) form. Figure 7 presents the example of such molecule: the probe 43 after the reaction with $O_2^{\bullet -}$ undergoes photoexcitation inducing a change in the distribution of the electronic charge, that increases the ability to form internal hydrogen bond and, subsequently, rapid phototautomerization from the low-fluorescent enol form (43-ox-E*) into the high-fluorescent, excited keto form (43-ox-K*).

Photoinduced electron Transfer (PeT) is extensively used in the development of fluorogenic off/on probes (Nagano 2010; De Silva et al. 1997). Such fluorescent probes are usually assembled from the receptor part (responsible for sensing) that is covalently linked to the fluorophore part acting as a reporter. The fluorophore can be excited by light and can exhibit increased fluorescence as an effect of the chemical changes arising from the reaction of the receptor with ROS. The selectivity depends on the receptor that should be designed to react with a specific ROS.

The fluorescence is an effect of radiative transfer of the electrons from the lowest unoccupied molecular orbital (LUMO) of excited fluorophore (fluorophore which absorbed light) to their highest occupied molecular orbital (HOMO) (Fig. 8b). However, a competitive process, PeT is able to quench the fluorescence, and the probe exists at non-emissive state. This situation is graphically represented in Fig. 8a and c. There are two possible mechanisms for this phenomenon. If the fluorophore is

Table 4 Examples of molecular precursors of the fluorescent probes for ROS detection based on deprotection of fluorophore that leads to increase of emission

No.	Compound	Ex/em wavelengths (nm)	Detected ROS	Applications
26	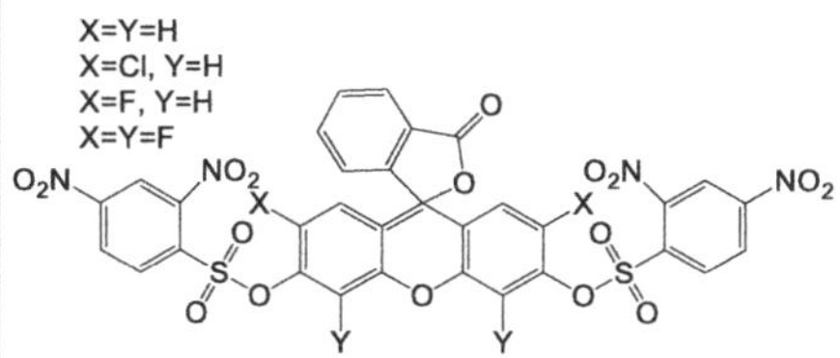 **Bis(2,4-dinitrobenzenesulfonyl) fluorescein**	485/515 511/540 (X = Y=F)	$O_2^{\bullet -}$	This probe provides a highly specific fluorescent response toward $O_2^{\bullet -}$ over other ROS and is capable of the real-time detection of $O_2^{\bullet -}$ generated in living cells with a high degree of selectivity (Maeda et al. 2005).
27	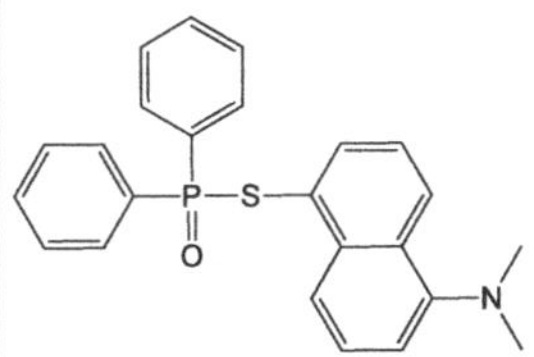 **TPP** based on 5-(dimethylamino)- naphthalene-1-thiol	345/470	$O_2^{\bullet -}$	TPP was successfully applied in two-photon microscopy for imaging endogenous superoxide anion radicals in living cells and tissues (Chen et al. 2019).
27	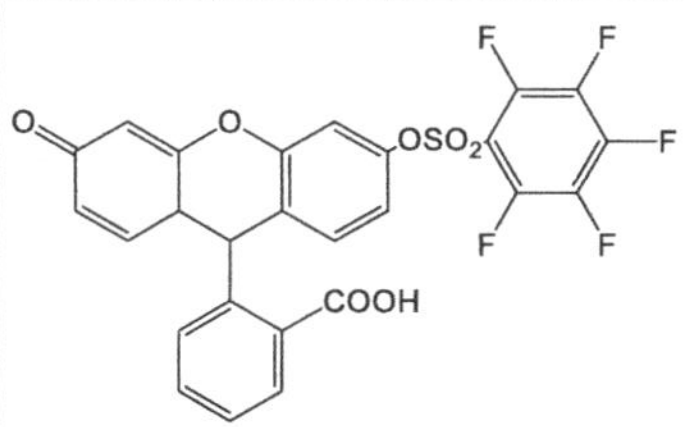 **Pentafluorobenzenesulfonyl fluorescein**	485/530	H_2O_2	Acetyl derivatives of this fluorescent probe have been applied to the detection of H_2O_2 in algal cells. Measurements of cell-derived H_2O_2 and monitoring the dynamic functions of oxidative stress, not only in algal cells, but also in phagocytes and vascular endothelium cells. The probe is active in cellular systems without loss of selectivity (Maeda et al. 2004).

(continued)

Table 4 (continued)

No.	Compound	Ex/em wavelengths (nm)	Detected ROS	Applications
28	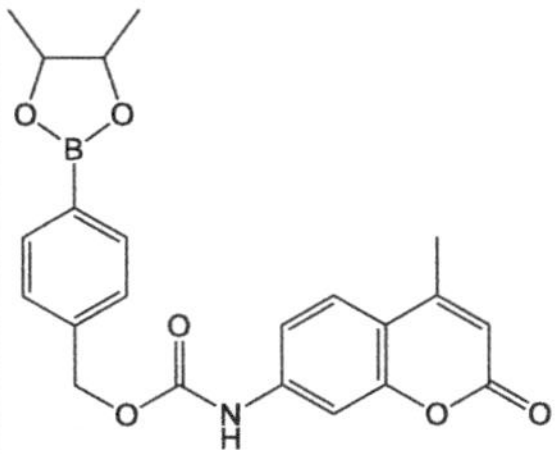 **Aminocoumarin masked by butanediol ester of Dobz derivative**	348/440	H_2O_2	This probe was used to measure H_2O_2 under alkaline conditions, no reported biological applications (Lo and Chu 2003).
29	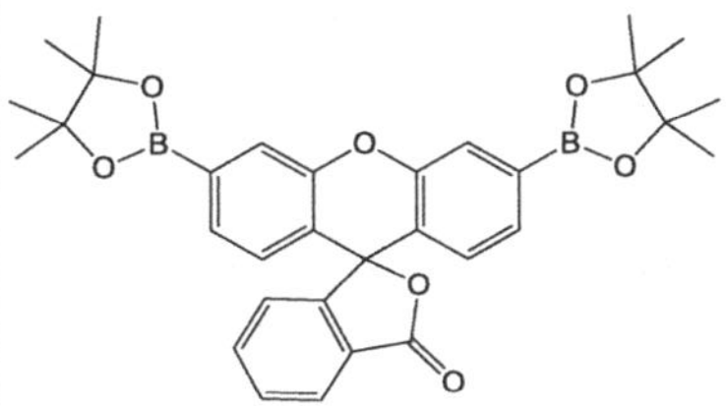 **Peroxyfluor-1 (PF1)**	450/510	H_2O_2	This kind of dyes is able to cross the biomembranes and can be used to measure even nanomolar changes in intracellular H_2O_2 concentration in cells, including hippocampal neurons from primary culture, using confocal microscopy and two-photon fluorescence microscopy (Chang et al. 2004).
30	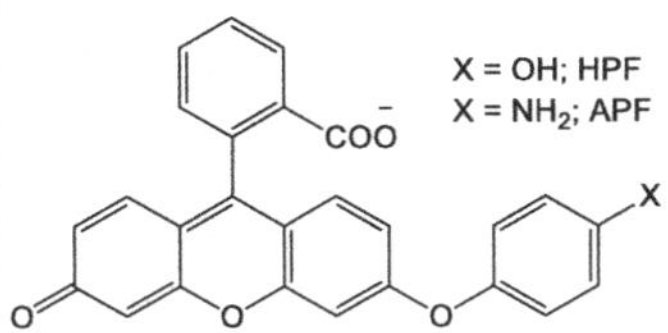 **HPF and APF**	500/520	$HO^•$	Detection of $HO^•$ or HOCl in either cellular or non-cellular systems. Utilizing HPF and APF together allows visualizing of ^{-}OCl generation in stimulated neutrophils (Setsukinai et al. 2003).

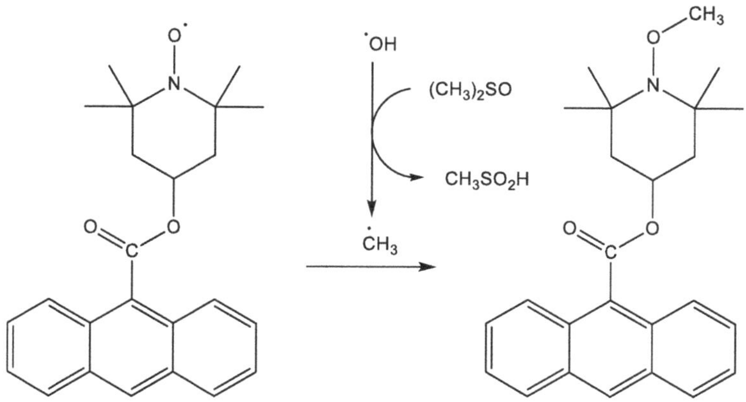

Fig. 6 Indirect detection of hydroxyl radical by 4-(9-anthroyloxy)-2,2,6,6-tetramethylpiperidine-1-oxyl

the electron acceptor, the fluorescence is quenched via electron transfer from the HOMO of the receptor (quencher) to the semi-occupied HOMO level of the excited chromophore, as visualized in Fig. 8a. Second mechanism is observed when the fluorophore is an electron donor, thus the electron is promoted to LUMO level and subsequently transferred from the LUMO of the photo-excited fluorophore to LUMO level of the quencher (receptor) (Fig. 8c) efficiently reducing the fluorescence of the entire molecule by inhibition of emission from the excited fluorophore.

The efficiency of PeT depends on the difference between the one-electron redox potential for the electron donor in the PeT process and the one-electron redox potential for the electron acceptor (Krumova et al. 2009). That makes almost any receptor to be viable as a part of the fluorogenic probe as long as it is connected to the fluorophore with suitable redox potential. Other parameters typically expected for good fluorescent probes are also important, like high extinction coefficient, quantum yield of fluorescence, high Stokes Shifts, good photostability, and emission in the visible or near-infrared range of the spectrum. This last feature helps to prevent cellular damage and increases tissue penetration if a fluorogenic probe is to be used in cell studies.

An example of fluorogenic lipophilic ROS sensors exploiting PeT phenomenon are the series of compounds developed by Cosa et al. (Oleynik et al. 2007; Krumova et al. 2012; Greene et al. 2017; Krumova et al. 2013; Greene et al. 2018) as a fluorogenic antioxidants mimicking the structure and activity of common antioxidants. The mechanism of action is presented in Fig. 9. B-TOH is a two segment receptor-reporter type probe that is non-emissive in the reduced state but reaction with radicals disables PeT and renders the molecule emissive.

Chromanol moiety with radical trapping activity similar to α-tocopherol is the receptor reacting with the radicals present in lipid phase (usually, lipid peroxyl

Table 5 Overview of some commonly used fluorogenic **[off/on]** fluorescent probes for ROS detection based on intramolecular electron and charge transfers that lead to emission increase and their applications

No.	Compound	Ex/em wavelengths (nm)	Detected ROS	Type of action	Applications
31	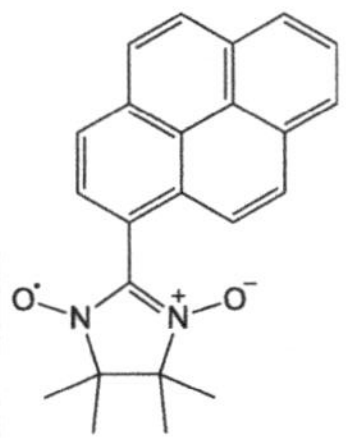 **4,4,5,5-tetramethyl-2-(1-pyryl)-2-imidazoline-1-oxyl-3-oxide (FNO)**	340/440	$O_2^{\bullet -}$	Nitroxide radical quenching	Studies of $O_2^{\bullet -}$ dynamics and the antioxidant status of biological systems (Medvedeva et al. 2004).
32	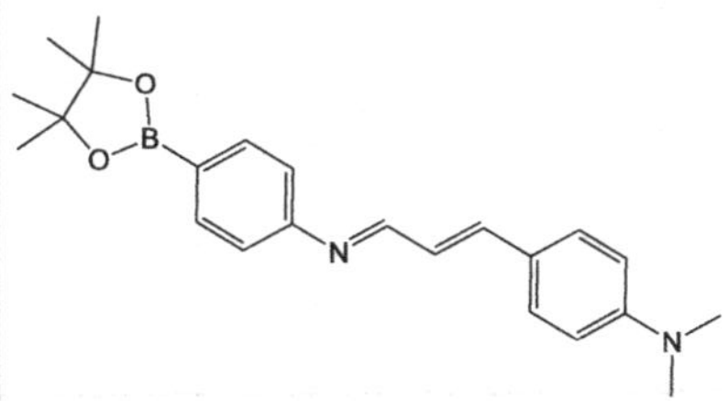 **Dimethylaminocinnamaldehyde linked arylboronate based probe**	400/470 (EtOH)	H_2O_2	TICT	Used for detection H_2O_2 in prostate cancer cells (Kumar et al. 2012).

33	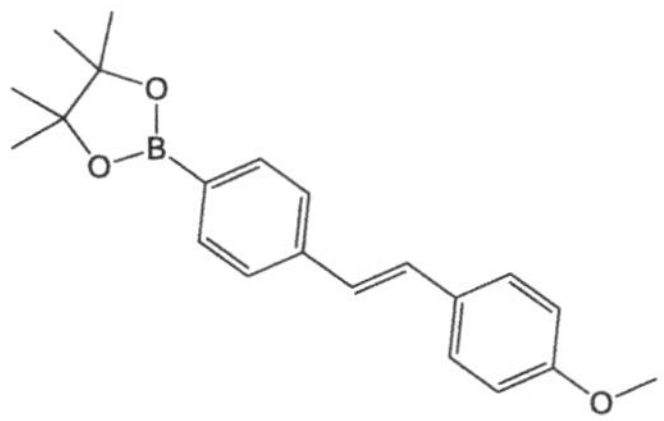 **MSTBPin**	330/360	H_2O_2	ICT	Used for sensing the intracellural H_2O_2. The whole family of similar boronate probes (i.e. DSTBPin, DAPOX-BPin) was presented by Lampard et al. (Lampard et al. 2018) and reviewed by Wu et al. (Wu et al. 2019a)
34	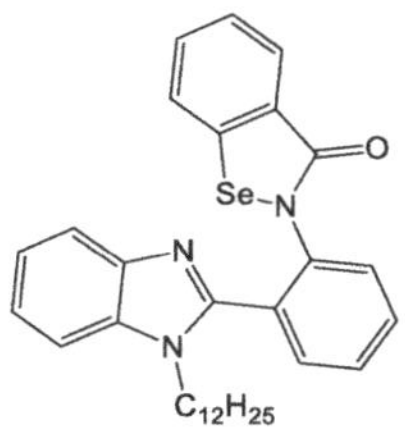 **D-HMSe**	330/476	H_2O_2	PeT	Sensing and monitoring H_2O_2 in water based systems. Authors suggested that probe might be used as a good sensor for use in food or environmental science (Liao et al. 2014).
35	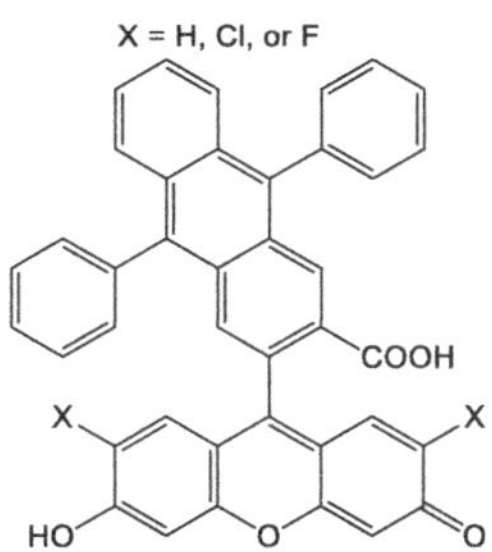 **9-[2-(3-Carboxy-9,10-diphenyl)anthryl]-6-hydroxy-3H-xanthen-3-ones (DPAXs)**	495/515	$^1O_2$	PeT	The first chemical traps for singlet oxygen that allowed to use fluorescence detection for measurements. Some DPAX derivatives can permeate the biomembranes (Umezawa et al. 1999).

(continued)

Table 5 (continued)

No.	Compound	Ex/em wavelengths (nm)	Detected ROS	Type of action	Applications
36	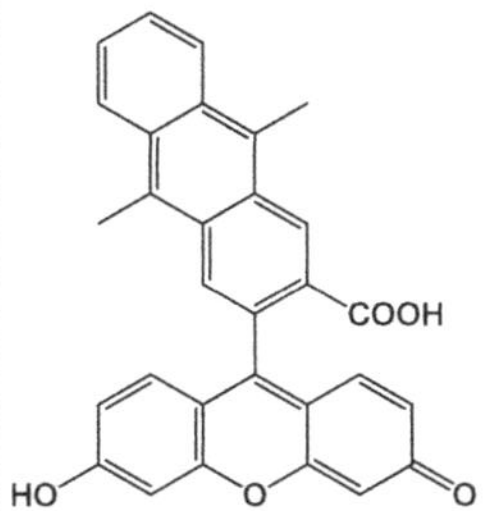 **9-[2-(3-Carboxy-9,10-dimethyl)anthryl]-6-hydroxy-3H-xanthen-3-one (DMAX)**	495/515	$^1O_2$	PeT	DMAX is less hydrophobic than DPAXs, therefore is more appropriate for detection of $^1O_2$ in biological systems (Tanaka et al. 2001).
37	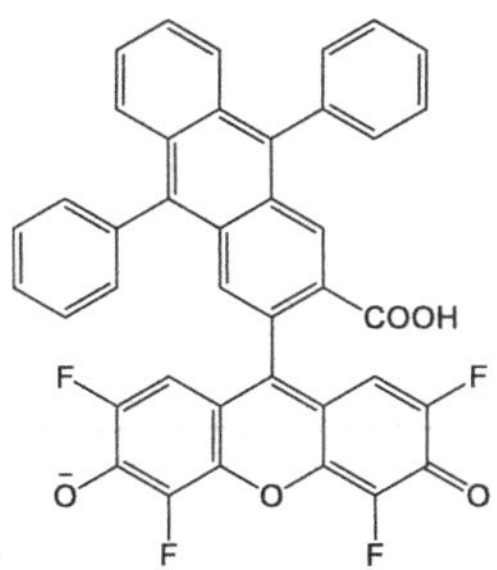 **Aarhus sensor Green (ASG)**	420/537 (D_2O)	$^1O_2$	PeT	Intracellular (as it is non-cytotoxic) singlet oxygen probe in mammalian cells (Pedersen et al. 2014).

38	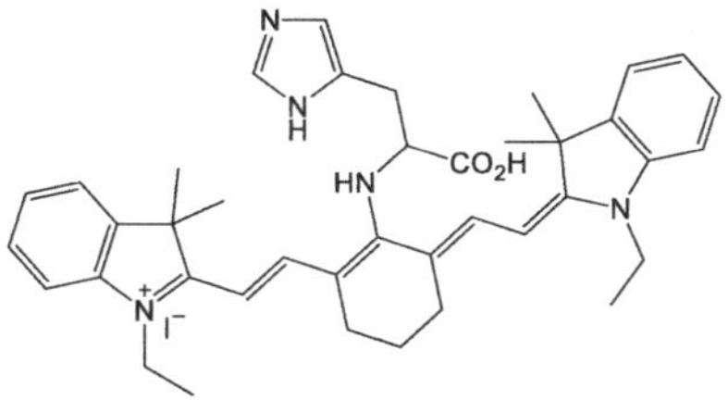**Tricarbocyanine-histidine based near-infrared fluorescent probe His-Cy**	754/794	$^1O_2$	PeT	The probe was applied for imaging $^1O_2$ generation in phorbol myristate acetate (PMA)- stimulated RAW264.7 cells (Xu et al. 2011).
39	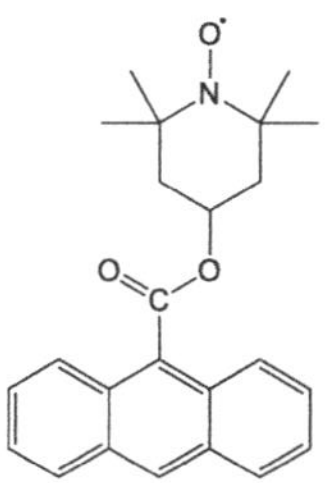**4-(9-Anthroyloxy)-2,2,6,6-tetramethylpiperidine-1-oxyl**	377/427	$HO^{\bullet}$	Nitroxide radical quenching	Indirect detection of $HO^{\bullet}$ via reaction with reaction witch DMSO, the probe can be monitored either by fluorescence or by EPR spectroscopy (Yang and Guo 2001).
40	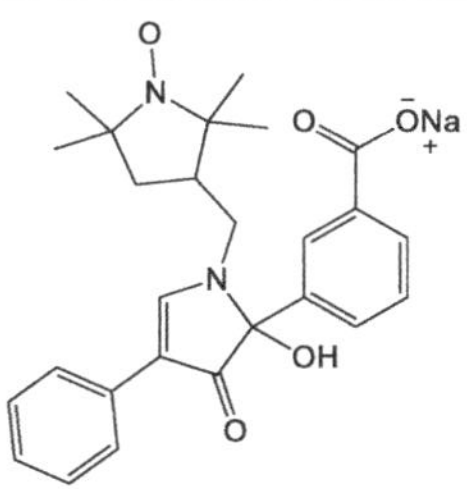**Fluorescamine-linked nitroxide**	384/485	$HO^{\bullet}$, $O_2^{\bullet-}$	Nitroxide radical quenching	Used for detection of $HO^{\bullet}$ and $O_2^{\bullet-}$ generated in stimulated neutrophils (Pou et al. 1993) or $HO^{\bullet}$ during redox cycling of quinone containing anticancer agents (Li et al. 2000).

(continued)

Table 5 (continued)

No.	Compound	Ex/em wavelengths (nm)	Detected ROS	Type of action	Applications
41	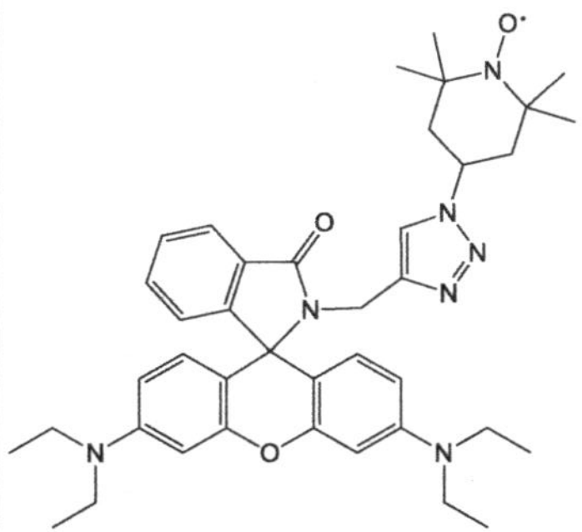 **Rhodamine nitroxide based fluorescent probe**	560/588	$HO^{\bullet}$	Nitroxide radical quenching	This probe was used as marker of $HO^{\bullet}$ produced in human retinal pigment epithelial cells under oxidative stress. In addition, the probe has also been applied to imaging $HO^{\bullet}$ in cancer cells (Yapici et al. 2012).
42	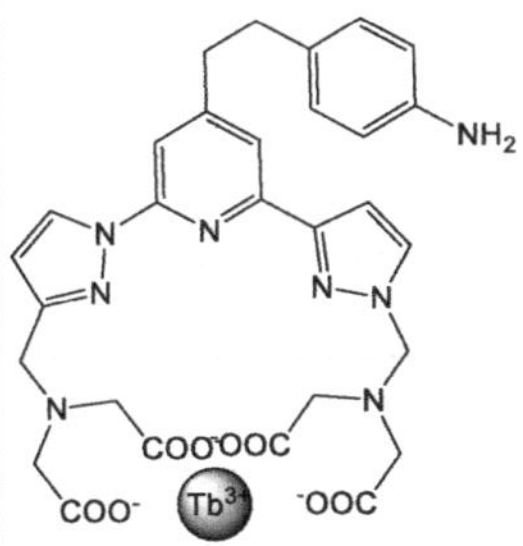 **Tb complex-based, luminescent probe ($BMPTA\text{-}TB^{3+}$)**	320/545	$HO^{\bullet}$	PeT	Other ROS do not affect the detection of $HO^{\bullet}$, indicating that the probe has high selectivity towards $HO^{\bullet}$. The probe was employed to detect $HO^{\bullet}$ in HeLa cells (Cui et al. 2011).

43	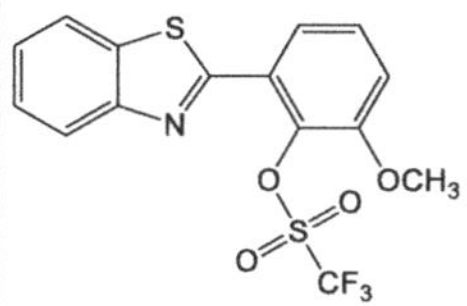 **HMBT-LW**	310/460	$O_2^{\bullet -}$	ESIPT	This probe was developed to rapidly detect low concentrations of $O_2^{\bullet -}$. And it demonstrates great selectivity and sensitivity towards it. No biological applications were presented (Wu et al. 2019b).
44	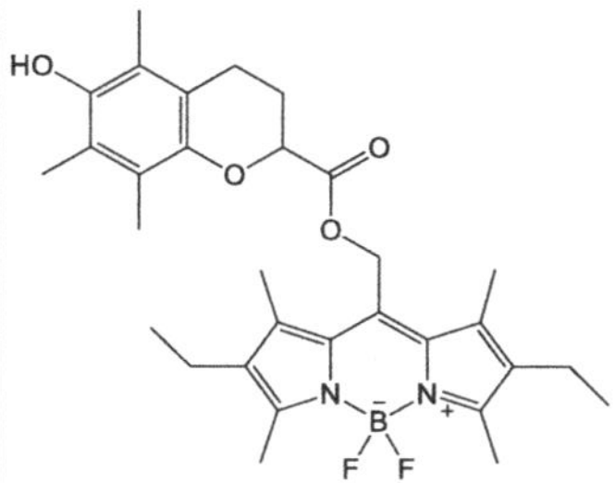 **B-TOH**	549/555 (hexane)	$ROO^{\bullet}$	PeT	Monitoring the lipid peroxyl radicals (peroxidation extent) in model homogenous systems, lipid membranes and in living cells. Can be used in fluorescent microscopy (Oleynik et al. 2007; Krumova et al. 2012; Greene et al. 2017).
45	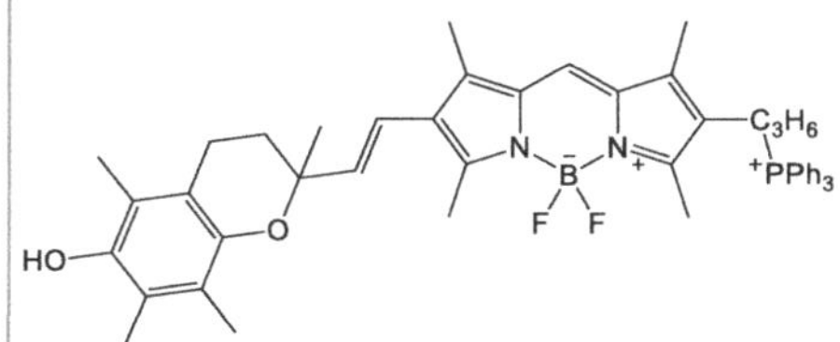 **Mito-BODIPY-TOH**	536/582 (MeCN)	$ROO^{\bullet}$	PeT	Mito-BODIPY-TOH enables real-time monitoring of lipid peroxyl radical generation and antioxidant depletion within the inner mitochondrial membrane (Krumova et al. 2013).

(continued)

Table 5 (continued)

No.	Compound	Ex/em wavelengths (nm)	Detected ROS	Type of action	Applications
46	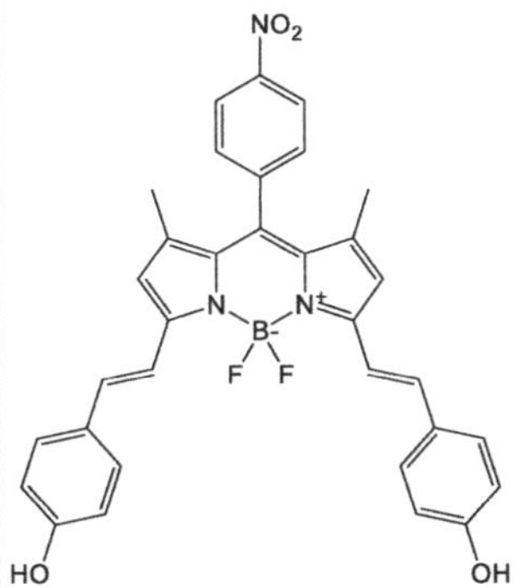 **Hydroxycinnamyl derived BODIPY (NB-2)**	575/610, 674 (MeOH)	$ROO^{\bullet}$	Probably PeT	Detection of peroxyl radicals in homogenous solution and in micelles (Kusio et al. 2020).

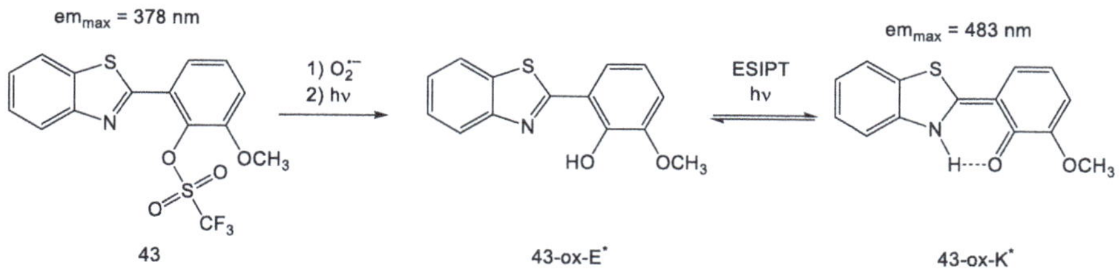

Fig. 7 Mechanism of action of the ESIPT-based fluorescent probe HMBT-LW for detecting $O_2^{•-}$ (Wu et al. 2019b)

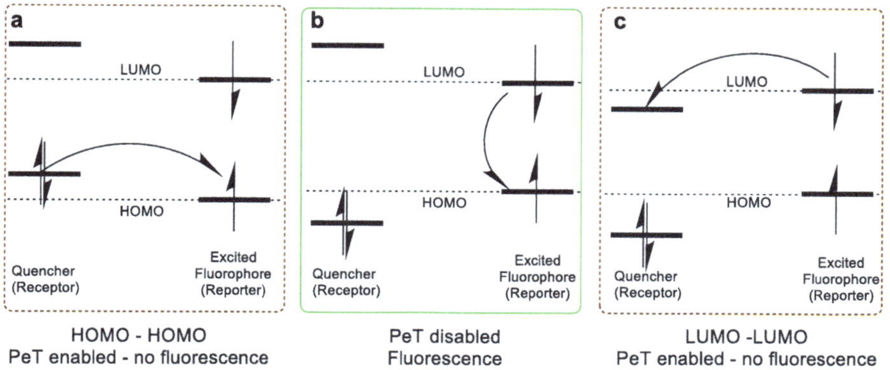

Fig. 8 Scheme of action mechanism of Photoinduced electron Transfer (PeT). (**a**) Electron transfer from quencher (receptor) HOMO energy level to fluorophore (reporter) HOMO energy level disabling fluorescence. (**b**) Fluorescence phenomenon occurring normally due to disabled PeT. (**c**) Electron transfer from LUMO energy level of the fluorophore to LUMO energy level of quencher preventing the occurrence of normal fluorescence phenomenon

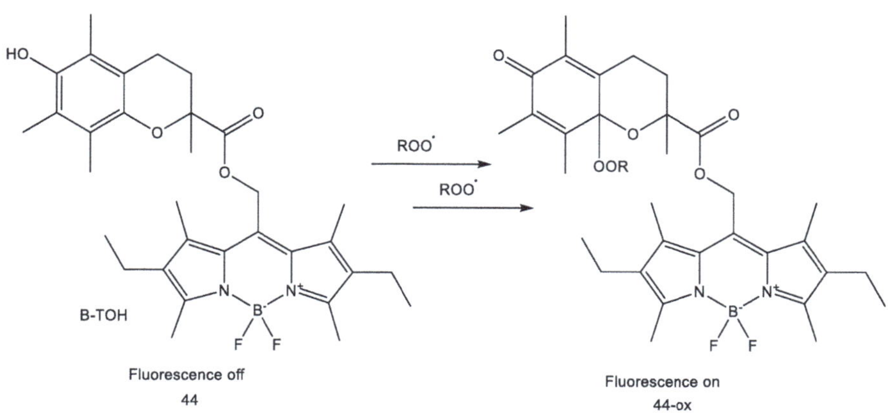

Fig. 9 Model of action BODIPY-tocopherol (B-TOH) peroxide fluorescent probe invented by Cosa et al.

radicals). Such reaction causes deactivation of PeT and is immediately indicated as enhancement of fluorescence emitted by reporter part, namely 4,4-difluoro-4-bora-3a,4a-diaza-s-indacene (*abbr.* BODIPY). BODIPY dyes are small molecules strongly absorbing UV and emitting relatively sharp fluorescence peaks with small Stokes shifts and high quantum yields. They are relatively insensitive to the solvent polarity and pH, and reasonably stable at physiological conditions (Loudet and Burgess 2007). Up to ten-fold emission enhancement is observed for this compound during reaction with peroxyl and alkoxyl radicals in homogenous solutions (Oleynik et al. 2007). Fluorescent probes designed and developed in the Cosa group contain not only chromanol, but also parts resembling di-*o*,*o*-*tert*-butylphenol (Krumova et al. 2009), vitamin K (Belzile et al. 2016) and non-polar parts making the fluorescent probe suitable to detect mitochondrial ROS (Krumova et al. 2013). Recently, a new BODIPY-based phenolic sensor for radicals has been presented by our group (Kusio et al. 2020). We decided to attach phenol moieties in different sides of BODIPY. This probe (named NB-2, see Table 5) comprises two receptor segments (4-hydroxycinnamyl moieties). Oxygen uptake measurements of the rate of lipid oxidation in the presence of NB-2 indicates moderate chain-breaking antioxidant properties, the rate constant for a reaction with these radicals (k_{inh}) is $1000 \pm 100\ M^{-1}\ s^{-1}$, one order of magnitude smaller than for the reaction of PMHC (analogue of -tocopherol). Such reactivity is sufficient for sensing peroxyl radicals, and importantly, as ten times less reactive than tocopherol, thus, being not too reactive, NB-2 is a promising fluorescent probe for monitoring the oxidation status of the systems containing other antioxidants. NB-2 can report the progress of reaction with peroxyl radicals via fluorescence intensity enhancement (off/on switch marker). The examples of fluorogenic probes acting via the mechanism of intramolecular electron and charge transfers are presented in Table 5.

The presence of a phenolic moiety as a reporter part that competes with other radical trapping agents can sometimes be problematic because the fluorescent probe might affect the redox equilibria when the probe acts as an antioxidant and, additionally, can synergistically interact with other antioxidants/co-antioxidants. A different approach, inspired by BODIPY with attached fatty acids used as membrane probes, (Johnson et al. 1991) was proposed by Drummen et al. (Drummen and van Liebergen 2002) Fig. 10a presents the structure and mechanism of action of C11-BODIPY$^{581/591}$. This probe has the sensitivity to oxidation comparable to that of endogenous fatty acyl, the reaction of this probe with peroxyl radical causes shift of fluorescence from 595 to 520 nm, due to reduced conjugation when peroxyl radical is added.

C11-BODIPY$^{581/591}$ can serve as a marker of peroxidation in fluorescence and confocal microscopy and became a popular compound for monitoring peroxidation in cell culture. Naguib (Naguib 2000b) proposed a modification of 47a in order to obtain highly lipophilic BODIPY$^{665/676}$ used for monitoring the antioxidant activity of series of carotenoids (with tocopherol and Trolox as calibrating antioxidants). It is worth to note that comparing to C11-BODIPY$^{581/591}$, a longer wavelength (665 nm) is the additional advantage when strongly absorbing samples are studied. Another advantage is that BODIPY$^{665/676}$ is more sensitive towards peroxyl radicals and the

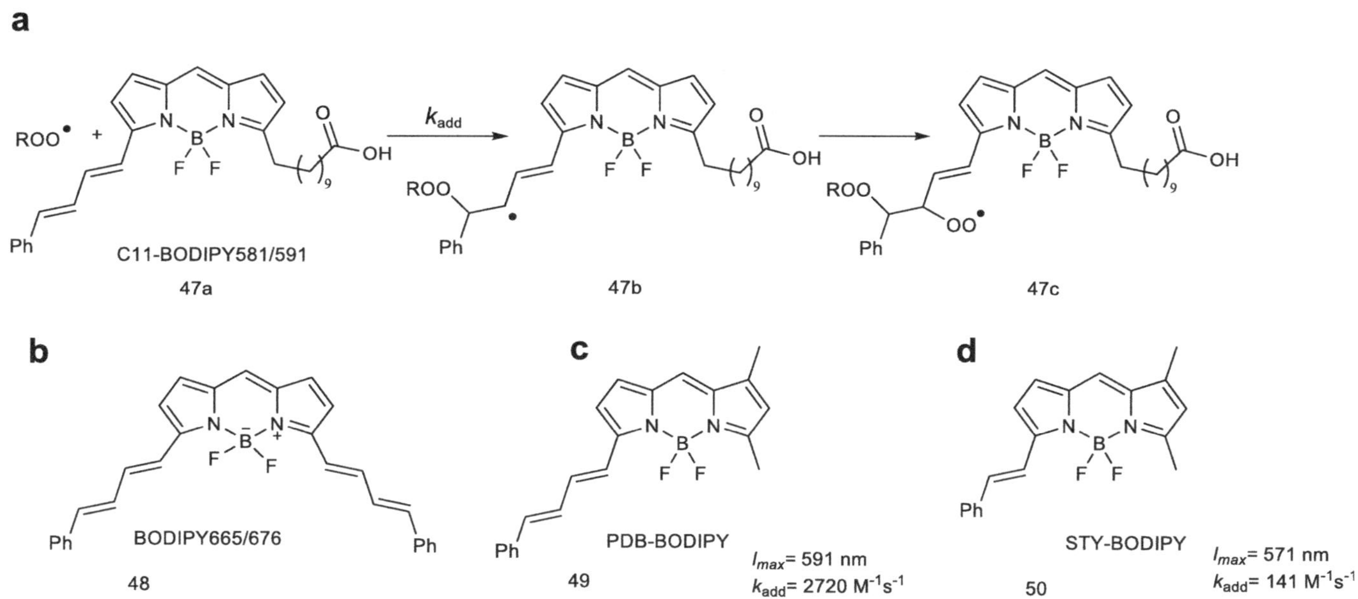

Fig. 10 (**a**) Scheme of propagation steps in the autoxidation of C11-BODIPY 581/591. (**b–d**) Structures of BODIPY 665/676, PDB-BODIPY, and STY-BODIPY

Stokes shift is 5 nm bigger than for C11-BODIPY (Naguib 2000b). $BODIPY^{665/676}$ became popular FP used for studies of mechanism of oxidation of lipids in complex multiphase food systems, with recent observations of inter-micellar transfer of radicals and pro-oxidants (Li et al. 2020), and quantitative spatiotemporal mapping of lipid and protein oxidation in emulsified food systems (Yang et al. 2020). Laguerre, Lecomte and Villeneuve in their review (Laguerre et al. 2007) grouped the methods utilising C11- $BODIPY^{581/591}$, $BODIPY^{665/676}$, parinaric acid, and fluorescein derivatives as the methods based on the measurements of substrate loss. Apart from fluorescein (that is a phenol), the mechanism of action of those FPs is the same, since they mimic the polyunsaturated lipids, participate in propagation process and do not exhibit any chain-breaking activity.

The idea of a compound that would be used directly for the kinetic studies of peroxidation was further developed by Pratt, Haidasz and Van Kessel (Haidasz et al. 2016). As the conventional methodology of kinetic studies in non-polar solvents employs peroxidation of styrene or cumene, they modified C11-$BODIPY^{581/591}$ into STY-BODIPY and PDB-BODIPY (Fig. 10c and d), both reacting with peroxyl radicals in the same way as shown in Fig. 10a. These "neutral" fluorescent probes co-participate in the propagation process without breaking the kinetic chain of peroxidation. The probes give clear and strong absorbance, thus, the spectrophotometry instead of spectrofluorimetry was successfully employed for monitoring the oxidation course. The rate constants determined for the reaction of $ROO^{\bullet}$ with PDB-BODIPY and with STY-BODIPY were determined as 141 $M^{-1}s^{-1}$ and 2720 $M^{-1}s^{-1}$, respectively. Therefore, peroxidation of styrene doped with PDB-BODIPY (or cumene doped with STY-BODIPY) in the presence of chain-breaking antioxidant could be used for determination of the inhibition rate constant for the reaction of $ROO^{\bullet}$ with antioxidant (k_{inh}), as competitive to the reaction of $ROO^{\bullet}$ with the probe. The proposed methodology was validated for series of phenolic antioxidants as well as diarylamine (Haidasz et al. 2016). In the same work, the authors proved that STY-BODIPY can be applied for determination of k_{inh} for peroxidation of THF in water inhibited by water-soluble antioxidants. Recently, Pratt and co-workers used STY-BODIPY and PDB-BIDIPY for studies of antioxidant action of series of putative antioxidants in phospholipid liposomal environment. In their experimental setup, they switched the way of detection from spectrophotometry to spectrofluorimetry as better suited for micro-plate based assay for oxidation performed in non-homogeneous lipid/water liposomal system (Shah et al. 2019).

Intensive research on the aetiology and consequences of oxidative stress were the driving force for the development of many fluorescent probes for sensing and imaging the species which usually initiate peroxidation (like $O_2^{\bullet-}$, $^{\bullet}OH$, $^1O_2$, H_2O_2) and mediate the kinetic chain reaction (like $LOO^{\bullet}$). Some of the probes found the application in the studies of natural and synthetic antioxidants and their role as key redox gatekeepers in the cell. Interestingly, a relatively large number of probes for sensing H_2O_2 contrasts with rather small number of FPs dedicated to lipid or alkyl hydroperoxides. Perhaps, in life sciences and in food sciences detection of ROS responsible for initiation or being markers of advanced peroxidation was more important (and easier) than detection of peroxyl radicals. Oxidation status of the sample/system can be monitored via secondary markers of oxidation (malondialdehyde adducts, volatile compounds, hexanal and hydroxyalkenals, peroxide number, etc.) but the above mentioned methods, although very useful, seem to be an input (initiation) and output (secondary products) for the black box model cannot be understood without detailed kinetic studies. Historically, the kinetic studies of autoxidation of organic materials (hydrocarbons, lipids, fats, polymers) were based on the measurements of the rate of the process defined as the rate of the oxygen consumption or the rate of production of hydroperoxides. The amount of hydroperoxides was monitored by laborious iodometric titration of periodically collected samples hydrocarbons, later replaced by chromatographic analysis or by monitoring the conjugated dienes (of unsaturated fatty acids). The problem of quantification of hydroperoxides still is crucial for the safety and quality assessment in food chemistry, pharmacy, polymer chemistry. The new approach to this problem was inspired by the reduction of hydroperoxides by triphenylphosphine: ROOH + $PPh_3 \rightarrow$ ROH, but one of the phenyls in PPh_3 was replaced by the fluorophore (coumarin) see Fig. 11a (Hanthorn et al. 2012). Thus, the reaction can be visualized as ten-fold enhancement of emission ($\Phi_{ox}/\Phi_{red} = 10.2$). The proposed probe gives reliable results when applied for the kinetic studies of autoxidation of 7-dehydrocholesterol at 37 °C and hexadecane at 160 °C (samples were collected periodically from the reactor, then cooled to 25 °C, diluted with tert-amyl alcohol, loaded into 96-well microplate and treated with fluorescent probe dissolved in CH_3CN (Hanthorn et al. 2012).

In the last part of this chapter we would like to mention the application of FPs for monitoring autoxidation processes proceeding at high temperatures and relevant to some industrial processes, when saturated hydrocarbons are additionally contacted with metal surfaces. At high temperatures hydroperoxides primarily formed during autoxidation decompose to myriads of products with aldehydes, ketones and carboxylic acids. As less volatile and stable, acids can be regarded as markers of autoxidation at higher temperature, when hydroperoxides of saturated hydrocarbons undergo rapid decomposition. An example of such fluorescent probe with coumarin as reporter and tertiary amine moiety as the receptor of protons is presented in Fig. 11b. In neutral (non-protonated) form, the fluorescence of coumarin is efficiently quenched because of PeT, but protonation of amine moiety results in an efficient, 38-fold increase of fluorescence ($\lambda_{ex} = 315$ nm, $\lambda_{em} = 395$ nm) (Shah and Pratt 2016). The probe does not react with hydroperoxides that would be an advantage when simple acid-base equilibrium is to be monitored. However, oxidation of hydrocarbons generates hydroperoxides which consecutively decompose to carbonyl and carboxyl compounds, therefore, the knowledge about the level of both, hydroperoxides and acids, is more relevant to the total oxidation status of the sample. Both coumarin-based fluorescent probes, for hydroperoxides (51a) and for acids (52a), have a similar spectrum, thus, they cannot be used in the same sample. However, Pratt and co-workers elaborated methodology for such simultaneous analysis employing coumarin-triarylphosphine (sensor of ROOH) and BODIPY based sensor for acids, 53a ($\lambda_{ex} = 495$ nm, $\lambda_{em} = 500$ nm), see Fig. 11c (Shah

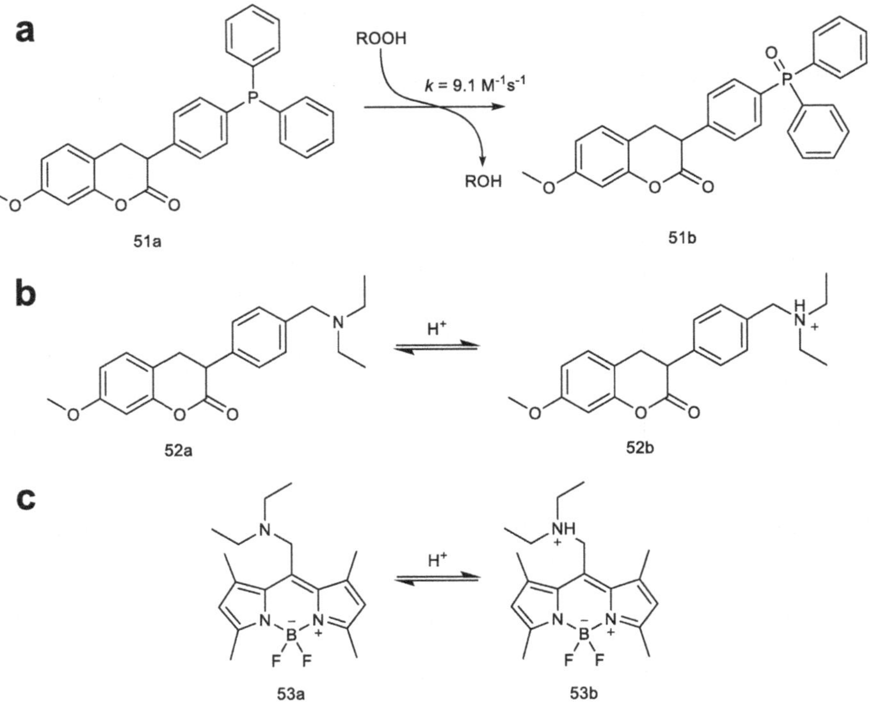

Fig. 11 (**a**) Reaction of fluorogenic coumarin-phosphine with alkyl(lipid) hydroperoxide (Hanthorn et al. 2012). (**b**) Coumarin derived tertiary amine in neutral and protonated form (Shah and Pratt 2016). (**c**) The relevant equilibrium of the BODIPY–trialkylamine and its conjugate acid

and Pratt 2016). A combination of those two probes in one assay to be used in the fully automated analysis (96-well microplate with automated reagent dispenser and microplate reader) is a promising alternative to laborious iodometric and acid-base titrations.

3.4 Conclusions

Fluorescent probes constructed on the functionalised BODIPY fluorophore have many advantages over the non-BODIPY probes. They can be tailored to specific needs, depending on the functional groups or bigger fragments (phenols or amines) acting as receptors responsible for selective sensing of ROS or by addition of polar / non-polar side-groups to modify the overall polarity of the molecule. This feature is crucial for the localisation of the mitochondrial fluorescent probe Mito-BODIPY-TOH in Table 5. Together with the clear and strong analytical signal, the BODIPY-based probes have the great potential to become the next generation of tools to be applied in life sciences, medical diagnostics, but also in pharmacy and food chemistry, whenever the ROS should be detected and quantitatively monitored, from the studies of oxidative stress and its consequences to the antioxidant action and oxidative degradation of non-biological materials.

Acknowledgements Financial support from the National Science Centre, Poland (NCN grant No. 2018/31/B/ST4/02354) is gratefully acknowledged.

References

Ahmad MH, Sahar A, Hitzmann B (2017) Fluorescence spectroscopy for the monitoring of food processes. In: Hitzmann B (ed) Measurement, modeling and automation in advanced food processing. Springer International Publishing, Berlin, pp 121–151. https://doi.org/10.1007/978-3-319-60111-3

Akasaka K, Ohrui H (2000) Development of phosphine reagents for the high-performance liquid chromatographic-fluorometric determination of lipid hydroperoxides. J Chromatogr A 881 (1–2):159–170

Akasaka K, Suzuki T, Ohrui H, Meguro H (1987) Study on aromatic phosphines for novel fluorometry of hydroperoxides (ii) -the determination of lipid hydroperoxides with diphenyl-1-pyrenylphosphine. Anal Lett 20(5):797–807

Atik S, Singer LA (1978) Fluorescence quenching by nitroxyl radicals in micellar environments. A useful probe for studying micelle-substrate interactions. J Am Chem Soc 100(10):3234–3235

Banwell CN, EM MC (1994) Fundamentals of molecular spectroscopy. McGraw-Hill, London

Barja G (2002) The quantitative measurement of H2O2 generation in isolated mitochondria. J Bioenerg Biomembr 34(3):227–233

Bartoli GM, Galeotti T, Azzi A (1977) Production of superoxide anions and hydrogen peroxide in Ehrlich ascites tumour cell nuclei. BBA – General Subjects 497(2):622–626

Baruah M, Qin W, Vallée RAL, Beljonne D, Rohand T, Dehaen W, Boens N (2005) A highly potassium-selective ratiometric fluorescent indicator based on BODIPY azacrown ether excitable with visible light. Org Lett 7(20):4377–4380

Belzile M-N, Godin R, Durantini AM, Cosa G (2016) Monitoring chemical and biological electron transfer reactions with a fluorogenic vitamin k analogue probe. J Am Chem Soc 138 (50):16388–16397

Bindokas VP, Jordán J, Lee CC, Miller RJ (1996) Superoxide production in rat hippocampal neurons: selective imaging with hydroethidine. J Neurosci 16(4):1324–1336

Brouwer AM (2011) Standards for photoluminescence quantum yield measurements in solution (IUPAC technical report). Pure Appl Chem 83(12):2213–2228

Bystryak S, Goldiner I, Niv A, Nasser AM, Goldstein L (1995) A homogeneous immunofluorescence assay based on dye-sensitized Photobleaching. Anal Biochem 225(1):127–134

Caldefie-Chézet F, Walrand S, Moinard C, Tridon A, Chassagne J, Vasson MP (2002) Is the neutrophil reactive oxygen species production measured by luminol and lucigenin chemiluminescence intra or extracellular? Comparison with DCFH-DA flow cytometry and cytochrome c reduction. Clin Chim Acta 319(1):9–17

Cao G, Alessio HM, Cutler RG (1993) Oxygen-radical absorbance capacity assay for antioxidants. Free Radic Biol Med 14(3):303–311

Chang MCY, Pralle A, Isacoff EY, Chang CJ (2004) A selective, cell-permeable optical probe for hydrogen peroxide in living cells. J Am Chem Soc 126(47):15392–15393

Chen X, Wang F, Hyun JY, Wei T, Qiang J, Ren X, Shin I, Yoon J (2016) Recent progress in the development of fluorescent, luminescent and colorimetric probes for detection of reactive oxygen and nitrogen species. Chem Soc Rev 45(10):2976–3016

Chen L, Cho MK, Wu D, Kim HM, Yoon J (2019) Two-photon fluorescence probe for selective monitoring of superoxide in live cells and tissues. Anal Chem 91(22):14691–14696

Corey EJ, Taylor WC (1964) A study of the peroxidation of organic compounds by externally generated singlet oxygen molecules. J Am Chem Soc 86(18):3881–3882

Costa D, Fernandes E, Santos JLM, Pinto DCGA, Silva AMS, Lima JLFC (2007) New noncellular fluorescence microplate screening assay for scavenging activity against singlet oxygen. Anal Bioanal Chem 387(6):2071–2081

Crosby DG, Berthold RV (1962) Fluorescence spectra of some simple coumarins. Anal Biochem 4 (5):349–357

Crow JP (1997) Dichlorodihydrofluorescein and dihydrorhodamine 123 are sensitive indicators of peroxynitrite in vitro: implications for intracellular measurement of reactive nitrogen and oxygen species. Nitric Oxide 1(2):145–157

Cui G, Ye Z, Chen J, Wang G, Yuan J (2011) Development of a novel terbium(III) chelate-based luminescent probe for highly sensitive time-resolved luminescence detection of hydroxyl radical. Talanta 84(3):971–976

De Silva AP, Gunaratne HQN, Gunnlaugsson T, Huxley AJM, McCoy CP, Rademacher JT et al (1997) Signaling recognition events with fluorescent sensors and switches. Chem Rev 97 (5):1515–1566

Dorta E, Fuentes-Lemus E, Speisky H, Lissi E, López-Alarcón C (2018) Evaluation of the antioxidant capacity of food samples: a chemical examination of the oxygen radical absorbance capacity assay. In: Apak R, Capanoglu E, Shahidi F (eds) Measurement of antioxidant activity and capacity: recent trends and applications. Wiley, Oxford, pp 39–55

Drummen GPC (2012) Fluorescent probes and fluorescence (microscopy) techniques — illuminating biological and biomedical research. Molecules 17(12):14067–14090

Drummen GPC, van Liebergen LCM (2002) Op den Kamp JAF, Post JA. C11-BODIPY581/591, an oxidation-sensitive fluorescent lipid peroxidation probe: (micro)spectroscopic characterization and validation of methodology. Free Radic Biol Med 33(4):473–490

Drummen GPC, Op den Kamp JAF, Post JA (1999) Validation of the peroxidative indicators, cis-parinaric acid and parinaroyl-phospholipids, in a model system and cultured cardiac myocytes. Biochimica et Biophysica Acta (BBA) - Molecular and Cell Biology of Lipids 1436(3):370–382

Entradas T, Waldron S, Volk M (2020) The detection sensitivity of commonly used singlet oxygen probes in aqueous environments. J Photochem Photobiol B Biol 204:111787

Fahrni CJ, O'Halloran TV (1999) Aqueous coordination chemistry of quinoline-based fluorescence probes for the biological chemistry of zinc. J Am Chem Soc 121(49):11448–11458

Gao JJ, Xu KH, Tang B, Yin LL, Yang GW, An LG (2007) Selective detection of superoxide anion radicals generated from macrophages by using a novel fluorescent probe. FEBS J 274 (7):1725–1733

Gao X, Ding C, Zhu A, Tian Y (2014) Carbon-dot-based ratiometric fluorescent probe for imaging and biosensing of superoxide anion in live cells. Anal Chem 86(14):7071–7078

Godin R, Liu H-W, Cosa G (2014) Ambient condition oxidation in individual liposomes observed at the single molecule level. Chem Sci 5(6):2525–2529

Gomes A, Fernandes E, Lima JLFC (2005) Fluorescence probes used for detection of reactive oxygen species. J Biochem Biophys Methods 65(2–3):45–80

Gomes A, Fernandes E, Lima JLFC (2006) Use of fluorescence probes for detection of reactive nitrogen species: a review. J Fluoresc 16(1):119–139

Green JA, Singer LA (1974) Di-tert-butyl nitroxide as a convenient probe for excited singlet states. Pyrene luminescence. J Am Chem Soc 96(9):2730–2733

Greene LE, Lincoln R, Cosa G (2017) Rate of lipid peroxyl radical production during cellular homeostasis unraveled via fluorescence imaging. J Am Chem Soc 139(44):15801–15811

Greene LE, Lincoln R, Cosa G (2018) Spatio-temporal monitoring of lipid peroxyl radicals in live cell studies combining fluorogenic antioxidants and fluorescence microscopy methods. Free Radic Biol Med 128(February):124–136

Grynkiewicz G, Poenie M, Tsien RY (1985) A new generation of Ca2+ indicators with greatly improved fluorescence properties. J Biol Chem 260(6):3440–3450

Haidasz EA, Van Kessel ATM, Pratt DA (2016) A continuous visible light spectrophotometric approach to accurately determine the reactivity of radical-trapping antioxidants. J Org Chem 81 (3):737–744

Halliwell B, Gutteridge JM (2015) Free radicals in biology and medicine. Oxford University Press, Oxford

Halliwell B, Whiteman M (2004) Measuring reactive species and oxidative damage in vivo and in cell culture: how should you do it and what do the results mean? Br J Pharmacol 142 (2):231–255

Hanthorn JJ, Haidasz E, Gebhardt P, Pratt DA (2012) A versatile fluorescence approach to kinetic studies of hydrocarbon autoxidations and their inhibition by radical-trapping antioxidants. Chem Commun 48(81):10141–10143

Henderson LM, Chappell JB (1993) Dihydrorhodamine 123: a fluorescent probe for superoxide generation? Eur J Biochem 217(3):973–980

Huang D, Ou B, Hampsch-Woodill M, Flanagan JA, Deemer EK (2002) Development and validation of oxygen radical absorbance capacity assay for lipophilic antioxidants using randomly methylated β-cyclodextrin as the solubility enhancer. J Agric Food Chem 50 (7):1815–1821

Iuliano L, Piccheri C, Coppola I, Praticò D, Micheletta F, Violi F (2000) Fluorescence quenching of dipyridamole associated to peroxyl radical scavenging: a versatile probe to measure the chain breaking antioxidant activity of biomolecules. Biochim Biophys Acta Gen Subj 1474 (2):177–182

Jablonski A (1933) Efficiency of anti-stokes fluorescence in dyes. Nature 131(3319):839–840

Johnson ID, Kang HC, Haugland RP (1991) Fluorescent membrane probes incorporating dipyrrometheneboron difluoride fluorophores. Anal Biochem 198(2):228–237

Keston AS, Brandt R (1965) The fluorometric analysis of ultramicro quantities of hydrogen peroxide. Anal Biochem 11(1):1–5

Kooy NW, Royall JA, Ischiropoulos H, Beckman JS (1994) Peroxynitrite-mediated oxidation of dihydrorhodamine 123. Free Radic Biol Med 16(2):149–156

Krumova K, Cosa G (2013) Fluorogenic probes for imaging reactive oxygen species. Photochemistry 41:279–301

Krumova K, Oleynik P, Karam P, Cosa G (2009) Phenol-based lipophilic fluorescent antioxidant indicators: a rational approach. J Org Chem 74(10):3641–3651

Krumova K, Friedland S, Cosa G (2012) How lipid unsaturation, peroxyl radical partitioning, and chromanol lipophilic tail affect the antioxidant activity of α-tocopherol: direct visualization via high-throughput fluorescence studies conducted with fluorogenic α-tocopherol analogues. J Am Chem Soc 134(24):10102–10113

Krumova K, Greene LE, Cosa G (2013) Fluorogenic α-tocopherol analogue for monitoring the antioxidant status within the inner mitochondrial membrane of live cells. J Am Chem Soc 135 (45):17135–17143

Kumar M, Kumar N, Bhalla V, Sharma PR, Qurishi Y (2012) A charge transfer assisted fluorescent probe for selective detection of hydrogen peroxide among different reactive oxygen species. Chem Commun 48(39):4719–4721

Kusio J, Sitkowska K, Konopko A, Litwinienko G (2020) Hydroxycinnamyl derived BODIPY as a lipophilic fluorescence probe for peroxyl radicals. Antioxidants 9(1):88

Kuypers FA, van den Berg JJ, Schalkwijk C, Roelofsen B (1987) Op den Kamp JA. Parinaric acid as a sensitive fluorescent probe for the determination of lipid peroxidation. Biochim Biophys Acta 921(2):266–274

Lageweg W, Steen I, Tager JM, Wanders RJA (1991) A fluorimetric assay for acyl-CoA synthetase activities. Anal Biochem 197(2):384–388

Laguerre M, Lecomte J, Villeneuve P (2007) Evaluation of the ability of antioxidants to counteract lipid oxidation: existing methods, new trends and challenges. Prog Lipid Res 46(5):244–282

Laguerre M, Bily A, Birtić S (2020) Chapter 7 – lipid oxidation in food. In: Galanakis CM (ed) Lipids and edible oils. Academic Press, Cambridge, pp 243–287
Lakowicz JR (2006) Principles of fluorescence spectroscopy. Springer, Berlin, pp 1–954
Lampard EV, Sedgwick AC, Sun X, Filer KL, Hewins SC, Kim G et al (2018) Boronate-based fluorescence probes for the detection of hydrogen peroxide. ChemistryOpen 7:262
Li B, Blough NV, Gutierrez PL (2000) Trace detection of hydroxyl radicals during the redox cycling of low concentrations of diaziquone: a new approach. Free Radic Biol Med 29 (6):548–556
Li P, Zhang W, Li K, Liu X, Xiao H, Zhang W et al (2013) Mitochondria-targeted reaction-based two-photon fluorescent probe for imaging of superoxide anion in live cells and in vivo. Anal Chem 85(20):9877–9881
Li P, McClements DJ, Decker EA (2020) Application of flow cytometry as novel technology in studying lipid oxidation and mass transport phenomena in oil-in-water emulsions. Food Chem 315:126225
Liao Y-X, Li K, Wu M-Y, Wu T, Yu X-Q (2014) A selenium-contained aggregation-induced "turn-on" fluorescent probe for hydrogen peroxide. Org Biomol Chem 12(19):3004–3008
Lindqvist C, Nordström T (2001) Generation of hydroxyl radicals by the antiviral compound phosphonoformic acid (foscarnet). Pharmacol Toxicol 89(1):49–55
Liu H, Zhang L, Chen J, Zhai Y, Zeng Y, Li L (2013) A novel functional imidazole fluorescent ionic liquid: simple and efficient fluorescent probes for superoxide anion radicals. Anal Bioanal Chem 405(29):9563–9570
Lo LC, Chu CY (2003) Development of highly selective and sensitive probes for hydrogen peroxide. Chem Commun 21:2728–2729
Loudet A, Burgess K (2007) BODIPY dyes and their derivatives: syntheses and spectroscopic properties. Chem Rev 107(11):4891–4932
Maeda H, Fukuyasu Y, Yoshida S, Fukuda M, Saeki K, Matsuno H et al (2004) Fluorescent probes for hydrogen peroxide based on a non-oxidative mechanism. Angewandte Chemie – International Edition. 43(18):2389–2391
Maeda H, Yamamoto K, Nomura Y, Kohno I, Hafsi L, Ueda N et al (2005) A design of fluorescent probes for superoxide based on a nonredox mechanism. J Am Chem Soc 127(1):68–69
Makrigiorgos GM, Baranowska-Kortylewicz J, Bump E, Sahu SK, Berman RM, Kassis AI (1993) A method for detection of hydroxyl radicals in the vicinity of biomolecules using radiation-induced fluorescence of coumarin. Int J Radiat Biol 63(4):445–458
Makrigiorgos GM, Kassis AI, Mahmood A, Bump EA, Savvides P (1997) Novel fluorescein-based flow-cytometric method for detection of lipid peroxidation. Free Radic Biol Med 22 (1–2):93–100
Maulik G, Kassis AI, Savvides P, Makrigiorgos GM (1998) Fluoresceinated phosphoethanolamine for flow-cytometric measurement of lipid peroxidation. Free Radic Biol Med 25(6):645–653
Medvedeva N, Martin VV, Weis AL, Likhtenshten GI (2004) Dual fluorophore-nitronyl probe for investigation of superoxide dynamics and antioxidant status of biological systems. J Photochem Photobiol A Chem 163(1):45–51
Meng L, Wu Y, Yi T (2014) A ratiometric fluorescent probe for the detection of hydroxyl radicals in living cells. Chem Commun 50(37):4843–4845
Minta A, Kao JPY, Tsien RY (1989) Fluorescent indicators for cytosolic calcium based on rhodamine and fluorescein chromophores. J Biol Chem 264(14):8171–8178
Mohanty JG, Jaffe JS, Schulman ES, Raible DG (1997) A highly sensitive fluorescent micro-assay of H2O2 release from activated human leukocytes using a dihydroxyphenoxazine derivative. J Immunol Methods 202(2):133–141
Nagano T (2010) Development of fluorescent probes for bioimaging applications. Proc Jpn Acad Ser B Phys Biol Sci 86(8):837–847
Naguib YM (1998) A fluorometric method for measurement of peroxyl radical scavenging activities of lipophilic antioxidants. Anal Biochem 265(2):290–298

Naguib YMA (2000a) A fluorometric method for measurement of oxygen radical-scavenging activity of water-soluble antioxidants. Anal Biochem 284(1):93–98

Naguib YMA (2000b) Antioxidant activities of astaxanthin and related carotenoids. J Agric Food Chem 48(4):1150–1154

Noguchi N, Numano R, Kaneda H, Niki E (1998) Oxidation of lipids in low density lipoprotein particles. Free Radic Res 29(1):43–52

Noomnarm U, Clegg RM (2009) Fluorescence lifetimes: fundamentals and interpretations. Photosynth Res 101(2–3):181–194

Ohyashiki T, Nunomura M, Katoh T (1999) Detection of superoxide anion radical in phospholipid liposomal membrane by fluorescence quenching method using 1,3-diphenylisobenzofuran. Biochim Biophys Acta Biomembr 1421(1):131–139

Okimoto Y, Watanabe A, Niki E, Yamashita T, Noguchi N (2000) A novel fluorescent probe diphenyl-1-pyrenylphosphine to follow lipid peroxidation in cell membranes. FEBS Lett 474 (2–3):137–140

Oleynik P, Ishihara Y, Cosa G (2007) Design and synthesis of a BODIPY-α-tocopherol adduct for use as an off/on fluorescent antioxidant indicator. J Am Chem Soc 129(7):1842–1843

Olojo RO, Xia RH, Abramson JJ (2005) Spectrophotometric and fluorometric assay of superoxide ion using 4-chloro-7-nitrobenzo-2-oxa-1,3-diazole. Anal Biochem 339(2):338–344

Ou B, Hampsch-Woodill M, Prior RL (2001) Development and validation of an improved oxygen radical absorbance capacity assay using fluorescein as the fluorescent probe. J Agric Food Chem 49(10):4619–4626

Ou B, Chang T, Huang D, Prior RL (2013) Determination of total antioxidant capacity by oxygen radical absorbance capacity (ORAC) using fluorescein as the fluorescence probe: first action 2012.23. J AOAC Int 96(6):1372–1376

Palamakumbura AH, Trackman PC (2002) A fluorometric assay for detection of lysyl oxidase enzyme activity in biological samples. Anal Biochem 300(2):245–251

Pastor I, Esquembre R, Micol V, Mallavia R, Mateo CR (2004) A ready-to-use fluorimetric biosensor for superoxide radical using superoxide dismutase and peroxidase immobilized in sol-gel glasses. Anal Biochem 334(2):335–343

Pedersen SK, Holmehave J, Blaikie FH, Gollmer A, Breitenbach T, Jensen HH et al (2014) Aarhus sensor green: a fluorescent probe for singlet oxygen. J Org Chem 79(7):3079–3087

Peng X, Du J, Fan J, Wang J, Wu Y, Zhao J, Sun S, Xu T (2007) A selective fluorescent sensor for imaging Cd^{2+} in living cells. J Am Chem Soc 129(6):1500–1501

Pou S, Huang YI, Bhan A, Bhadti VS, Hosmane RS, Wu SY et al (1993) A fluorophore-containing nitroxide as a probe to detect superoxide and hydroxyl radical generated by stimulated neutrophils. Anal Biochem 212(1):85–90

Prior RL, Hoang H, Gu L, Wu X, Bacchiocca M, Howard L et al (2003) Assays for hydrophilic and lipophilic antioxidant capacity (oxygen radical absorbance capacity (ORACFL)) of plasma and other biological and food samples. J Agric Food Chem 51(11):3273–3279

Qu X, Kirschenbaum LJ, Borish ET (2000) Hydroxyterephthalate as a fluorescent probe for hydroxyl radicals: application to hair melanin. Photochem Photobiol 71(3):307–313

Ruch W, Cooper PH, Baggiolini M (1983) Assay of H2O2 production by macrophages and neutrophils with homovanillic acid and horse-radish peroxidase. J Immunol Methods 63 (3):347–357

Schaich K (2008) Co-oxidations of oxidizing lipids: Reactions with proteins. In: Kamal-Eldin A, Min D (eds) Lipid oxidation pathways, 2nd edn. AOCS Press, Champaign, pp 183–274

Setsukinai K, Urano Y, Kakinuma K, Majima HJ, Nagano T (2003) Development of novel fluorescence probes that can reliably detect reactive oxygen species and distinguish specific species. J Biol Chem 278(5):3170–3175

Shah R, Pratt DA (2016) Determination of key hydrocarbon autoxidation products by fluorescence. J Org Chem 81(15):6649–6656

Shah R, Farmer LA, Zilka O, Van Kessel ATM, Pratt DA (2019) Beyond DPPH: Use of fluorescence-enabled inhibited autoxidation to predict oxidative cell death rescue. Cell Chemical Biol 26(11):1594–607.e7

Sherman WR, Robins E (1968) Fluorescence of substituted 7-hydroxycoumarins. Anal Chem 40 (4):803–805

Singh A, McIntyre NR, Koroll GW (1978) Photochemical formation of metastable species from 1,3-diphenylisobenzofuran. Photochem Photobiol 28(4–5):595–601

Soh N (2006) Recent advances in fluorescent probes for the detection of reactive oxygen species. Anal Bioanal Chem 386(3):532–543

Spiteller G (2006) Peroxyl radicals: inductors of neurodegenerative and other inflammatory diseases. Their origin and how they transform cholesterol, phospholipids, plasmalogens, polyunsaturated fatty acids, sugars, and proteins into deleterious products. Free Radic Biol Med 41 (3):362–387

Staniek K, Nohl H (1999) H_2O_2 detection from intact mitochondria as a measure for one-electron reduction of dioxygen requires a non-invasive assay system. Biochim Biophys Acta Bioenerg 1413(2):70–80

Tai C, Gu X, Zou H, Guo Q (2002) A new simple and sensitive fluorometric method for the determination of hydroxyl radical and its application. Talanta 58(4):661–667

Tanaka K, Miura T, Umezawa N, Urano Y, Kikuchi K, Higuchi T et al (2001) Rational design of fluorescein-based fluorescence probes. Mechanism-based design of a maximum fluorescence probe for singlet oxygen. J Am Chem Soc 123(11):2530–2536

Tang B, Zhang L, Hu J-X, Li P, Zhang H, Zhao Y-X (2004a) Indirect determination of superoxide anion radical in the plant of red sage based on vanillin-8-aminoquinoline with fluorescence. Anal Chim Acta 502(1):125–131

Tang B, Zhang L, Zhang LL (2004b) Study and application of flow injection spectrofluorimetry with a fluorescent probe of 2-(2-pyridil)-benzothiazoline for superoxide anion radicals. Anal Biochem 326(2):176–182

Tang B, Zhang L, Geng Y (2005) Determination of the antioxidant capacity of different food natural products with a new developed flow injection spectrofluorimetry detecting hydroxyl radicals. Talanta 65(3):769–775

Tikekar RV, Johnson A, Nitin N (2011) Fluorescence imaging and spectroscopy for real-time, in-situ characterization of interactions of free radicals with oil-in-water emulsions. Food Res Int 44(1):139–145

Uluata S, Durmaz G, McClements DJ, Decker EA (2021) Comparing DPPP fluorescence and UV based methods to assess oxidation degree of krill oil-in-water emulsions. Food Chem 339:127898

Umezawa N, Tanaka K, Urano Y, Kikuchi K, Higuchi T, Nagano T (1999) Novel fluorescent probes for singlet oxygen. Angewandte Chemie – International Edition 38(19):2899–2901

Usui Y (1973) Determination of quantum yield of singlet oxygen formation by photosensitization. Chem Lett 1973:743–744

Wang H, Joseph JA (1999) Quantifying cellular oxidative stress by dichlorofluorescein assay using microplate reader. Free Radic Biol Med 27(5–6):612–616

Wilkinson F, Brummer JG (1981) Rate constants for the decay and reactions of the lowest electronically excited singlet state of molecular oxygen in solution. J Phys Chem Ref Data 10 (4):809–999

Wu L, Sedgwick AC, Sun X, Bull SD, He X-P, James TD (2019a) Reaction-based fluorescent probes for the detection and imaging of reactive oxygen, nitrogen, and sulfur species. Acc Chem Res 52(9):2582–2597

Wu L, Liu L, Han HH, Tian X, Odyniec ML, Feng L et al (2019b) ESIPT-based fluorescence probe for the ratiometric detection of superoxide. New J Chem 43:2875

Würth C, Geißler D, Behnke T, Kaiser M, Resch-Genger U (2015) Critical review of the determination of photoluminescence quantum yields of luminescent reporters. Anal Bioanal Chem 407 (1):59–78

Xiao H, Parkin KL (2002) Antioxidant functions of selected allium thiosulfinates and s-alk(en)yl-l-cysteine sulfoxides. J Agric Food Chem 50(9):2488–2493
Xu K, Wang L, Qiang M, Wang L, Li P, Tang B (2011) A selective near-infrared fluorescent probe for singlet oxygen in living cells. Chem Commun 47(26):7386–7388
Yan EB, Unthank JK, Castillo-Melendez M, Miller SL, Langford SJ, Walker DW (2005) Novel method for in vivo hydroxyl radical measurement by microdialysis in fetal sheep brain in utero. J Appl Physiol 98(6):2304–2310
Yang XF, Guo XQ (2001) Investigation of the anthracene-nitroxide hybrid molecule as a probe for hydroxyl radicals. Analyst 126(10):1800–1804
Yang S, Verhoeff AA, Merkx DWH, van Duynhoven JPM, Hohlbein J (2020) Quantitative spatiotemporal mapping of lipid and protein oxidation in mayonnaise. Antioxidants 9(12):1–13
Yapici NB, Jockusch S, Moscatelli A, Mandalapu SR, Itagaki Y, Bates DK et al (2012) New rhodamine nitroxide based fluorescent probes for intracellular hydroxyl radical identification in living cells. Org Lett 14(1):50–53
Yuan M, Li Y, Li J, Li C, Liu X, Lv J, Xu J, Liu H, Wang S, Zhu D (2007) A colorimetric and fluorometric dual-modal assay for mercury ion by a molecule. Org Lett 9(12):2313–2316
Zeng L, Miller EW, Pralle A, Isacoff EY, Chang CJ (2006) A selective turn-on fluorescent sensor for imaging copper in living cells. J Am Chem Soc 128(1):10–11
Zhou M, Diwu Z, Panchuk-Voloshina N, Haugland RP (1997) A stable nonfluorescent derivative of resorufin for the fluorometric determination of trace hydrogen peroxide: applications in detecting the activity of phagocyte NADPH oxidase and other oxidases. Anal Biochem 253 (2):162–168

Lipid Oxidation in Meat Systems: Updated Means of Detection and Innovative Antioxidant Strategies

Trinidad Pérez-Palacios and Mario Estévez

1 Introduction

Advancing in the food science field requires a detailed assessment of the chemistry fundamentals of food components and their reactivity during handling, storage and processing. Lipids are crucial components in meat systems as lipid quantity and composition influence sensory properties (appearance, juiciness…) and nutritional value of meat and meat products (Gandemer 2002). The impact of fatty acid composition of muscle lipids on meat quality lingers as a hot topic in the meat science field (Wood et al. 2008). Moreover, the lipid components of meat systems are highly reactive and lipid oxidation has straightforward consequences on the odor and flavor of meat products. Additionally, the toxicity, mutagenicity and carcinogenicity of malondialdehyde (MDA) as a major lipid oxidation product, is well known (Gandemer 2002; Esterbauer et al. 1993).

Lately, the assessment of the severity and nature of lipid oxidative reactions has progressed from the optimization and application of simple and routine methods to highly sophisticated procedures to identify and quantify specific oxidation products and pathways. Regarding the former, classic methods such as the thiobarbituric-reactive substances (TBARS) or the headspace isolation and subsequent analysis of volatile lipid oxidation products, have been optimized into speedier and more accurate procedures (Ganhão et al. 2011; Grotta et al. 2017). Regarding the latter, the application of innovative and more complex approaches has enabled the gain of further insight into the chemistry fundamentals and provided many additional benefits. For instance, non-destructive procedures have been developed to perform fast and accurate analysis of the quantity, composition and oxidation state of meat lipids with the purpose of predicting their influence on meat quality. Some of the

T. Pérez-Palacios · M. Estévez (✉)
Iprocar Research Institute, University of Extremadura, Cáceres, Spain
e-mail: mariovet@unex.es

C. Bravo-Diaz (ed.), *Lipid Oxidation in Food and Biological Systems*,
https://doi.org/10.1007/978-3-030-87222-9_4

technologies applied include Near-Infrared Spectroscopy (NIRS), Hyperspectral Imaging (HSI), Nuclear Magnetic Resonance (NMR), Magnetic Resonance Imaging (MRI) or Electron Spin Resonance (ESR), among others (Antequera et al. 2021). Furthermore, the arrival of versatile mass spectrometric equipments based on exact mass technology has enabled the onset of accurate and large-scale studies of muscle lipids (*lipidomics*) including their involvement in muscle biology and redox reactions (Hu et al. 2017; Paradiso et al. 2018; Züllic et al. 2020).

The remarkable impact of lipid oxidation on meat quality and safety emphasizes the necessity of applying antioxidant solutions to counteract these reactions and their negative effects. The recent advances achieved in the antioxidant protection of meat and meat products have been enabled by the usage of novel sources of antioxidants and by the development of effective and innovative means of application. Furthermore, understanding the chemistry fundamentals has also contributed to target the oxidation threats and neutralize specific oxidation routes, oxidation products and their noxious effects. Antemortem antioxidant solutions include managing animal feeding to minimize the onset of oxidative stress in the living animal and hence, contribute to enhance the oxidative stability of the muscle tissue upon slaughter. Assorted and synergist dietary strategies can be applied including formulating feeds antioxidant additives such as tocopherols (Romero et al. 2021), phytogenics and phenolic-rich plant materials (Selim et al. 2021; Romero et al. 2021), sulphur-amino acids (methionine) (Estévez et al. 2020), and certain micronutrients such as selenium (Surai and Fisinin 2015; Batorska et al. 2017) and magnesium (Estévez and Petracci 2019). Upon slaughter, antioxidant solutions can be directly applied to meat and meat products by using a protective packaging system (i.e. vacuum packing) (Soladoye et al. 2015) or by formulating with ingredients and additives of antioxidant potential (Trigo et al. 2020). Considerable advances have been made in the meat industry in regards to the incorporation of the so-called natural antioxidants (phytochemicals with antioxidant properties) by using assorted technological approaches such as marination, pulverization, injection, edible films etc. (Domínguez et al. 2018, 2019).

The present book chapter reviews the most recent advances in the analysis of lipid and lipid oxidation products and the most innovative strategies to control such reactions in meat and meat products.

2 Lipid Oxidation in Meat Products: Innovative Means of Analysis

Methodologies for the analysis of lipid oxidation can be classified as a function of the type of oxidation product (primary or secondary) aimed to be detected and quantified. For routine analyses, most meat scientists show preference towards the analysis of MDA or volatile compounds (as secondary lipid oxidation products) over the assessment of primary oxidation products such as hydroperoxides and

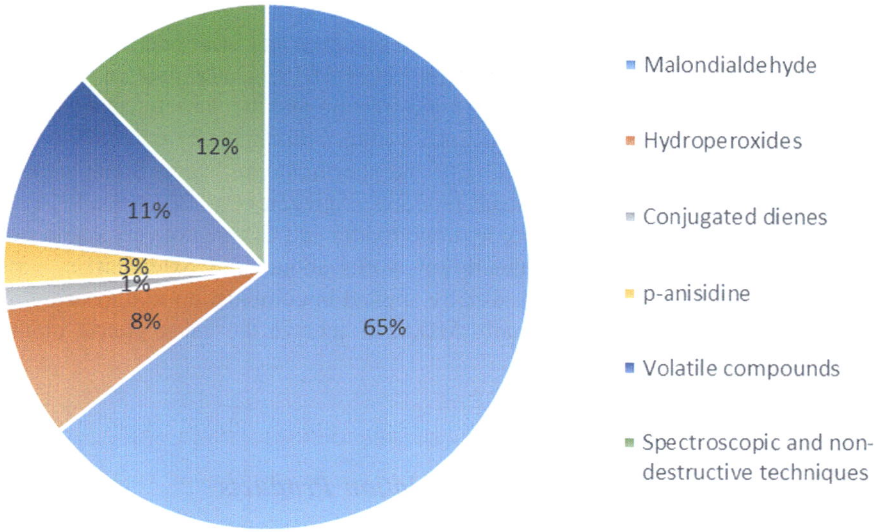

Fig. 1 Estimation of the extent of analysis of different lipid oxidation products in meat in the last five years (2017–2021)

conjugated dienes (CD). Moreover, the depletion of unsaturated fatty acids is not usually monitored to assess lipid oxidation, as such decrease may be only remarkable and meaningful at the latest stage of the lipid oxidation process (Guyon et al. 2016). Besides, there is an increasing interest in evaluating the capability of other techniques, mainly spectroscopic and non-destructive, to analyze lipid oxidation in meat products. Based on a scientific search in the last five years (2017–2021) using as key words "lipid oxidation" and "meat", Fig. 1 shows an estimation of the extent of analyses of these oxidation products to determine lipid oxidation in meat and its derivatives. After the analysis of MDA, the spectroscopic and non-destructive techniques and the volatile compounds analysis have been mainly applied to assess the lipid oxidation in meat products, following in decreasing order by the determinations of hydroperoxides, p-anisidine and CD.

2.1 Assessment of Primary Oxidation Products

The Peroxide Value (PV) is a method to measure the content of hydroperoxides and it is based on the oxidizing power of the hydroperoxides on iodine or iron. Two principal PV methods have been applied in recent studies: the iodometric titration (Hes et al. 2017; Rasinka et al. 2019; Pérez-Andrés et al. 2020), based on the titration of iodine oxidized upon reaction with hydroperoxides, and the spectrophotometric method ferric-xylenol orange (FOX) that uses ferrous ion (Mir et al. 2017; Chen et al. 2021). In both cases, a previous lipid extraction is required. In comparison to the iodometric titration method, FOX presents some advantages, for being simpler, less susceptible to atmospheric oxygen, more sensitive and offer better correlation with other oxidation parameters (Domínguez et al. 2019). Determination of CD is also initiated with the extraction of the lipid fraction and the subsequent measurement by ultraviolet spectrophotometry at 234 nm. Although CD measurement is faster and simpler than PV, its sensitivity and specificity are lower. Some authors have lately used CD in different meat products (Agregán et al. 2018; Dalle Zotte et al. 2018; Wazir et al. 2019). Since hydroperoxides and CD are not very stable and rapidly decompose, in most of the recent works about lipid oxidation in meat products, PV and CD methods are usually applied in combination to a measure of secondary oxidation products, mainly MDA, to achieve the entire course of the oxidation process.

2.2 Assessment of Secondary Oxidation Products

2.2.1 Malonaldehyde

The TBARS test is, without doubt, the most commonly used method to assess secondary lipid oxidation phenomenon in meat and meat products. It is based on the reaction between MDA and the thiobartituric acid (TBA) that gives a colorimetric complex with a maximum absorbance at 532–535 nm, measured with UV-Vis spectrophotometer. Nevertheless, it should be noted that other oxidation products, such as aldehydes, also react with TBA. Due to this, the TBARS test is used to determine the lipid oxidation status, rather than to quantify MDA (Domínguez et al. 2019). The test initiates with the extraction of MDA, which has been commonly carried out by distillation (Tarladgis et al. 1960) or homogenization with acids (Witte et al. 1970). With both extraction procedures, artifacts can be formed, due to high temperature that degrades existing hydroperoxides to secondary oxidation products, when applying distillation, or the extraction of colored components in the other case. This may lead to an overestimation of the TBARS values (Estévez et al. 2009). Besides, in cured meat products, the extraction of residual nitrite may react with MDA, causing an underestimation of the levels of TBARS (Gray and Monahan 1992). Different strategies have been proposed to solve these drawbacks, such as removing the interfering compounds by solid-phase extraction (Raharjo et al. 1992) or the addition of sulfanilamide to these cured products (Lee et al. 2017a). Wazir et al. (2019) have recently used a TBARS assay kit that is advertised for not requiring heating and reducing the interferences generated at high temperatures. There are not substantial differences between classical TBARS method and the assay kit. In the latter, samples are homogenized with distilled water and then added a TBA reagent, which consists of TBA and an appropriate acid catalyst. Moreover, several authors have tried to improve the TBARS procedure by evaluating different variables, such as MDA + TBA incubation conditions, type of sample or solvent extractions. Ganhão et al. (2011) suggested carrying out the reaction between

MDA and TBA at room temperature for 15–24 h. Grotta et al. (2017) recommended (1) avoiding stored or frozen samples and (2) the use of 7% trichloracetic acid (TCA) with antioxidants to avoid further oxidation during extraction. In fact, the most recent studies aimed to assess lipid oxidation in meat products by means of TBARS, have implemented the use of TCA (Baldi et al. 2019; Cheng et al. 2019; Costa et al. 2020; Sobral et al. 2020; Auriema et al. 2020; Wang et al. 2021). Also, to solve the biases from the reaction of MDA with TBA and other compounds, Reitznerová et al. (2017) proposed the derivatization of MDA with 2,4-dinitrophenylhydrazine (DNPH) and a subsequent analysis by reverse-phase high-performance liquid chromatography (RP-HPLC) with diode array detector (DAD). These authors found an overestimation of MDA of around 25% when using the TBARS test in comparison to the chromatographic method. A similar derivatization procedure has been carried out by Goethals et al. (2020) to measure MDA in commercial luncheon meat products by means of HPLC-UV/VIS at 310 nm. Other authors have used HPLC in combination with different detectors such as UV (Mousa and Al-Khateeb 2017), DAD (Mäkinen et al. 2020), and fluorescence (Gruffat et al. 2021). The use of ultraperformace liquid chromatography (UPLC) with DAD or fluorescence detector has also been proposed to determine MDA in meat samples and avoid the overestimation of the TBARs method (Bertolín et al. 2019).

2.2.2 Lipid-Derived Volatiles

Analysis of lipid-derived volatile compounds in meat products is mainly carried out to evaluate lipid oxidation processes and to assess their implication on the flavor and odor (Domínguez et al. 2019). Firstly, volatile compounds are extracted from the samples, with the headspace solid phase microextraction (SPME) being one of the most popular technique, since it is simple, solvent free and has high sensitivity. Once extracted, the volatile compounds are analyzed by chromatographic means. For that, the preferred analytical method is gas chromatography coupled to mass spectrometry (GC-MS). Recent studies on volatile compound determination in meat samples have focused on the optimization of the extraction conditions and the quantification procedure. Mansur et al. (2018) evaluated and compared the performance of two different types of fibers in the headspace extraction of volatile compounds in different pork and beef cuts, concluding that carboxen/polydimethylsiloxane fibers showed major extraction yields. The SPME conditions for extraction of beef volatiles were recently optimized by Wang et al. (2018a) who established an extraction time of 50 min at 80 °C followed by desorption for 2 min. Being aware of the lack of control over changes related to fiber adsorption-desorption or to detector sensitivity, Bueno et al. (2019) proposed the selection and addition of target compounds as internal standards. Specifically, methyl 2-methylbutyrate and 2,6-dichloroanisole were selected in this study with beef samples to control the quantification procedure despite of changes in the fiber, the column, the sample or the oxidation level. Some researchers have applied advanced MS detectors, such as selected ion flow tube

(SIFT)-MS (Flores et al. 2013; Carrapiso et al. 2015), which was proposed as an accurate tool for the analysis of volatiles in real time conditions, or time-of-flight (TOF)-MS (Gamero-Negrón et al. 2015; Wang et al. 2018b), which improves resolution and detection of polar volatile compounds.

However, the application of these innovative technologies in meat products has been scarce during the last years. Aldehydes are the most abundant family of volatile compounds derived from lipid oxidation. Although unsaturated aldehydes, such propenal, 4-heptenal, 2,4-heptadienal, 2-octenal or 2-nonenal, may be used as lipid oxidation indicators, saturated aldehydes are commonly selected for their higher stability, with hexanal being the major lipid oxidation marker (Barriuso et al. 2013). Hexanal is normally determined by SPME-GC-MS, as above explained, but it has also been analyzed by means of UPLC-DAD with a previous sample derivatization with DNPH in sulfuric acid (Vilarninho et al. 2018). It is worth emphasizing that the analysis of volatiles may not only provide information on the extent of lipid oxidation but also on the aromatic profile of the samples. On this line, many lipid-derived volatiles are odor-active compounds and the analysis of the full volatiles profile may be valuable to investigate the onset of off-flavors or of pleasant odor notes in meat and meat products (Domínguez et al. 2019).

2.3 Advanced Methodologies for Lipid Oxidation Assessment: Non-destructive Methods

The above revised methods to determine primary and secondary lipid oxidation in meat, though accurate, are time and solvent consuming, tedious and, in most cases, require the destruction of the samples. Thus, they do not fulfil the requirements demanded by meat industries for on-line methodologies and real-time results (Kucha et al. 2018). Because of this, there is an increasing interest in the evaluation of the capability of non-destructive techniques based on spectra and/or images for the analysis of quality parameters of meat products, such as Couple-Charges Devices cameras, Computed Tomography, NIRS, HSI, NMR or MRI (Antequera et al. 2021). In the case of lipid oxidation measurements, recent studies have mainly focused on HSI and NIRS, there being some recent work on NMR and MRI. It is worth to mention the great efforts for improving the data analysis when using these non-destructive technologies to evaluate meat products (Kucha et al. 2018). HSI collects and processes information from across the electromagnetic spectra. Aheto et al. (2019) have combined spectra and image data from of HSI in pork samples processed under different conditions to predict TBARS and PV. These authors applied Partial Least Square Regression (PLSR) models and found better results with fusion information of spectra and images for TBARS (R_p^2 (determination coefficient of prediction) = 0.865, RMSEP (root mean square error prediction) = 0.994, RPD (residual predictive deviation) = 2.280) and PV (R_p^2 = 0.899, RMSEP = 0.966, RPD = 2.314) than with models based exclusively on full wavelengths (R_p^2 = 0.832, RMSEP = 1.713, RPD = 2.141, for TBARS; R_p^2 = 0.857, RMSEP = 1.197, RPD = 2.023, for PV). Infrared spectroscopy measures the broad overtone and combination bands of some of the fundamental vibrations, being mid-infrared (MIR) and NIRS the most applied in food (Xu et al. 2015). The combination of HIS and NIRS to evaluate the lipid oxidation (measured as TBARs) in frozen pork has been recently investigated. Cheng et al. (2019) have developed a calibration model based on two scatter correction factors from Multiplicative scatter correction (MSC), achieving accurate results (R_p^2 = 0.926, RMSEP = 0.036). Jáuregui-López et al. (2020) aimed to determine TBARS by means of MIR in horse meat by applying MSC and PLSR. The prediction models obtained from the MIR spectra resulted less consistent than those achieved by the TBARS method. The authors attributed these results to the use of too width regression treatments that no delimits the range of the spectrum coinciding with the wavelengths in which many fatty acid bonds can be identified.

The capability of both Raman and ESR technologies for characterize the lipid oxidation process during freeze-thaw in beef has been evaluated by Chen et al. (2018). Results have shown significant correlations between TBARS and PV and these techniques, which are popular for their advantages of rapidness and online monitoring. Auriema et al. (2021) have revealed that NMR is an efficient technique to determine lipid oxidation in meat, since it achieves the identification of oxidation products (double allylic groups) not detected by conventional methods. These authors also noted the correlation in the appearance of these compounds with the increase of trans fatty acids. Very recently, Carvalho et al. (2021) have carried out a study aimed the analysis of chicken breast affected by white striping myopathy by MRI. The MRI images of the chicken breast were computationally analyzed, obtaining a vector of features that were correlated with physico-chemical characteristics by means of prediction techniques of data mining. Accurate values for correlation coefficient (R = 0.807–0.811), Mean Absolute Error (MAE = 0.013–0.018) and Weighted Absolute Percent Error (WAPE = 0.106–0.157) were obtained for TBARS prediction when applying Multiple Linear Regression (MLR) as predictive technique.

It is worth denoting that some of these advanced methods have not only enabled a more accurate and non-destructive assessment of the extent of lipid oxidation, they have further facilitated the understanding of the redox chemistry behind the oxidative damage to lipids, the oxidation pathways and mechanisms. As aforementioned, a better comprehension of this fundamental chemistry has greatly enabled the development of innovative means to counteract lipid oxidation and their negative consequences. The following section concisely covers some of these recent advances.

3 Lipid Oxidation in Meat Products: Innovative Antioxidant Solutions

The description of the assorted and serious consequences of the onset of oxidative reactions in meat and meat products, with some of them already reported, fully justify the application of antioxidant solutions and, hence, inhibit or alleviate the impact of such reactions on meat quality. These negative effect of lipid and protein oxidation includes deterioration of sensory properties (flavor, texture, appearance, among others) (Estévez 2011; Bekhit et al. 2013), loss of nutritional value (loss of essential nutrients, impaired digestibility of proteins) (Soladoye et al. 2015) and abiotic threats owing to the formation of toxic compounds (MDA, oxidized aminoacids) (Estévez and Xiong 2019). On this line, it is essential to identify the potential threat caused by the onset of the oxidative reactions in a particular meat system and envisage the potential benefits that such antioxidant strategy should eventually display. According to the existing literature, oxidative reactions can be counteracted at each of the stages between the farm and the fork, as lipid and protein oxidation takes place all the way from the livestock (≈ oxidative stress) to the ready-to-eat meat product and beyond, during the digestion in consumer's gastrointestinal tract (Estevez 2015). Hence, the application of comprehensive antioxidant strategies and inclusive solutions are interesting to control oxidative stress in the living animal, inhibit oxidative degradation of lipids and proteins during meat storage and processing and finally, control further oxidative deterioration of meat components during digestion that may lead to formation of potentially toxic compounds (Estévez 2011). In the following lines, antioxidant solutions for meat and meat products will be categorized in two groups, namely, those applied ante-mortem (at the farm), and those applied post-mortem (at the meat processing plant). Regardless of the strategy and the stage of the food chain, only innovative strategies shown to display benefits in terms of protection of meat quality, will be included.

3.1 Ante-Mortem Antioxidant Solutions

The oxidative stress occurred in animals' tissues is not only responsible for the depletion of production rates, impairment of animal welfare and the onset of pathological conditions; in vivo oxidation stress has an impact on meat quality (Fellenberg and Speisky 2006; Estevez 2015). Scientific evidence indicate that oxidative stress can be counteracted by avoiding the exposure of animals to environmental and dietary sources of such conditions. Among the former, we may cite extreme temperatures (particularly heat), poor ventilation and unproper handling that leads to physical stress and adrenergic responses (Fellenberg and Speisky 2006; Lykkesfeldt and Svendsen 2007; Estevez 2015). The intake of oxidized lipids is a major source of dietary oxidative stress though mycotoxins and other substances of toxicological concern may also promote in vivo oxidation (Estevez 2015). Evidence

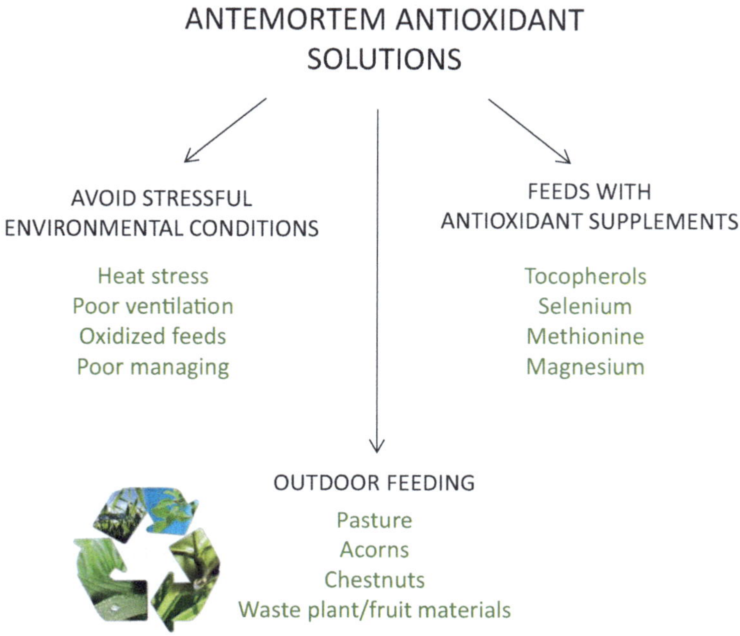

Fig. 2 Summary of most relevant ante-mortem antioxidant solutions for meat and meat products

also shows that oxidative stress and its negative consequences can be avoided or minimized through dietary strategies (Fig. 2). The efficiency of enhancing the oxidative stability of meat animals by supplementing tocopherols and, to a lesser extent, carotenoids, have been known for decades. The supplementation with 100–200 mg tocopherol/kg feed is a popular practice and have been found to protect meat from mammals (pork and beef) and poultry, against oxidation (Estévez 2011; Romero et al. 2021). While α-tocopherol can be supplemented in feeds, using natural sources of tocopherols such as pasture (Descalzo and Sancho 2008), acorns (Ventanas et al. 2007), chestnuts (Pugliese et al. 2009) and olive leaf (Botsoglou et al. 2012), to feed animals, has also been reported as an efficient strategy to deliver tocopherols and increase the oxidative stability of meat animals.

Dietary selenium (Se) is known to increase the concentration of glutathione peroxidase (GPx) in muscle and hence, protect this tissue against oxidative stress. The jointly work of GPx with catalase (CAT), glutathione reductase (GR), superoxide dismutase (SOD) and glutathione (GSH) leads to an effective neutralization of the pro-oxidant actions of ROS in muscle tissue. While Se has been classically supplemented as inorganic Se (as selenite or selenate), organic Se such as Se-yeast or other selenomethionine (SeMet) preparations (Surai and Fisinin 2015) have been reported to be more efficient. Profuse literature supports the use of 2-hydroxy-4-methylselenobutanoic acid (HMSeBA), a hydroxyl analog of SeMet, as a highly bioavailable source of SeMet (Jlali et al. 2014). The benefits of supplementation of organic Se have been documented for poultry meat (Estevez 2015), pork (Batorska et al. 2017), and beef (O'Grady et al. 2001). Other microelements and minerals such as copper and magnesium have been tested for their beneficial effects against oxidative stress (Estevez 2015). The latter, in particular, was recently found to strengthen the endogenous antioxidant defences of broilers supplemented with 0.3% of MgO and that was reflected in preserved water-holding capacity and reduced incidence of breast myopathies (Estévez and Petracci 2019). Sulphur amino acids have recently been regarded as "more than building blocks" owing to their recognized antioxidant effects when supplemented to the livestock (Estévez et al. 2020). Delivering DL-Met at supranational levels (>0.38%) to broilers and pigs has benefits in terms of meat quality such as color protection and enhanced water-holding capacity (Castellano et al. 2015; Zhai et al. 2016). The hydroxyl analog of DL-Met (DL-2-hydroxy-(4-methylthio) butanoic acid, DLHMB) can also be used for a better absorption, faster metabolization and more effective antioxidant protection (Estévez et al. 2020).

Finally, phytogenics have also been proposed to be used in animal feeding not only to protect against oxidative stress, but also to improve gut health, animal welfare, immune response and protect against biotic threats (Estevez 2015). Phytogenics are defined as molecular components from plant kingdom (e.g. roots, fruits, nuts, leafs, and their essential oils and extracts) with proven bioactivity. Such bioactivity is due to the occurrence of phytochemicals such as phenolic diterpenes, phenolic acids and polyphenols. Yet, there is limited evidence of significant accumulations of dietary phytochemicals (or their metabolites) in muscle tissue and hence, a straightforward antioxidant action of such phytogenics on meat systems is a topic under dispute. As redox-active species, some phytochemicals may be able to promote the endogenous antioxidant defences in livestock (Selby-Pham et al. 2017; Estévez et al. 2020). Such strengthening, via modulation of gene expression, could have benefits in terms of muscle redox status and meat quality (Lee et al. 2017b). Recent applications of dietary phytogenics to livestock have been reported by Moroney et al. (2015) (Laminariaceae seaweed); Beghelli et al. (2014) (Lamiaceae herbs); Aditya et al. (2018) (grape pomace); Tayengwa et al. (2020) (dried citrus pulp); Saleh et al. (2018) (pomegranate pomace), and Cimmino et al. (2018) (olive by-products), among many others. Further comprehensive information on the dietary application of phytochemicals in meat animals and its effects on meat quality can be found in the recent review by Serra et al. (2021).

3.2 Post-Mortem Antioxidant Solutions

Further to antioxidant solutions applied at the farm, additional strategies can be directly applied to meat and meat products to minimize the negative consequences of the onset of oxidative reactions during meat handling, cold storage, industrial processing, and culinary preparations (Fig. 3).

Vacuum packaging is regarded as a classic but effective antioxidant strategy to protect fresh meat against lipid and protein oxidation (Soladoye et al. 2015). Given

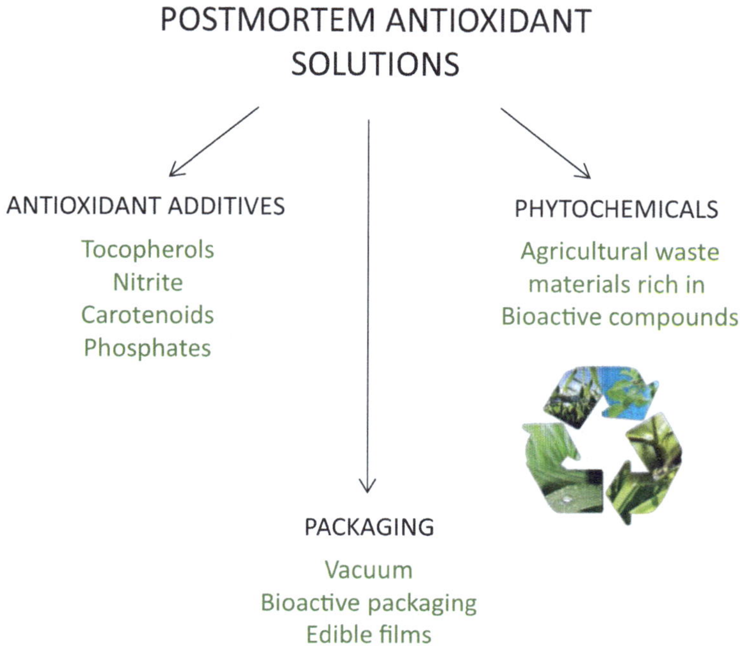

Fig. 3 Summary of most relevant post-mortem antioxidant solutions for meat and meat products

the recognized role of oxygen on the formation of ROS and the propagation of oxidative reactions, oxygen exclusion from the headspace of meats has been reported as highly efficient while high-oxygen packaging shows the opposite effect (Łopacka et al. 2016; Morcuende et al. 2020).

Certain additives commonly used in meat products' formulation such as nitrite, ascorbate and phosphates, and ingredients such as assorted spices, seasonings and non-muscle proteins and peptides, are largely known for protecting meat and meat products against oxidation (Xiong 2017). Yet, current consumer's concerns over some of the aforementioned additives urges the meat industry to apply innovative antioxidant solutions. The so-called "natural antioxidants" (those derived from plant kingdom) can be an excellent alternative to classic antioxidant additives for two reasons. First, waste materials from agricultural production such as seeds, pomaces, residues etc. are rich in phytochemicals with assorted bioactivities of indisputable interest in the meat industry (Fierascu et al. 2020). Promoting circular economy is not only a feasible and admirable purpose; the advantages of using these waste materials as sources of "natural antioxidants" are well documented (Trigo et al. 2020). Second, the design of "clean label" products by replacing standard additives by phytochemicals with versatile bioactivities is recently attracting considerable attention (Estévez 2021). This is particularly pertinent in the current scenario of consumer's mistrust in additives with known antioxidant activity (nitrite, phosphates, among others). Muíño et al. (2017) designed lamb patties with a phenolic-rich olive oil waste material and observed significant antioxidant protection. Aquilani et al. (2018) employed grape seeds and chestnuts as sources of phytochemicals with antioxidant potential in Cinta Senese dry-fermented sausages with added nitrites. Alirezalu et al. (2019) evaluated the ability of plant pigments, such as betanin, phyllocactin and betanidin to replace the antioxidant effects of nitrite in frankfurters. More recently, Martillanes et al. (2020) formulated clean-label pork burger patties with rice bran extract, which were subsequently subjected with high-hydrostatic pressure. Nanoencapsulated cinnamic aldehyde has been tested as replacement of the antioxidant and antimicrobial effects of nitrite in cooked sausages (Karim et al. 2020). Yet, the application of certain phenolic-rich materials as inhibitors of oxidative reactions in meat products is limited by the occurrence of distinctive and intense flavors from the plant that may not match the sensory properties of particular meat products. This can be avoided by using deodorized extracts or essential oils from rosemary or sage, for instance (reviewed by Estévez 2021).

The application of phytochemicals with antioxidant properties may also be justified in ready-to-eat meat products in which the extent of oxidation is such that the sensory and safety properties of these ultra-processed foods, are compromised. Examples of this approach can be found in literature using berries (Utrera et al. 2015), chestnuts (Ferreira et al. 2017), sage and laurel (Akcan et al. 2017) and tea plants (Wojtasik-Kalinowska et al. 2021) among many others. Some of the benefits of using such plant materials in ultra-processed meat products include reduction of rancidity in burger patties and meatballs (Ferreira et al. 2017; Akcan et al. 2017), improved texture in cooked sausages (Estévez et al. 2005) and inhibited formation of toxic compounds such as MDA in cooked beef products (Van Hecke et al. 2017).

Along with the search for new, sustainable and environmentally friendly sources of "natural antioxidants", great efforts have also been made in designing means of application to a large variety of meat products. The technological approach for the addition of "natural antioxidants" to meat products vary depending on the type of meat product (whether it is ground or intact, for instance) the physicochemical properties of the plant material (odor, polarity, stability…) and the antioxidant protection we aim to achieve (whether it is against lipid or protein oxidation…). Homogenization of the plant material or crude extract with the meat and the other ingredients is easy and inexpensive and hence, the most common procedure in ground products such as burger patties (Ganhão et al. 2011), meatballs (Suniati and Purnomo 2019), liver and emulsions (Estévez et al. 2005), or dry-cured sausages (Ozaki et al. 2021). Conversely, the incorporation of phytochemicals to intact meats and meat products such as steaks or cooked hams, requires other means such as pulverization (Morcuende et al. 2020), marination (Arcanjo et al. 2019) or injection (Armenteros et al. 2016). Other option is the addition of the natural antioxidants into edible films (Ribeiro et al. 2021). Using this antioxidant solution, Akcan et al. (2017) reported significant reduction of lipid oxidation in ready-to-eat beef cooked meatballs treated with whey protein isolate-based edible films containing natural antioxidant extracts from laurel (*Laurus nobilis* L.) and sage (*Salvia officinalis*). Using bioactive packaging materials impregnated with "natural antioxidants" has also been proposed to protect muscle foods against oxidation (Domínguez et al. 2018). One of the main benefits of this approach, as compared to the direct addition of antioxidants to meat products, is that active materials may act as a source of antioxidants to be released to the meat products at controlled rates during storage (Domínguez et al. 2018). Successful applications of these antioxidant solutions are found in literature. Baldin et al. (2016) were able toreduce lipid oxidation of fresh pork sausages during 15 days of storage by addition of microencapsulated jabuticaba extract. Ghaderi-Ghahfarokhi et al. (2017) also observed that encapsulated cinnamon essential oil effectively retarded lipid oxidation and improved the consumer acceptability of beef patties.

4 Future Perspectives

The future of lipid oxidation analysis and control relies on the research at the frontier of knowledge and "OMICS" techniques have attracted considerable attention. Lipidomics can be described as "the science of the large-scale determination of individual lipid species". Although the field of lipidomics has not progressed as much as other omics disciplines, due to the lipidome complexity and analysis challenges, it has experimented important advances in the last decade (Züllic et al. 2020). Within lipidomics, "oxidomics" may be included as a sophisticated and advanced tool aimed to comprehensively study the full profile of oxidation products ("oxidome") in a given food product (Hu et al. 2017). Paradiso et al. (2018) have studied the network of oxidation pathways and pattern of oxidation lipid products in highly-purified olive oil added with increasing amounts of free fatty acids. These authors highlighted the valuable support of oxidomics to increase the understanding of uncertain oxidation concerns. However, no oxidomics studies has been developed in meat products yet. The progress in the field of the application of plant antioxidants in meat products will also need further research at the frontier of knowledge. The design of justified, dependable and harmless antioxidant solutions for the meat industry can only be made by a comprehensive understanding the molecular basis of the oxidative reactions and the molecular redox-properties of phytochemicals. The upcoming antioxidant strategies in meat products will surely obey two essential requirements: sustainability and safety. In the current scenario of growing concern on the environment, the application of green and safe antioxidant solutions based on phytochemicals will be a sensible choice. Consumers preference for clean-label and minimally processed muscle foods will plausibly be consolidated in the new future.

References

Aditya S, Ohh S-J, Ahammed M, Lohakare J (2018) Supplementation of grape pomace (*Vitis vinifera*) in broiler diets and its effect on growth performance, apparent total tract digestibility of nutrients, blood profile, and meat quality. Anim Nutr 4(2):210–214

Agregán R, Franco D, Carballo J, Tomasevic I, Barba FJ, Gómez B, Muchenje V, Lorenzo JM (2018) Shelf life study of healthy pork liver pâté with added seaweed extracts from *Ascophyllum nodosum, Fucus vesiculosus* and *Bifurcaria bifurcate*. Food Res Int 112:400–411

Aheto JH, Huang X, Tian X, Ren V, Bonah E, Alenyorege EA, Lv R, Dai C (2019) Combination of spectra and image information of hyperspectral imaging data for fast prediction of lipid oxidation attributes in pork meat. J Food Process Eng 42:13225

Akcan T, Estévez M, Serdaroğlu M (2017) Antioxidant protection of cooked meatballs during frozen storage by whey protein edible films with phytochemicals from *Laurus nobilis* L. and *Salvia officinalis*. LWT – Food Sci Technol. 77:323–331

Alirezalu K, Hesari J, Nemati Z, . . ., Barba FJ, Lorenzo JM (2019) Combined effect of natural antioxidants and antimicrobial compounds during refrigerated storage of nitrite-free frankfurter-type sausage. Food Res Int. 120:839–850

Antequera T, Caballero D, Grassi S, Uttaro B, Perez-Palacios T (2021) Evaluation of fresh meat quality by hyperspectral imaging (HSI), nuclear magnetic resonance (NMR) and magnetic resonance imaging (MRI): a review. Meat Sci 172:108340

Aquilani C, Sirtori F, Flores M, . . ., Lebret B, Pugliese C (2018) Effect of natural antioxidants from grape seed and chestnut in combination with hydroxytyrosol, as sodium nitrite substitutes in Cinta Senese dry-fermented sausages. Meat Sci. 145:389–398

Arcanjo NMO, Morcuende D, Andrade MJ, . . ., Madruga MS, Estevez M (2019) Bioactivities of wine components on marinated beef during aging. J Funct Foods. 57:19–30

Armenteros M, Morcuende D, Ventanas J, Estévez M (2016) The application of natural antioxidants via brine injection protects Iberian cooked hams against lipid and protein oxidation. Meat Sci 116:253–259

Auriema BE, Vicente J, de Carvalho MG, Castro RN, Luchese RH, Mathias SP (2020) Correlation between nuclear magnetic resonance and traditional method to evaluate the lipid oxidation of emulsified chicken meat products with fat replacement by green banana biomass. J Food Process Preserv 45:15277

Auriema BE, Vicente J, de Carvalho M, Castro RN, Rosa HL, Mathias SP (2021) Correlation between nuclear magnetic resonance and traditional method to evaluate the lipid oxidation of emulsified chicken meat products with fat replacement by green banana biomass. J Food Process Preserv 45:15277

Baldi G, Chauhan SS, Linden N, Dunshea FR, Hopkins DL, Rossi CAS, Dell'Orto V, Ponnampalam EN (2019) Comparison of a grain-based diet supplemented with synthetic vitamin E versus a lucerne (alfalfa) hay-based diet fed to lambs in terms of carcass traits, muscle vitamin E, fatty acid content, lipid oxidation, and retail colour of meat. Meat Sci 148:105–112

Baldin JC, Michelin EC, Polizer YJ, . . ., Fernandes AM, Trindade MA (2016) Microencapsulated jabuticaba (*Myrciaria cauliflora*) extract added to fresh sausage as natural dye with antioxidant and antimicrobial activity. Meat Sci. 118:15–21

Barriuso B, Astiasarán I, Ansorena D (2013) A review of analytical methods measuring lipid oxidation status in foods: a challenging task. Eur Food Res Technol 236(1):1–15

Batorska M, Więcek J, Rekiel A (2017) Influence of organic vs inorganic source and different dietary levels of selenium supplementation in diets for growing pigs on meat quality. J Elem 22:653–662

Beghelli D, Bailetti L, Cavallucci C, . . ., Olivi L, Polidori P (2014) Lipid metabolism, antioxidant status and meat sensorial profile in pigs supplemented with plant extracts (*Origanum vulgare* L. and/or *Rosmarinus officinalis*). Progress in Nutr. 6(4):263–268

Bekhit AE-DA, Hopkins DL, Fahri FT, Ponnampalam EN (2013) Oxidative processes in muscle systems and fresh meat: sources, markers, and remedies. Comp Rev Food Sci Food Saf 12:565–597

Bertolín JR, Joy M, Blanco M (2019) Malondialdehyde determination in raw and processed meat products by UPLC-DAD and UPLC-FLD. Food Chem 298:125009

Botsoglou E, Govaris A, Fletouris D, Botsoglou N (2012) Lipid oxidation of stored eggs enriched with very long chain n-3 fatty acids, as affected by dietary olive leaves (*Olea europea* L.) or α-tocopheryl acetate supplementation. Food Chem 134:1059–1068

Bueno M, Resconi VC, Campo MM, Ferreira V, Escudero A (2019) Development of a robust HS-SPME-GC-MS method for the analysis of solid food samples. Analysis of volatile compounds in fresh raw beef of differing lipid oxidation degrees. Food Chem 281:49–56

Carrapiso AI, Noseda B, García C et al (2015) SIFT-MS analysis of Iberian hams from pigs reared under different conditions. Meat Sci 104:8–13

Carvalho L, Perez-Palacios T, Caballero D, Antequera T, Madruga MS, Estévez M (2021) Computer vision techniques on magnetic resonance images for the non-destructive classification and quality prediction of chicken breasts affected by the white-striping myopathy. J Food Eng 306:110633

Castellano R, Perruchot M-H, Conde-Aguilera JA, van Milgen J, Collin A, Tesseraud S, Mercier Y, Gondret F (2015) A methionine deficient diet enhances adipose tissue lipid metabolism and alters anti-oxidant pathways in young growing pigs. PLoS ONE 10(7):e0130514

Chen Q, Xie Y, Xi J, Guo Y, Qian H, Cheng Y, Chen Y, Chen Y, Yao W (2018) Characterization of lipid oxidation process of beef during repeated freeze-thaw by electron spin resonance technology and Raman spectroscopy. Food Chem 243:58–64

Chen X, Luo J, Lou A, Wang Y, Yang D, Shen QW (2021) Duck breast muscle proteins, free fatty acids and volatile compounds as affected by curing methods. Food Chem 338:128138

Cheng W, Sorensen KM, Engelsen SB, Sun D-W, Pu H (2019) Lipid oxidation degree of pork meat during frozen storage investigated by near-infrared hyperspectral imaging: effect of ice crystal growth and distribution. J Food Eng 263:311–319

Cimmino R, Barone CMA, Claps S, ..., Campanile G, Neglia G (2018) Effects of dietary supplementation with polyphenols on meat quality in Saanen goat kids. BMC Vet Res. 14 (1):181

Costa FS, Cabral AR, Silva SL, Silva MAI, Henrique W, Mazlli MR, Baldi FS, Mueller LF, Ferrinhho AM, Corte RRPS, Pereira ASC (2020) Effects of n-3 and n-6 feeding sources on the quality and lipid oxidation of meat from feedlot-finished *Bos indicus* steers. Meat Sci 161:107966

Dalle Zotte A, Cullere M, Tasoniero G, Gerencsér Z, Szendrő Z, Novelli E, Matics Z (2018) Supplementing growing rabbit diets with chestnut hydrolyzable tannins: effect on meat quality and oxidative status, nutrient digestibilities, and content of tannin metabolites. Meat Sci 146:101–108

Descalzo AM, Sancho AM (2008) A review of natural antioxidants and their effects on oxidative status, odor and quality of fresh beef produced in Argentina. Meat Sci 79:423–436

Domínguez R, Barba FJ, Gómez B, ..., Santos EM, Lorenzo JM (2018) Active packaging films with natural antioxidants to be used in meat industry: a review. Food Res Int. 113:93–101

Domínguez R, Pateiro M, Gagaoua M, Barba FJ, Zhang W, Lorenzo JM (2019) A comprehensive review on lipid oxidation in meat and meat products. Antioxidants. 8:429

Esterbauer H, Muskiet F, Horrobin DF (1993) Cytotoxicity and genotoxicity of lipid-oxidation products. Am J Clin Nutr 57:779S–786S

Estévez M (2011) Protein carbonyls in meat systems: a review. Meat Sci 89(3):259–279

Estevez M (2015) Oxidative damage to poultry: from farm to fork. Poult Sci 94:1368–1378

Estévez M (2021) Critical overview of the use of plant antioxidants in the meat industry: opportunities, innovative applications and future perspectives. Meat Sci. In press

Estévez M, Petracci M (2019) Benefits of magnesium supplementation to broiler subjected to dietary and heat stress: Improved redox status, breast quality and decreased myopathy incidence. Antioxidants. 8(10):456

Estévez M, Xiong Y (2019) Intake of oxidized proteins and amino acids and causative oxidative stress and disease: recent scientific evidences and hypotheses. J Food Sci 84(3):387–396

Estévez M, Ventanas S, Cava R (2005) Protein oxidation in frankfurters with increasing levels of added rosemary essential oil: effect on color and texture deterioration. J Food Sci 70(7):c427–c432

Estévez M, Morcuende D, Ventanas S (2009) Determination of oxidation. In: Nollet LML, Toldrá F (eds) Handbook of muscle foods analysis. CRC Press, Boca Raton, FL, USA, pp 221–240. isbn: 9781420045291

Estévez M, Geraert P-A, Liu R, . . ., Mercier Y, Zhang W (2020) Sulphur amino acids, muscle redox status and meat quality: more than building blocks – invited review. Meat Sci. 163:108087

Fellenberg MA, Speisky H (2006) Antioxidants: their effects on broiler oxidative stress and its meat oxidative stability. World Poult Sci J 62:53–70

Ferreira VCS, Morcuende D, Hérnandez-López SH, . . ., Silva FAP, Estévez M (2017) Antioxidant extracts from acorns (*Quercus ilex* L.) effectively protect ready-to-eat (RTE) chicken patties irrespective of packaging atmosphere. J Food Sci. 82(3):622–631

Fierascu RC, Sieniawska E, Ortan A, Fierascu I, Xiao J (2020) Fruits by-products – a source of valuable active principles A short review. Front Bioeng Biotech 8:319

Flores M, Olivares A, Dryahina K, Španěl P (2013) Real time detection of aroma compounds in meat and meat products by SIFT-MS and comparison to conventional techniques (SPME-GC-MS). Current Anal Chem 9:622–630

Gamero-Negrón R, Sánchez del Pulgar J, Cappellin L et al (2015) Immune-spaying as an alternative to surgical spaying in Iberian × Duroc females: effect on the VOC profile of dry-cured shoulders and dry-cured loins as detected by PTR-ToF-MS. Meat Sci 110:169–173

Gandemer G (2002) Lipids in muscles and adipose tissues, changes during processing and sensory properties of meat products. Meat Sci 62:309–321

Ganhão R, Estévez M, Morcuende D (2011) Suitability of the TBA method for assessing lipid oxidation in a meat system with added phenolic-rich materials. Food Chem 126:772–778

Ghaderi-Ghahfarokhi M, Barzegar M, Sahari MA, Gavlighi HA, Gardini F (2017) Chitosan-cinnamon essential oil nano-formulation: application as a novel additive for controlled release and shelf life extension of beef patties. Int J Biol Macromol 102:19–28

Goethals S, Van Hecke T, Vossen E, Vanhaeckeb L, Van Camp J, De Smet S (2020) Commercial luncheon meat products and their in vitro gastrointestinal digests contain more protein carbonyl compounds but less lipid oxidation products compared to fresh pork. Food Res Int 136:109585

Gray JI, Monahan FJ (1992) Measurement of lipid oxidation in meat and meat products. Trends Food Sci Technol 3:315–319

Grotta L, Castellani F, Palazzo F, Naceur Haouet M, Martino G (2017) Treatment optimisation and sample preparation for the evaluation of lipid oxidation in various meats through TBARs assays before analysis. Food Anal Met 10:1870–1880

Gruffat D, Bauchart D, Thomas A, Parfita E, Durand D (2021) Fatty acid composition and oxidation in beef muscles as affected by ageing times and cooking methods. Food Chem 343:128476

Guyon C, Meynier A, de Lamballerie M (2016) Protein and lipid oxidation in meat: a review with emphasis on high-pressure treatments. Trends Food Sci Technol 50:131–143

Hes M, Szwengiel A, Dziedizic K, Thanh-Blicharz JL, Kmiecik D, Gorecka D (2017) The effect of buckwheat hyll extract on lipid oxidation in frozen-stored meat products. J Food Sci 82:882–889

Hu C, Wang M, Han X (2017) Shotgun lipidomics in substantiating lipid peroxidation in redox biology: methods and applications. Redox Biol 12:946–955

Jáuregui-López I, Zulategi F, Beriain MJ, Sarriés MV, Beruete M, Insausti K (2020) Lipid and protein oxidation marker compounds in horse meat determined by MIR spectroscopy. Foods. 9:1828

Jlali M, Briens M, Rouffineau F, Geraert P-A, Mercier Y (2014) Evaluation of the efficacy of 2-hydroxy-4-methylselenobutanoic acid on growth performance and tissue selenium retention in growing pigs. J Anim Sci 92:182–188

Karim M, Fathi M, Soleimanian-Zad S (2020) Nanoencapsulation of cinnamic aldehyde using zein nanofibers by novel needle-less electrospinning: production, characterization and their application to reduce nitrite in sausages. J Food Eng 288:110140

Kucha C, Liu L, Ngadi MO (2018) Non-destructive spectroscopic techniques and multivariate analysis for assessment of fat quality in pork and pork products: a review. Sensors 18:377–400

Lee JH, Alford L, Kannan G, Kouakou B (2017a) Curing properties of sodium nitrite in restructured goat meat (chevon) jerky. Int J Food Prop 20:526–537

Lee MT, Lin WC, Yu B, Lee TT (2017b) Antioxidant capacity of phytochemicals and their potential effects on oxidative status in animals – a review. As-Austral J Anim Sci 30(3):299–308

Łopacka J, Półtorak A, Wierzbicka A (2016) Effect of MAP, vacuum skin-pack and combined packaging methods on physicochemical properties of beef steaks stored up to 12 days. Meat Sci 119:147–153

Lykkesfeldt J, Svendsen O (2007) Oxidants and antioxidants in disease: oxidative stress in farm animals. Vet J 173:502–511

Mäkinen S, Hellström J, Mäki M, Korpinen R, Mattila PH (2020) Bilberry and sea buckthorn leaves and their SubcriticalWater extracts prevent lipid oxidation in meat products. Foods. 9:265–279

Mansur AR, Lee HJ, Choi H-K, ..., Jang HW, Nam TG (2018) Comparison of two commercial solid-phase microextraction fibers for the headspace analysis of volatile compounds in different pork and beef cuts. J Food Process Pres. 42:e13746

Martillanes S, Ramírez R, Amaro-Blanco G, ..., Gil MV, Delgado-Adámez J (2020) Effect of rice bran extract on the preservation of pork burger treated with high pressure processing. J Food Proc Pres. 44(1):e14313

Mir SA, Masoodi FA, Raja J (2017) Influence of natural antioxidants on microbial load, lipid oxidation and sensorial quality of rista-a traditional meat product of India. Food Bioesci 20:79–87

Morcuende D, Vallejo-Torres C, Ventanas S, Martínez SL, Ruiz SC, Estévez M (2020) Effectiveness of sprayed bioactive fruit extracts in counteracting protein oxidation in lamb cutlets subjected to a high-oxygen MAP. Foods 9(11):1715

Moroney NC, O'Grady MN, Robertson RC, Stanton C, O'Doherty JV, Kerry JP (2015) Influence of level and duration of feeding polysaccharide (laminarin and fucoidan) extracts from brown seaweed (*Laminaria digitata*) on quality indices of fresh pork. Meat Sci 99:132–141

Mousa RMA, Al-Khateeb LA (2017) Influence of binary and ternary mixtures of spices on the inhibition of lipid oxidation and carcinogenic heterocyclic amines in fried hamburger patties. J Food Process Pres 41:12976

Muíño I, Díaz MT, Apeleo E, ..., Lauzurica S, Fuente JDL (2017) Valorisation of an extract from olive oil waste as a natural antioxidant for reducing meat waste resulting from oxidative processes. J Clean Prod. 140:924–932

O'Grady MN, Monahan FJ, Fallon RJ, Allen P (2001) Effects of dietary supplementation with vitamin E and organic selenium on the oxidative stability of beef. J Anim Sci 79:2827–2834

Ozaki MM, Munekata PES, Jacinto-Valderrama RA, ..., Lorenzo JM, Pollonio MAR (2021) Beetroot and radish powders as natural nitrite source for fermented dry sausages. Meat Sci. 171:108275

Paradiso VM, Pasqualone A, Summo C, Caponio F (2018) An "omics" approach for lipid oxidation in foods: the case of free fatty acids in bulk purified olive oil. Eur J Lipid Sci Technol 120:1800102

Pérez-Andrés JM, Cropotova J, Harrison SM, Brunton NP, Cullen PJ, Rustad T, Tiwari BK (2020) Effect of cold plasma on meat cholesterol and lipid oxidation. Foods. 9:1786

Pugliese C, Sirtori F, D'Adorante S, Parenti S, Rey A, Lopez-Bote C, Franci O (2009) Effect of pasture in oak and chestnut groves on chemical and sensorial traits of cured lard of Cinta Senese pigs. Italian J Anim Sci 9:131–142

Raharjo S, Sofos JN, Schmidt GR (1992) Improved speed, specificity, and limit of determination of an aqueous acid extraction thiobarbituric acid-cig method for measuring lipid peroxidation in beef. J Agric Food Chem 40:2182–2185

Rasinka E, Rutkowska J, Czarnicka-Skubina E, Tambor K (2019) Effects of cooking methods on changes in fatty acid contents, lipid oxidation and volatile compounds of rabbit meat. LWT-Food Sci Technol 110:64–70

Reitznerová A, Uleková M, Nagy J, Marcinčák S, Semjon B, Čertík M, Klempová T (2017) Lipid peroxidation process in meat and meat products: a comparison study of malondialdehyde determination between modified 2-thiobarbituric acid spectrophotometric method and reverse-phase high-performance liquid chromatography. Molecules 22:1988

Ribeiro AM, Estevinho BN, Rocha F (2021) Preparation and incorporation of functional ingredients in edible films and coatings. Food BioprocTechnol 14(2):209–231

Romero C, Nardoia M, Arija I, . . ., Prodanov M, Chamorro S (2021) Feeding broiler chickens with grape seed and skin meals to enhance α-and γ-tocopherol content and meat oxidative stability. Antioxidants. 10(5):699

Saleh H, Golian A, Kermanshahi H, Mirakzehi MT (2018) Antioxidant status and thigh meat quality of broiler chickens fed diet supplemented with α-tocopherolacetate, pomegranate pomace and pomegranate pomace extract. Ital J Anim Sci 17(2):386–395

Selby-Pham SNB, Cottrell JJ, Dunshea FR, . . ., Bennett LE, Howell KS (2017) Dietary phytochemicals promote health by enhancing antioxidant defence in a pig model. Nutrients. 9(7):758

Selim S, Seleiman MF, Hassan MM, Saleh AA, Mousa MA (2021) Impact of dietary supplementation with *moringa oleifera* leaves on performance, meat characteristics, oxidative stability, and fatty acid profile in growing rabbits. Animals 11(2):248. 1-16

Serra V, Salvatori G, Pastorelli G (2021) Dietary polyphenol supplementation in food producing animals: effects on the quality of derived products. Animals 11(2):401. 1-44

Sobral MMC, Casal S, Faria MA, Cunha SC, Ferreira IMPLVO (2020) Influence of culinary practices on protein and lipid oxidation of chicken meat burgers during cooking and in vitro gastrointestinal digestion. Food Chem Toxicol 141:111401

Soladoye OP, Juárez ML, Aalhus JL, Shand P, Estévez M (2015) Protein oxidation in processed meat: mechanisms and potential implications on human health. Comp. Rev. Food Sci. Food Saf. 14:106–122

Suniati FRT, Purnomo H (2019) Goroho (*Musa acuminafe*, sp) banana flour as natural antioxidant source in indonesian meatball production. Food Res 3(6):678–683

Surai PF, Fisinin VI (2015) Selenium in pig nutrition and reproduction: boars and semen quality – a review. Asian Australas. JAnim Sci 28:730–746

Tarladgis BG, Watts BM, Younathan MT, Dugan L Jr (1960) A distillation method for the quantitative determination of malonaldehyde in rancid foods. J Am Oil Chem Soc 37:44–48

Tayengwa T, Chikwanha OC, Dugan MER, Mutsvangwa T, Mapiye C (2020) Influence of feeding fruit by-products as alternative dietary fibre sources to wheat bran on beef production and quality of Angus steers. Meat Sci 161:107969

Trigo JP, Alexandre EMC, Saraiva JA, Pintado ME (2020) High value-added compounds from fruit and vegetable by-products–characterization, bioactivities, and application in the development of novel food products. Crit Rev Food Sci Nutr 60(8):1388–1416

Utrera M, Morcuende D, Ganhão R, Estévez M (2015) Role of phenolics extracting from *Rosa canina* L. on meat protein oxidation during frozen storage and beef patties processing. Food biopro. Technol. 8(4):854–864

Van Hecke T, Ho PL, Goethals S, De Smet S (2017) The potential of herbs and spices to reduce lipid oxidation during heating and gastrointestinal digestion of a beef product. Food Res Int 102:785–787

Ventanas S, Ventanas J, Tovar J, García C, Estévez M (2007) Extensive feeding versus oleic acid and tocopherol enriched mixed diets for the production of Iberian dry-cured hams: effect on chemical composition, oxidative status and sensory traits. Meat Sci 77:246–256

Vilarninho F, Andrade M, Buonocore GG, Stanzione M, Vaz MF, Silva AS (2018) Monitoring lipid oxidation in a processed meat product packaged with nanocomposite poly(lactic acid) film. Eur Polym J 98:362–367

Wang W, Feng X, Zhang D et al (2018a) Analysis of volatile compounds in Chinese dry-cured hams by comprehensive two-dimensional gas chromatography with high-resolution time-of-flight mass spectrometry. Meat Sci 140:14–25

Wang X, Zhu L, Han Y et al (2018b) Analysis of volatile compounds between raw and cooked beef by HS-SPME–GC–MS. J. Food Process. Pres. 42:13503

Wang Z, Tu J, Zhou H, Lu A, Xu B (2021) A comprehensive insight into the effects of microbial spoilage, myoglobin autoxidation, lipid oxidation, and protein oxidation on the discoloration of rabbit meat during retail display. Meat Sci 172:108359

Wazir H, Chay SY, Zarei M, Hussin FS, Mustapha NA, Zunairah W, Ibadullah W, Saari N (2019) Efects of storage time and temperature on lipid oxidation and protein co-oxidation of low-moisture shredded meat products. Antioxidants 8:486

Witte VC, Krause GF, Bailey ME (1970) A new extraction method for determining 2-thiobarbituric acid values of pork and beef during storage. J Food Sci 35:582–585

Wojtasik-Kalinowska I, Onopiuk A, Szpicer A, Wierzbicka A, Półtorak A (2021) Frozen storage quality and flavor evaluation of ready to eat steamed meat products treated with antioxidants. CYTA – J Food 19(1):152–162

Wood JD, Enser M, Fisher AV, Nute GR, Sheard PR, Richardson RI, Hughes SI, Whittington FM (2008) Fat deposition, fatty acid composition and meat quality: a review. Meat Sci 78:343–358

Xiong YL (2017) Inhibition of hazardous compound formation in muscle foods by antioxidative phytophenols. Ann N Y Acad Sci 1398:37–46

Xu J-L, Riccioli C, Sun D-W (2015) An overview on nondestructive spectroscopic techniques for lipid and lipid oxidation analysis in fish and fish products. Compr Rev Food Sci Food Saf 14:466–477

Zhai W, Schilling MW, Jackson V, Peebles ED, Mercier Y (2016) Effects of dietary lysine and methionine supplementation on Ross 708 male broilers from 21 to 42 days of age (II): breast meat quality. J Appl Poult Res 25(2):212–222

Züllic T, Trötzmüller M, Köfeier H (2020) Lipidomics from sample preparation to data analysis: a primer. Anal. Bioana.L Chem 412:2191–2209

Part II
Lipid Oxidation and Inhibition in Different Systems

The Underrecognized Role of the Hydroperoxyl ($HOO^•$) Radical in Chain Propagation of Lipids and its Implication in Antioxidant Activity

Yafang Guo and Riccardo Amorati

1 Introduction

Degradation of organic molecules under atmospheric O_2 is due to a radical-chain reaction named autoxidation, or peroxidation. The two names of this reaction derive from the fact that it is characterized by an apparently spontaneous occurrence under mild conditions, and that (hydro)peroxides are the main first products of this process (Burton and Ingold 1986). The mechanism of peroxidation is well established, and is based on a radical chain mechanism propagated by carbon-centred (alkyl, $R^•$) and oxygen-centred (alkylperoxyl, $ROO^•$) radicals, as shown in Scheme 1 (Zielinski and Pratt 2017).

Hydroperoxides (ROOH) decompose through radical and non-radical mechanisms (Zielinski and Pratt 2017) to a variety of potentially toxic electrophilic carbonyl compounds such as acrolein, 4-hydroxy-trans-2-nonenal and malondialdehyde (Vieira et al. 2017). Formation of ROOH in turn renders the material more susceptible to oxidation because ROOH is a substrate of initiation events such as the Fenton chemistry (Zielinski and Pratt 2017). Besides the well-established role of $R^•$ and $ROO^•$ radicals, the hydroperoxyl radical ($HOO^•$) has been recently discovered as an important player in the propagation and termination of lipid peroxidation. Although the formation of $HOO^•$ during peroxidation was recognized long time ago, (Howard and Ingold 1967a; Denisov 1996) the role of $HOO^•$ radical has been appreciated in its full importance only after the pioneering works by Foti and Ingold on the antioxidant effect of γ-terpinene (Foti and Ingold 2003). As summarized in Scheme 1, and explained in this chapter, the $HOO^•$ radical can potentially impact on both propagation and termination steps of the autoxidation reaction (see Scheme 1), as effect of its peculiar chemical characteristics. Notably,

Y. Guo · R. Amorati (✉)
Department of Chemistry "G. Ciamician" University of Bologna, Bologna, Italy
e-mail: Riccardo.amorati@unibo.it

C. Bravo-Diaz (ed.), *Lipid Oxidation in Food and Biological Systems*,
https://doi.org/10.1007/978-3-030-87222-9_5

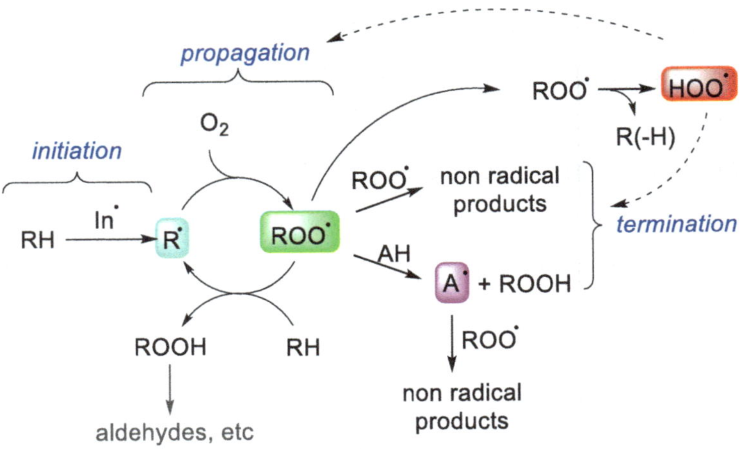

Scheme 1 Role of hydroperoxyl (HOO•) radicals in the autoxidation of an organic substrate (RH), initiated by In• radicals, in the presence of antioxidants (AH)

HOO• is formed typically by the fragmentation of ROO• radicals of specific oxidizable substrates, which therefore represent a possible source of this radical see Sect. 3). An enhancing effect of HOO• on the activity of several phenolic and nonphenolic antioxidants has been recently uncovered (see Sect. 4). In this chapter, we will show the mechanisms that lead to HOO• formation during the lipid peroxidation, and how the HOO• presence impacts on the self-termination of peroxyl radicals and on the efficacy of antioxidants. As the role of HOO• is still relatively undiscovered, we will also provide suggestions for future research to fully exploit the rich potentialities of HOO• chemistry.

2 The HOO• Radical

2.1 Biochemical Role

HOO• (hydroperoxyl radical), also known as perhydroxyl radical, is the protonated form of superoxide $O_2^{\bullet-}$. Radical chemistry of HOO•/$O_2^{\bullet-}$ has been implicated in a wide variety of processes, such as metal catalyzed oxidation, (Zheng et al. 2015; Chang et al. 2013) environmental chemistry, (Kumar and Francisco 2015; Anglada et al. 2015) and cell biology (Campomanes et al. 2015). $O_2^{\bullet-}$ generated by the reduction of molecular oxygen, $O_2 + e^- \rightarrow O_2^{\bullet-}$ is the predominant form of superoxide in neutral water, (Sawyer and Valentine 1981; Lian et al. 1974) while HOO• ($pKa = 4.7$) predominates at low pH (Warren et al. 2010; Bielski 1978; Behar et al. 1970). The HOO•/$O_2^{\bullet-}$ couple works as the precursor of many other reactive oxygen species (ROS), for example, their disproportionation yields hydrogen peroxide and oxygen, (Bielski and Allen 1977) and H_2O_2 can generate HO• radicals through the Fenton reaction. In the context of the reactive oxygen species (ROS), hydroperoxyl radicals are believed to play an important role in the oxidation of biomembranes. In fact, $O_2^{\bullet -}$ very poorly interacts with polyunsaturated fatty acids and proteins (McCord and Fridovich 1969; Weisiger and Fridovich 1973) and it is also unreactive towards phenolic antioxidants such as catechols and α-tocopherol (Nanni et al. 1980). The hydroxyl radical (HO•), because of its high reactivity, (Sawyer and Valentine 1981) indistinctly reacts with organic molecules at nearly diffusion-controlled rate, and its short lifetime (~10^{-9} sec) prevents it from reaching the inner layer of lipid membrane to attack the unsaturated double bonds (Czapski 1984). Although the low p*K*a of HOO• reduces its abundance in the cytoplasm (pH 7.4), and less than 1% of $[O_2^{\bullet -}]$ is present in pronated form, HOO• is more hydrophobic and reactive than $O_2^{\bullet -}$ (Min and Ahn 2005). The pH value in proximity of the charged membranes may be several units lower than in the bulk cytoplasm. It should be also kept in mind that HOO• has no charge, so it can easily pass into or cross the lipid membrane, providing its action as lipid peroxidation initiator, by abstracting hydrogen atoms like alkylperoxyl radicals do, in contrast with $O_2^{\bullet -}$, which is a poor in hydrogen abstraction reactions (Ingold 2003). The HOO• radical is capable to react with polyunsaturated lipids to generate pentadienyl type radicals, (Bielski et al. 1983) however in membranes and liposomes HOO• can efficiently trigger lipid peroxidation only if hydroperoxides are already present (Aikens and Dix 1993). This aspect is indeed very important to explain the behavior of HOO• in a complex system and deserves further investigations. However, it should be noted that $O_2^{\bullet -}$/HOO• radicals don't have only negative effects, due to the potential physiological functions such as their role as antibacterial (Davies 1995; Sundaresan et al. 1995; Rosen et al. 1995) and inhibition on the growth and spread of cancer cells (Huang et al. 2000).

2.2 Physico-Chemical Characteristics of HOO•

Oxidizing Properties Superoxide ($O_2^{\bullet -}$) is a weak H-atom transfer oxidizer because of the relatively small strength of the $^-O_2$-H bond, whose free energy is reported as 81.6 kcal/mol in water (Warren et al. 2010). The hydroperoxyl radical is instead a better H-atom abstractor as the strength of the HOO-H bond is larger and similar to that of organic hydroperoxides. The Bond Dissociation Enthalpy, BDE, of HOO-H is 87.5 kcal/mol in the gas phase, while those of CH_3OO-H and ^tBuOO-H are 87.8 and 84.3 kcal/mol, respectively (Warren et al. 2010).

Reducing Properties The unusual chemistry of hydroperoxyl radical is deeply connected to its hydrogen donating ability. The BDE (•OO-H) is 49.2 kcal/mol, (Warren et al. 2010) that is notably lower than the O-H BDE of phenolic and non-phenolic antioxidants (typically 77 to 82 kcal/mol), (Warren et al. 2010; Amorati et al. 2011) as well as of most semiquinones radical (Fattahi et al. 2005). This demonstrates from a thermodynamic point of view the possibility to regenerate antioxidants from their radicals ($A^•$) or from an oxidized quinone (Q) by $HOO^•$ through the reactions 1 and 2.

$$A^• + HOO^• \rightarrow AH + O_2 \quad (1)$$

$$Q + HOO^• \rightarrow QH^• + O_2 \quad (2)$$

In Scheme 2, a BDE ladder showing the comparison between the BDE($^•$OO-H) and that of species relevant for the chemistry of peroxidation and antioxidant activity is reported. All compounds that can form a bond stronger than 49.2 kcal/mol with a H atom can potentially be reduced by $HOO^•$, although the reaction will be more efficient if the strength of the newly formed bond is much higher than this value.

It has been shown that the H-atom transfer from $HOO^•$, like those reported in Eqs. (1 and 2), are slowed down by H-bonding with solvent, as shown in Scheme 3a (Foti et al. 2005). The regeneration of the antioxidant radical A• by HOO• is also reduced by radical-radical combination, that causes the loss of the antioxidant activity (Scheme 3b) (Cedrowski et al. 2016). Interestingly, the radicals of nitroxides ($R_2NO^•$) or diphenylamines ($Ar_2N^•$) have a small tendency to add $HOO^•$ and $ROO^•$ radicals, and this is related to their superior antioxidant activity compared to phenols in situations where regeneration by $HOO^•$ plays an important role (Poon et al. 2020).

2.3 Formation of $HOO^•$

The generation of $HOO^•$ radical during the autoxidation of different classes of organic substrates is based on an intramolecular 1,4-H-atom transfer (intra-1,4-HAT) as exemplified in Scheme 4.

Besides the obvious requirement of the availability of an abstractable H-atom in 4-position with respect to the terminal oxygen of the peroxyl group, many structural characteristics of the oxidizable substrate can facilitate the $HOO^•$ release. These characteristics are summarized below, divided by the functional group undergoing autoxidation.

Alcohols The α-C-H group in primary and secondary alcohols (**1**), see Scheme 5, is preferentially attacked by radicals, owning to the adjacent hydroxyl group which lowers its BDE (Denisov and Afanas'ev 2005). The α-hydroxy alkyl radical **2** (also named ketyl radical) reacts with O_2 at diffusion rate to afford the α-hydroxy-peroxyl radical **3** which decomposes to a carbonyl compound (**4**) and $HOO^•$ (Denisov and Afanas'ev 2005; Valgimigli et al. 2008).

Amines Like the oxygen atom in alcohols, nitrogen in amines lowers the BDE of α-C-H by resonance with the unpaired electron. Thus, the α-position of alkyl-substituted amines (**5, 9**) is strongly activated toward H-atom abstraction, forming aminoalkyl radicals (**6, 10**) that, in the presence of O_2, form the corresponding α-amino peroxyl radicals (**7, 11**). In the case of primary and secondary amines (**5**),

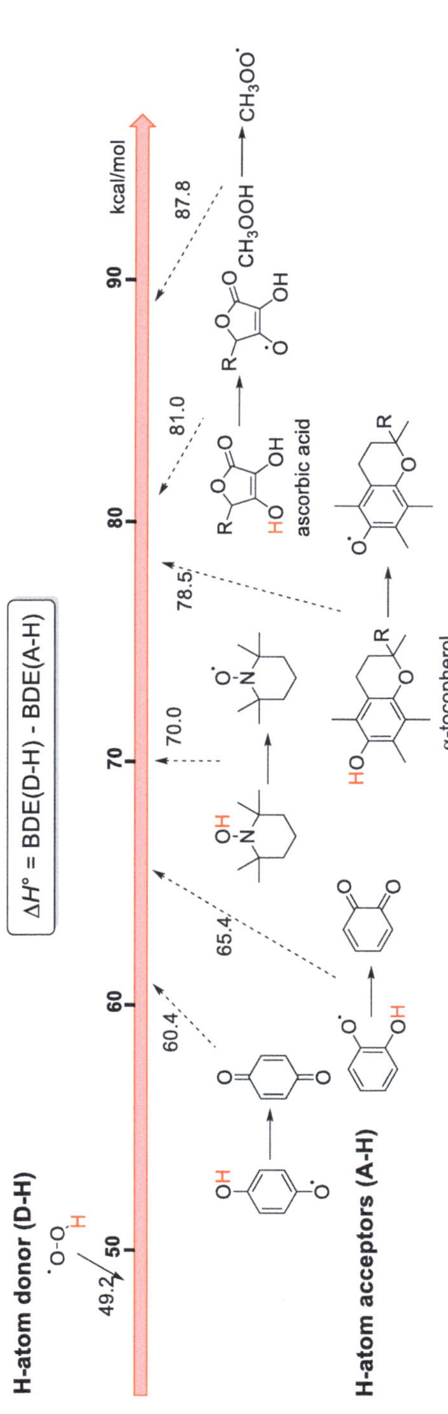

Scheme 2 Enthalpy values in kcal/mol for the donation of a H-atom from HOO• to various acceptors. The reaction is exothermic for BDE(A-H) > BDE(D-H)

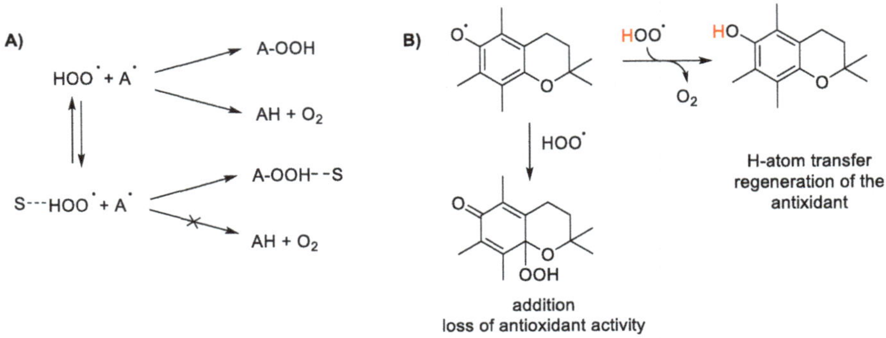

Scheme 3 Mechanism leading to a decreased regeneration of antioxidants by HOO•. (**a**) Interaction with a H-bond accepting solvent. (**b**) Competition between H-atom transfer and addition

Scheme 4 General mechanism of HOO• elimination from alkyl peroxyl radicals

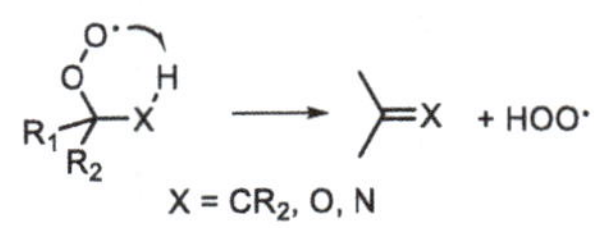

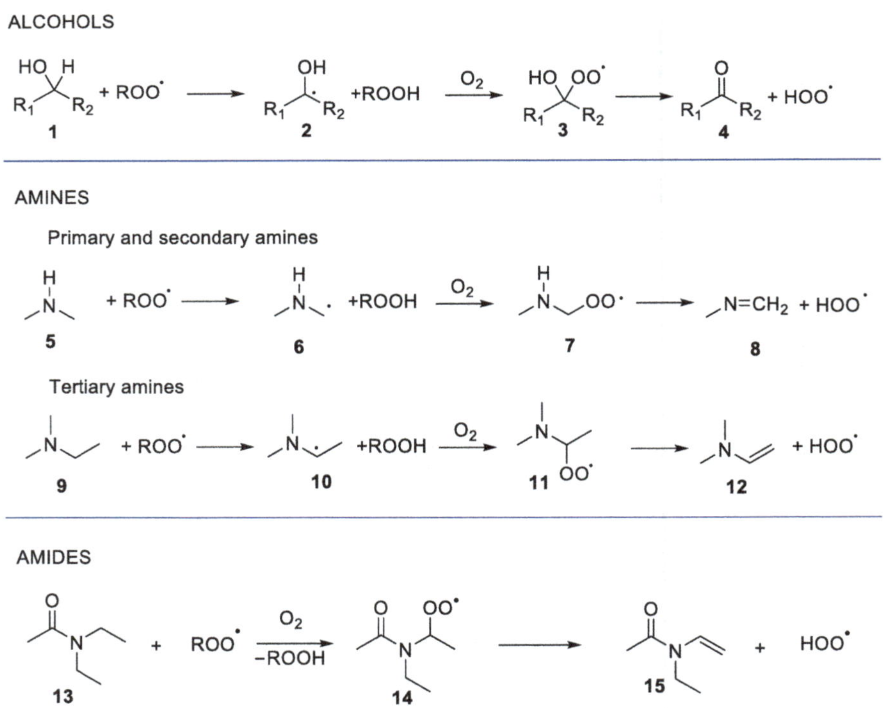

Scheme 5 Generation of HOO• during the autoxidation of alcohols, amines and amides

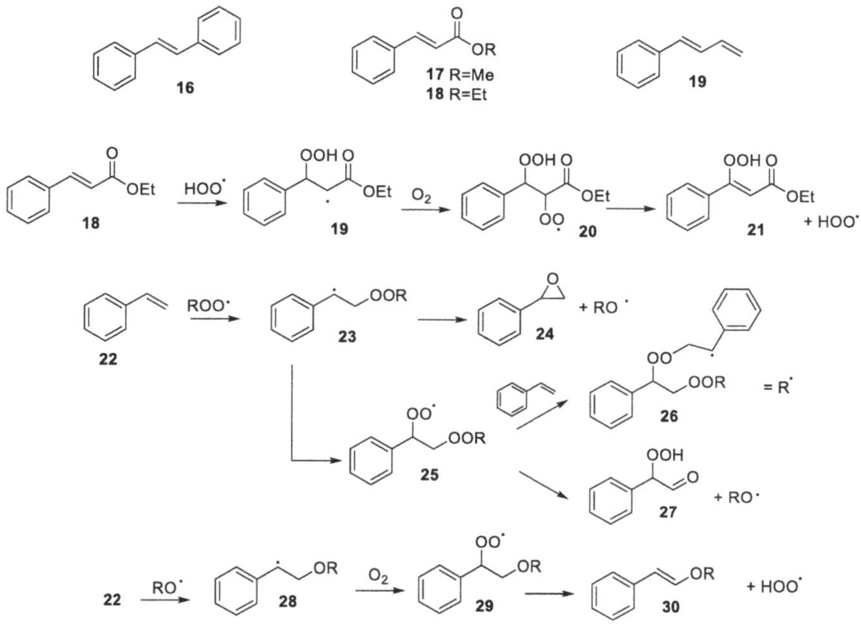

Scheme 6 Mechanism proposed for the formation of HOO• during the autoxidation of alkenes

the aminoperoxyl radical **7** decomposes to HOO• and an imine **8** (Schumperli et al. 2012). In the case of tertiary alkylamines having at least one CH_2CH_2 substituent, such in the case of compound **9**, HOO• is generated with great efficiency clearly indicating that the aminoalkene **12** is the product of the reaction (see Scheme 5) (Howard and Yamada 1981).

Amides N-alkyl amides are less reactive toward ROO• than amines because the lone pair of the N atom is shared with the carbonyl and thus it is less available for the stabilization of the alkyl radical. The peroxyl radical of primary amides is stable and does not decompose to afford HOO•. On the other hand, N,N-diethylacetamide (**13**) was reported to yield HOO• during its autoxidation, reasonably because of the decomposition of the peroxyl radical **14** to afford the alkene **15**, by following a mechanism similar to that observed in the case of trialkylamines (Scheme 5) (Aleksandrov 1980).

Alkenes Generation of HOO• during the autoxidation of alkenes is often complex and includes the formation of many oxidized products including epoxides, alkyl hydroperoxides and (poly)peroxides. Pliss reported that the substituted alkenes **16**–**19**, having no abstractable C-H groups, are *quantitatively* oxidized at 323 K with AIBN (azobisisobutyronitrile) initiator to hydroperoxides with a radical-chain carried on by HOO•. The proposed reaction sequence is reported in Scheme 6 for the case of **18** taken as an example. The authors suggest that the hydroperoxyl radical adds to the C=C double bond to yield an alkyl radical (**19**) which reacts with O_2 to

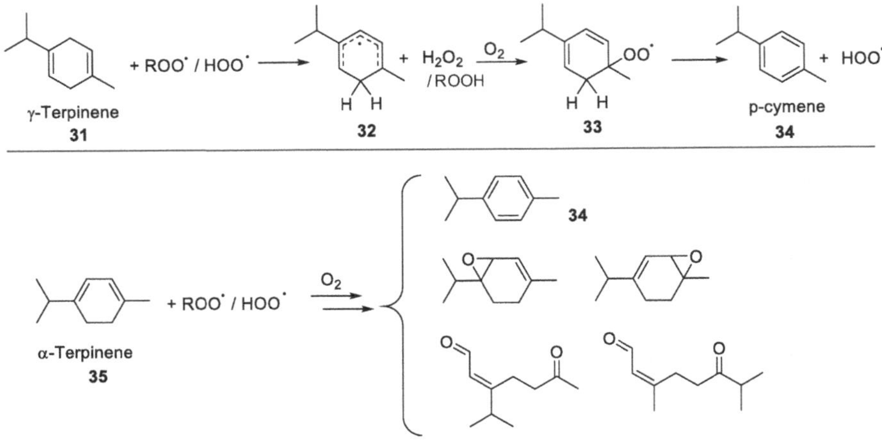

Scheme 7 Products formed by the autoxidation of γ- and α-terpinene

form the peroxyl radical **20**. Then, 1,4-HAT in **20** affords the unsaturated hydroperoxide **21**, that was isolated and partially identified by ^{1}H-NMR and IR spectra (Pliss et al. 2017).

Pratt has recently reported that HOO• is formed during the autoxidation of styrene, cyclooctene (343 K) and of hexadecene (373 K) initiated by di-*tert*-butylperoxide (Harrison et al. 2018). Based on theoretical calculations, a mechanism featuring a chain reaction involving alkoxyl (RO•), alkylperoxyl (ROO•) and hydroperoxyl radicals was proposed, as reported in Scheme 6 for the case of styrene (**22**). During the autoxidation of **22**, which is known to occur mainly through the intermediates **23**, **25** and **26**, some RO• radicals are formed *via* the intramolecular cyclization of the alkyl radical **23** to afford the epoxide **24**. In turn, alkoxyl radicals add to the double bond of **22** forming **28** that, after reaction with O_2, affords the peroxyl radical **29**, which yields HOO• and the alkoxy-substituted alkene **30** by intra-1,4-HAT. Interestingly, the generation of HOO• by intra-1,4-HAT in the peroxyl radical **25** is impossible, because it would rather produce aldehyde **27** and an alkoxyl radical. On the basis of this mechanism, it can be expected that HOO• is produced during the autoxidation of many different kinds of olefins which have epoxides among the possible oxidized products.

1,4-Cyclohexadiene derivatives such as γ-terpinene (**31**), a common terpene found in the essential oils of oregano and lemon (see Scheme 7), represent a unique class of alkenes as they produce high yields of HOO• during their autoxidation (Foti and Ingold 2003). γ-Terpinene reacts with radicals exclusively by H-atom transfer from the bis-allylic CH_2 groups forming a highly stabilized alkyl radical (such as **32**) which reacts with O_2 to yield a peroxyl radical with conjugated double bonds (such as **33**). Then, HOO• is released by the by intra-1,4-HAT mechanism, whose driving force is the formation of an aromatic product, p-cymene (**34**) (Sortino et al. 2003). Other pro-aromatic compounds such as 1,4-cyclohexadiene itself, methyl-1,4-cylohexadiene, 1,4-dihydronaphthalene and 9,10-dihydroanthracene quantitatively generate HOO• during their autoxidation (Howard and Ingold 1967a; Baschieri et al. 2019).

At variance with γ-terpinene, the chain propagation of its isomer α-terpinene (**35**), having conjugated double bonds, is only partially propagated by HOO• (Vardanyan and Denisov 1971). The autoxidation of **35** yields mainly epoxides and diketones, the latter presumably formed by the decomposition of alkyl hydroperoxides. p-Cymene **34**, which is formed by an intra-1,4-HAT from the peroxyl radical of α-terpinene, accounts only for about 20% of the products of the autoxidation, implying that HOO• is involved only to a small extent in chain propagation (Rudbäck et al. 2012).

2.4 *Kinetics of Intra-1,4-HAT*

Selected experimental and calculated values of the rate constants for the HOO• elimination reaction from different peroxyl radicals are reported in Table 1 (Harrison et al. 2018; Sortino et al. 2003; Bothe et al. 1977). By comparing entries 1 and 2 to 3 and 4, it can be noticed that the reaction is fast in the case of X = O (see Scheme 4), in line with the well-known transition-state stabilization effect of the H-bonding between the donor and the acceptor (Valgimigli et al. 2008). In general, HOO• elimination is favoured by the presence of substituents on the double bond of the product (compare entry 1 to 2).

The possibility to form an aromatic product, such as in the case of cyclohexadiene derivatives (see entry 5) greatly boosts the HOO• elimination (Sortino et al. 2003). The HAT nature of the mechanism is confirmed by the fact that this reaction slows down when replacing the transferred H with a D atom (kinetic deuterium effect, KDE). This effect is exceptionally large in the case of the transfer from a C-H bond (see entry 4) because in this case HAT rate is enhanced by quantum tunnelling (Harrison et al. 2018).

3 Antioxidant Effects in LOO•/HOO• Mediated Chain Propagation of Lipids

The unexpected antioxidant activity of γ-terpinene (**31**), a common constituent of *origanum* and other essential oils, disclosed a peculiar effect of HOO• on the lipid peroxidation reaction. γ-Terpinene is an easily oxidizable terpene that is converted under O_2 into aromatic p-cymene **34** with the production of H_2O_2 (see Scheme 7). If the autoxidation of linoleic acid is performed in the presence of small amounts of **31**, instead of increasing the oxidation rate as it would be expected given its high oxidizability, **31** causes a paradox antioxidant effect (Foti and Ingold 2003). This

Table 1 Rate constant and Kinetic Deuterium Effect (KDE) for the elimination of hydroperoxyl from peroxyl radicals by intra-1,4-HAT

		k (s^{-1})	KDE	References
1	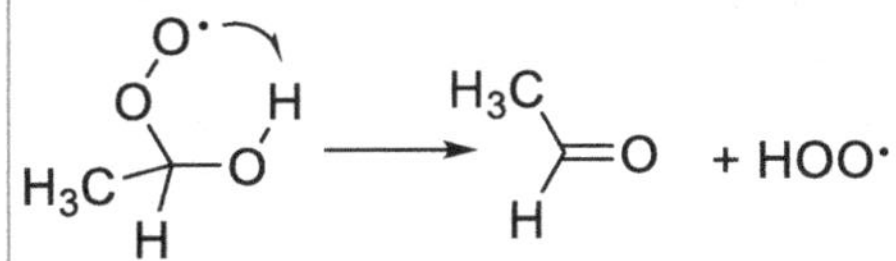	52^{a}		Bothe et al. (1977)
2	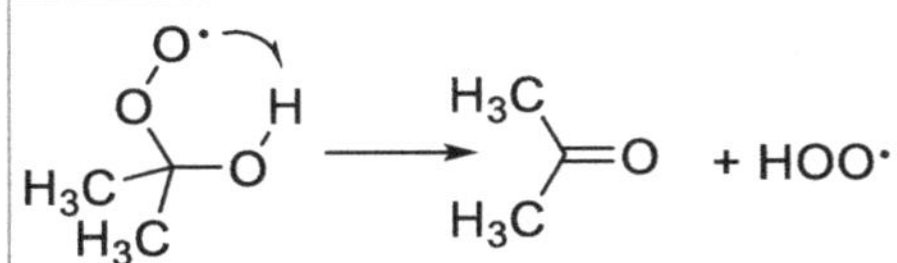	665^{a}	3.5	Bothe et al. (1977)
3	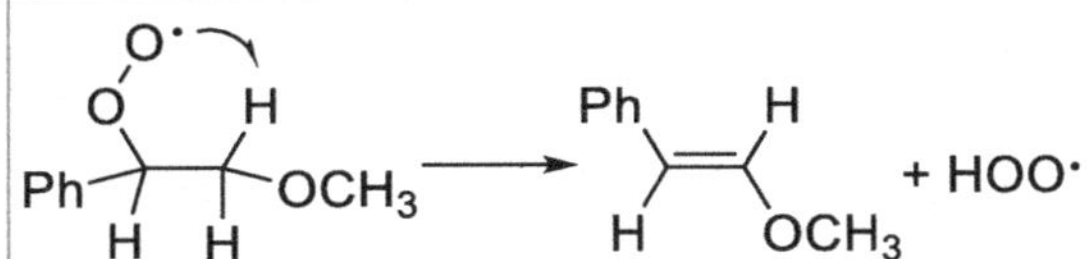	4.4^{b}		Harrison et al. (2018)
4	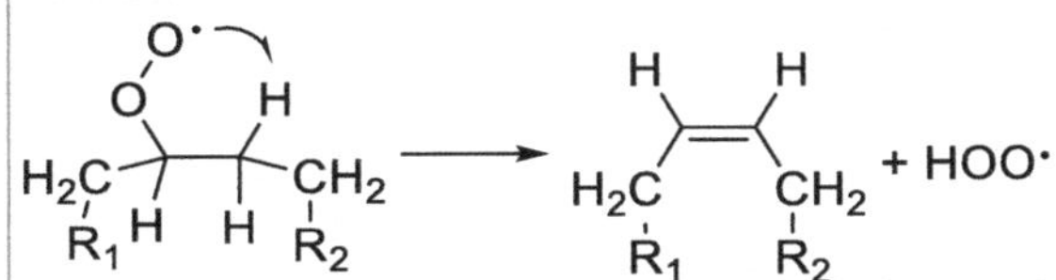	13^{b}	1145	Harrison et al. (2018)
5	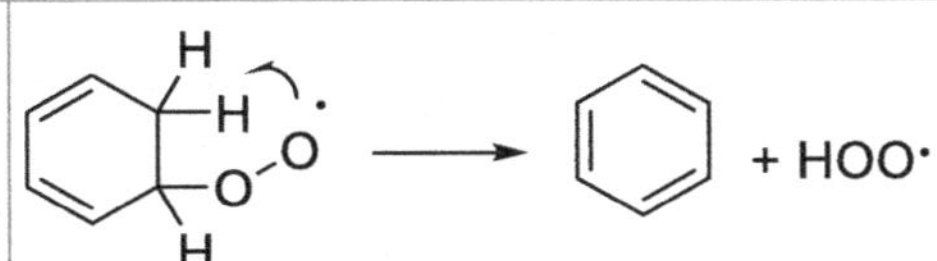	4×10^{4c}		Sortino et al. (2003)

[a] Experimental, 22 °C, water
[b] Calculated, 70 °C
[c] Experimental, acetonitrile, 20 °C

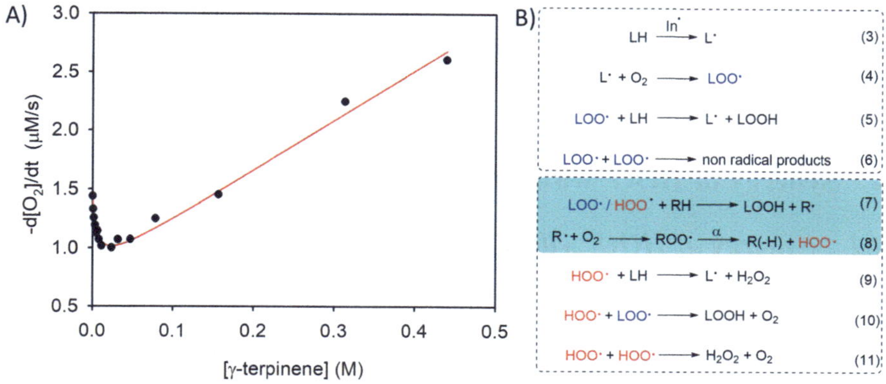

Fig. 1 (**a**) Rate of O_2 consumption during the autoxidation of styrene in the presence of increasing amounts γ-terpinene ([styrene] = 4.2 M, solvent PhCl, 30 °C, initiator AIBN 0.05 M). (**b**) Kinetic scheme for the co-oxidation of a lipid (LH) and HOO• precursor (RH)

effect can be evidenced also when mixing **31** to other organic substrates. In the example reported in Fig. 1, the rate of autoxidation of styrene, experimentally determined by measuring the O_2 consumption, (Unpublished data n.d.) is reported as function of the concentration of **31**. Styrene is a substrate used frequently in autoxidation studies because its behavior is similar to that of linoleic acid (Amorati and Valgimigli 2015).

The addition of **31** gradually reduces the O_2 consumption, reaching the maximum effect (i.e. minimum O_2 consumption) at [**31**] ≈ 0.02 M, while at [**31**] > 0.15 M, the antioxidant effect disappears. This result can be explained on the basis of reactions 3–11, reporting the autoxidation of a generic lipid substrate LH initiated by In• radicals in the presence of a second oxidizable molecule RH able to produce HOO• (Amorati and Pedulli 2008). The addition of RH causes the transformation of a fraction of LOO• into HOO•, that depends on RH concentration, on the rate constant of the reaction LOO• + RH and on the efficiency of HOO• formation by ROO• (α), as reported in reactions 7 and 8. As explained in Sect. 2.1, HOO• is different from ROO• because it undergoes fast termination thanks to the ability to form O_2 after H-atom donation. Typical rate constants for these processes are k_{10} and $k_{11} = 10^8$–$10^9\ M^{-1}\ s^{-1}$, whereas $10^6 < k_6 < 10^8\ M^{-1}\ s^{-1}$ for primary and secondary peroxyl radicals, and $10^3 < k_6 < 10^4\ M^{-1}\ s^{-1}$ for tertiary ones (Foti and Ingold 2003; Howard and Ingold 1967b). As a consequence, reactions 10 and 11 override the relatively slow self-termination reaction of LOO• (reaction 6), leading to an overall chain-breaking effect and to a slowdown of the peroxidation rate. Clearly, RH autoxidation contributes to the overall O_2 consumption and to hydroperoxide formation. If RH is less oxidizable than LH, such as in the cases of some amines used as inhibitors of cumene autoxidation, (Howard and Yamada 1981) the antioxidant effect is present at every RH concentrations. On the other hand, in the case of more oxidizable co-substrates (RH) such as **31**, at high tested RH concentrations the O_2 consumption is almost entirely caused by the autoxidation of RH, and the inhibition effect is restricted to a narrow RH concentration range.

4 Synergy of $HOO^•$ with Antioxidants

4.1 Synergy of $HOO^•$ with Phenols and Aromatic Amines

Recent results have shown that molecules able to generate $HOO^•$ radicals increase the duration of the induction period of phenolic chain-breaking antioxidants. Addition of a small amount of **31** to various oxidizable substrates such as stripped sunflower oil, squalene and styrene prolonged the duration of the inhibition of α-tocopherol (see Fig. 2) (Guo et al. 2021a).

The mechanism consists of the reaction of $ROO^•$ with **31** to form $HOO^•$ radicals, which are able to regenerate the radicals of the antioxidants through the reactions reported in Scheme 8 (Cedrowski et al. 2016). In the case of catechol derivatives such as **39**, the active antioxidant can be also regenerated from the quinone **41** (see Sect. 4.2) (Guo et al. 2021a).

Aromatic amines provide cyclic termination with peroxyl radicals in the presence of $HOO^•$ sources, such as alcohols, amines and cyclohexadienes (Denisov 1996). Pratt and coworkers have demonstrated that in liposome certain aromatic amines can be regenerated by reacting with $O_2^{•-}$ present in the aqueous phase (Poon et al. 2020).

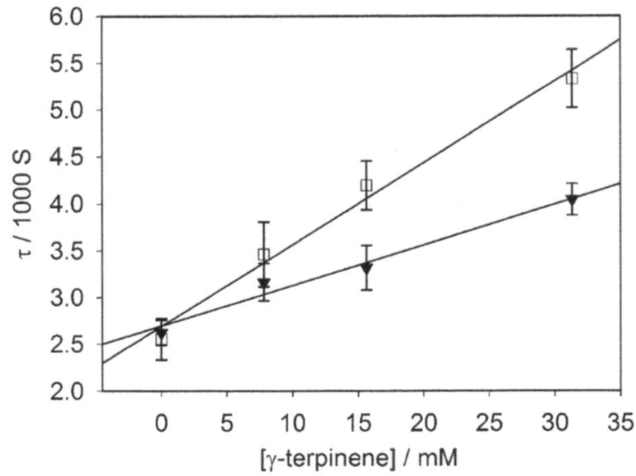

Fig. 2 Duration of the inhibition periods (τ) produced by 2.5 μM of the synthetic α-tocopherol analog 2,2,5,7,8-pentamethyl-6-chromanol in the oxidation of sunflower oil (square) and squalene (triangle) at 30 °C as function of the concentration of γ-terpinene. Data from ref. (Guo et al. 2021a)

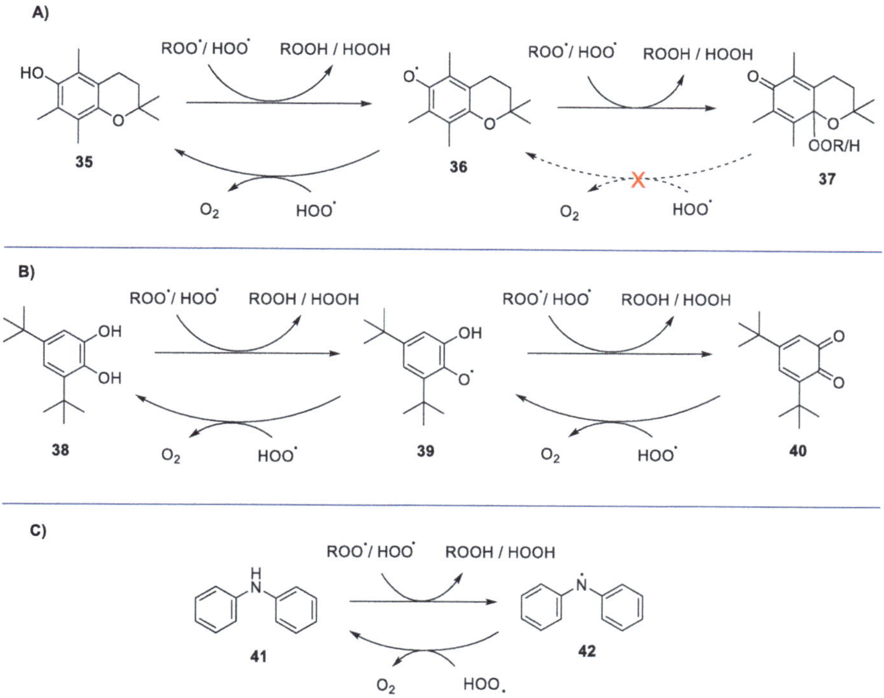

Scheme 8 Reaction pathways for the reaction of HOO• and ROO• with (**a**) the analog of α-tocopherol **35**, (**b**) the catechol derivative **38** and (**c**) a representative aromatic amine **41**

4.2 Synergy of HOO• with Nitroxides and Quinones

Molecules without antioxidant activity can behave as autoxidation inhibitors in the presence of HOO• radicals. The nitroxide TEMPO (see Scheme 9) is a stable radical endowed with excellent antioxidant activity in water (Griesser et al. 2018) and in organic solvent in the presence of weak acids (Griesser et al. 2018; Amorati et al. 2010). Under these conditions, nitroxides trap ROO• by donating an electron to the radical, which is then protonated by the solvent or by the added acid (Genovese et al. 2021). Nitroxides are instead poor inhibitors of the autoxidation of organic substrates in non-protic systems. We have shown that the addition of 1,4-cyclohexadiene dramatically increases the antioxidant activity of the nitroxide TEMPO, by the onset of the catalytic cycle reported in Scheme 9a. The duration of the inhibition is exceptionally long if compared to that provided by phenols, because the cyclic mechanism is interrupted only by the reaction of alkyl radicals with the nitroxide with alkyl radicals, whose concentration in the presence of O_2 is small (Baschieri et al. 2018a). The catalytic antioxidant activity of nitroxides has been also proven in the case of the autoxidation of styrene, cyclooctene and hexadecane, under

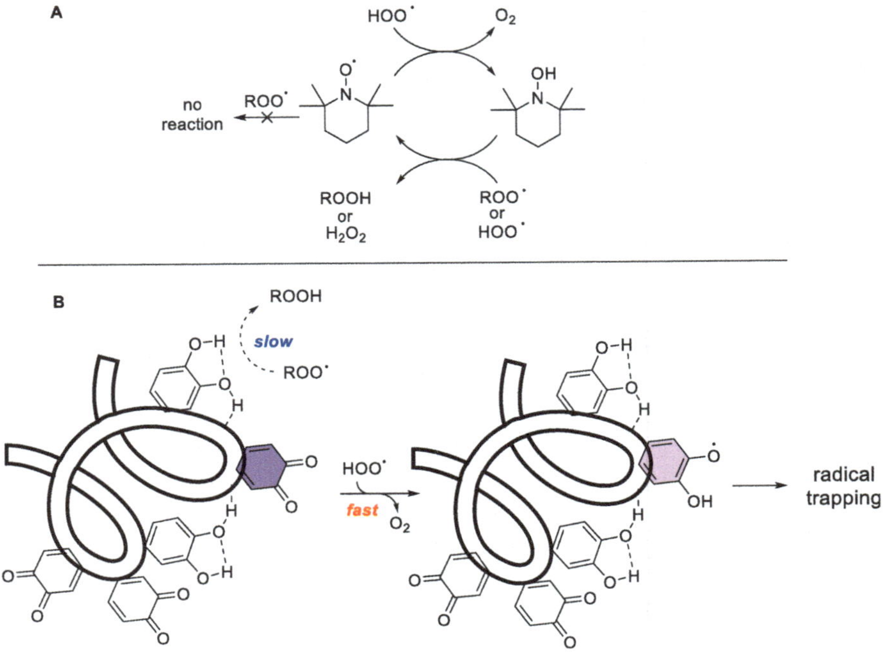

Scheme 9 Catalytic activity of TEMPO (**a**) and of melanins (**b**) with HOO• radicals

conditions in which the chain carrying radicals are both ROO• and HOO• (Poon et al. 2020).

Quinones also behave as inhibitors of the autoxidation of organic compounds in the presence of HOO• radicals and display catalytic activity, that is, multiple chains are terminated by each quinone molecule (Denisov 1969; Vardanyan et al. 1972; Degtyareva et al. 1979). However, at variance with nitroxides, the inhibition duration is shorter because the semiquinone radicals react also by addition with ROO•/HOO• radicals. Interestingly, *ortho*-benzoquinones have a higher antioxidant effect than *para*-benzoquinones, because the HAT from HOO• to the quinone is more exothermic for the *ortho* than for the *para* isomer (Guo et al. 2021b). The good catalytic ROO•/HOO• radical trapping activity of *ortho*-quinones explains why many polymeric phenol-based materials rich of quinone moieties such as polydopamine and melanin display antioxidant activity (see Scheme 9b) (Guo et al. 2021b).

Scheme 10 Trapping of HOO• by a monophenolic antioxidant

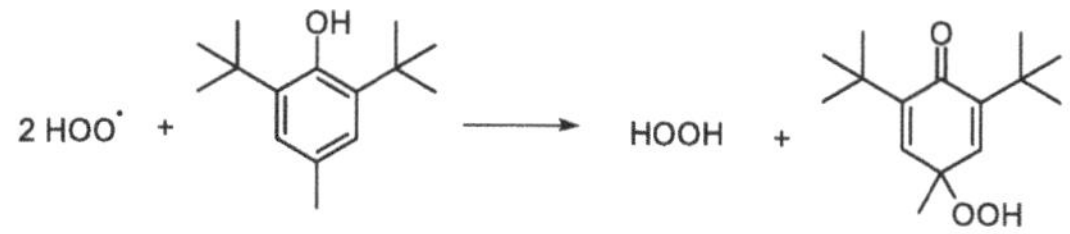

5 Methods to Detect HOO• Formation

The involvement of HOO• in the autoxidation of organic substrates can be inferred from both product and kinetic studies. As shown in Sect. 2.3, the intra-1,4-HAT reaction of peroxyl radicals affords characteristic products: alkenes, aromatic compounds, and in the case of alcohols and amines, carbonyl compounds and imines, whose presence represents an indirect proof of HOO• formation. Hydrogen peroxide is produced by HOO• recombination and/or HAT from the substrate. The HOO• radical can be trapped by using mono-phenolic antioxidants, to afford a peroxidic adduct shown in Scheme 10, which can be quantified by chromatography (Howard and Ingold 1967a; Howard and Yamada 1981).

The kinetic methods represent a quick and reliable way to reveal the presence of HOO•. They are mainly based on the use of antioxidants able to react only with HOO• and not with ROO• radicals, such as nitroxides, which however are selective toward HOO• only in aprotic solvents and in the absence of acids, and quinones (see Sect. 4.2). Another property that can be exploited is the ability of HOO• to prolong the inhibition of antioxidants. Therefore, the HOO• presence can be deduced by comparing the stoichiometry of radical trapping of an antioxidant in different oxidizable substrates, by using a substrate which does not produce HOO•, such as cumene, as reference (Poon et al. 2020).

6 Conclusions

In this chapter, we have provided an account of the recent studies of the role of HOO• in the autoxidation of lipids and other oxidizable substrates in homogeneous systems. Many aspects of the HOO• chemistry are still unexplored, such as the influence of HOO•/$O_2^{\bullet -}$ on the autoxidation of emulsions and liposomes and the capability of molecules occurring in oxidizable materials, like phospholipids and phytosterols, to produce HOO•. The hydrophilicity of HOO• and its deprotonation to $O_2^{\bullet -}$ is expected to play a fundamental role to "export the radical" from the lipid phase to the aqueous phase, reducing the radical chain propagation (Bowry and Stocker 1993). At the same time, $O_2^{\bullet -}$ and HOO• are able to regenerate the antioxidants present at the interface, thus providing an increase of the radical trapping stoichiometry (Poon et al. 2020). The interplay between HOO• and ROO• radicals is expected to play a role to explain the antioxidant behaviour of metal and organic nanomaterials (Valgimigli et al. 2018; Amorati and Mollica 2019; Baschieri et al. 2018b). In conclusion, we believe that a better clarification of the HOO• role in radical chain processes will provide a deeper comprehension of the mechanisms of lipid peroxidation, and will allow the rational development of novel antioxidant systems.

Acknowledgements Y.G. acknowledges a fellowship from China Scholarship Council (CSC:201706790016). R.A. Thanks Dr. Mario Foti for helpful suggestions.

References

Aikens J, Dix TA (1993) Hydrodioxyl (perhydroxyl), peroxyl, and hydroxyl radical-initiated lipid peroxidation of large unilamellar vesicles (liposomes): comparative and mechanistic studies. Arch Biochem Biophys 305(2):516–525

Aleksandrov AL (1980) β-CH regeneration of antioxidants in N,N-diethylacetamide. Bull Acad Sci USSR Div Chem Sci 29(12):1896–1901

Amorati R, Mollica F (2019) Oximetry methods for evaluating the activity of nanoantioxidants. (minireview). Bulg Chem Commun 51(Special Issue A):289–294

Amorati R, Pedulli GF (2008) Do garlic-derived allyl sulfides scavenge peroxyl radicals? Organic & biomolecular chemistry. Org Biomol Chem 6(6):1103–1107

Amorati R, Valgimigli L (2015) Advantages and limitations of common testing methods for antioxidants free radical research. Free Radic Res 49(5):633–649

Amorati R, Pedulli GF, Pratt DA, Valgimigli L (2010) TEMPO reacts with oxygen-centered radicals under acidic conditions. Chem Commun 46:5139–5141

Amorati R, Pedulli GF, Valgimigli L (2011) Kinetic and thermodynamic aspects of the chain-breaking antioxidant activity of ascorbic acid derivatives in non-aqueous media. Org Biomol Chem 9:3792–3800

Anglada M, Martins-Costa M, Francisco JS, Ruiz-Ljpez MF (2015) Interconnection of reactive oxygen species chemistry across the interfaces of atmospheric, environmental, and biological processes. Acc Chem Res 48:575–583

Baschieri A, Valgimigli L, Simone G, DiLabio GA, Romero-Montalvo E, Amorati R (2018a) Extremely fast hydrogen atom transfer between nitroxides and $HOO^{•}$ radicals and implication for catalytic coantioxidant systems. J Am Chem Soc 140:10354–10362

Baschieri A, Del Secco B, Zaccheroni N, Valgimigli L, Amorati R (2018b) The role of onium salts in the pro-oxidant effect of gold nanoparticles in lipophilic environments. Chem Eur J 24 (36):9113–9119

Baschieri A, Amorati R, Valgimigli L (2019) Sambri L. 1-Methyl-1,4-cyclohexadiene as a traceless reducing agent for the synthesis of catechols and hydroquinones. J Org Chem 84 (21):13655–13664

Behar D, Czapski G, Rabani J, Dorfman LM, Schwartz HA (1970) The acid dissociation constant and decay kinetics of the perhydroxyl radical. J Phys Chem 74(17):3209–3213

Bielski BHJ (1978) Reevaluation of the spectral and kinetic properties of $HOO^{·}$ and $O2^{-}$ free radicals. Photochem Photobiol 28(4–5):645–649

Bielski BHJ, Allen AO (1977) Mechanism of the disproportionation of superoxide radicals. J Phys Chem 81(11):1048–1050

Bielski BH, Arudi RL, Sutherland MW (1983) A study of the reactivity of HO_2/O_2^{-} with unsaturated fatty acids. J Biol Chem 258:4759–4761

Bothe E, Behrens G, Shulte-Frohlinde DZ (1977) Mechanism of the first order decay of 2-hydroxy-propyl-2-peroxyl radicals and of $O_2^{•-}$ formation in aqueous solution. Z Naturforsch B 32:886–889

Bowry VW, Stocker R (1993) Tocopherol-mediated peroxidation. The prooxidant effect of vitamin E on the radical-initiated oxidation of human low-density lipoprotein. J Am Chem Soc 115 (14):6029–6044
Burton GW, Ingold KU (1986) Vitamin E: Application of the principles of physical organic chemistry to the exploration of its structure and function. Acc Chem Res 19:194–201
Campomanes P, Rothlisberger U, Alfonso-Prieto M, Rovira C (2015) The molecular mechanism of the catalase-like activity in horseradish peroxidase. J Am Chem Soc 137(34):11170–11178
Cedrowski J, Litwinienko G, Baschieri A, Amorati R (2016) Hydroperoxyl radicals (HOO•): vitamin E regeneration and H-bond effects on the hydrogen atom transfer. Chem Eur J 22 (46):16441–16445
Chang CR, Yang XF, Long B, Li J (2013) Water-promoted mechanism of alcohol oxidation on an au (111) surface: understanding the catalytic behavior of bulk gold. ACS Catal 3(8):1693–1699
Czapski G (1984) Reaction of ·OH. Method Enzymol 105(984):209–215
Davies KJ (1995) Oxidative stress: the paradox of aerobic life. Biochem Soc Symp 61:1–31
Degtyareva TG, Denisov ET, Martem'yanov VS, Badretdinova LY (1979) The activity of poly-atomic alcohols in reactions with peroxide radicals. Bull Acad Sci USSR Div Chem Sci 28 (6):1139–1144
Denisov ET (1969) Mechanism of the inhibiting effect of quinone in the chain reaction of oxidation of isopropyl alcohol. Russ Chem Bull 18:282–285
Denisov ET (1996) Cyclic mechanisms of chain termination in the oxidation of organic compounds. Russ Chem Rev 65:505–520
Denisov ET, Afanas'ev IB (2005) Oxidation of alcohols and ethers. In: Oxidation and antioxidants in organic chemistry and biology. Taylor and Francis, Boca Raton, FL
Fattahi A, Kass SR, Liebman JF, Matos MAR, Miranda MS, Morais VMF (2005) The enthalpies of formation of o-, m-, and p-benzoquinone: gas-phase ion energetics, combustion calorimetry, and quantum chemical computations combined. J Am Chem Soc 127:6116–6122
Foti MC, Ingold KU (2003) Mechanism of inhibition of lipid peroxidation by γ-terpinene, an unusual and potentially useful hydrocarbon antioxidant. J Agric Food Chem 51:2758–2765
Foti MC, Sortino S, Ingold KU (2005) New insight into solvent effects on the formal HOO• + HOO• reaction. Chem Eur J 11:1942–1948
Genovese D, Baschieri A, Vona D, Baboi RE, Mollica F, Prodi L, Amorati R, Zaccheroni N (2021) Nitroxides as building blocks for nanoantioxidants. ACS Appl Mater Interfaces 13 (27):31996–32004
Griesser M, Shah R, Kessel ATV, Zilka O, Haidasz EA, Pratt DA (2018) The catalytic reaction of nitroxides with peroxyl radicals and its relevance to their cytoprotective properties. J Am Chem Soc 140(10):3798–3808
Guo Y, Baschieri A, Amorati R, Valgimigli L (2021a) Synergic antioxidant activity of γ-terpinene with phenols and polyphenols enabled by hydroperoxyl radicals. Food Chem 345:128468
Guo Y, Baschieri A, Mollica F, Valgimigli L, Cedrowski J, Litwinienko G, Amorati R (2021b) Hydrogen atom transfer from HOO• to ortho-quinones explains the antioxidant activity of polydopamine. Angew Chem Int Ed 60:15220 –15224
Harrison KA, Haidasz EA, Griesser M, Pratt DA (2018) Inhibition of hydrocarbon autoxidation by nitroxide-catalyzed cross-dismutation of hydroperoxyl and alkylperoxyl radicals. Chem Sci 9:6068–6079
Howard JA, Ingold KU (1967a) Absolute rate constants for hydrocarbon autoxidation. V. the hydroperoxy radical in chain propagation and termination. Can J Chem 45:785–792
Howard JA, Ingold KU (1967b) Absolute rate constants for hydrocarbon autoxidation. VI. Alkyl aromatic and olefinic hydrocarbons. Can J Chem 45:793–802
Howard JA, Yamada T (1981) Absolute rate constants for hydrocarbon autoxidation. 31. Autoxidation of cumene in the presence of tertiary amines. J Am Chem Soc 103:7102–7106
Huang P, Feng L, Oldham EA, Keating MJ, Plunkett W (2000) Superoxide dismutase as a target for the selective killing of cancer cells. Nature 407:390–395

Ingold KU (2003) Reactions of water-soluble Alkylperoxyl radicals and superoxide with DNA, lipoproteins and phospholipid vesicles: the role played by electrostatic forces. Curr Med Chem 10:2631–2642

Kumar M, Francisco JS (2015) Red-light-induced decomposition of an organic peroxy radical: a new source of the $HOO^{\bullet}$ radical. Angew Chem Int Ed 127:15937–15940

Lian YA, Meisel D, Czapski G (1974) The redox potential of the O_2^- system in aqueous media. Israel J Chem 12(4):891–895

McCord JM, Fridovich I (1969) Superoxide dismutase an enzymic function for erythrocuprein (hemocuprein). J Biol Chem 244:6049–6055

Min B, Ahn DU (2005) Mechanism of lipid peroxidation in meat and meat products -a review. Food Sci Biotechnol 14(1):152–163

Nanni JEJ, Stallings MD, Sawyer DT (1980) Does superoxide ion oxidize catechol, alpha-tocopherol, and ascorbic acid by direct electron transfer? J Am Chem Soc 102:4481–4485

Pliss EM, Machtin VA, Grobov AM, Pliss RE, Sirick AV (2017) Kinetics and mechanism of radical-chain oxidation of 1,2-substituted ethylene and 1,4-substituted butadiene-1,3. Int J Chem Kinet 49(3):173–181

Poon J-F, Zilka O, Pratt DA (2020) Potent ferroptosis inhibitors can catalyze the cross-dismutation of phospholipid-derived peroxyl radicals and hydroperoxyl radicals. J Am Chem Soc 142:14331–14342

Rosen GM, Pou S, Ramos CL, Cohen MS, Britigan BE (1995) Free radicals and phagocytic cells. FASEB J 9:200–209

Rudbäck J, Bergström MA, Börje A, Nilsson U, Karlberg A-T (2012) α-Terpinene, an antioxidant in tea tree oil, autoxidizes rapidly to skin allergens on air exposure. Chem Res Toxicol 25:713–721

Sawyer DT, Valentine JS (1981) How super is superoxide? Acc Chem Res 14:393–400

Schumperli MT, Hammond C, Hermans I (2012) Reactivity of a-amino-peroxyl radicals and consequences for amine oxidation chemistry. Chem Phys 14:11002–11007

Sortino S, Petraliaa S, Foti MC (2003) Absolute rate constants and transient intermediates in the free-radical-induced peroxidation of γ-terpinene, an unusual hydrocarbon antioxidanty. New J Chem 27:1563–1567

Sundaresan M, Yu Z-X, Ferrans VJ, Irani K, Finkel T (1995) Requirement for generation of H_2O_2 for platelet-derived growth factor signal transduction. Science 270:296–299

Unpublished data

Valgimigli L, Amorati R, Fumo MG, DiLabio GA, Pedulli GF, Ingold KU, Pratt DA (2008) The unusual reaction of semiquinone radicals with molecular oxygen. J Org Chem 73:1830–1841

Valgimigli L, Baschieri A, Amorati R (2018) Antioxidant activity of nanomaterials. J Mater Chem B 6:2036–2051

Vardanyan RL, Denisov ET (1971) Regeneration of inhibitors in oxidized 1,3-cyclohexadienes. Bull Acad Sci USSR Div Chem Sci. 12:2681–2683

Vardanyan RL, Denisov ET, Zozulya VI (1972) Measurement of the rate constants of the reaction of peroxy radicals with alcohols by the interlinked oxidation method with a selective inhibitor. Bull Acad Sci USSR Div Chem Sci. 21(3):572–574

Vieira SA, Zhang G, Decker EA (2017) Biological implications of lipid oxidation products. J Am Oil Chem Soc 94:339–351

Warren JJ, Tronic TA, Mayer JM (2010) Thermochemistry of proton-coupled electron transfer reagents and its implications. Chem Rev 110(12):6961–7001

Weisiger RA, Fridovich I (1973) Mitochondrial superoxide dismutase site of synthesis and intramitochondrial localization. J Biol Chem 248:4793–4796

Zheng Y, Jiao Y, Jaroniec M, Qiao SZ (2015) Advancing the electrochemistry of the hydrogen-evolution reaction through combining experiment and theory. Angew Chem Int Ed 54(1):52–65

Zielinski ZAM, Pratt DA (2017) Lipid peroxidation: kinetics, mechanisms, and products. J Org Chem 82:2817–2825

Location, Orientation and Buoyance Effects of Radical Probes as Studied by EPR

Carolina Aliaga and Marcos Caroli Rezende

1 Introduction

1.1 Location and Orientation of Probes and Antioxidants in Heterogeneous Systems

The assessment of the antioxidant activity of natural products of potential pharmacological or dietary value (Nogueira et al. 2014; Thippeswamy et al. 2020) and, in particular, common vegetables (Thippeswamy et al. 2020), fruits (Özcan et al. 2020) and food extracts is a subject of permanent interest. Global activities of antioxidants (AOs) are frequently reported, with little concern for the distribution of the active compounds in the extracts, or even for their homogeneity. As a result, authors are often confronted with different evaluations of the same sample, depending on the methods employed in the extraction process (Kumar et al. 2011).

Heterogeneous systems are often found in plant and food extracts and emulsions. This is often neglected by many authors, that thus overlook important aspects of the AO-probe interaction, such as their solubility in a micro-heterogeneous sample and, consequently, of their location. Comparisons and conclusions are drawn based exclusively on structural factors (Montoro et al. 2005), ignoring the lipophilicities of the reacting species (Ahmadi et al. 2007; Almajano et al. 2008; López-Alarcón et al. 2005; Yeşilyurt et al. 2008) and leading to results that are not accurate and cannot be generalized. Ambiguous or even contradictory results can be obtained

C. Aliaga (✉)
Facultad de Química y Biología, Universidad de Santiago de Chile, Santiago, Chile

Centro para la Nanociencia y la Nanotecnología CEDENNA, Santiago, Chile
e-mail: carolina.aliaga@usach.cl

M. C. Rezende
Facultad de Química y Biología, Universidad de Santiago de Chile, Santiago, Chile

C. Bravo-Diaz (ed.), *Lipid Oxidation in Food and Biological Systems*,
https://doi.org/10.1007/978-3-030-87222-9_6

from the same sample, depending on the method and/or probe employed for measuring its AO activity.

When measuring AO-activities, the location of the antioxidant or of the probe in a micro-heterogeneous system has long been recognized as an important issue (Cui et al. 2015). This has given rise to considerable effort in the past decades to measure the distribution of antioxidants in micelles (Aspée et al. 2017; Parmar et al. 2011; Teixeira et al. 2005) and emulsions (Losada-Barreiro et al. 2020; Costa et al. 2020; Wang et al. 2020; Raimúndez-Rodríguez et al. 2019; Meireles et al. 2019; Freiría-Gándara et al. 2018; Ferreira et al. 2018; Keller et al. 2016; Romsted and Zhang 2004), either by direct observation (Ye et al. 2019) or by indirect determination based on kinetic protocols (Aspée et al. 2017; Costa et al. 2020; Raimúndez-Rodríguez et al. 2019; Romsted and Zhang 2004).

In the following sections, the importance of different factors in the assessment of antioxidant activities will be addressed. In particular, besides the AO or probe location, the importance of their orientation in a micro-heterogeneous system will be stressed, with a novel interpretation of cut-off effects based on this factor. Structural requirements, based on their amphiphobic nature, for AOs or probes to exhibit this behavior will also be discussed.

1.2 Cut-off Effects and Probe Orientation

In the assessment of antioxidant activities of series of analogous compounds, a puzzling behavior that is often encountered is the "polar paradox" or "cut-off effect". The effect is not limited to antioxidants. It also appears in evaluations of pharmacologically active compounds, where structure-activity relationships are sought (Balgavý and Devínsky 1996).

The paradox can be summarized by stating that "polar compounds are more effective in non-polar media" (Boroski et al. 2018; Noon et al. 2020). Thus, the polar paradox runs contrary to the expectation that, for a series of compounds of increasing hydrophobicity, the observed activity in a nonpolar medium should increase with that property (Mateos et al. 2015). What is often observed is an initial increase in activity with the increased hydrophobicity of the series, up to a cut-off point where this activity starts to diminish (Laguerre et al. 2009).

Cut-off effects have been observed when measuring the AO activities of alkyl gallates in emulsions (González et al. 2015), the antioxidant efficacies of rutin and rutin esters in oil and in emulsions (Lue et al. 2017), in the effectiveness of *n*-alcohols as anesthetics, (Zapata-Morin et al. 2020) or in the toxicity of N-alkylimidazolium ionic liquids (Shao et al. 2018, 2019; Kusumahastuti et al. 2019). The paradox is so ubiquitous that examples where it is not followed are presented as noteworthy (Martinović et al. 2019).

The competition between surfactant properties with structural features in a series of active compounds has been invoked as a possible source of the effect (González et al. 2015). Structural aspects related with the behavior of an active compound in the interface of a micro-heterogeneous system provide a possible explanation for its effect in such media (González et al. 2015; Costa et al. 2017). Kinetic methods (Dar et al. 2017) or molecular dynamic simulations (Zapata-Morin et al. 2020) have been used to gain insight into the position of an antioxidant or an active species in a micro-heterogeneous environment.

EPR spectroscopy constitutes a convenient tool for locating a radical probe in a micro-heterogeneous environment (Aliaga et al. 2009a). Besides being convenient probes for the antioxidant activity of phenols, the 2,2,6,6-tetramethylpyperidineoxyl (TEMPO) radical and its derivatives have been the object of partition studies in micellar systems (Aliaga et al. 2009b, 2016a) and emulsions (Krudopp et al. 2015).

1.3 A Methodology Based on EPR Spectroscopy to Evaluate the Partitioning of Nitroxide Probes in Micelles

Though very useful as a measure of the hydrophobicity of different radical probes and antioxidants, log *P* values based on the octanol/water partitioning coefficient provide only an approximate and indirect way of identifying the site of reaction of a probe in a micelle or an emulsion. These micro-heterogeneous systems are more complex than the water-ethanol two-phase system, offering more than two micro-domains for the reacting species.

As mentioned above, one advantage of working with nitroxide radicals is their versatility as medium-dependent probes. With a view to exploiting the sensitivity of their EPR spectra in different solvents, the partitioning of five different 4-substituted TEMPO derivatives in aqueous solutions of reduced Triton XR-100 was measured with the aid of their spectrum *g*-factor (Aliaga et al. 2012).

The series of nitroxide radicals included, besides the unsubstituted TEMPO, derivatives substituted at the 4-position with amido, sulfonamide and carboxyl groups (see Scheme 1 for their structures).

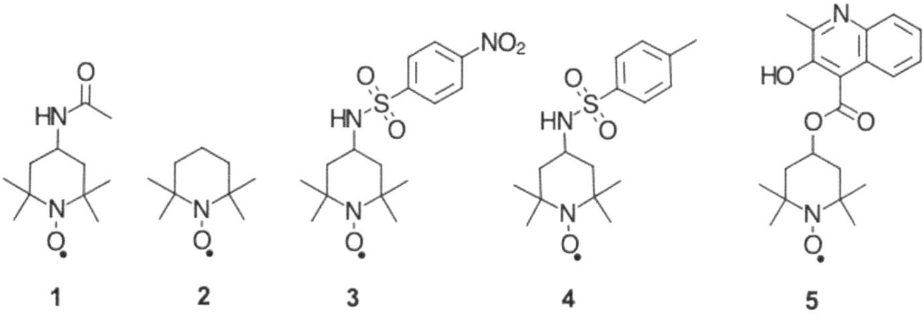

Scheme 1 Series of nitroxide radicals employed in the determination of partitioning constants in micellar solutions of reduced Triton XR-100, by means of the *g*-factors of their EPR spectra (Aliaga et al. 2012)

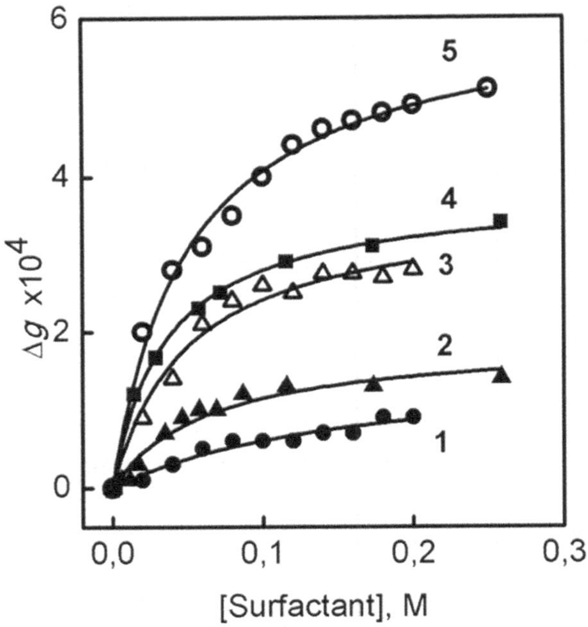

Fig. 1 Variations of the probe *g*-factor with the increasing concentration of added reduced Triton XR-100 surfactant. Numbers identify the probes shown in Scheme 1 (Aliaga et al. 2012). Fitting curves were drawn with the aid of Eq. 1, from which partitioning constants *K* were determined. Reprinted C. Aliaga et al. *Magn. Reson. Chem.* 2012, 50, 779–78. Copyright © 2012 John Wiley & Sons, Ltd

Since quenching measurements in a micellar system or an emulsion can be performed with the aid of the probe EPR spectra, variations of their *g*-factor in the same spectra offer an immediate and reliable way of determining their micro-environment in such media.

Adding increasing amounts of a surfactant to an aqueous solution of the probe changed the *g*-factor of its EPR spectrum, as the probe was incorporated into the more hydrophobic micellar core. The initial g_w value in water increased with the surfactant concentration reaching a plateau value g_m which corresponded to its full incorporation into the micelle (see Fig. 1).

$$\Delta g = \frac{(g_m - g_w) \cdot K \cdot [\text{Surfactant}]}{(1 + \mathrm{K} \cdot [\text{Surfactant}])} \tag{1}$$

Values of *K* thus obtained were in good agreement with values obtained from variations of other EPR spectral parameters, such as the line intensities of the first or third signals of the nitroxide triplet, or the linewidth of its third signal. In fact, the smallest deviations were observed with the method based on *g*-factor variations, thus validating this protocol for a reliable determination of the partitioning of nitroxide radicals in micro-heterogeneous systems.

The experiment also showed that log *P* values, though useful for a preliminary assessment of the hydrophobicities of nitroxide probes and their partitioning in a micellar system, can sometimes depart significantly from results obtained by direct measurements of the system. Thus, Table 1 compares *K* values for the TEMPO

Table 1 Comparison of partitioning constants *K* of compounds **1–5**, obtained by fitting the experimental data of Fig. 1 with Eq. 1, with log *P* values estimated by summing up fragment contributions

TEMPO derivative[a]	K (M^{-1})	log P[b]
1	6.3 ± 2.1	0.81
2	15.7 ± 3.5	1.78
3	20.0 ± 3.4	2.21
4	28.0 ± 1.8	3.02
5	19.8 ± 1.9	2.92

[a]Numbering corresponds to structures shown in Scheme 1
[b]Estimated with the ICM software based on fragment contributions

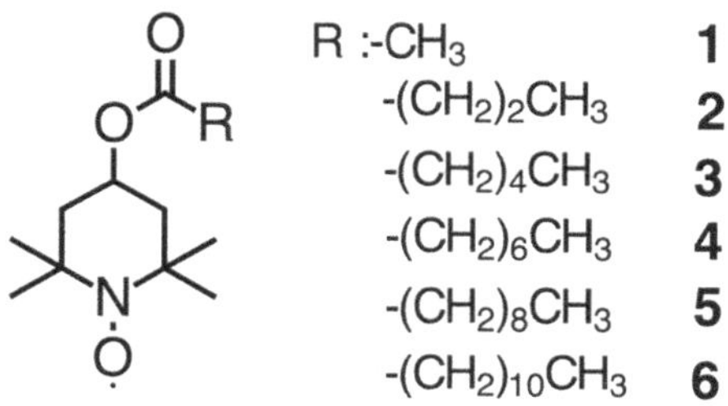

Scheme 2 Structure of 4-acyloxyTEMPO derivatives

derivatives **1–5** of Scheme 1, obtained by the above methodology, with their log *P* values estimated by fragment contributions.

An inspection of Table 1 reveals a significant inconsistency regarding compound **5**, with a partitioning constant *K* similar to that of compound **3** and rather different from that of **4**. Log *P* values predict a different behavior, with compound **4** and **5** exhibiting similar hydrophobicities. In fact, in spite of possessing an extended aromatic system, probe **5** probably exists in solution in an equilibrium between a neutral and a zwitterionic form, due to its 3-hydroxyquinolinyl fragment. This should make this compound more hydrophilic than anticipated by its heteroaromatic moiety, in agreement with its *K* value in Table 1, close to that of the more hydrophilic derivative **3**.

These last considerations validated the described methodology for the determination of partitionings of radical nitroxide probes in micelles and microemulsions. Nevertheless, it was also clear that, if a comparison of TEMPO derivatives was to be made in a structure-activity relationship (SAR) study in these systems, care should be taken to ensure a common substitution pattern for the probe series, so as to avoid any other structural effects not ascribable to pure hydrophobicity.

In a more systematic effort towards validation of the methodology based on *g*-factor measurements, 4-Alkanoyloxy TEMPO derivatives (Scheme 2) offer the additional feature of a variable lipophilicity modulated by their 4-substituent, so that a series of analogous radical probes with increasing hydrophobicity can be employed for the assessment of antioxidants in micro-heterogeneous media.

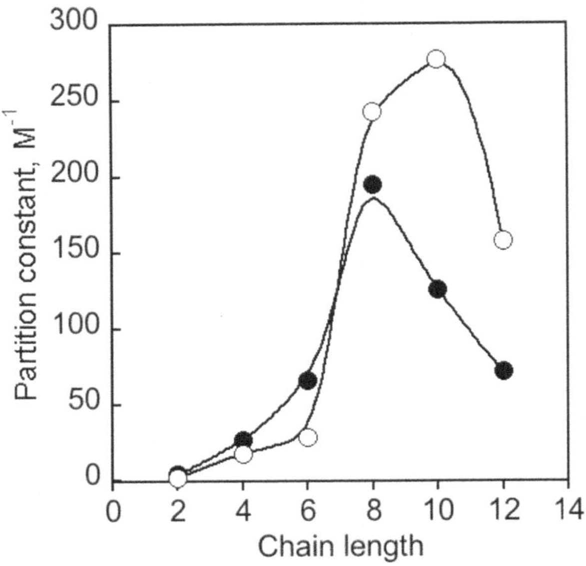

Fig. 2 Variations of the partition constants K_{EPR} (*filled circle*) and of pyrene-quenching constants K_{Fluo} (*open circle*) of 4-substituted TEMPO probes in a Triton-XR 100 micellar medium with the chain-length of the 4-acyloxy substituent. Reprinted C. Aliaga et al. *Food Chemistry*, **2016**, *192*, 395–401. Copyright 2015 Elsevier Ltd

When the partition constants K_{EPR}, or the pyrene-quenching constants K_{Fluo}, for a series of 4-substituted TEMPO derivatives with increasing lipophilicities were measured in aqueous Triton-XR 100 micelle, cut-off patterns were observed with the increasing chain-length of the radical 4-substituent (Fig. 2) (Aliaga et al. 2016a).

Molecular dynamic simulations of these probes in the micellar milieu reveals the origin of this cut-off effect: 4-substituted TEMPO derivatives insert and orient themselves in different ways in the micellar system, as their 4-substituent chain-length increases (Fig. 3).

The effect of orientation of the TEMPO probe in the micellar interface is more clearly seen in the scheme of (Fig. 4 left). The molecular dynamic simulation also reveals a cut-off pattern when the average distance between the NO· fragment of the radical probes and the micellar center is plotted against the 4-substituent chain-length (Fig. 4 right).

2 Cut-off Patterns and the Location of Probe/Antioxidant in Heterogeneous Media

The cut-off behavior of antioxidants reported by many groups when assessing antioxidant activities can therefore be explained by the relative location and orientation of the probe and the antioxidant in a micro-heterogeneous environment. If the same set of analogous probes – radical 4-substituted TEMPO derivatives of variable lipophilicity – measures the activity of two different antioxidants, a hydrophobic and

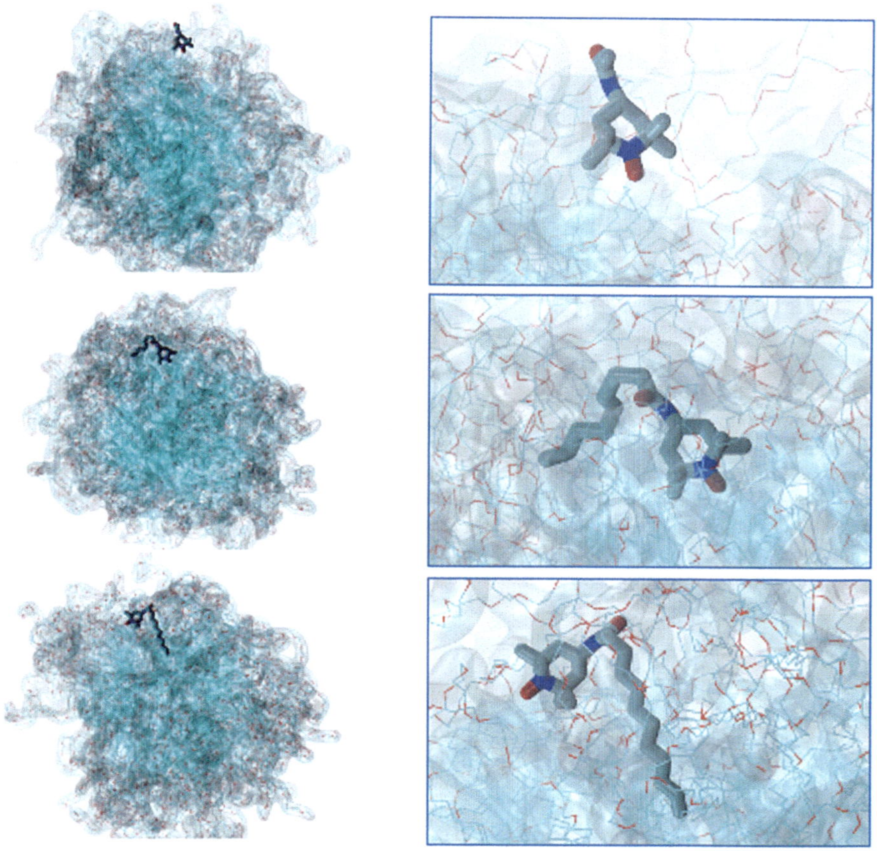

Fig. 3 Molecular simulation of the mode of insertion of 4-alkanoyl TEMPO radicals of variable lipophilicity into a micelle of Triton-XR 100 in water. Left pictures show the probe location in the water-micelle interface, with its orientation in this interface in the enlarged pictures (right). Reprinted C.Aliaga et al. *Food Chemistry*, **2016**, *192*, 395–401. Copyright 2015 Elsevier Ltd

a hydrophilic phenol in a micellar environment, what kind of cut-off patterns are to be expected from their interactions with the probes?

The answer to this question, illustrated in Fig. 5, gives rise to a novel protocol, that characterizes the location of a phenolic antioxidant in a micellar medium.

Hydrophilic antioxidants will give rise to a convex cut-off curve, whereas AOs that are more hydrophobic than the employed set of radical probes will generate concave curves. If a family of phenols of increasing hydrophobicity is evaluated in a micellar medium by this standard probe set, the resulting cut-off curves will gradually change from a convex to a concave pattern. This can be seen in Fig. 6, for AOs whose hydrophobicity increases in the order glutathione$<$ascorbate anion$<$*p*-cresol$<\alpha$-tocopherol (Aliaga et al. 2018).

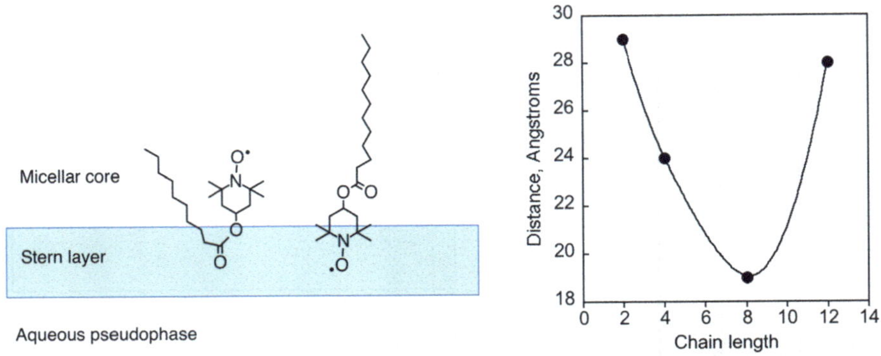

Fig. 4 Left: schematic representation of the orientation of 4-substituted TEMPO derivatives in a micellar environment. Depending on the hydrophobicity of the 4-substituent, the N-O· fragment is exposed to and assesses different micro-environments (Aliaga et al. 2016b) (Reprinted with permission from C.Aliaga et al. *Magnetic Resonance in Chemistry* 2016, 54, 870–873. Copyright 2016 John Wiley and Sons Ltd.). Right: variation of the average distance between the oxygen atom of the N-O· fragment of TEMPO derivatives and the center of a Triton-XR 100 micelle, estimated by molecular simulation. Reprinted from C. Aliaga et al. *Food Chemistry* 2016, 192, 395–401. Copyright 2015 Elsevier Ltd

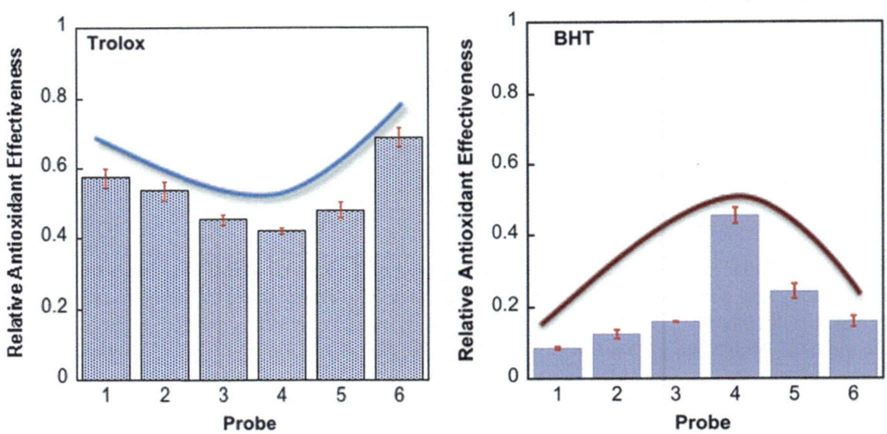

Fig. 5 Different cut-off patterns obtained by assessing the AO activity of a hydrophilic (Trolox) and a hydrophobic (BHT) phenolic antioxidant by the same set of probes (4-alkanoyl TEMPO radicals, with increasing substituent chain-length C-1 to C-12, Scheme 1) in aqueous micellar Triton-X 100 system (Aliaga et al. 2016c). (Reproduced from Aliaga C. et al. *Food Chemistry*, 2018, *245*,240–245. Copyright 2018 Elsevier Ltd.)

The described protocol for the location of the antioxidant/probe pair in a micro-heterogeneous medium is not limited to TEMPO probes. In a study of the fluorescence quenching of α-tocopherol by gallic acid (GA) and its propyl, octyl and dodecyl esters in oil-in-water emulsions, (Wang et al. 2020) the binding constants Ka ($\times 10^4$ M^{-1}) between the fluorescent probe and the gallic acid quenchers were

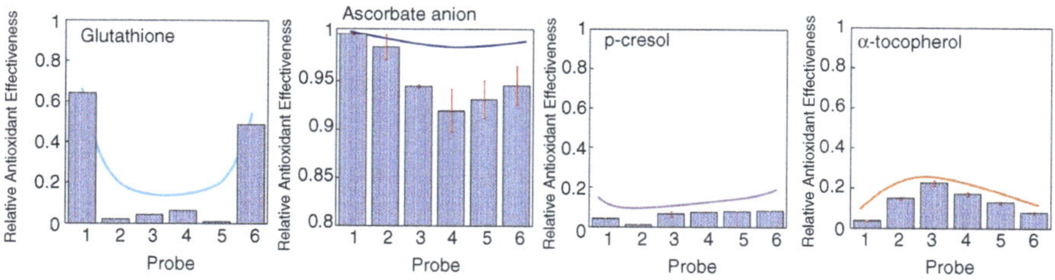

Fig. 6 Variation of the cut-off pattern with the increased hydrophobicity of the antioxidant: convex to concave curves are obtained as the AOs hydrophobicity increases, when its activity is evaluated by the standard 4-substituted TEMPO probes in aqueous Triton-XR 100. Reproduced from C. Aliaga et al. *Food Chemistry* 2018, *245*, 240–245. Copyright 2018 Elsevier Ltd

calculated. The Ka values increased sharply from the hydrophilic GA (0.06) to the more hydrophobic GA propyl ester (5.78), decreasing again for GA octyl (1.66) and GA dodecyl esters (0.48). This concave cut-off pattern is similar to that of Fig. 5 for α-tocopherol. Binding of the hydrophilic gallic acid with α-tocopherol is not very efficient in the emulsion and increases when the acid is converted to its more hydrophobic propyl ester. However, increasing the ester chain-length has the effect of tilting the quenching phenolic fragment towards a more hydrophilic environment, thus reducing its binding to the hydrophobic α-tocopherol. The observed cut-off pattern is thus a consequence of a change in orientation of the GA quenchers in the emulsion.

3 Amphiphobic Structure and the Diving-Swan Analogy

If the reactivity of a set of analogous probes, or of a family of antioxidants of increasing hydrophobicity, when measured in a micro-heterogeneous medium, can give rise to a cut-off behavior, do these compounds exhibit structural features responsible for this behavior?

As shown in Figs. 3 and 4, the answer is affirmative. The 4-alkanoyl TEMPO radicals (Scheme 2) are composed of two end-fragments, a constant piperidineoxyl and a 4-alkanoyl substituent of variable lipophilicity, both of which compete for the hydrophobic micellar core. This flexible arrangement of two fragments in the same molecule that compete for a hydrophobic microenvironment generates an *amphiphobic* structure and determines its variable orientation. A typically amphiphobic molecule, the 4-substituted TEMPO derivative, can adopt different conformations, when positioned in the interface of two micro-environments of different hydrophobicities (Fig. 7). As the size of its 4-substituent chain-length R increases, it is increasingly drawn towards the more hydrophobic micro-environment, tilting the N-oxide fragment towards a more hydrophilic region. An analogy can be made with a diving swan that plunges its neck towards the bottom of

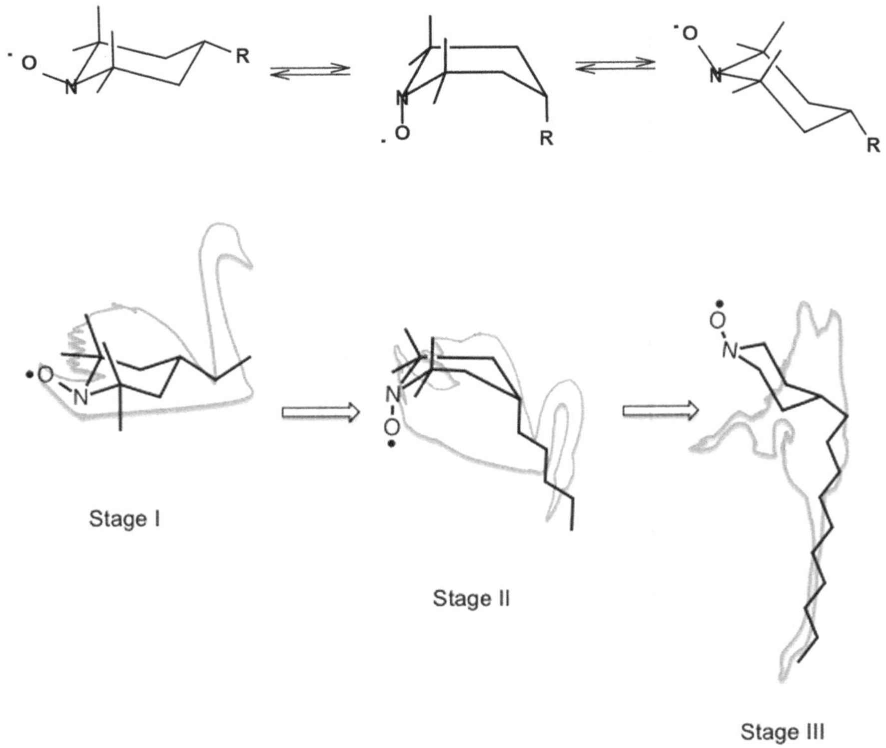

Fig. 7 The analogy between an amphiphobic N-oxide probe in the interface of two micro-environments and a "molecular" diving-swan floating on a "hydrophobic pond". Paradoxically, in Stage III the N-oxide fragment is raised to a more hydrophilic region, as the size of the hydrophobic "neck" increases, and the "molecular swan" plunges its head into the bottom of the pond. Reproduced from C.Aliaga et al. *Food Chemistry* **2019**, *279*, 288–293. Copyright 2018 Elsevier Ltd

a pond, assuming different conformations, depicted as Stages I-III (Lopez de Arbina et al. 2019).

The progression from Stage I to Stage III of Fig. 7 should be observed not only for the set of 4-substituted TEMPO derivatives, but for other amphiphobic probes or antioxidants. As an example, the antioxidant activity of caffeic acid (CA) and a series of its alkyl esters in soybean oil-in-water emulsions was measured by monitoring the formation of conjugated dienes with time, in the absence and in the presence of these AOs (Meireles et al. 2019; Costa et al. 2017). The antioxidant activity of these compounds increased from CA to its esters, reaching a maximum with the octyl caffeate, and decreasing again with the more hydrophobic hexadecanoate ester. Figure 8 depicts the changes in orientation of these AOs, as the ester chain-length increases. The caffeic acid and its esters act as a "molecular diving-swan" in the interface that separates the aqueous from the hydrophobic micro-environments. From the rather hydrophilic region that solvates the free

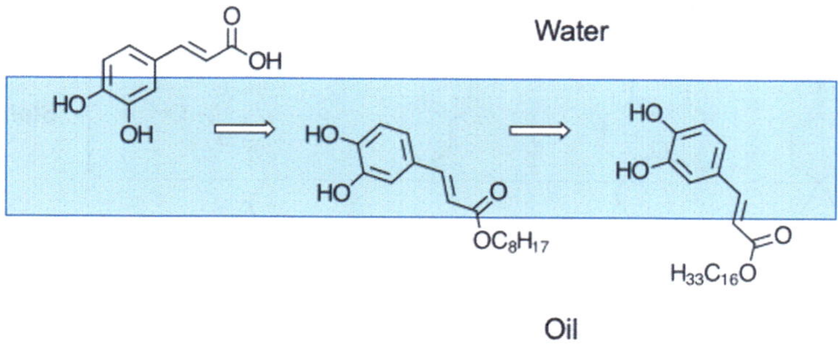

Fig. 8 Changes in orientation of caffeic acid and its esters in the interface of an oil-in-water emulsion (Costa et al. 2017), as a result of the amphiphobic structure of these antioxidants

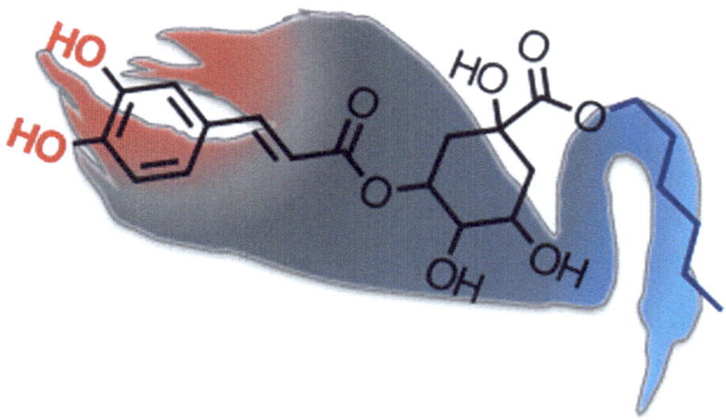

Fig. 9 The "swan-like" amphiphobic structure of the alkyl esters of chlorogenic acid, responsible for the cut-off behavior of their antioxidant activity in olive oil-in-water emulsions (Meireles et al. 2019)

carboxylic group of caffeic acid, its alkyl esters plunge into a more hydrophobic region, as their alkyl chain increases. However, for the very large hexadecyl chain, the "swan's" almost vertical "neck" has the effect of tilting upwards its phenolic "tail".

Similar cut-off curves were obtained when the antioxidant activity of a series of alkyl gallates was measured in fish oil-in-water emulsions, (Costa et al. 2020) or in the inhibition of liposome peroxidations (Durand et al. 2019).

Another family of amphiphobic antioxidants, alkyl esters of chlorogenic acid also exhibit a cut-off behavior in olive oil-in-water emulsions (Meireles et al. 2019). Their "swan-like" amphiphobic structure is easily recognizable in their diphenolic "tail", reminiscent of caffeic acid and its derivatives, its flexible, polyhydroxylated body and its alkyl-elongated ester "neck" (Fig. 9).

Thus, as expected, for three series of esters of chromancarboxylic, caffeic and gallic acid, all of them phenolic antioxidants with increasingly hydrophobic ester chains, cut-off curves are obtained in a micellar environment, when their antioxidant activity is assessed with the standard TEMPO probes (Lopez de Arbina et al. 2019). As expected, these curves evolve from a convex to a concave pattern, indicating that the antioxidant esters plunge into increasingly more hydrophobic environments, as their "necks" increase in size. However, for these compounds, although an evolution

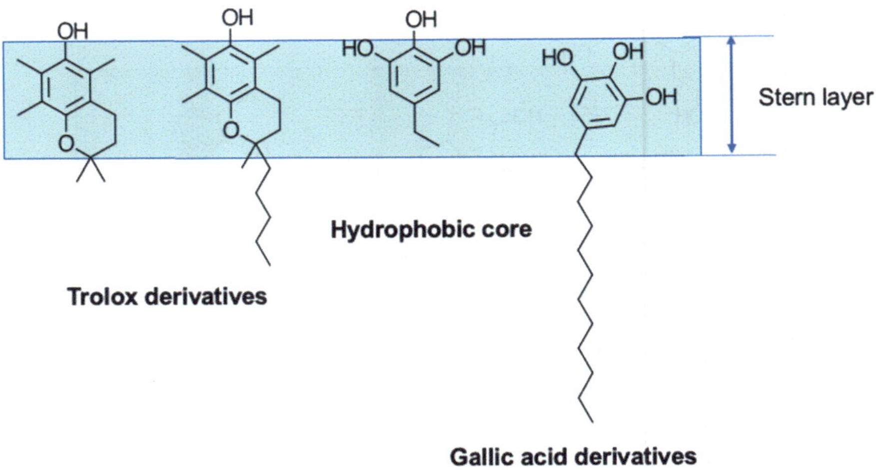

Fig. 10 Orientation of esters of chromancarboxylic (Trolox) and of gallic acid esters with increasingly hydrophobic chains in the Stern layer of a micellar system. The more hydrophilic phenolic fragment is situated in or near the aqueous layer and acts as a "buoy", preventing the molecular plunge into the hydrophobic core

from Stage I to Stage II of Fig. 7 is observed, the paradoxical situation of Stage III is not attained. The reason for this is due to a "buoyance effect" of the phenolic hydroxyl groups of these thee acid esters, which firmly retains this rather hydrophilic fragment of these molecules in or near the aqueous region of the micellar system. This "buoyance" effect is illustrated in Fig. 10, that depicts the orientation of these antioxidants in the micellar interface.

Interestingly, the flexibility of the hydrophobic "swan's neck" seems to have some effect on the AO activity of these amphiphobic molecules. The activity of two series of vitamin E homologs as singlet oxygen-quenching species were measured in micellar solutions (Mukai et al. 2018).The two homologs differed in the degree of insaturation of their side chain: α-,β-. and γ-tocopherols (Toc) had a completely saturated chain, while their homologs α-,β-. and γ- tocotrienols (Toc-3) had an unsaturated isoprenoid or farnesyl side chain. Toc-3 proved in all cases better quenchers in the micellar medium than the corresponding Toc homologs, a result that was rationalized by the authors by the bent and shorter chain of the former, when compared with the larger side-chain of the latter. These differences, according to the authors, would favor a greater mobility of the Toc-3 homologs, responsible for their greater reactivity. An alternative view, based on the "diving-swan" analogy, could invoke the shorter and more rigid "necks" of the Toc-3 molecules, as responsible for positioning their corresponding phenolic "tails" in a slightly less hydrophilic microenvironment than their Toc analogs (Fig. 11).

Cut-off patterns associated with series of amphiphobic molecules can be found for rather different systems. Even when a detailed description of their mode of action is missing, the paradoxical effect of these molecules can be sought in their changing

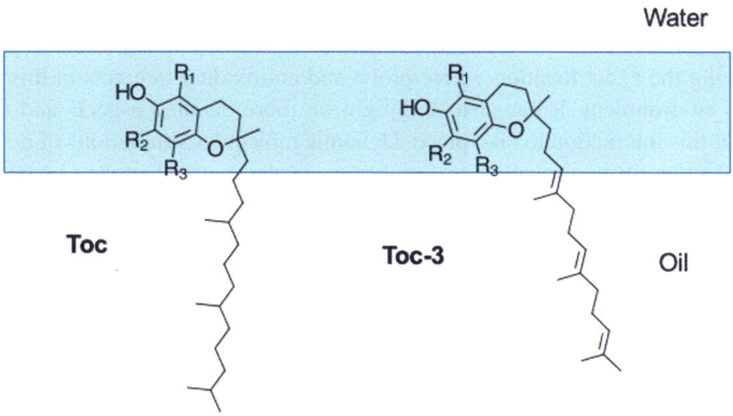

Fig. 11 Possible explanation for the different reactivities of Toc and Toc-3 in the interface of a micellar system (Mukai et al. 2018), according to the "diving-swan" analogy. The fully saturated sidechain of Toc would be more capable than the analogous rigid and shorter chain of Toc-3 to tilt the phenolic hydroxide to a slightly less hydrophobic micro-environment

orientation in a receptor site. As an example, the acute toxicity of the ionic liquids 1-alkyl-3-methylimidazolium nitrates of varying chain-length to earthworms presents a cut-off pattern (Shao et al. 2017). Though probably of a very complex nature, this effect could be associated with the changing orientation of the imidazolium ring, as the length of the 1-alkyl substituent increases.

4 Summary and Perspectives

The paradoxical behavior of antioxidants in food and biological systems has long been identified and characterized (Porter 1993; Porter et al. 1989), and various explanations for this "polar paradox" have been put forward by distinct groups. Since the critical overview of the polar paradox by Shahidi and Zhong (Shahidi and Zhong 2011), ten years have elapsed, and a more detailed understanding of its causes have emerged. The recognition of the heterogeneous environment that is the scenario of these paradoxical findings led to the need of determining the exact location of the antioxidant or probe in these systems (Aliaga et al. 2016a) More subtle aspects of the probe-antioxidant interactions have emerged, requiring a more detailed identification of their reactive groups and or their proximity in a micro-region. Even when situated in the same micro-environment, phenolic antioxidants react differently with a substrate, depending on the proximity of the reactive fragments of the pair. Thus, the conclusion of a study on the comparative inhibition of lipid peroxidation by resveratrol, piceid, BHT and vitamin E is that "in the presence of lipid peroxidation, resveratrol, piceid and BHT can react more quickly with the peroxyl radical than vit. E for distance motives" (Fabris et al. 2008).

The use of a series of 4-alkanoyl TEMPO derivatives as radical probes for the assessment of antioxidants in micellar media has not only confirmed the importance of knowing the exact location where probe and antioxidant interacts in this heterogeneous environment. It has also shed light on more detailed aspects and requirements for this interaction to take place. Dynamic molecular simulations of how these probes are present in a micellar system have revealed a new factor to be taken into account, when examining probe/AO interactions: the orientation of the probe in the micellar interface, leading to slightly different micro-environments for the reacting N-oxide fragment. Thus, these changes in the probe orientation could be held responsible for the cut-off effects frequently observed in these systems. This hypothesis was confirmed by varying in a systematic way series of analogous antioxidants, or by replacing the micellar system by oil-in-water emulsions.

The choice of the 4-substituted TEMPO probes as references for the assessment of different AOs has proved rather fruitful, because they have also suggested main structural requirements for such changes of orientation to take place.

The competition between two end-groups of similar hydrophilicity or hydrophobicity for a particular region in a micro-heterogeneous environment gives rise to different orientations of a probe or substrate, resulting in different reactivities in this medium. The cut-off effects frequently observed in many heterogeneous systems (Aliaga et al. 2016c) can thus be ascribed to the amphiphobic nature of these probes or antioxidants, and their variable orientation in these environments. Examples of such behavior can be deduced from data in the literature that have been presented in the past without any interpretation based on structural features. As an example, in a study of the solubilization of some phenolic antioxidants in pluronic micelles (Parmar et al. 2011), the authors suggested different orientations for the studied hydroxyphenyl carboxylic acids in the micelle (Fig. 12). Notice that the location of the reactive phenolic hydroxyl groups depends on the orientation of the acidic antioxidant, and this orientation is the result of a competition between two hydrophilic end-groups, the carboxylic $-CO_2H$ and the phenolic OH fragments. This is a nice example of a change in orientation of related antioxidants due to an amphiphobic (or amphiphilic) competition.

The possibility of modifying the orientation of an amphiphobic molecule by systematically changing the hydrophobicity of one of its end-groups thus leads to cut-off patterns of behavior. The form of the obtained curves can also be used as a protocol to identify the location of an antioxidant in a heterogeneous milieu. The resulting "diving-swan" analogy describes the different stages of probe insertion into the hydrophobic core of a micelle and explains their paradoxical reactivity in such a medium. Interestingly, as shown in the previous sections, different families of widely employed antioxidants that give rise to cut-off effects in micellar systems or microemulsions possess a "swan-like" structure. This observation allows a generalization to be made regarding a large number of apparently unrelated systems, that share common structural features.

Thus, information gathered from EPR spectra, molecular-dynamics simulations and kinetic measurements, allied with the fortunate choice of a family of amphiphobic probes, have led to a novel explanation of the paradoxical behavior

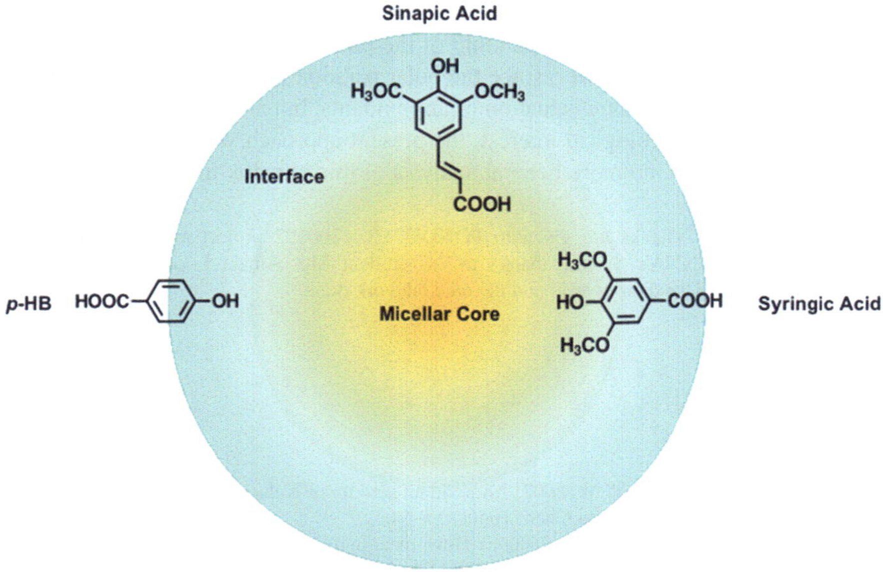

Fig. 12 Schematic representation of the location of phenolic antioxidants *p*-hydroxybenzoic (*p*-HB), sinapic and syringic acid in a pluronic micelle. Adapted with permission from Parmar A. et al. Colloids and Surfaces B: Biointerfaces **2011**, 86, 319–326. Copyright 2011 Elsevier B.V

of antioxidants in micro-heterogeneous media. More interesting perhaps, from the point of view of future applications, this novel framework suggests interpretations for other pharmacological or biological phenomena not directly related with antioxidants and their behavior, that exhibit the same polar paradox. Series of compounds with an amphiphobic structure that exhibit a cut-off behavior are candidates for systematic investigations based on their possible changes of orientation in a heterogeneous biological environment. In this respect, molecular simulations constitute a theoretical tool that can be more frequently utilized to confirm or dismiss these hypotheses. As mentioned before, they have for the first time stressed the importance of the probe orientation in these media. They can also provide new explanations for the cut-off behavior of other families of biologically active compounds. This is the case, for example, of the anesthetic effect of *n*-alkanols, which exhibit a cut-off pattern of activity. The alcohol interactions with a lipid membrane were investigated with molecular dynamic simulations, leading to a detailed description of these interactions, and relating them with their paradoxical activity (Zapata-Morin et al. 2020). Though not amenable to the interpretation developed in the previous sections, this example of an explanation for the cut-off behavior in biological systems with the aid of molecular dynamic simulations illustrates the wide scope of this interesting and puzzling phenomenon. Probably, it cannot be reduced to a single explanation or interpretation, but constitute a macroscopic evidence for the interplay of different factors and interactions at a microscopic level. The approach described in this chapter, emphasizing the proximity of the reacting groups of a probe/antioxidant pair, and of the probe and/or AO orientations to modulate their interaction, represents a contribution to the understanding of the polar paradox in the assessment of antioxidant activities. However, since the polar paradox is a more general phenomenon, found not only in the evaluation of antioxidants, but also in a larger variety of systems with pharmacological interest, the present approach will also find applications in the future in the more general analysis of this paradox in these systems.

Acknowledgements Thanks are given to FONDECYT 1200192 project and Basal financing AFB180001 (CEDENNA). Special thanks to our student Mg. Amaia López de Arbina and Ms. Camila Pastenes for their helpful and devoted labwork done.

References

Ahmadi F, Kadivar M, Shahedi M (2007) Antioxidant activity of Kelussia odoratissima Mozaff. In model and food systems. Food Chem 105(1):57–64

Aliaga C, Rezende MC, Arenas A (2009a) How meaningful is the assessment of antioxidant activities in microheterogeneous media? Food Chem [Internet] 113(4):1083–1087

Aliaga C, Rezende MC, Arenas A (2009b) How meaningful is the assessment of antioxidant activities in microheterogeneous media? Food Chem 113(4):1083–1087

Aliaga C, Torres P, Silva F (2012) A simple method for the determination of the partitioning of nitroxide probes in microheterogeneous media. Magn Reson Chem 50(12):779–783

Aliaga C, Bravo-Moraga F, Gonzalez-Nilo D, Márquez S, Lühr S, Mena G et al (2016a) Location of TEMPO derivatives in micelles: subtle effect of the probe orientation. Food Chem 192:395–401

Aliaga C, Rezende C, Mena G (2016b) The effect of micellization on the EPR spectra tetramethylpiperidinoxyl (TEMPO) radicals. Magn Reson Chem 54(11):870–873

Aliaga C, Lopez de Arbina A, Rezende MC (2016c) "Cut-off" effect of antioxidants and/or probes of variable lipophilicity in microheterogeneous media. Food Chem 206:119–123

Aliaga C, López de Arbina A, Pastenes C, Rezende MC (2018) Antioxidant-spotting in micelles and emulsions. Food Chem 245:240–245

Almajano MP, Carbó R, Jiménez JAL, Gordon MH (2008) Antioxidant and antimicrobial activities of tea infusions. Food Chem 108(1):55–63

Aspée A, Aliaga C, Maretti L, Zúñiga-Núñez D, Godoy J, Pino E et al (2017) Reaction kinetics of phenolic antioxidants toward Photoinduced Pyranine free radicals in biological models. J Phys Chem B 121(26):6331–6340

Balgavý P, Devínsky F (1996) Cut-off effects in biological activities of surfactants. Adv Colloid Interf Sci 66:23–63

Boroski M, Aguiar AC, Rotta EM, Bonafe EG, Valderrama P, Souza NE et al (2018) Antioxidant activity of herbs and extracted phenolics from oregano in canola oil. Int Food Res J 25(6): 2444–2452

Costa M, Losada-Barreiro S, Paiva-Martins F, Bravo-Díaz C (2017) Physical evidence that the variations in the efficiency of homologous series of antioxidants in emulsions are a result of differences in their distribution. J Sci Food Agric 97(2):564–571

Costa M, Losada-Barreiro S, Bravo-Díaz C, Vicente AA, Monteiro LS, Paiva-Martins F (2020) Influence of AO chain length, droplet size and oil to water ratio on the distribution and on the activity of gallates in fish oil-in-water emulsified systems: emulsion and nanoemulsion comparison. Food Chem 310:125716–125723

Cui L, Mcclements DJ, Decker EA (2015) Impact of phosphatidylethanolamine on the antioxidant activity of α-tocopherol and trolox in bulk oil. J Agric Food Chem 63(12):3288–3294

Dar AA, Bravo-Diaz C, Nazir N, Romsted LS (2017) Chemical kinetic and chemical trapping methods: unique approaches for determining respectively the antioxidant distributions and interfacial molarities of water, counter-anions, and other weakly basic nucleophiles in association colloids. Curr Opin Colloid Int Sci 32:84–93

Durand E, Zhao Y, Ruesgas-Ramón M, Figueroa-Espinoza MC, Lamy S, Coupland JN et al (2019) Evaluation of antioxidant activity and interaction with radical species using the vesicle conjugated Autoxidizable Triene (VesiCAT) assay. Eur J Lipid Sci Technol 121(5):1–8

Fabris S, Momo F, Ravagnan G, Stevanato R (2008) Antioxidant properties of resveratrol and piceid on lipid peroxidation in micelles and monolamellar liposomes. Biophys Chem 135(1–3): 76–83

Ferreira I, Costa M, Losada-Barreiro S, Paiva-Martins F, Bravo-Díaz C (2018) Modulating the interfacial concentration of gallates to improve the oxidative stability of fish oil-in-water emulsions. Food Res Int 112:192–198

Freiría-Gándara J, Losada-Barreiro S, Paiva-Martins F, Bravo-Díaz C (2018) Enhancement of the antioxidant efficiency of gallic acid derivatives in intact fish oil-in-water emulsions through optimization of their interfacial concentrations. Food Funct 9(8):4429–4442

González MJ, Medina I, Maldonado OS, Lucas R, Morales JC (2015) Antioxidant activity of alkyl gallates and glycosyl alkyl gallates in fish oil in water emulsions: relevance of their surface active properties and of the type of emulsifier. Food Chem 183:190–196

Keller S, Locquet N, Cuvelier ME (2016) Partitioning of vanillic acid in oil-in-water emulsions: impact of the tween®40 emulsifier. Food Res Int 88:61–69

Krudopp H, Sönnichsen FD, Steffen-Heins A (2015) Partitioning of nitroxides in dispersed systems investigated by ultrafiltration, EPR and NMR spectroscopy. J Colloid Interface Sci 452:15–23

Kumar A, Chauhan PK, Bhardwaj VS, Kumar R, Tyagi A (2011) In vitro antioxidant & phytochemical investigations of ethanolic extracts of viola serpens & morus nigrav. J Chem Pharm Res 3(4):166–171

Kusumahastuti DKA, Sihtmäe M, Kapitanov IV, Karpichev Y, Gathergood N, Kahru A (2019) Toxicity profiling of 24 L-phenylalanine derived ionic liquids based on pyridinium, imidazolium and cholinium cations and varying alkyl chains using rapid screening *Vibrio fischeri* bioassay. Ecotoxicol Environ Saf 172:556–565

Laguerre M, López Giraldo LJ, Lecomte J, Figueroa-Espinoza M-C, Baréa B, Weiss J et al (2009) Chain length affects antioxidant properties of Chlorogenate esters in emulsion: the cutoff theory behind the polar paradox. J Agric Food Chem 57(23):11335–11342

Lopez de Arbina A, Losada-Barreiro S, Rezende MC, Vidal M, Aliaga C (2019) The location of amphiphobic antioxidants in micellar systems: the diving-swan analogy. Food Chem 279:288–293

López-Alarcón C, Aspée A, Henríquez C, Campos AM, Lissi EA (2005) Interaction and reactivity of urocanic acid towards peroxyl radicals. Redox Rep 10(4):227–234

Losada-Barreiro S, Sova M, Mravljak J, Saso L, Bravo-Díaz C (2020) Synthesis, in vitro antioxidant properties and distribution of a new cyanothiophene-based phenolic compound in olive oil-in-water emulsions. Antioxidants 9(7):1–16

Lue BM, Sørensen ADM, Jacobsen C, Guo Z, Xu X (2017) Antioxidant efficacies of rutin and rutin esters in bulk oil and oil-in-water emulsion. Eur J Lipid Sci Technol 119(4):1–15

Martinović N, Poklar Ulrih N, Abramovič H (2019) Sinapic acid and its derivatives increase oxidative stability in different model lipid systems. Eur J Lipid Sci Technol 121(4):1–10

Mateos R, Madrona A, Pereira-Caro G, Domínguez V, Cert RM, Parrado J et al (2015) Synthesis and antioxidant evaluation of isochroman-derivatives of hydroxytyrosol: structure-activity relationship. Food Chem 173:313–320

Meireles M, Losada-Barreiro S, Costa M, Paiva-Martins F, Bravo-Díaz C, Monteiro LS (2019) Control of antioxidant efficiency of chlorogenates in emulsions: modulation of antioxidant interfacial concentrations. J Sci Food Agric 99(8):3917–3925

Montoro P, Braca A, Pizza C, De Tommasi N (2005) Structure-antioxidant activity relationships of flavonoids isolated from different plant species. Food Chem 92(2):349–355

Mukai K, Ishikawa E, Ouchi A, Nagaoka S-I, Abe K, Suzuki T et al (2018 Jun 1) Measurements of singlet oxygen-quenching activity of vitamin E homologs and palm oil and soybean extracts in a micellar solution. Lipids 53(6):601–613

Nogueira CCR, Paixão ICNDP, Teixeira VL (2014) Antioxidant activity of natural products isolated from red seaweeds. Nat Prod Commun 9(7):1031–1036

Noon J, Mills TB, Norton IT. The use of natural antioxidants to combat lipid oxidation in O/W emulsions. Journal of Food Engineering [Internet]. 2020; 281:110006–19. Available from: https://doi.org/10.1016/j.jfoodeng.2020.110006

Özcan MM, Al Juhaimi F, Ahmed IAM, Uslu N, Babiker EE, Ghafoor K (2020) Effect of microwave and oven drying processes on antioxidant activity, total phenol and phenolic compounds of kiwi and Pepino fruits. J Food Sci Technol 57(1):233–242

Parmar A, Singh K, Bahadur A, Marangoni G, Bahadur P (2011) Interaction and solubilization of some phenolic antioxidants in Pluronic® micelles. Colloids Surf B: Biointerfaces 86(2): 319–326

Porter WL (1993) Paradoxical behavior of antioxidants in food and biological systems. Toxicol Ind Health 9(1–2):93–122

Porter WL, Black ED, Drolet AM (1989) Use of polyamide oxidative fluorescence test on lipid emulsions: contrast in relative effectiveness of antioxidants in bulk versus dispersed systems. J Agric Food Chem 37(3):615–624

Raimúndez-Rodríguez EA, Losada-Barreiro S, Bravo-Díaz C (2019) Enhancing the fraction of antioxidants at the interfaces of oil-in-water emulsions: a kinetic and thermodynamic analysis of their partitioning. J Colloid Interface Sci 555:224–233

Romsted LS, Zhang J (2004) Determining antioxidant distributions in model food emulsions: development of a new kinetic method based on the pseudophase model in micelles and opaque emulsions. Progr Colloid Polym Sci 123:182–187

Shahidi F, Zhong Y (2011) Revisiting the polar paradox theory: a critical overview. J Agric Food Chem 59(8):3499–3504

Shao Y, Du Z, Zhang C, Zhu L, Wang J, Wang J (2017) Acute toxicity of imidazole nitrate ionic liquids with varying chain lengths to earthworms (*Eisenia foetida*). Bull Environ Contam Toxicol 99(2):213–217

Shao Y, Wang J, Du Z, Li B, Zhu L, Wang J (2018) Toxicity of 1-alkyl-3-methyl imidazolium nitrate ionic liquids to earthworms: the effects of carbon chains of different lengths. Chemosphere 206:302–309

Shao Y, Wang J, Wang J, Du Z, Li B, Zhu L et al (2019) Oxidative stress and genotoxic effects in earthworms induced by five imidazolium bromide ionic liquids with different alkyl chains. Chemosphere 227:570–579

Teixeira S, Siquet C, Alves C, Boal I, Marques MP, Borges F et al (2005) Structure-property studies on the antioxidant activity of flavonoids present in diet. Free Radic Biol Med 39(8):1099–1108

Thippeswamy TG, Vidyashree HK, Hemavathi, Kengaiah J (2020) Antioxidant activity in methanolic extract of selected vegetables. Int J Pharm Pharm 17(2):151–160

Wang YL, Wu C, Zhou XY, Zhang MR, Chen Y, Nie SP et al (2020) Combined application of gallate ester and α-tocopherol in oil-in-water emulsion: their distribution and antioxidant efficiency. J Dispers Sci Technol 41(6):909–917

Ye L, Pham-Mondala A, Li J, Joseph P, Nahas R, Michel-Salaun F (2019) Using confocal microscopy to estimate the distribution of natural antioxidants in poultry meal and extruded kibbles. Eur J Lipid Sci Technol 121(9):1–7

Yeşilyurt V, Halfon B, Öztürk M, Topçu G (2008) Antioxidant potential and phenolic constituents of Salvia cedronella. Food Chem 108(1):31–39

Zapata-Morin PA, Sierra-Valdez FJ, Ruiz-Suárez JC (2020) The cut-off effect of n-alcohols in lipid rafts: a lipid-dependent phenomenon. J Mol Graph Model 101:1–7

Lipid Antioxidants: More than Just Lipid Radical Quenchers

K. M. Schaich

1 Introduction

When the food industry selects compounds to inhibit lipid oxidation, phenolic compounds are always the first in line because they have long been recognized for their ability to quench radicals rapidly and efficiently while causing few complicating side reactions. At the same time, use of phenolic antioxidants in foods is not straightforward. Not all phenolic compounds are equally effective in slowing or stopping lipid oxidation, some are not effective at all, and under some conditions phenolic compounds can even become pro-oxidant. To complicate matters still further, phenols do more than just quenching radicals to inhibit lipid oxidation and its effects. Phenols have multiple actions that affect total oxidative stability (+ or −) of any *system* in which they are located, particularly foods, biological tissues, and personal care products, including (in addition to radical scavenging):

- metal chelating
- singlet oxygen quenching
- carbonyl complexation
- protein complexation and enzyme inhibition

These additional pathways extend protection past radical scavenging but also may interfere or compete with other stabilizing reactions.

Now add to this still another layer of complexity—differences in reactions and reactivity of phenol vs quinone forms of [phenolic] antioxidants. While a system may start out with only phenols, these are oxidized in the process of radical quenching and eventually product quinones accumulate and can exert different influences on system stabilization. Are these anti or pro-oxidant?

K. M. Schaich (✉)
Department of Food Science, Rutgers University, New Brunswick, NJ, USA
e-mail: schaich@sebs.rutgers.edu

C. Bravo-Diaz (ed.), *Lipid Oxidation in Food and Biological Systems*,
https://doi.org/10.1007/978-3-030-87222-9_7

It is easy to take the most simplistic stabilization approach and focus only on radical quenching to slow the rate of oxidation as measured by decreasing levels of hydroperoxides. However, nearly everyone working with antioxidants has encountered some mystifying results where the compounds added were ineffective, exerted effects beyond lipid oxidation, changed sensory or physical properties in unexplained ways, added browning, and more. Such situations are where it is critical to look beyond radical quenching to other actions of antioxidants that stabilize or destabilize systems in different ways.

Natural (poly)phenols appear to participate in alternate reactions much more than the traditional synthetic phenols BHA and BHT, perhaps because the structures are more complex with multiple sites for reaction and with mixed solubilities in oil and aqueous phases. Current emphasis on replacing synthetic phenolic antioxidants with natural compounds thus makes it particularly important to recognize the complexity of antioxidant actions, to begin looking routinely for their multiple actions, and to distinguish unique reactions in different materials. To stimulate broader consideration of antioxidant protections and disruptions in systems beyond quenching of lipid radicals, this chapter provides a general overview of phenolic antioxidant actions at all three stages of lipid oxidation—initiation, propagation, and termination—and beyond in interactions with proteins. The chapter then closes with a new view for interconnections between lipids, proteins, and phenolic antioxidants in the oxidative stability of total systems.

2 Prevention of Initiation

2.1 Metal Complexation and Cycling

Phenols complex metals very effectively through combinations of phenol moieties or phenols and carbonyls in adjacent positions on the same ring (a and b, respectively) or in *peri* positions of two fused rings (c and d, respectively) (Hider et al. 2001).

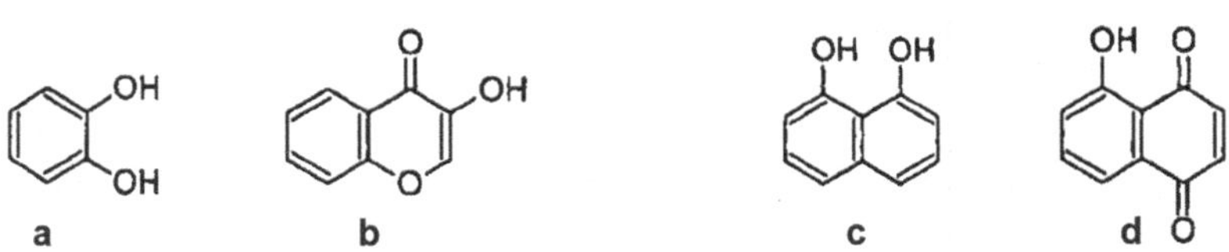

a b c d

Metals may be complexed to a single phenol (monodentate, 1:1) or chelated by multiple phenolic groups (bidentate 2:1 or higher 3:1, etc.) depending on the available orbitals in the metal, and the metal binding energies and reactivities vary with each (Hider et al. 2001).

1:1 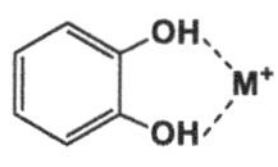

2:1 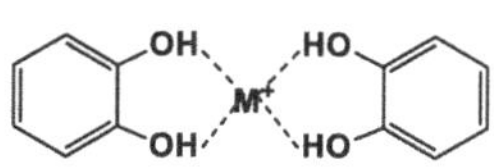

For example, quercetin forms 1:1, 2:1, and 3:1 complexes with iron. Three quercetin molecules are required to fill all the orbitals and fully inactivate a single atom of iron, but the strongest iron binding and hence most efficient removal of the metal from solution occurs in 2:1 complexes (Ren et al. 2008; Leopoldini et al. 2006). Di- and tri-dentate complexes dominate near neutral pH; 1:1 complexes are most common at acid pH where ligands are released (Perron and Brumaghim 2009). Similarly, catechin and its congeners catechol, methyl gallate, epigallocatechin, epicatechin gallate, and epigallocatechin gallate form 2:1 and 3:1 complexes with Fe^{3+} at pH 7 (Jovanovic et al. 1998).

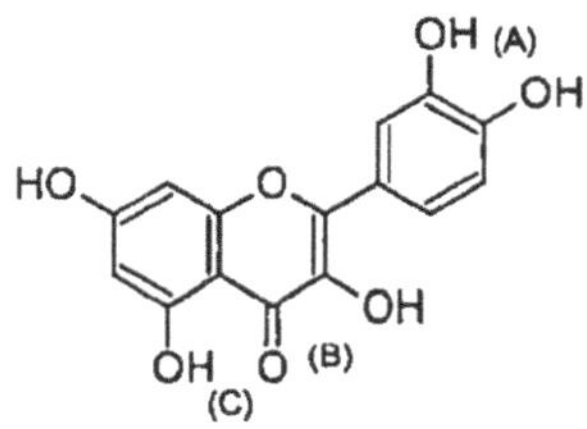

Multiple metal binding sites can be present in polyphenols, as shown in the structure above. Strength and preference for binding is A>B>C (Hider et al. 2001). Metals can be bound in two positions of a single polyphenol, either A+B or A+C, but not B and C (Hider et al. 2001). Types of metals bound by polyphenols include redox-active iron$^{(2+ \text{ and } 3+)}$ (Ren et al. 2008; Jovanovic et al. 1998; Leopoldini et al. 2004) and copper$^{(1+ \text{ and } 2+)}$ commonly found in foods (Md et al. 2005) as well as aluminum^{3+} (Zhang et al. 2016), chromium^{3+} (Chrysochoou and Reeves 2017), and tin^{2+} (Wehrer et al. 1984) that may be encountered in processing and packaging of foods.

Metal complexing/chelation has both physical and chemical mechanisms for inhibiting oxidation. Physically, binding moves metals out of solution and blocks some orbitals from electron transfer, thereby limiting reaction. However, the inhibition efficiency depends on the degree of ligand formation. Full chelation with all metal orbitals complexed does inactivate metals, but any open orbitals remain active. Iron in 2:1 complexes retains sufficient accessibility to catalyze reactions with small reactive oxygen species but is largely blocked from larger molecules such as lipids. Iron in 1:1 complexes is open to approach from other molecules so its activity is determined by the complex redox potential and phase localization. More important than physical association, though, chemical complexing and chelation alters metal redox potentials (generally lower) so that they become less active as an oxidizing agent. For example, the standard redox potential for Fe^{3+}/Fe^{2+} is 0.77 V vs the normal hydrogen electrode. Complexation by EDTA (ethylenediamine tetra-acetic

Table 1 Reduction Potentials of phenolic ligands and their metal complexes as determined by cyclic voltammetry (modified from Mahal et al. (2005); used with permission)

pH	Phenolic cmpd	$E°$ (V) vs NHE	Metal complex	$E°$ (V) vs NHE
7	propylgallate	0.42	Fe^{2+}	0.45
7			Mn	0.038
7	catechol	0.442	Cu	0.277
10.5	resveratrol	0.5, 0.74	Cu	0.3
9	curcumin	0.38	Cu	0.022
7/7.5	chlorogenic acid	0.4	Cu	0.251
7	rutin	0.475	Cu	0.325
7	3-OH anthranilic acid	0.58	Fe^{2+}	0.33

acid) drops the potential to 0.12 V where it is a good reducing agent but poor initiator. This indeed is the key inhibitory effect of metal complexation.

Redox potentials of phenol-metal complexes have seldom been measured. A few available values are reported in Table 1 (Mahal et al. 2005).

While these physical and chemical mechanisms are straightforward for standard chelators, metal complexes of phenols and polyphenols present an enigma that must be considered when using these antioxidants—they can inhibit initiation of lipid oxidation but with equal probability they can also promote it. First, polyphenols have limited and mixed solubilities in water versus oils, and the distribution between the two phases varies with the phenol structure (Schwarz et al. 1996). Unlike EDTA which moves metals out of the oil phase into the water phase, polyphenols shift metal phase distribution in still undetermined ways, so polyphenols may move metals to regions where they are more active as pro-oxidants rather than less. This includes moving metals to interfaces. Coupled with shifting of metal localization is the facility with which phenols undergo redox cycling, together with reactivity of both oxidized and reduced metals in the complexes. As noted above, complexing metals decreases the metal redox potential and its oxidizing power. At the same time, the metal becomes a stronger reducing agent so reacts more readily with oxygen or species such as hydroperoxides, ROOH, to generate radicals and return to its oxidized state. Redox cycling is largely responsible for pro-oxidant behavior observed for polyphenols, particularly at low phenol concentrations where 1:1 and 2:1 phenol-metal complexes are most likely (Jiang et al. 2015).

How redox cycling works is shown simplistically in Fig 1. When diphenols bind Fe^{3+}, H atoms are released and a semiquinone radical cation is formed (Rx 1). The released H atoms drop the pH, decreasing metal binding constants and redox potentials (Perron and Brumaghim 2009). Electron transfer from the Fe^{3+} generates a semiquinone radical (Rx 2) which then can reduce either oxygen to superoxide anions (Rx 3) or another Fe^{3+} to Fe^{2+} (Rx 4), yielding the quinone as the final product. Binding of Fe^{2+} by diphenols (Rx 5) is weak but is followed by rapid oxidation to regenerate the Fe^{3+} complex, semiquinone radical, and more $O_2^{-\bullet}$ (Rx 6) (Perron and Brumaghim 2009; Jiang et al. 2015).

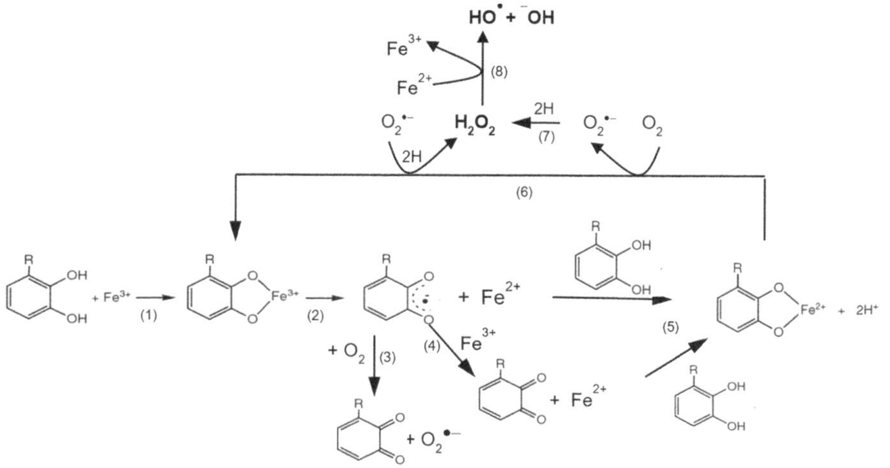

Fig. 1 Potential pro-oxidant actions resulting from redox cycling of Fe^{2+} and Fe^{3+} complexed by phenolic compounds

Redox cycling of metals has several ripple effects. One is that, when oxygen is present, even trace amounts of metals can be used over and over to generate large amounts of $O_2^{-\bullet}$/HOO$^\bullet$ and then H_2O_2 (Rx 7) and very reactive HO$^\bullet$ (Rx 8) (at the same time regenerating Fe^{3+} to restart the cycle) until all the phenolic complexer has been oxidized. This greatly increases the radical load which antioxidants must inactivate while at the same time decreasing the amount of phenol available for radical quenching (Dimitrić Marković et al. 2011). The product quinones remain active and mediate separate reactions with radicals, oxidation products, and proteins, as will be discussed in Sect. 3.1.2.

2.2 *Singlet Oxygen Quenching*

Singlet oxygen ($^1O_2$) is a powerful lipid oxidation catalyst produced by interaction of Type 2 photosensitizers with oxygen. It is particularly important as a catalyst because it adds indiscriminately to all carbons of all double bonds, generating intact hydroperoxides at multiple positions on fatty acids without an induction period.

Carotenoids are perhaps the best-known quenchers of $^1O_2$. One general explanation traditionally given for this effect has been the presence of many double bonds that compete with lipids for $^1O_2$ addition. $^1O_2$ adds preferentially to the extended conjugated double bond system so, in effect, carotenoids commit suicide to protect lipids. These actions consume the antioxidants making less available for subsequent radical quenching.

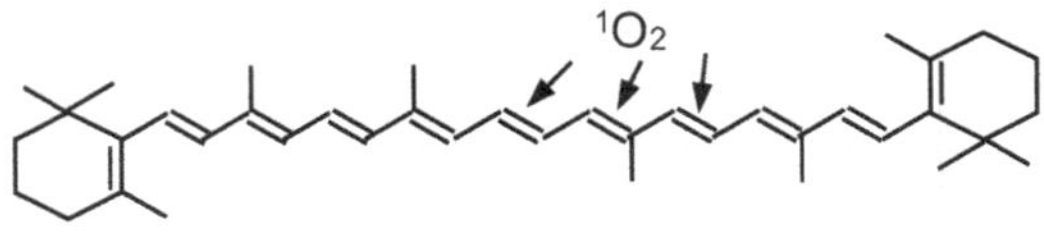

That being said, carotenoid loss is not always observed in the presence of light. A second even more active mechanism has been identified in some carotenoids, particularly in β-carotene: energy transfer to return singlet oxygen to its ground state, i.e. $^1O_2 \rightarrow {}^3O_2$. This is an extremely efficient action. In model systems, more than 250 moles of $^1O_2$ were quenched per mole of β-carotene without carotenoid loss. $^1O_2$ reaction rate was diffusion-controlled, 6.6×10^9 L $M^{-1}s^{-1}$ (Foote and Denny 1968).

Tocopherols also efficiently quench singlet oxygen and the same two mechanisms described for carotenoids have been reported. Fahrenholtz et al (Fahrenholtz et al. 1974) showed that 120 mols of $^1O_2$ were deactivated ($^1O_2 \rightarrow {}^3O_2$) before α-tocopherol was destroyed, indicating excitation deactivation. Rates were very fast, $2.5\text{–}5.0 \times 10^8$ L $M^{-1}s^{-1}$. However, $^1O_2$ addition to tocopherol double bonds to form 8-hydroperoxy-tocopherone (Toc-OOH) (Rx 1) has been documented as well (Trebst 2003). Toc-OOH then decomposes to α-tocopheryl-quinone.

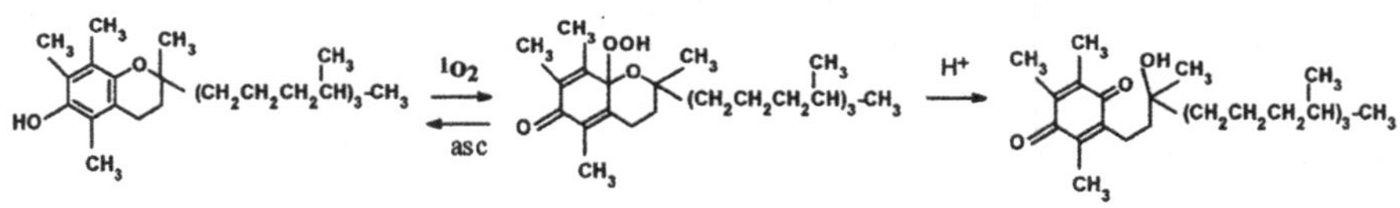

α–Tocopherol 8-Hydroperoxy-tocopherone α–Tocopheryl quinone

(1)

Polyphenols do quench $^1O_2$ but at rates generally lower than those of tocopherols, e.g. $k = 10^4\text{–}10^6$ L $M^{-1}s^{-1}$ for flavonoids (Tournaire et al. 1993). The same physical and chemical mechanisms of quenching are active, with ($^1O_2 \rightarrow {}^3O_2$) deactivation mediated by the presence of a catechol moiety on the B ring, and chemical addition directed by presence of a hydroxyl group activating the double bond in the C ring. Flavanones and catechins are chemically inert towards $^1O_2$.

3 Effects on Propagation

3.1 Inhibition by Radical Scavenging

3.1.1 Reactions of Phenol Forms of Antioxidants

Radicals can be quenched by adding a hydrogen atom and an electron to them, either together or separately (Rx 2).

$$R(O)O^{\bullet} + \text{Phe-OH} \xrightarrow[\text{or } H^{+}/e^{-}]{H^{\bullet}} R(O)OH + \text{Phe-O}^{\bullet} \quad (2)$$

For decades the potent antioxidant behavior of phenols has been attributed to two features: weak bond dissociation energies of the phenolic H bond release H atoms readily for transfer to oxyl radicals, and the resulting phenoxyl free electron delocalizes over the entire aromatic ring, drastically reducing reactivity. Consequently, the phenoxyl radicals formed when lipid and other radicals are quenched do not propagate radical chains. However, studies of why and how some phenol structures more effectively enhanced radical quenching uncovered multiple modes by which electrons and protons can be transferred from antioxidants to radicals, and these modes provide flexibility for actions of phenols with different structures as well as in different solvents or reaction milieu.

It is now well recognized that phenolic compounds also quench radicals by electron transfer which proceeds in parallel to H atom transfer or in sequence with it. A few antioxidants such as curcumin have multiple sites that act independently to provide both H atom and electron transfer. The end products are the same either way, so why are multiple quenching mechanisms important? Kinetics are much faster for electron transfer, while H atom transfer is thermodynamically favored. Preference for one mechanism or the other varies with the structures of the phenolic compound and perhaps more importantly, the medium, but is independent of radical structure (Lucarini et al. 1998; Valgimigli et al. 1995). Rates of the two mechanisms are strongly and oppositely affected by solvent, as will be explained below. Solvent effects become particularly important when polyphenols partition between aqueous and lipid phases and act by different mechanisms in each. Thus, understanding behavior of natural phenolic compounds used as antioxidants in multiphase systems such as foods—and choosing the best antioxidant to fit the system—requires knowledge of how antioxidant action changes with phenol structure and with localization in materials being stabilized. These mechanistic considerations also apply when designing assays for testing antioxidant activity and when running radical quenching reactions in different solvents.

Given space limitations, this chapter can only provide a brief overview of the differences in hydrogen atom versus electron transfer in radical quenching by phenolic compounds.

HAT: H Atom Transfer In this long-recognized mechanism, a hydrogen atom is transferred intact with its bonding electron (i.e., as a radical) from the phenols to an abstracting radical, quenching it and leaving a phenoxyl radical on the antioxidant. Phe-O$^{\bullet}$ then reacts with a second target radical (ROO$^{\bullet}$/RO$^{\bullet}$) to generate a variety of non-radical products, including adducts and decomposition products.

$$ROO^{\bullet}/RO^{\bullet} + \text{Phe-}\mathbf{OH} \rightarrow RO\mathbf{OH}/R\mathbf{OH} + \text{Phe-O}^{\bullet} \quad (3)$$

$$ROO^{\bullet}/RO^{\bullet} + \text{Phe-O}^{\bullet} \rightarrow \text{non-radical products} \quad (4)$$

Importantly, in HAT reactions the H and electron transfer occurs together to or from the same orbital of the same bond of the antioxidant, to the same orbital and bond of the target radical. No ionization, even partial, is involved so this is the dominant mechanism in lipids and hydrophobic and aprotic solvents (Litwinienko and Ingold 2007). The main phenol property driving this mechanism is the bond dissociation energy (BDE) of the phenol O–H. The weaker the bond, the easier it is to release the H atom intact. Structural features that decrease BDE and enhance HAT include electron donor groups (–OH, –OR, –CH=CH–, –R) in *ortho* and *para* position relative to the phenol group (Wright et al. 2001). Most efficient H-donors have dihydroxy functionality in a monophenol or the B ring of polyphenols such as gallic acid, caffeic acid, quercetin, and catechins.

Just as influential, however, is solvent: solvents that are strongly hydrogen bonding or that increase viscosity depress HAT action of antioxidants by blocking the H atom release and slowing movement away from the phenol. Under these conditions, phenolic compounds shift to electron transfer if they have structural capability. On the other side of the story, HAT is not affected by changes in pH and ionization state of the solvent. Both features allow HAT to proceed facilely in lipid phases.

Thermodynamically, HAT is the preferred mode of radical quenching by phenolic compounds. Even so, on a kinetic basis it is a relatively slow and diffusion-controlled process since the H atom must physically move between molecules. Thus, HAT can be outranked by rapid electron transfer with some polyphenols and under some reaction conditions (see next section).

PCET: Proton Coupled Electron Transfer Simplistically, PCET can be thought of as HAT in ionizable solvents. This mode is functionally equivalent to HAT except that electrons and protons transfer from different orbitals on the donor to different orbitals on the acceptor (Mayer 2004). The e^-/H^+ donor orbitals and e^-/H^+ acceptor orbitals interact electronically, enabling simultaneous transfer (Litwinienko and Ingold 2007). The process requires preassociation short-range nature of proton transfer so transfer occurs in three steps: (1) a hydrogen bond forms between the phenolic –OH and an O in the ($ROO^{\bullet}$) radical, (2) proton transfer occurs along the H-bond from the phenol to one lone pair of the free radical O atom while electron transfer occurs from a lone pair of the antioxidant to the SOMO (singly occupied molecular orbital) of the free radical, and (3) the phenoxyl radical and ROOH separate (Di Meo et al. 2013).

While the difference between HAT and PCET may seem to be quibbling with details, a major practical consequence is kinetic. PCET contributes to increases in reaction rates of some phenolic compounds in water compared to lipids or organic solvents and is the only mechanism active at low pH (Di Meo et al. 2013).

SET: Single Electron Transfer In contradistinction to HAT which transfers radicals, SET involves transfer of electrons and protons separately as *ions* from different orbitals. Which transfers first is determined by the ionization potential or energy (IP or IE) and proton dissociation enthalpy (PDE) of the phenol. Lower IE allows easier electron release, lower PDE facilitates release of H^+. The actual sequence of electron and proton transfer for any phenolic compound is influenced first by the relative IE and PDE values, and second by the solvent. Also, because electron transfer occurs via ions, electron transfer requires a solvent that at least partially ionizes and consequently is strongly dependent on pH; it slows to negligible in acid because ionization is repressed (Lemanska et al. 2001).

As noted above, electron transfer is very fast (10^8–10^9 $M^{-1}s^{-1}$) (Steenken and Neta 1982), in part because it is not diffusion limited as is H atom transfer (Litwinienko and Ingold 2007).

Electron transfer in any mode is supported by phenolic structure with extended delocalization and conjugation of pi electrons as well as by resonance established when ring adducts are conjugated with the aromatic system. Notably, these features are found in flavonoids with 2,3 double bonds plus 4-ketone group in the C ring together with *o* and *p* diphenols in the A ring (Leopoldini et al. 2004; Steenken and Neta 1982). Resveratrol, catechin, epicatechin, epigallocatechin, epicatechin gallate, epigallocatechin gallate, methyl gallate, quercetin, hydroxycinnamic acids, and tocopherol have all demonstrated significant electron transfer capabilities (Medina et al. 2007; Musialik et al. 2009; Muzolf et al. 2008).

A number of SET mechanisms have been proposed in the literature, varying mainly in the sequence of electron and H atom transfer and the origin and target locations for transfer. There is considerable inconsistency of definitions, descriptions, and terms cited in the literature, especially between scientific fields. There is also considerable argument over which mechanisms are active. Given space limitations of this chapter, only the SET variants most applicable to phenolic inhibition of lipid oxidation and oxygen radical are described briefly here. The intent is not to argue for or against any mechanism but rather to offer alternatives that explain variations in antioxidant behaviors with solvent (e.g. in antioxidant activity assays or model oxidation studies) and phenol structure.

SPLET: Single Electron Transfer, Sequential Proton Loss Electron Transfer This mode of electron transfer comes into play when reactions occur in solvents that form H bonds with the phenolic –OH. Under such conditions, radical quenching is greatly impeded for phenols capable only of HAT (Litwinienko and Ingold 2007) yet is greatly enhanced with phenolic compounds capable of electron transfer (Foti et al. 2004). The quenching rate increases because now both HAT and SET are active simultaneously, with the electron transfer being much more rapid. Three reaction steps are involved (Litwinienko and Ingold 2007):

1. Ionization of the phenolic compound Phe-OH and transfer of the phenolic H to the solvent forms a phenoxy anion,

$$\text{Phe-OH} \rightarrow \text{Phe-O}^- + \text{H}^+ \ldots \text{S} \tag{5}$$

2. Electron transfer from the radical to the phenolic anion generates a phenoxyl radical and converts the target radical to an anion,

$$\text{Phe-O}^- + \text{RO(O)}^\bullet \rightarrow \text{Phe-O}^\bullet + \text{RO(O)}^- \quad (6)$$

3. Proton transfer from the solvent to the target anion forms the product ROOH or ROH.

$$\text{RO(O)}^- + \text{H}^+ \rightarrow \text{RO(O)H} \quad (7)$$

Structural requirements for SPLET include: (1) the radical must be electron deficient, as in DPPH$^\bullet$ and ROO$^\bullet$ (Litwinienko and Ingold 2007; Musialik et al. 2009), (2) the phenolic group must have a readily abstractable and fairly acidic (low pK_a) H atom, as can be found in the C7 –OH in the A ring and the B ring diphenols of quercetin and other flavonoids (Musialik et al. 2009; Amorati et al. 2017) as well as in tocopherols (Bakhouche et al. 2015), and (3) the solvent must be able to support partial or full ionization of the phenol (Litwinienko and Ingold 2007). Methanol supports weak ionization, non-acidic water supports full ionization, and hydrocarbons do not support ionization at all, so this mechanism primarily describes phenolic actions in polar phases. Although most data on SPLET has been collected by reacting phenols with the stable radicals DPPH$^\bullet$, the mechanism applies equally to peroxyl radical ROO$^\bullet$ quenching (Amorati et al. 2016; Valgimigli et al. 1999).

HAT and SPLET mechanisms are not mutually exclusive. The relative contributions of each depends on the phenol structure and the solvent. Polyphenols, in particular, may quench a given radical by multiple mechanisms, each at different sites. For example, quercetin reacts by HAT through its B ring –OH groups in organic solvents but by SPLET at the acidic 7–OH on the A ring in ionizing solvents (Musialik et al. 2009).

For more information on SPLET, the reader is referred to an extensive body of research from Foti, Litwinienko, Valgimigli, Amorati and others in the laboratory of Keith Ingold (Litwinienko and Ingold 2007; Amorati et al. 2016; Litwinienko and Ingold 2003; Valgimigli et al. 2009; Litwinienko and Ingold 2004; Litwinienko and Ingold 2005).

SET-PT: Single Electron Transfer Followed by Proton Transfer In this mode, the phenolic compound first transfers an electron to the target radical, converting it to an anion and becoming a radical cation in the process.

$$\text{Phe-OH} + \text{ROO}^\bullet \rightarrow \text{Ar-OH}^{\bullet +} + \text{ROO}^- \quad (8)$$

The proton is then released and transferred to the target anion. In the common case of peroxyl radicals, this forms the expected hydroperoxide (Leopoldini et al. 2004; Wright et al. 2001).

$$\text{Ar-OH}^{\bullet +} \rightarrow \text{Ar-O}^\bullet + \text{H}^+ \quad (9)$$

$$H^+ + ROO^- \rightarrow ROOH \tag{10}$$

Much of the early literature describing electron transfer by phenolic antioxidants cited this mechanism and it provides reasonable explanation for radical quenching in neat lipid phases. Nevertheless, whether lipids support weak ionization and electron transfer remains an open question. One would think that the hydrocarbon chains preclude ionization, and this may be true for saturated fatty acids and esters. However, double bonds have enough polarity to support some ionization, and the unsaturated acyl chains of phospholipids contribute to formation of ion channels at interfaces (Cordero-Morales and Vásquez 2018). Tests in our laboratory find that phenolic compounds showing strong and dominant electron transfer in methanol also exhibit highest inhibition of oxidation in acetonitrile and lipid phases while compounds that are strongly HAT in methanol inhibit lipid oxidation only slowly in lipid phases and acetonitrile where HAT should be dominant (K.M. Schaich, unpublished data). Thus, there is reason to question whether electron transfer by phenols can occur in unsaturated lipids in contrast to standard dogma that electron transfers occur only in aqueous and polar phases and HAT occurs in lipid phases. Detailed research is needed to address this issue and understand which actions are possible in lipid phases. The difference is important in interpreting behavior of different (poly)phenols in various antioxidant activity assays and in different food and reaction systems.

3.1.2 Reactions of Quinone Forms of Antioxidants

Lipid radical quenching is generally considered to stop with reactions of phenols. However, phenols oxidize in the radical scavenging process, yielding quinones as major oxidation products. Not as much is known about quinone reactions, but they do contribute to radical flow to both inhibit and produce radicals.

Inhibition of Lipid Oxidation by Alkyl Radical Addition Quinones stop radical chains mostly by radical addition. Phenols react with alkyl radicals only slowly ($k \sim 10^2$–10^4 $M^{-1}s^{-1}$), strongly preferring oxyl radicals ($k = 10^8$–10^9 $M^{-1}s^{-1}$) (Franchi et al. 1999). In contrast, alkyl radicals readily add to quinones of any structure, such as are found in hydroquinone, tocopherol, chlorogenic acid, quercetin, flavonols with vicinal diphenols in B ring, catechol, and catechins. Alkyl radical additions are fast enough to be competitive with H abstractions, e.g. benzoquinone adds 5-hexenyl radicals at a rate of 2×10^7 L $m^{-1}s^{-1}$ under nitrogen (Citterio et al. 1979) in degassed acetonitrile-acetic acid and adds hydroxyethyl radicals in undegassed cyclohexane 1.9×10^8 L $m^{-1}s^{-1}$ (Maroz and Brede 2003).

As long as open positions are available on the ring, radicals add to the ring ortho to one carbonyl, as shown here for addition of hydroxyethyl radicals to benzoquinone (Maroz and Brede 2003). The resulting semiquinone radical can react with another alkyl radical to form a dimer addition product, RQR (Denisov 2006).

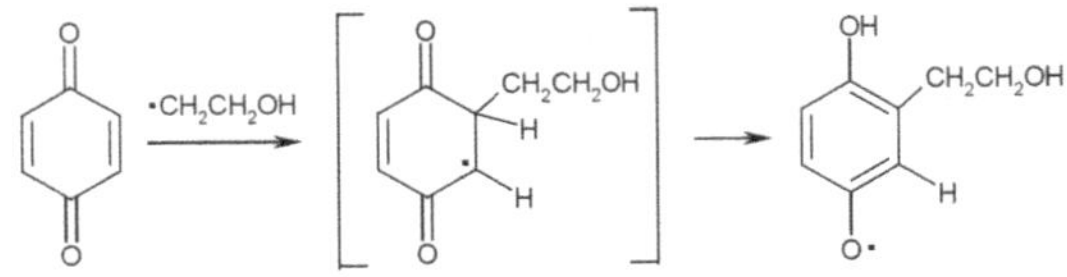

(11)

When ring positions are blocked, as in many natural polyphenols, radicals can add to the oxygen (Kumli et al. 2006), as shown in the structure below. Additions to *o*-quinones are preferred over *p*-quinones (Denisov 2006).

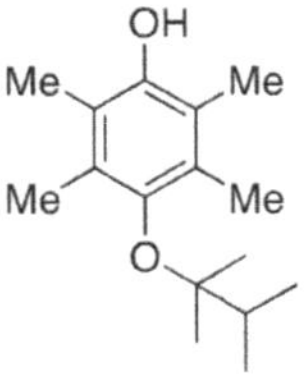

This action provides a pathway for synergism between quinone and hydroquinone forms of antioxidants: the hydroquinone quenches electrophilic peroxyl radicals by H atom transfer while its oxidation product quinone can continue to inhibit lipid oxidation by covalently trapping nucleophilic alkyl radicals. Alkyl radical addition to quinones may play a particularly important role under the low oxygen conditions in which oils and foods are packaged and stored.

Inhibition of Lipid Oxidation by Electron Transfer Quinones are very susceptible to redox reactions: they can be oxidized to radical cations ($SQ^{\bullet+}$) by reagents with high redox potentials such as the hydroxyl radicals ($HO^{\bullet}$) that are produced in the presence of metals and during heating of oils and foods,

$$Q + {}^{\bullet}OH \rightarrow SQ^{\bullet+} + H_2O \tag{12}$$

or reduced to semiquinone radical anions ($SQ^{\bullet-}$) by superoxide anion ($O_2^{-\bullet}$), thiols, ascorbic acid, and similar reducing agents (Maroz and Brede 2003).

$$Q + O_2^{-\bullet}/HOO^{\bullet} \leftrightarrow O_2 + SQ^{-\bullet} \tag{13}$$

Quinone cycling of superoxide anion back to oxygen can contribute significantly to oxidative stability by preventing accumulation of $O_2^{-\bullet}/HOO^{\bullet}$ which dismutates to H_2O_2 that, in turn, decomposes to the extremely reactive hydroxyl radical, $HO^{\bullet}$ (see Fig. 1).

The intermediate semiquinone anions ($SQ^{-\bullet}$) are most important as antioxidants because they reduce oxygen to $O_2^{-\bullet}$ (Rx 14), $O_2^{-\bullet}$ to H_2O_2 (Rx 15), and peroxyl radicals to anions (Rx 16), reducing production and accumulation of initiating $LOO^{\bullet}$ and hence limiting the overall primary radical load on available antioxidants. In the presence of transition metals, however, Rxs 14 and 15 can become pro-oxidant.

$$SQ^{-\bullet} + O_2 \leftrightarrow Q + Q_2^{-\bullet} \tag{14}$$

$$SQ^{-\bullet} + O_2^{-\bullet} \xrightarrow{2H^+} Q + H_2O_2 \tag{15}$$

$$SQ^{-\bullet} + ROO^{\bullet} \rightarrow Q + ROO^{-} \tag{16}$$

Enhancement of Lipid Oxidation: Radical Production by Electron Transfer Quinones have the potential for becoming pro-oxidant at high pO_2, especially when ascorbic acid (Asc) is present, as occurs in most phenolic extracts from fruits. Under these conditions, reducing activity of the ascorbic acid maintains a supply of semiquinone radical anions in equilibrium with the quinones and this establishes a redox cycle that continuously reduces oxygen to reactive oxygen species (Li et al. 2012). The same activation occurs when quinones are reduced enzymatically in vivo (Metodiewa et al. 1999).

$$Asc^{-} + Q \longrightarrow Asc^{\bullet} + SQ^{-\bullet} \xrightarrow{O_2} Q + O_2^{-\bullet} \xrightarrow{O_2^{-\bullet}} H_2O_2 \longrightarrow HO^{\bullet} + {}^{-}OH \tag{17}$$

3.2 *Shifts in Propagation Pathways*

3.2.1 Modification of Primary Lipid Oxidation Products

Phenol Reactions that Increase LOOH

Inhibition of LOO$^{\bullet}$ β-Scission and Addition to Double Bonds Initial peroxyl radicals formed in lipid oxidation have more options for reaction than merely abstracting H atoms from neighboring molecules to form hydroperoxides. Competing with H abstraction are β-elimination of the peroxyl oxygen to regenerate the original alkyl radical, addition to double bonds to form peroxydimers that decompose to epoxides and alkoxyl radicals (Rx 18), internal rearrangements to epidioxides, and dismutation of the LOO$^{\bullet}$ (Schaich 2020; Schaich 2005). All of these alternate reactions are constantly active, especially when H donors are limited (as in low lipid concentrations) or not readily available (as in all cis unsaturated fatty acids) so LOO$^{\bullet}$ does not immediately find an H partner. Oxidation continues but by pathways other than H abstraction, generating different products while hydroperoxides form at much lower levels (Schaich 2020). When only peroxide values are measured, lipid oxidation appears to be low. Phenols shift the competition in this mix. Because they donate H atoms much faster than do lipids (Jovanovic et al. 1992; Erben-Russ et al. 1987) (Table 2), they can block both the loss of LOO$^{\bullet}$ oxygen by β-scission (Weenan and Porter 1982; Porter and Wujek 1987) and the formation of epoxides by LOO$^{\bullet}$ addition to double bonds (Rx 18a) (K.M. Schaich, unpublished

Table 2 Rate constants for alternate reactions of lipid peroxyl (LOO$^\bullet$) and alkoxyl (LO$^\bullet$) radicals compared to H abstraction by phenolic compounds (adapted from Schaich (2005); used with permission)

		LOO$^\bullet$	RO$^\bullet$	HO$^\bullet$
H abstraction, LH	nonpolar organic	<1–400 $M^{-1}s^{-1}$	10^4–10^7 M^{-1} s^{-1}	
	polar, aqueous		10^6–10^8 L $M^{-1}s^{-1}$	
Tocopherol	**aqueous**	**10^7 [c]**		**0.1–2 × 10^8 $M^{-1}s^{-1}$ [f]**
	nonpolar organic	**10^6 [d]**	**10^9 [e]**	
H abstraction, LOOH	nonpolar organic	600 $M^{-1}s^{-1}$	2.5 × 10^8 $M^{-1}s^{-1}$	
	polar, aqueous	NA	NA	
Internal cyclization[a]	nonpolar organic	10^1–10^3 s^{-1}	10^4–10^5 s^{-1}	
	polar, aqueous	NA	NA	
Addition to double bonds	nonpolar organic	NA	10^4–10^8 $M^{-1}s^{-1}$	
	polar, aqueous	NA	NA	
β-scission[b]	oleate	1–8 s^{-1}	10^3–10^5 s^{-1} org	
	linoleate	27–430 s^{-1}	10^4–10^5 s^{-1} polar org	
			10^6–10^7 s^{-1} aq	
Dismutation	nonpolar organic	10^6–10^9 L $M^{-1}s^{-1}$	10^9–10^{10} $M^{-1}s^{-1}$	
	polar, aqueous	10^7–10^8 L $M^{-1}s^{-1}$	NA	
	oleate-OO$^\bullet$	10^6 $M^{-1}s^{-1}$		

NA not available
[a]LOO$^\bullet$ cyclization to epidioxides, LO$^\bullet$ cyclization to epoxides
[b]Scission of O_2 from LOO$^\bullet$, scission of the acyl chain on either side of LO$^\bullet$ to release alkanes and carbonyls
[c](Simic 1981)
[d](Warren and Mayer 2010)
[e](Snelgrove 2000)
[f]Multiple phenol donors (Bors and Michel 1999)

data), increasing the formation of LOOH and toc-peroxyl radical adducts (Rx 18b) (see LO(O)$^\bullet$ and L$^\bullet$ Add to Phenols section below). Now when only peroxide values are measured to quantitate lipid oxidation, this effect makes phenols *appear* to be pro-oxidant.

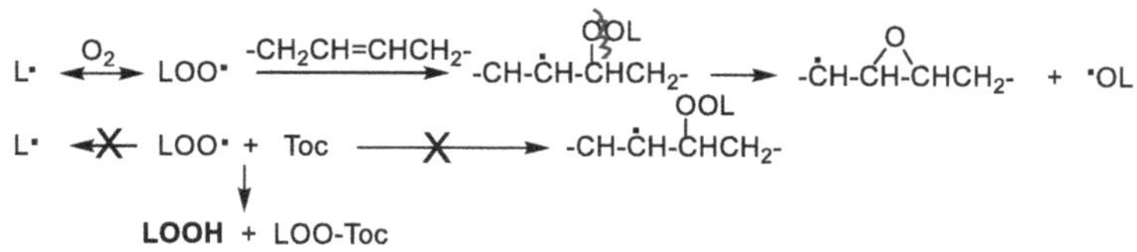

(18)

(18a)

(18b)

An interesting side effect of β-scission of oxygen is a shift in conjugated double bond configuration of linoleic acid from trans-cis to trans-trans (Porter et al. 1981). The presence of phenolic antioxidants prevents this isomerization.

Stabilization of LOOH by H bonding with Phe-OH Phenols also establish H-bonds with hydroperoxides (Mäkinen and Hopia 2000), stabilizing these initial products. While this slows progression into rapid propagation by faster reacting alkoxyl radicals, it also exerts an apparent pro-oxidant effect when only peroxide values are measured.

It seems reasonable to expect that enhancement of LOOH by blocking alternate reactions of $LOO^•$ and by stabilizing LOOH very likely contribute to apparent pro-oxidant activity at high phenol concentrations. However, more information is needed to establish concentration dependence of these effects, as well as what happens to other lipid oxidation products in the process so that procedures can be developed to distinguish these actions of antioxidants in differential product distributions.

Phenol Reactions that Decrease LOOH

$LO(O)^•$ and $L^•$ Add to Phenols As noted several times in this chapter, phenolic antioxidants do not just quench radicals in lipid oxidation—they have multiple mechanisms for controlling radical chains reactions and these shift in importance with reaction conditions. Thus, antioxidant mechanisms most active in oils during room temperature storage can be replaced with other actions during heating, when lipids are dispersed in dry systems, when they are organized in bilayers or membranes, or when they oxidized in model systems in different solvents. This section focuses on phenol removal of radicals by forming covalent adducts with them.

Pulse radiolysis studies show that at high pH $LOO^•$ addition to kaempferol and quercetin to form covalent adducts is very fast ($k > 10^7$ L $mol^{-1}s^{-1}$) and addition to corresponding phenoxyl radicals is even faster ($k > 10^8$ L $mol^{-1}s^{-1}$) (Erben-Russ et al. 1987). While it may be argued that reactions under these extreme conditions are not representative of normal conditions (reviewer comments) or that lipid radical addition to phenolic rings should not be significant due to steric hindrance from the long lipid chains (Kamal-Eldin and Appelqvist 1996), there is, nevertheless, substantial evidence for addition of lipid radicals and other bulky peroxyl and alkoxyl radicals to both mono- and poly- phenolic compounds (Amorati et al. 2016; Yamauchi 2007).

Early studies first demonstrated the phenomenon in reactions of 2,6-di-tert-butyl-p-cresol (butylated hydroxytoluene, BHT) with t-butyl peroxyl radicals (Campbell and Coppinger 1952; Tsuchiya et al. 1983) and peroxyl radicals from decomposition of AIBN (azoisobutyronitrile) (Boozer et al. 1955). That the stoichiometry was about two peroxyl radicals quenched per phenol (Horswill et al. 1966; Horswill and Ingold 1966) and peroxyl radical-phenol adducts accounted for half of the products (Boozer et al. 1955) led to the proposal that two reactions occur in sequence (Rx 19): (1) BHT quenches the first $ROO^{\bullet}$ by donating an H atom to form ROOH and a phenoxyl radical that becomes resonance-stabilized over the aromatic ring, and (2) a second $ROO^{\bullet}$ adds to a radical site in the ring, in this case at C4.

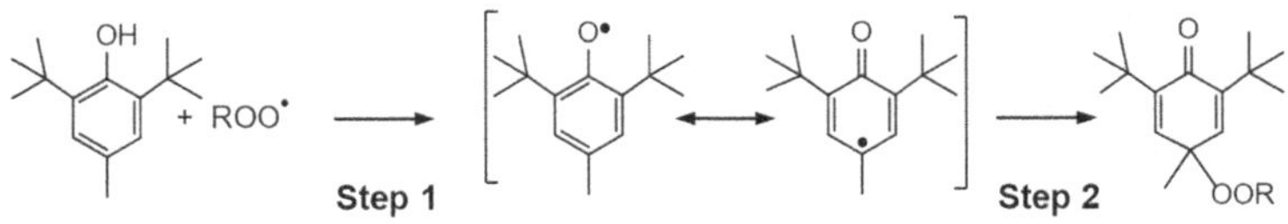

(19)

With increasing phenolic complexity, radical adducts of tocopherols formed via phenoxyl radicals in the same way (Rx 20) have been documented extensively. Alkyl radicals add to phenoxyl radicals $TO^{\bullet}$ (Urano et al. 1977; Gardner et al. 1972) (Rx 20a, left), but peroxyl radicals add to resonance radical ($T^{\bullet}$) at position 8a (Rx 20a, right) as has been shown in γ-tocopherol (Yamauchi et al. 1990a), δ-tocopherol (Yamauchi et al. 1990b), and α-tocopherol (Yamauchi et al. 1995). Alkoxyl radicals also add at C-8a of tocopherol (Altwicker 1967). This pattern has been demonstrated with peroxyl radicals from the azo initiator 2,2′-azobis (2,4-dimethylvaleronitrile) (AMVN) (Yamauchi et al. 1990a, 1990b), t-butyl hydroperoxide (Tsuchiya et al. 1983), alkyl radicals (Urano et al. 1977; Urano and Matsuo 1976), and alkyl and oxyl radicals from oxidizing lipids (Yamauchi 2007; Gardner et al. 1972; Yamauchi et al. 1995; Liebler et al. 1991) in model systems and the comparable products have been detected in natural materials (Yamauchi 2007).

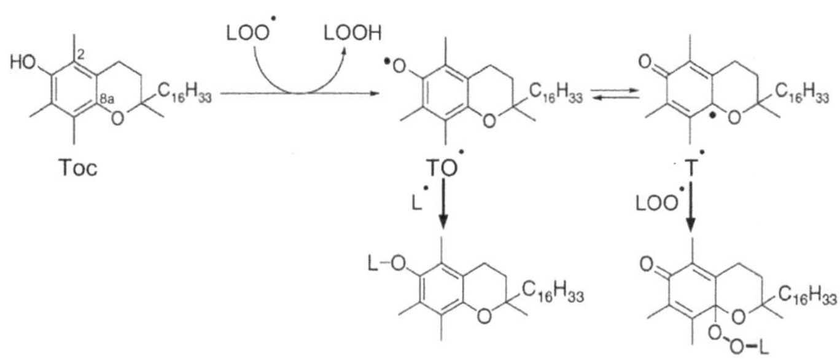

(20)

(20a)

Addition at C-2 (ortho to the phenol) occurs to a lesser extent (Rx 21). Stable adducts from methyl radicals have been observed (Urano et al. 1977; Urano and

Matsuo 1976), but peroxyl adducts here transform rapidly to other products (Kamal-Eldin and Appelqvist 1996; Yamauchi et al. 1996).

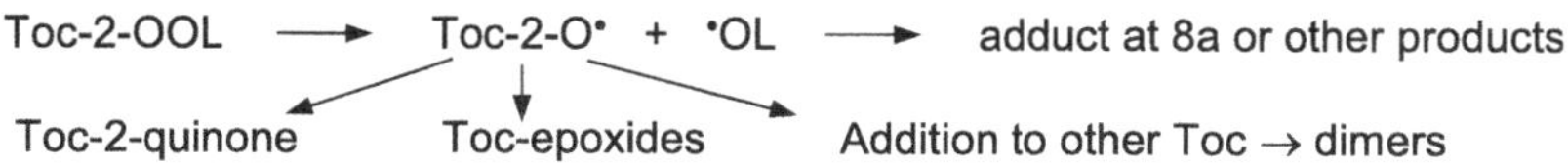

(21)

How much lipid radical addition to phenols contributes to product mixes is determined by competition between H abstraction and the various addition reactions, and this in turn is influenced strongly by reaction conditions including physical matrix, temperature, oxygen content, and solvent. The differences described below point out the importance of not expecting antioxidants to act by the same mechanisms in all types of solutions, food systems, or biological materials. For example, when linoleic acid was oxidized with α-tocopherol in a monolayer on silicic acid at 80 °C, 40% of products were 1:1 chromanol adducts (shown below) resulting from addition of the resonance-stabilized radical at C-2 in the aromatic ring to a double bond of the fatty acid, followed by ring closure with the phenolic O (Porter et al. 1971). The authors proposed that the solid system did not restrict reactivity but greatly reduced mobility of both lipid and phenol. Steric restrictions or physical separation nearly eliminated formation of normal tocopherol quinones and dimers but facilitated radical reactions at less favorable sites.

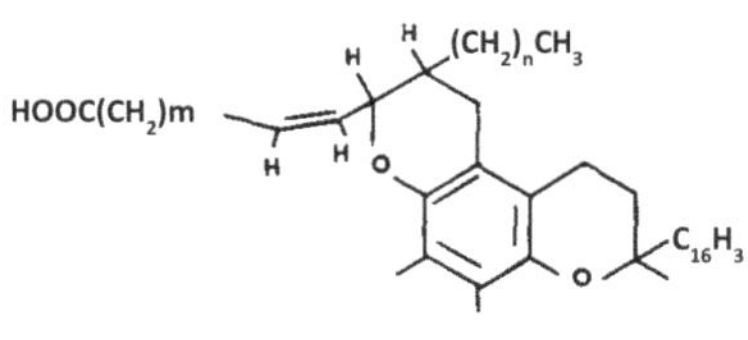

In contrast, when tocopherol was reacted with radicals in AMVN-stimulated phospholipid bilayers in liposomes (Liebler et al. 1991), the principal product was lipid peroxyl adducts at 8a but a second major product was 2,3-epoxy,8a-hydroperoxy tocopherol (Rx 22). The authors attributed this to autoxidation of the tocopherol because AMVN loss did not correlate. However, a more reasonable explanation for this product is AMVN stimulation of lipid oxidation in the bilayers, then *lipid* peroxyl radicals add to the C2 double bond, followed by O–O scission to form the epoxide and a radical on C8a. Addition of oxygen and abstraction of an H atom from another tocopherol then forms the lipid hydroperoxide (Rx 20). An analogous addition of LOO$^{\bullet}$ to lipid double bonds (see Rx 18) is responsible for early production of epoxides at high levels during lipid oxidation (Schaich 2020).

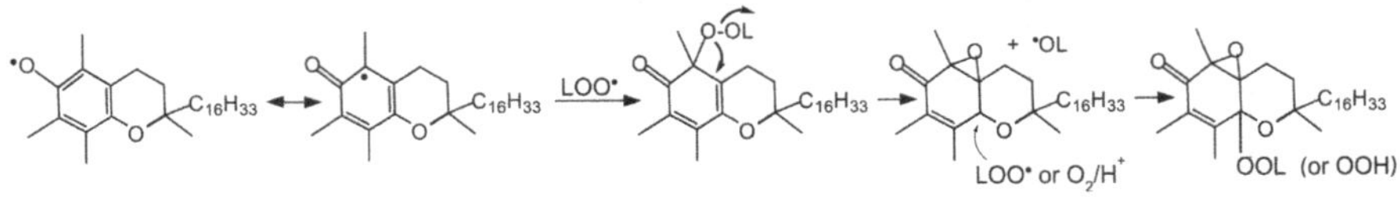

(22)

The 2,3-epoxy,8a-hydroperoxy product was observed again in bulk methyl linoleate (ML) oxidized with α-tocopherol in air at different temperatures. At 37 °C, dominant products were a mixture of the epoxy-hydroperoxy tocopherol (Rx 22) and the 8a LOO$^•$ adduct (Rx 20a, right) (Yamauchi et al. 1995), but at 60 °C the major products were adducts formed by addition of a ML-C13$^•$ or –C9$^•$ radicals to the tocopherol –OH (structure shown below). Similar alkyl radical adducts (mixtures of positional and geometric isomers) were dominant products when MLOO$^•$ and MLO$^•$ radicals were reacted with α-tocopherol under low or no oxygen; these disappeared in the presence of air as phenol oxidation to quinones became the favored reaction (Gardner et al. 1972).

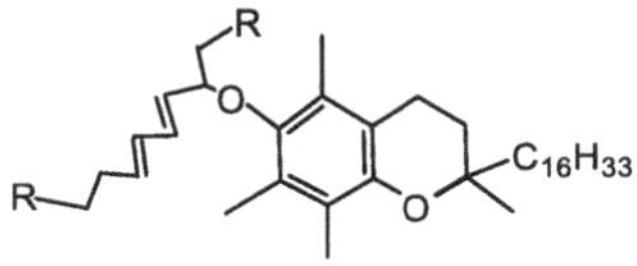

Polyphenols also add lipid radicals. Genistein reacted with ROO$^•$ from AMVN in acetonitrile 50 C 3 hrs, found 3 different adducts from LOO$^•$ and LO$^•$ at 1′ in the B ring (Arora et al. 2000).

Finally, while adduct formation reduces measured hydroperoxides, classifying this as an antioxidant process, the resulting formation of quinones is a two-edged sword. Quinones react with carbon-centered radicals, R$^•$ in general or L$^•$ in lipid acyl chains, at rates fast enough to be competitive with O_2 addition to form peroxyl radicals, so especially under low oxygen conditions, quinones can stop the radical chain at the L$^•$ stage and be strongly synergistic with its phenol form (Mellors and Tappel 1966). At the same time, the presence of reducing agents such as ascorbic acid or phenolic compounds with lower redox potentials cycle the quinone backwards to the semiquinone then to the parent phenol. Again, under low oxygen this can be a protective action, regenerating the phenol and extending its useful antioxidant lifetime. However, as shown in Section 3.1.2, under high pO_2 and in the presence of ascorbic acid or phenols with low redox potential, these quinones may become pro-oxidant. This is perhaps still another mechanism that contributes to pro-oxidant effects observed at high antioxidant concentrations.

Several additional questions must be raised about the impact of lipid radical complexes with antioxidants. Do the adducts have any toxic potential? Do the quinone forms contribute to any browning condensations or other reactions, either via phenolic browning or via reactions with proteins? Do they have any flavor impacts in foods? These issues need to be investigated as use of natural antioxidants increases.

4 Alteration of Products

4.1 *Formation of Adducts with Lipid Carbonyls*

Secondary products of lipid oxidation seldom receive attention in antioxidant strategies, presumably under the assumption that if radicals are controlled, these will not be formed. This is a short-sighted viewpoint since antioxidants are never perfectly effective and secondary products have critical effects on sensory properties, nutritional value, functionalities, and potential toxicity of oils and foods. Lipid carbonyls provide key off-odors and flavors that are sensory cues indicating rancidity for consumers. Dicarbonyls are also critical intermediates and commitment points for several damaging reactions including browning reactions (Zamora and Hidalgo 2005) and Strecker degradation (Zamora 2011) and are also important stress inducers in vivo (Rabbani and Thornalley 2015). Importantly, phenols also block formation of toxic compounds evolving from carbonyl-amine interactions, e.g. heterocyclic aromatic amines (Zamora et al. 2013; Lee et al. 2020). Thus, secondary products need to be controlled along with lipid radicals. Interestingly, phenolic compounds can provide this function by forming covalent adducts with carbonyls.

Recognition that phenol trapping of carbonyls could be an active mechanism for protection in biological systems and in foods began when epicatechin was observed to block Maillard reactions of sugars (Totlani and Peterson 2006). Detailed chemical analyses revealed that the epicatechin reacted with aldehyde forms of sugars as well as with secondary aldehydic products of the browning reactions and Strecker degradations. After determining the structures of the complexes, the authors proposed that reaction most likely occurred via electrophilic substitution at C6 or C8 of the epicatechin A ring:

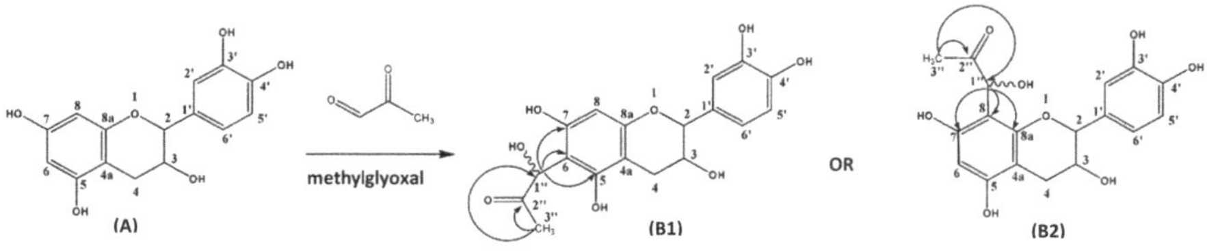

(23)

Understanding the potential importance of removing aldehydes from reaction flows, they also proposed that complexation of aldehydes, including those from oxidized lipids, may be a critical fourth pathway by which phenols limit oxidations.

The idea caught on and other studies investigating structural requirements (both phenol and aldehyde) for reaction soon followed. A large number of natural phenolic compounds with different structures were screened to determine which chemical features were necessary for reaction with aldehydes, using acrolein and 4-hydroxynonenal (HNE) as models (Zhu et al. 2009a).

$$\text{Acrolein}: \ CH_2 = CH - CHO \qquad \text{HNE}: \ CH_3(CH_2)_4CH(OH)CH = CH - CHO$$

Reacting 0.5 mM aldehyde with 0.1–0.2 mM phenol in phosphate buffer pH 7, only flavan-3-ols, theaflavins, cyanomaclurin, and dihydrochalcones effectively trapped acrolein and hydroxynonenal, and they did this by working as sacrificial nucleophiles. The critical structural feature was the phloroglucinol moiety in the A ring. The most reactive phenolic compound was phloretin.

Phloroglucinol

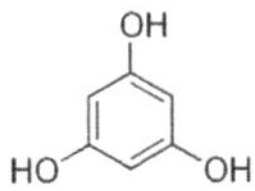

Phloretin

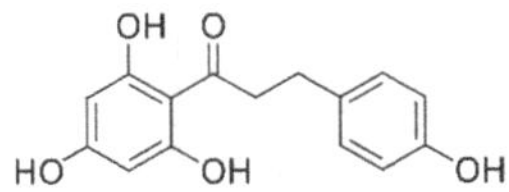

Phloretin has three electron-donating HO– groups in the A ring, two of which are in *meta* configuration relative to the main phenol. This generates electron-rich centers on the unsubstituted carbons, which facilitates substitution from electrophiles such as aldehydes. Absence of an electron-withdrawing C=O on the C ring is an additional requirement in flavonoids, and substitution on any ring –OH also reduces activity. In all reactive polyphenols, the main addition sites were in the A ring at positions 6 and 8 (Zhu et al. 2009a).

Using resorcinol as a simpler model (single ring meta diphenols), Hidalgo and Zamora confirmed that *m* phenol configuration is an *absolute requirement* for reaction with carbonyls and went on to identify the reason and mechanism. With the –OH groups at 1 and 3 in *m* diphenols, electron delocalization activates positions 2, 4, and 6 for loss of a proton. This gives *m*-diphenols two sites for reaction with aldehydes (Hidalgo and Zamora 2014). They further showed that the presence of groups increasing nucleophilicity of the phenol enhance aldehyde trapping while groups increasing steric hindrance without increasing nucleophilicity depress carbonyl trapping (Hidalgo et al. 2017a). Taken together, these two studies suggest that *m*-diphenols such as resorcinols and polyphenols with *m*-diphenols in the A ring (e.g. quercetin, catechins) should be capable of reacting with lipid carbonyls.

The preference for meta configuration of ring hydroxyls relative to the primary phenol is quite interesting. Section 3.1.1 explained how ortho and para phenols were usually most active as radical scavengers and meta phenols were relatively inactive. The reverse appears to be true for carbonyl complexation. Rather than being a detriment, such opposite actions provide a new mechanism for synergism between phenolic compounds with different structures.

A broad range of carbonyls derived from lipid oxidation react with compounds having *m*-phenol structures. In saturated aldehydes, the aromatic double bond alpha to the phenol group adds to the aldehyde carbonyl, forming a hydroxyl derivative (Rx 24) (Hidalgo et al. 2017a; Zamora and Hidalgo 2018).

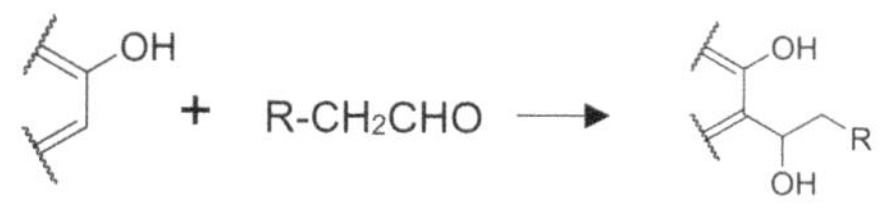

(24)

In unsaturated aldehydes, phenols add to the alkenal C=C double bond either through the phenol –OH or the phenol double bond at 4, as shown in Fig. 2 (Hidalgo and Zamora 2014). Four classes of adducts were isolated from reactions of resorcinol congeners: 2H-chromenols, chromandiols, chromanols, and dihydropyrano[3,2-g] chromenes. The dominant product class was 2H-chromenols, but these were unstable—they participated in browning reactions and polymerized. Chromandiols were the most stable products but were produced at much lower concentrations (Hidalgo and Zamora 2014). This observation raises a very important issue to be remembered in monitoring oxidation—*products detected reflect what is most stable, not all that were produced.*

Hydroxyalkenals reacted with the same phenols produce similar products—chromane-2,7-diols and 2H-chromen-7-ols—plus dihydrobenzofuranols (Hidalgo and Zamora 2019; Zamora and Hidalgo 2016). Oxo-alkenals formed a number of adducts involving multiple addition sites on both the phenol and carbonyl compound. However, only one carbonyl was complexed and the other remained free and reactive, which probably accounted for the unstability of the adducts and tendency to rapidly polymerize (Hidalgo et al. 2018). Epoxyalkenals reacted rapidly with resorcinol by opening the epoxide ring and adding to one phenol group (Rx 25). The aromatic carbon α to the adducted –OH then adds to the β-carbon of the double bond

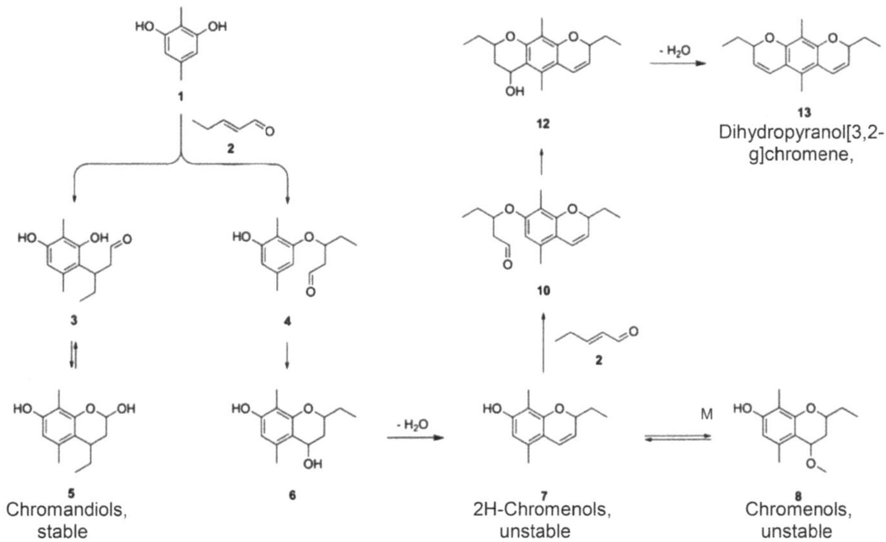

Fig. 2 Products from addition 2-pentenal addition to dimethyl resorcinol showing points of initial addition, cyclization, dehydration, and polymerization. Adapted from (Hidalgo and Zamora 2014); used with permission

in the alkenal to form a ring (Rx 25a, stage two). The products have two reactive groups—a phenolic –OH and a carbonyl from the alkenal, both with muted activity. The phenol retains some radical quenching, but more importantly, the remaining carbonyl has greatly reduced reactivity compared to the original epoxide (Zamora et al. 2017).

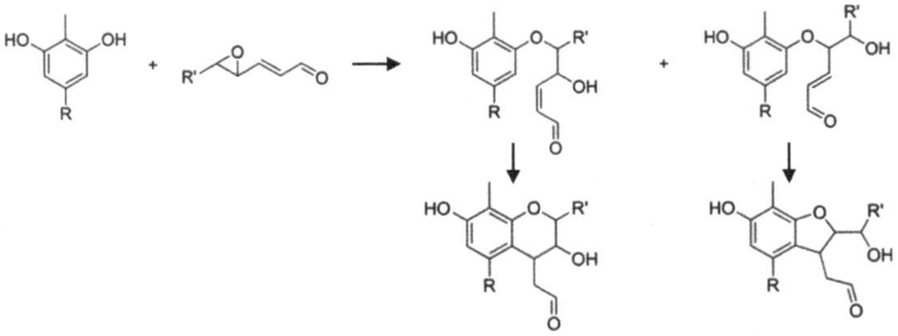

(25)

(25a)

Conditions Required for Phenol-Carbonyl Complexation As with carbonyl-amine condensations in Maillard reactions, formation of carbonyl-phenol adducts increases with pH, up to about pH 7-8; it also increases linearly with temperature (Hidalgo and Zamora 2019; Hidalgo et al. 2018). Interestingly, in foods this puts phenolic antioxidants in place to accept carbonyls under conditions where they are generated in highest concentrations. A particularly important point considering the heat requirement is that phenol reactions with lipid-derived carbonyls are faster than carbonyl-amine reactions (Hidalgo et al. 2017b), so this process removes problematic lipid oxidation products while leaving amino compounds intact to maintain protein functionality and radical quenching capability.

Complications of Phenol-Carbonyl Complexation Observations of aldehyde complexation raise attractive possibilities for deliberately directing phenolic complexation of undesirable carbonyls to control off-flavors and odors of lipid oxidation (Hidalgo et al. 2017a). However, it is instructive to remember that no chemistry occurs in isolation so here are some issues to keep in mind when targeting aldehydes with phenolic compounds.

Browning. Hidalgo and Zamora noted browning side reactions in their studies (Hidalgo and Zamora 2014). Another study of reactions between several simple aldehydes (acetaldehyde, glyoxylic acid, furfural, 5-hydroxy-methylfurfural) and polyphenols catechin, epicatechin, and malvidin 3-O-glucoside found colorless ethyl-linked flavanol oligomers and colored flavanol-anthocyanin adducts (Es-Safi et al. 2002). Is increased browning a trade-off of carbonyl blocking? Investigations to date have been qualitative rather than quantitative. More research is needed to determine the extent to which phenol carbonyl complexation contributes to browning and other color changes in any products where lipids are oxidizing.

Competition from carbonyls in other molecules. Side reactions with carbonyls in other molecules is a second complication of phenol-carbonyl complexation. For example, quercetin complexed aldehyde groups in starch, which altered starch functionality. Some antioxidant activity was retained but levels were much lower than with quercetin alone (Yong et al. 2020). Starches are known to "scalp" flavors by binding them in foods. Potential for antioxidant scalping by food molecules also needs to be investigated. Such action may provide one explanation of how antioxidants are often less effective in foods than in solution.

Removal of desirable as well as rancid flavors from aldehydes. Carbonyls from lipid oxidation are chiefly responsible for "rancid" off-flavors and odors and as such provide critical sensory cues for consumers. At the same time, low levels of compounds derived from lipid carbonyls also contribute some characteristic and desirable flavors. Research is needed to identify levels of lipid carbonyls that need to be removed to limit browning and formation of toxic products while also retaining flavors in foods.

Applications of Phenol-Carbonyl Complexation Application of quercetin successfully trapped toxic aldehydes acrolein and crotonaldehyde as well as pentenal spiked into oils during frying. The same adducts formed here were found in oils used for general frying of foods, indicating that phenols may reduce toxic carbonyls by complexation as well as slowing degradation of frying oils (Zamora et al. 2016). Similarly, catechins in green and black tea extracts almost completely eliminated levels of acrolein and malonaldehyde in oxidized seal blubber oil (Zhu et al. 2009b). However, complexation with acetaldehyde and propanal was only about half or less. More quantitative studies are needed to differentiate which lipid carbonyl products are preferentially trapped and under what conditions.

5 Phenol Effects on System Oxidation Beyond Lipids

Lipids oxidize alone only in oils. In foods and all other products, oxidizing lipids interact with other molecules in the system and they transfer oxidation to those molecules. Thus, except in oils, we really must address *system* oxidation rather than just lipid oxidation to stabilize the product. This structure holds for biological systems as well. In foods and living tissues, lipids are closely associated with proteins and the interactions between these two classes of molecules is extensive, particularly in oxidation. Oxidizing lipids co-oxidize proteins by transferring radicals and by reacting lipid epoxides and carbonyls with protein amino, sulfhydryl, and tryptophan groups (Schaich 2008). Protein quenching of lipid radicals stops chain reactions in lipids but the radical reactions don't stop—radicals move to proteins and initiate reactions there. Reactions of lipid epoxides and carbonyls with proteins moves lipid oxidation products out of the analytical stream and reduces lipid off-odors and flavors but replaces lipid degradation with protein degradation that affects protein functionality and overall food product qualities, including flavor, texture, and color. Thus, lipid and protein oxidation are inextricably linked in foods and biological tissues.

The concepts of protein co-oxidation and total system oxidation have been slow to catch on in the food industry, but they nevertheless are critically important for designing stabilization approaches and optimizing antioxidant effectiveness for *systems*, not just lipids. Proteins need protection against oxidation, just as do lipids, and fortunately proteins can get this protection from phenols, as do lipids. The problem is, what happens when phenols are protecting both molecules? Do phenol interactions with proteins divert radical quenching activity away from oxidizing lipids, and will such interaction add to or alleviate protein modification by lipids? The last section of this chapter outlines phenol-protein reactions (non-covalent, covalent, and chemical reactions not involving bonding) and discusses how these actions of phenols move antioxidant action beyond lipids to potentially contribute to total system stabilization.

5.1 Phenol Associations with Proteins via Noncovalent Interactions

Phenolic compounds link to proteins non-covalently through expected hydrogen bonds to free amino groups and thiols as well as pi (aromatic) bonding and hydrophobic associations with tryptophan and other aromatic amino acids (Rawel et al. 2001a; Rawel et al. 2005; Hasni et al. 2011; Kanakis et al. 2011). Less recognized are charge interactions between proteins and ring substituents, particularly phenolic acids and charged amines. Although reversible and sometimes weak, these non-covalent associations nevertheless play several important roles.

- They block access to and from protein groups, thus limiting access of lipid oxyl radicals and secondary carbonyl products (protective) to amino acids but also impeding H atom transfer from proteins to quench lipid radicals (potentially destructive). This action is analogous to inhibition of HAT radical quenching of phenols in H-bonding solvents.
- Non-covalent bonds at key residues are a critical first step for covalent interactions and their consequences: they disrupt protein tertiary structure to increase access of reactive amino acids to phenolic compounds and bring phenols into position for covalent binding. They also reconfigure secondary structures that control functions such as emulsification and gelling. Phenol associations with tryptophan are observed in almost every study and appear to the center pin of this action (Skrt et al. 2012)
- The balance between various non-covalent associations controls the molecular environment for phenol-protein interactions, varying with pH and amino acid dissociation state, with protein structure and amino acids available, and with phenolic structure that governs susceptibility of the ring to additions. This local

environment in large part explains why some proteins are more susceptible, why some phenolic compounds are more active, and why some reaction conditions in foods favor phenol complexation of proteins.

Non-covalent associations between phenols and proteins alone are sufficient to modify protein functionalities such as gelling, emulsification and foaming, and digestibility, and there is a reasonable body of literature addressing these changes. For the purposes of this chapter, however, the key issue is whether phenol-protein associations interfere with antioxidant activity. Although experiments have attempted to measure complexing effects, the question of activity remains open. Testing antioxidant activity in any of the *in vitro* screening tests such as DPPH and TEAC (Trolox equivalent antioxidant capacity) is highly questionable due to steric hindrance, solubility issues, and electron transfer dominant mechanisms (Schaich et al. 2015). Definitive studies are needed to distinguish whether non-covalent association with phenols protects proteins but derails radical quenching away from oxidizing lipids, whether the net effect is reduction in net system oxidation, or other more complex impacts result.

5.2 *Quinone Covalent Reactions with Proteins*

Nearly every study investigating phenol interactions with proteins has found that phenol oxidation to quinones is required for covalent binding of these two molecules. The partial charges that are established in the ring and on the quinone oxygens then prime the molecule for additions via Michael addition of RSH and RNH_2 to ring double bonds and via Schiff base condensation of RNH_2 with carbonyls of quinones (Rohn 2014). Both additions need the oxidized phenol (quinone or semiquinone) to temporarily accept protons while amines or thiols are adding to the ring.

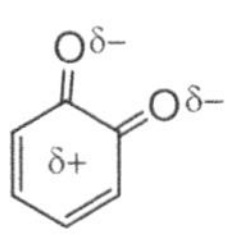

+ Nucleophile → Products (26)

Michael Addition of Protein Thiols and Amines to Aromatic Double Bonds Protein thiols (Rx 27) (Bassil et al. 2005) and amines (Rx 28) (Ozdal et al. 2013; Rawel et al. 2001b) add rapidly to the β-carbon of the double bond next to C-1 carbonyl of quinones. Because they are more strongly nucleophilic, thiols generally add with greater facility than amines (Li et al. 2016). Notice that the products retain the quinone function, and this can undergo further radical quenching or protein binding reactions.

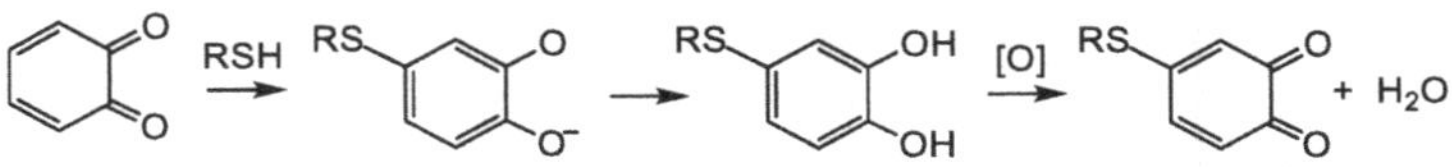
(27)

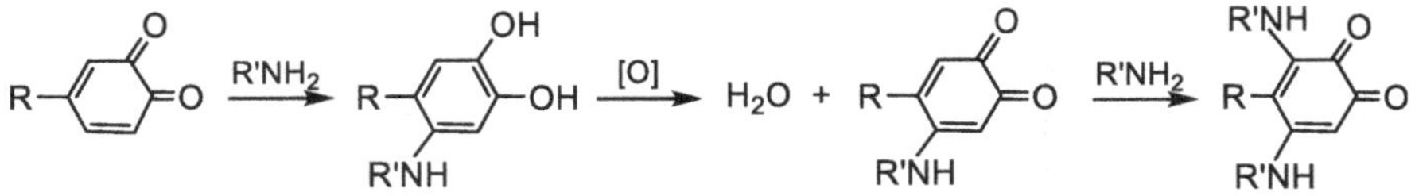
(28)

Schiff Base Condensation of Amines with Carbonyls in Quinones or Semiquinones Here the quinone provides carbonyls for condensation with amines in the Maillard reaction (Rx 29) (Bittner 2006).

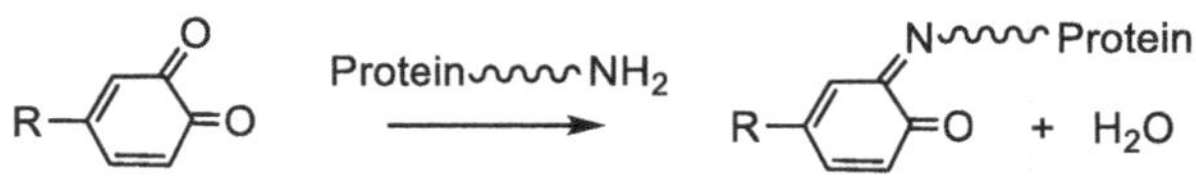
(29)

An important difference from complexation by Michael addition is that Schiff base formation blocks only one quinone carbonyl and leaves the other free. These modified residues then become prime target sites for lipid radical addition to open positions on the ring (Sect. 3.1.2), as well as for protein crosslinking via amine addition to the ring or to the carbonyl as described above.

5.3 *Chemical Reactions Not Involving Binding*

Protein amino, thiol, and tryptophan losses seldom match phenol binding so chemical reactions of phenolic compounds that transform both reactants but do not result in binding are very likely to always be present. These provide a background of amino acid damage that contributes to changes in protein structure and surface interactions that, in turn, alter functional properties as well as access to phenolic compounds.

Oxidation of Thiols The well-established ability of thiols such as cysteine and glutathione to reduce quinones and semiquinones to parent phenols (Cilliers and Singleton 1990; Wardman 1990) provides an important mechanism for recycling and regenerating antioxidant capability.

$$RSH + Q \longrightarrow RS^{\bullet} + SQ^{\bullet} \xrightarrow{RSH} QH_2 + RS^{\bullet} \qquad (30)$$

The two thiyl radicals ($RS^{\bullet}$) formed in the process then can dimerize to reform disulfide crosslinks (Cilliers and Singleton 1990),

$$2RS^{\bullet} \rightarrow RSSR \tag{31}$$

oxidize to sulfenic, sulfinic, and sulfonic acid oxides (Alcock et al. 2018),

$$RS^{\bullet} \rightarrow RSOH, RS(O)OH, RS(O)_2OH \tag{32}$$

or add as a radical to a semiquinone (Fujimoto and Masuda 2012). Sulfhydryl radical addition is actually faster than the Michael addition shown in Sect. 5.2 and provides important protection against disulfide crosslinking. At the same time, it uses up both phenols and thiols, which over time gradually degrades antioxidant reserves.

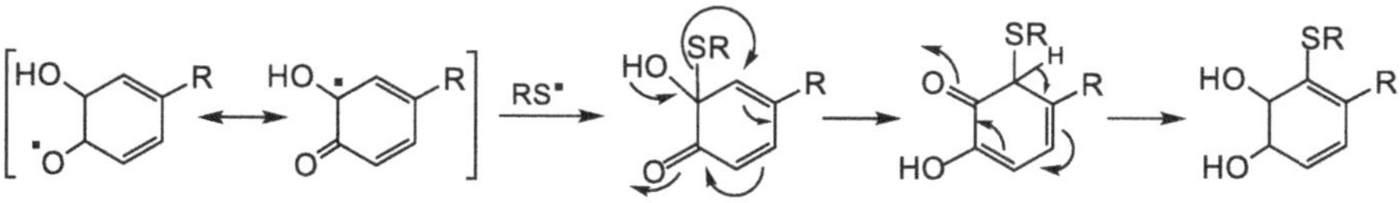

(33)

Radical-Mediated Oxidation of Amino Acids While quenching lipid radicals, phenols oxidize to phenoxyl radicals and also, in the presence of air, semiquinone radicals which reduce oxygen to H_2O_2 and very reactive hydroxyl radicals, $HO^{\bullet}$ (Kalyanaraman et al. 1985; Shendrik et al. 2012). When formed close to proteins, $HO^{\bullet}$ can attack amino acids by several mechanisms, leading to loss of lys, trp, his, arg, cys (Davies 1987).

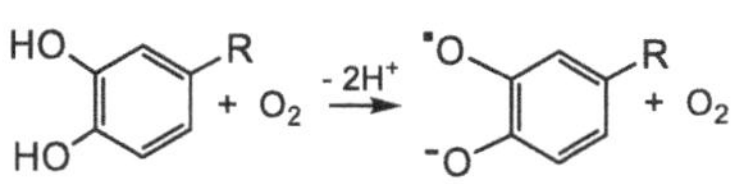

(34)

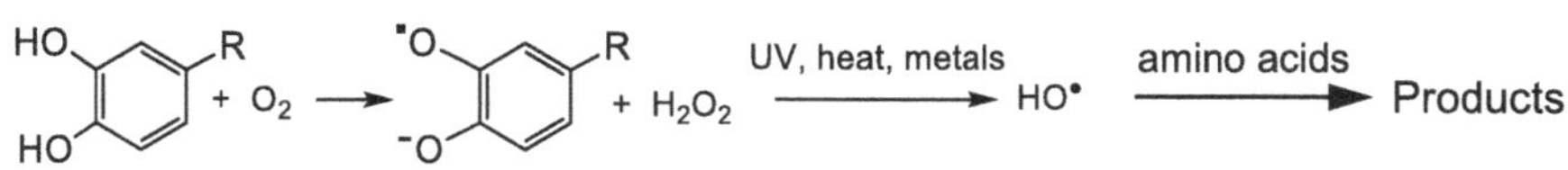

(35)

6 Summary: What Does This All Mean?

Everyone working with phenolic antioxidants has experienced situations where phenolic compounds did not work as expected, worked in oils but only poorly inhibited lipid oxidation when added to foods, exhibited much less than stoichiometric limiting of lipid oxidation, and sometimes just did not work at all. We have also encountered puzzling actions of antioxidants beyond lipids, as in blocking browning and protein crosslinking that did not correlate with inhibition of lipid oxidation (K.M. Schaich, unpublished data). Antioxidant action in oils or foods is nearly always evaluated by lipid hydroperoxide levels as a first approximation, and when changes in peroxide values don't explain system degradation, additional chemistry must be in play, i.e. phenols do more than just quench lipid radicals.

This chapter has presented evidence for actions of phenolic compounds that carry their antioxidant actions far beyond quenching of lipid oxyl radicals and formation of lipid hydroperoxides. While most of the evidence comes from model system experiments to simplify product analysis, studies are also coming forth to verify that multiple phenol actions also occur in oils, food systems, and even biological tissues. It is likely that alternate reactions of phenolic compounds that compete with and augment simple lipid oxyl radical quenching are probably always in play in different balances, contributing to the complex and often perplexing antioxidant behaviors of different phenols in different systems under different reaction systems.

When viewed all together, the multiple antioxidant pathways of phenolic compounds are internally cooperative and designed for maximum stabilization of natural systems. Oxidizing lipids start the process generating reactive peroxyl and alkoxyl radicals. Phenolic compounds quench these radicals and in the process are converted to a variety of products, one of which is the quinone form. Radical quenching generally increases with the number of phenolic –OH groups, but these moieties must be in ring positions ortho and para to the base phenol. Given this requirement, many phenolic compounds with *m*-diphenols and the A ring of flavonoids were thought to be poorly active as antioxidants. Now there is evidence that *m*-phenols contribute their own unique activity in forming adducts with aldehydes formed as secondary products in lipid oxidation, and they retain one phenol group that now has increased radical quenching capability. Hence, mixtures of *o,p*-phenols with *m*-phenols should give more complete reduction of lipid oxidation species, both primary and secondary. Such cooperation between phenols provides new definition to the frequent observation that application of total fruit or plant extracts are more effective in oxidation control than any single isolated component.

For a second level of cooperation, add proteins to this mix. Proteins provide important antioxidant protection in complex systems, both donating H atoms to quench lipid oxyl radicals and removing reactive epoxide and carbonyl products by reaction. When this occurs, lipid oxidation appears to be low in standard analyses, but oxidation has just moved to proteins instead, and although typical lipid rancid odors and flavors may be absent, they are replaced by other quality degradation in flavor, texture, and color associated with protein oxidation. Here, attempts to correlate system changes with lipid oxidation fail.

Now for a third level of cooperation, let us add the phenol-quinone pair. Explanation of why phenolic compounds are good antioxidants always specify low reactivity of the resulting phenolic radicals or formation of unreactive products, and quinones are totally discounted. Evidence presented in this chapter shows this exclusion is a mistake—quinones generated during early processes involving both lipid and proteins exert several important actions that contribute to total system stabilization, as well as some that accelerate destabilization. Starting with cooperation in lipid oxidation, quinones can compete with oxygen for $L^{\bullet}$, eliminating these initiating radicals by adding them to the aromatic ring. Semiquinone radical anions

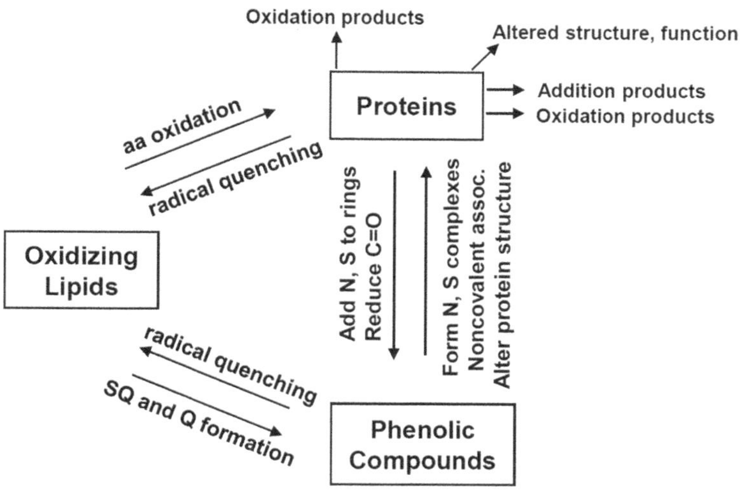

Fig. 3 Illustration of dependent yet competing interrelationships of oxidizing lipids, proteins, and phenolic antioxidants in stabilizing and destabilizing total system oxidation in complex materials. *aa* amino acid, *SQ* semiquinone, *Q* quinone

also reduce peroxyl radicals to anions. These two actions of quinones provide important limits at initial stages of lipid oxidation and reduce the radical load that must be combatted by phenols.

Moving beyond lipids to proteins, phenols quench amino acid radicals formed from reaction with lipid radicals, regenerating the original amino acids. They can also associate non-covalently with amine, sulfhydryl, and aromatic groups to modify reactivity of both protein and phenols groups. Does non-covalent association with phenols inhibit amino acid ability to transfer H atoms to lipids and quench radicals? To participate in co-oxidation and browning reactions? Does association with proteins diminish the amount of phenol available for inhibiting lipid oxidation? Answers to these questions are not yet known. Quinones come into play by covalently complexing protein amine and sulfhydryl groups. Since the bound protein groups are also those involved in quenching lipid radicals, does this action reduce protection for lipids while blocking co-oxidation of proteins? Or does this action occur parallel to lipid radical quenching? What is the net impact of all these interactions on total system oxidation in complex materials? The answers to these questions are also not yet known.

These multifaceted interactions between phenolic antioxidants, oxidizing lipids, and proteins are integrated and interdependent, as illustrated in Fig. 3, and must be considered together when designing stabilization strategies for oils and complex materials. Much research will be needed to determine relative kinetics of the various reactions and assess which reactions with lipids are parallel, synergistic, or antagonistic.

Acknowledgements This work was supported in part by USDA National Institute of Food and Agriculture Hatch Project 1008424 through the New Jersey Agricultural Experiment Station, Hatch Project NJ10157, and in part by USDA-NIFA grant 2017-67017-26465.

References

Alcock LJ, Perkins MV, Chalker JM (2018) Chemical methods for mapping cysteine oxidation. Chem Soc Rev 47(1):231–268

Altwicker ER (1967) The chemistry of stable phenoxy radicals. Chem Rev 67(5):475–531

Amorati R, Baschieri A, Morroni G, Gambino R, Valgimigli L (2016) Peroxyl radical reactions in water solution: a gym for proton-coupled electron-transfer theories. Chemistry 22(23): 7924–7934

Amorati R, Baschieri A, Cowden A, Valgimigli L (2017) The antioxidant activity of quercetin in water solution. Biomimetics 2(3):9

Arora A, Valcic S, Cornejo S, Nair M, Timmermann B, Liebler D (2000) Reactions of genistein with alkylperoxyl radicals. Chem Res Toxicol 13:638–645

Bakhouche K, Dhaouadi Z, Jaidane N, Hammoutène D (2015) Comparative antioxidant potency and solvent polarity effects on HAT mechanisms of tocopherols. Computat Theoret Chem 1060: 58–65

Bassil D, Makris DP, Kefalas P (2005) Oxidation of caffeic acid in the presence of l-cysteine: isolation of 2-S-cysteinylcaffeic acid and evaluation of its antioxidant properties. Food Res Int 38(4):395–402

Bittner S (2006) When quinones meet amino acids: chemical, physical and biological consequences. Amino Acids 30(3):205–224

Boozer CE, Hammond GS, Hamilton CE, Sen JN (1955) Air oxidation of hydrocarbons. II. The stoichiometry and fate of inhibitors in benzene and chlorobenzene. J Amer Chem Soc 77(12): 3233–3237

Bors W, Michel C (1999) Antioxidant capacity of flavanols and gallate esters: pulse radiolysis studies. Free Radical Biol Med 27(11/12):1413–1426

Campbell TW, Coppinger GM (1952) The reaction of t-butyl hydroperoxide with some phenols. J Amer Chem Soc 74(6):1469–1471

Chrysochoou M, Reeves K (2017) Reduction of hexavalent chromium by green tea polyphenols and green tea nano zero-valent iron (GT-nZVI). Bull Environ Contam Toxicol 98(3):353–358

Cilliers JJL, Singleton VL (1990) Caffeic acid autoxidation and the effects of thiols. J Agric Food Chem 38(9):1789–1796

Citterio A, Arnoldi A, Minisci F (1979) Nucleophilic character of alkyl radicals. 18. Absolute rate constants for the addition of primary alkyl radicals to conjugated olefins and 1,4-benzoquinone. J Org Chem 44(15):2674–2682

Cordero-Morales JF, Vásquez V (2018) How lipids contribute to ion channel function, a fat perspective on direct and indirect interactions. Curr Opin Struct Biol 51:92–98

Davies KAJ (1987) Protein damage and degradation by oxygen radicals. I. General aspects. J Biol Chem 262(20):9895–9901

Denisov ET (2006) Reactivity of quinones as alkyl radical acceptors. Kinet Catal 47(5):662–671

Di Meo F, Lemaur V, Cornil J, Lazzaroni R, Duroux J-L, Olivier Y et al (2013) Free radical scavenging by natural polyphenols: atom versus electron transfer. J Phys Chem A 117(10): 2082–2092

Dimitrić Marković JM, Marković ZS, Brdarić TP, Pavelkić VM, Jadranin MB (2011) Iron complexes of dietary flavonoids: combined spectroscopic and mechanistic study of their free radical scavenging activity. Food Chem 129(4):1567–1577

Erben-Russ M, Bors W, Saran M (1987) Reactions of linoleic acid peroxyl radicals with phenolic antioxidants: a pulse radiolysis study. Int J Rad Biol 52(3):393–412

Es-Safi NE, Cheynier V, Moutounet M (2002) Role of aldehydic derivatives in the condensation of phenolic compounds with emphasis on the sensorial properties of fruit-derived foods. J Agric Food Chem 50(20):5571–5585

Fahrenholtz SR, Doleiden FH, Trozzolo AM, Lamola AA (1974) On the quenching of singlet oxygen by α-tocopherol. Photochem Photobiol 20(6):505–509

Foote CS, Denny RW (1968) Chemistry of singlet oxygen. VII. Quenching by Beta-carotene. J Am Chem Soc 90(22):6233–6235

Foti MC, Daquino C, Geraci C (2004) Electron-transfer reaction of cinnamic acids and their methyl esters with the DPPH$^{\bullet}$ radical in alcoholic solution. J Org Chem 69(7):2309–2314

Franchi P, Lucarini M, Pedulli GF, Valgimigli L, Lunelli B (1999) Reactivity of substituted phenols toward alkyl radicals. J Amer Chem Soc 121(3):507–514

Fujimoto A, Masuda T (2012) Chemical interaction between polyphenols and a cysteinyl thiol under radical oxidation conditions. J Agric Food Chem 60(20):5142–5151

Gardner HW, Eskins K, Grams GW, Inglett GR (1972) Radical addition of linoleic hydroperoxides to α-tocopherol or the analogous hydroxychroman. Lipids 7(5):324–334

Hasni I, Bourassa P, Hamdani S, Samson G, Carpentier R, Tajmir-Riahi H-A (2011) Interaction of milk α- and β-caseins with tea polyphenols. Food Chem 126(2):630–639

Hidalgo FJ, Zamora R (2014) 2-Alkenal-scavenging ability of m-diphenols. Food Chem 160: 118–126

Hidalgo FJ, Zamora R (2019) Characterization of carbonyl–phenol adducts produced by food phenolic trapping of 4-hydroxy-2-hexenal and 4-hydroxy-2-nonenal. J Agric Food Chem 67(7):2043–2051

Hidalgo FJ, Aguilar I, Zamora R (2017a) Model studies on the effect of aldehyde structure on their selective trapping by phenolic compounds. J Agric Food Chem 65(23):4736–4743

Hidalgo FJ, Delgado RM, Zamora R (2017b) Protective effect of phenolic compounds on carbonyl-amine reactions produced by lipid-derived reactive carbonyls. Food Chem 229:388–395

Hidalgo FJ, Aguilar I, Zamora R (2018) Phenolic trapping of lipid oxidation products 4-oxo-2-alkenals. Food Chem 240:822–830

Hider R, Liu Z, Khodr H (2001) Metal chelation of polyphenols. Methods Enzymol 335:190–203

Horswill EC, Ingold KU (1966) The oxidation of phenols: I. The oxidation of 2,6-di-t-butyl-4-methylphenol, 2,6-di-t-butylphenol, and 2,6-dimethylphenol with peroxy radicals. Can J Chem 44(3):263–268

Horswill EC, Howard JA, Ingold KU (1966) The oxidation of phenols: III. The stoichiometries for the oxidation of some substituted phenols with peroxy radicals. Can J Chem 44(9):985–991

Jiang C, Garg S, Waite TD (2015) Hydroquinone-mediated redox cycling of iron and concomitant oxidation of hydroquinone in oxic waters under acidic conditions: comparison with iron–natural organic matter interactions. Environ Sci Technol 49(24):14076–14084

Jovanovic SV, Jankovic I, Josimovic L (1992) Electron-transfer reactions of alkylperoxy radicals. J Amer Chem Soc 114(23):9018–9021

Jovanovic SV, Simic MG, Steenken S, Hara Y (1998) Iron complexes of gallocatechins. Antioxidant action or iron regulation? J Chem Soc Perkin Trans 2:2365–2369

Kalyanaraman B, Felix CC, Sealy RC (1985) Semiquinone anion radicals of catechol(amine)s, catechol estrogens, and their metal ion complexes. Environ Health Perspect 64:185–198

Kamal-Eldin A, Appelqvist LA (1996) The chemistry and antioxidant properties of tocopherols and tocotrienols. Lipids 31(7):671–701

Kanakis CD, Hasni I, Bourassa P, Tarantilis PA, Polissiou MG, Tajmir-Riahi H-A (2011) Milk β-lactoglobulin complexes with tea polyphenols. Food Chem 127(3):1046–1055

Kumli E, Montermini F, Renaud P (2006) Radical addition to 1,4-benzoquinones: addition at O-versus C-atom. Org Lett 8(25):5861–5864

Lee S, Yim D, Lee D, Kim OY, Kang HJ, Kim HS et al (2020) Overview of the effect of natural products on reduction of potential carcinogenic substances in meat products. Trends Food Sci Technol 99(May):568–579

Lemanska K, Szymusiak H, Tyrakowska B, Zielinski R, Soffer AEMF, Rietjens IMCM (2001) The influence of pH on the antioxidant properties and the mechanisms of antioxidant action of hydroxyflavones. Free Radical Biol Med 31:869–881

Leopoldini M, Marino T, Russo N, Toscano M (2004) Antioxidant properties of phenolic compounds: H-atom versus electron transfer mechanism. J Phys Chem 108:4916–4922

Leopoldini M, Russo N, Chiodo S, Toscano M (2006) Iron chelation by the powerful antioxidant flavonoid quercetin. J Agric Food Chem 54(17):6343–6351

Li Y, Zhu T, Zhao J, Xu B (2012) Interactive enhancements of ascorbic acid and iron in hydroxyl radical generation in quinone redox cycling. Environ Sci Technol 46(18):10302–10309

Li Y, Jongberg S, Andersen ML, Davies MJ, Lund MN (2016) Quinone-induced protein modifications: kinetic preference for reaction of 1,2-benzoquinones with thiol groups in proteins. Free Radical Biol Med 97:148–157

Liebler DC, Kaysen KL, Burr JA (1991) Peroxyl radical trapping and autoxidation reactions of alpha-tocopherol in lipid bilayers. Chem Res Toxicol 4(1):89–93

Litwinienko G, Ingold KU (2003) Abnormal solvent effects on hydrogen atom abstractions. 1. The reactions of phenols with 2,2-diphenyl-1-picrylhydrazyl (DPPH$^•$) in alcohols. J Org Chem 68(9):3433–3438

Litwinienko G, Ingold KU (2004) Abnormal solvent effects on hydrogen atom abstraction. 2. Resolution of the curcumin antioxidant controversy. The role of sequential proton loss electron transfer. J Org Chem 69(18):5888–5896

Litwinienko G, Ingold KU (2005) Abnormal solvent effects on hydrogen atom abstraction. 3. Novel kinetics in sequential proton loss electron transfer chemistry. J Org Chem 70(22):8982–8990

Litwinienko G, Ingold KU (2007) Solvent effects on the rates and mechanisms of reaction of phenols with free radicals. Acc Chem Res 40(3):222–230

Lucarini M, Pedulli GF, Valgimigli L (1998) Do peroxyl radicals obey the principle that kinetic solvent effects on H-atom abstraction are independent of the nature of the abstracting radical? J Org Chem 63(13):4497–4499

Mahal HS, Kapoor S, Satpati AK, Mukherjee T (2005) Radical scavenging and catalytic activity of metal-phenolic complexes. J Phys Chem B 109(50):24197–24202

Mäkinen EM, Hopia AI (2000) Effects of alpha-tocopherol and ascorbyl palmitate on the isomerization and decomposition of methyl linoleate hydroperoxides. Lipids 35(11):1215–1223

Maroz A, Brede O (2003) Reaction of radicals with benzoquinone – addition or electron transfer? Rad Phys Chem 67(3):275–278

Mayer JM (2004) Proton-coupled electron transfer: a reaction chemist's view. Ann Rev Phys Chem 55:363–390

Md V, Kerman K, Tamiya E (2005) An electrochemical approach for detecting copper-chelating properties of flavonoids using disposable pencil graphite electrodes: possible implications in copper-mediated illnesses. Analyt Chim Acta 538(1):273–281

Medina I, Gallardo JM, Gonzalez MJ, Lois S, Hedges N (2007) Effect of molecular structure of phenolic families as hydroxycinnamic acids and catechins on their antioxidant effectiveness in minced fish muscle. J Agric Food Chem 55(10):3889–3895

Mellors A, Tappel AL (1966) Quinones and quinols as inhibitors of lipid peroxidation. Lipids 1(4): 282–284

Metodiewa D, Jaiswal AK, Cenas N, Dickancaité E, Segura-Aguilar J (1999) Quercetin may act as a cytotoxic prooxidant after its metabolic activation to semiquinone and quinoidal product. Free Radical Biol Med 26(1-2):107–116

Musialik M, Kuzmicz R, Paw Å,Owski TS, Litwinienko G (2009) Acidity of hydroxyl groups: an overlooked influence on antiradical properties of flavonoids. J Org Chem 74(7):2699-2709.

Muzolf M, Szymusiak H, Gliszczyńska-Świgło A, Rietjens IMCM, Tyrakowski B (2008) pH-Dependent radical scavenging capacity of green tea catechins. J Agric Food Chem 56(3): 816–823

Ozdal T, Capanoglu E, Altay F (2013) A review on protein–phenolic interactions and associated changes. Food Res Int 51(2):954–970

Perron NR, Brumaghim JL (2009) A review of the antioxidant mechanisms of polyphenol compounds related to iron binding. Cell Biochem Biophys 53(2):75–100

Porter NA, Wujek JS (1987) Allylic hydroperoxide rearrangement: β-scission or concerted pathway? J Org Chem 52:5085–5089

Porter WL, Levasseur LA, Henick AS (1971) An addition compound of oxidized tocopherol and linoleic acid. Lipids 6(1):1–8
Porter NA, Lehman LS, Weber BA, Smith KJ (1981) Unified mechanism for polyunsaturated fatty acid autoxidation. Competition of peroxy radical hydrogen atom abstraction, β-scission, and cyclization. J Am Chem Soc 103:6447–6455
Rabbani N, Thornalley PJ (2015) Dicarbonyl stress in cell and tissue dysfunction contributing to ageing and disease. Biochem Biophys Res Commun 458(2):221–226
Rawel HM, Kroll J, Hohl UC (2001a) Model studies on reactions of plant phenols with whey proteins. Food/Nahrung 45(2):72–81
Rawel HM, Kroll J, Rohn S (2001b) Reactions of phenolic substances with lysozyme — physicochemical characterisation and proteolytic digestion of the derivatives. Food Chem 72(1):59–71
Rawel HM, Meidtner K, Kroll J (2005) Binding of selected phenolic compounds to proteins. J Agric Food Chem 53(10):4228–4235
Ren J, Meng S, Lekka CE, Kaxiras E (2008) Complexation of flavonoids with iron: structure and optical signatures. J Phys Chem B 112(6):1845–1850
Rohn S (2014) Possibilities and limitations in the analysis of covalent interactions between phenolic compounds and proteins. Food Res Int 65:13–19
Schaich KM (2005) Lipid oxidation in fats and oils: theoretical aspects. In: Shahidi F (ed) Bailey's industrial fats and oil products, 6th edn. John Wiley, New York, pp 2681–2767
Schaich KM (2008) Co-oxidations of oxidizing lipids: reactions with proteins. In: Kamal-Eldin A, Min DB (eds) Lipid oxidation pathways, vol 2. CRC Press, Boca Raton, FL, pp 183–274
Schaich K (2020) Lipid oxidation: new perspectives on an old reaction. In: Shahidi F (ed) Bailey's industrial fats and oil products 7th edit edible oil and fat products: chemistry, properties, and safety aspects. John Wiley & Sons, Hoboken, NJ, pp 1–72
Schaich KM, Tian X, Xie J (2015) Hurdles and pitfalls in measuring antioxidant efficacy: a critical evaluation of ABTS, DPPH, and ORAC assays. J Funct Foods 14:111–125
Schwarz K, Frankel EN, German JB (1996) Partition behaviour of antioxidative phenolic compounds in heterophasic systems. Eur J Lipid Sci 98(3):115–121
Shendrik AN, Odaryuk ID, Kanibolotska LV, Kalinichenko EA, Tsyapalo AS, Beznos VV et al (2012) Radical formation during phenol oxidation in aqueous media. Int J Chem Kinet 44(6): 414–422
Simic MG (1981) Free radical mechanisms in autoxidation processes. J Chem Ed 58(2):125
Skrt M, Benedik E, Podlipnik Č, Ulrih NP (2012) Interactions of different polyphenols with bovine serum albumin using fluorescence quenching and molecular docking. Food Chem 135(4): 2418–2424
Snelgrove DW (2000) Kinetic and mechanistic studies on some one-electron and two-electron reactions. PhD Dissertation, Carleton University, Ottawa, CA.
Steenken S, Neta P (1982) One-electron redox potentials of phenols. Hydroxy- and aminophenols and related compounds of biological interest. J Phys Chem 86(18):3661–3667
Totlani VM, Peterson DG (2006) Epicatechin carbonyl-trapping reactions in aqueous Maillard systems: identification and structural elucidation. J Agric Food Chem 54(19):7311–7318
Tournaire C, Croux S, Maurette MT, Beck I, Hocquaux M, Braun AM et al (1993) Antioxidant activity of flavonoids: efficiency of singlet oxygen (1 delta g) quenching. J Photochem Photobiol B 19(3):205–215
Trebst A (2003) Function of β-carotene and tocopherol in photosystem II. Zeitschrift für Naturforschung C, J Biosci 58:609–620
Tsuchiya J, Niki E, Kamiya Y (1983) Oxidation of lipids. IV. Formation and reaction of chromanoxyl radicals as studied by electron spin resonance. Bull Chem Soc Jpn 56(1):229–232
Urano S, Matsuo M (1976) A radical scavenging reaction of α-tocopherol with methyl radicals. Lipids 11(5):380–383
Urano S, Yamanoi S, Hattori Y, Matsui M (1977) Radical scavenging reactions of alpha-tocopherol. II. The reaction with some alkyl radicals. Lipids 12(1):105–108
Valgimigli L, Banks JT, Ingold KU, Lusztyk J (1995) Kinetic solvent effects on hydroxylic hydrogen atom abstractions are independent of the nature of the abstracting radical. Two extreme tests using vitamin E and phenol. J Am Chem Soc 117:9971

Valgimigli L, Banks JT, Lusztyk J, Ingold KU (1999) Solvent effects on the antioxidant activity of vitamin E. J Org Chem 64(9):3381–3383
Valgimigli L, Amorati R, Petrucci S, Pedulli GF, Hu D, Hanthorn JJ et al (2009) Unexpected acid catalysis in reactions of peroxyl radicals with phenols. Angew Chem Int Edit 48(44):8348–8351
Wardman P (1990) Bioreductive activation of quinones: redox properties and thiol reactivity. Free Radical Res Commun 8(4-6):219–229
Warren JJ, Mayer JM (2010) Predicting organic hydrogen atom transfer rate constants using the Marcus cross relation. Proc Natl Acad Sci U S A 107(12):5282–5287
Weenan H, Porter NA (1982) Autoxidation of model membrane systems: cooxidation of polyunsaturated lecithins with steroids, fatty acids, and α-tocopherol. J Am Chem Soc 104:5216–5221
Wehrer C, Bindler F, Laugel P (1984) Interactions between polyphenols and stannous ions in canned fruits and vegetables: a quantitative approach. Dtsch Lebensm Rundsch 80(9):273–279
Wright JS, Johnson ER, DiLabio GA (2001) Predicting the activity of phenolic antioxidants: theoretical method, analysis of substituent effects, and application to major families of antioxidants. J Am Chem Soc 123:1173–1183
Yamauchi R (2007) Addition products of alpha-tocopherol with lipid-derived free radicals. Vitam Horm 76:309–327
Yamauchi R, Matsui T, Kato K, Ueno Y (1990a) Reaction products of γ-tocopherol with an alkylperoxyl radical in benzene. Agric Biol Chem 54(10):2703–2709
Yamauchi R, Matsui T, Miyake N, Kato K, Ueno Y (1990b) Reaction of δ-tocopherol with an alkylperoxyl radical. Agric Biol Chem 54(11):2993–2999
Yamauchi R, Kato K, Ueno Y (1995) Free-radical scavenging reactions of α-tocopherol during the autoxidation of methyl linoleate in bulk phase. J Agric Food Chem 43(6):1455–1461
Yamauchi R, Yagi Y, Kato K (1996) Oxidation of alpha-tocopherol during the peroxidation of dilinoleoylphosphatidylcholine in liposomes. Biosci Biotechnol Biochem 60(4):616–620
Yong H, Bai R, Bi F, Liu J, Qin Y, Liu J (2020) Synthesis, characterization, antioxidant and antimicrobial activities of starch aldehyde-quercetin conjugate. Intl J Biol Macromol 156:462–470
Zamora R, Hidalgo FJ (2005) Coordinate contribution of lipid oxidation and Maillard reaction to the nonenzymatic food browning. Crit Rev Food Sci Nutr 45(1):49–59
Zamora R, Hidalgo FJ (2011) The Maillard reaction and lipid oxidation. Lipid Technol 23(3): 59–62
Zamora R, Hidalgo FJ (2016) The triple defensive barrier of phenolic compounds against the lipid oxidation-induced damage in food products. Trends Food Sci Technol 54:165–174
Zamora R, Hidalgo FJ (2018) Carbonyl–phenol adducts: an alternative sink for reactive and potentially toxic lipid oxidation products. J Agric Food Chem 66(6):1320–1324
Zamora R, Alcón E, Hidalgo FJ (2013) Strecker-type degradation of phenylalanine initiated by 4-oxo-2-alkenals in comparison to that initiated by 2,4-alkadienals, 4,5-epoxy-2-alkenals, or 4-hydroxy-2-nonenal. J Agric Food Chem 61(43):10231–10237
Zamora R, Aguilar I, Granvogl M, Hidalgo FJ (2016) Toxicologically relevant aldehydes produced during the frying process are trapped by food phenolics. J Agric Food Chem 64(27):5583–5589
Zamora R, Aguilar I, Hidalgo FJ (2017) Epoxyalkenal-trapping ability of phenolic compounds. Food Chem 237:444–452
Zhang L, Liu R, Gung BW, Tindall S, Gonzalez JM, Halvorson JJ et al (2016) Polyphenol–aluminum complex formation: implications for aluminum tolerance in plants. J Agric Food Chem 64(15):3025–3033
Zhu Q, Zheng ZP, Cheng KW, Wu JJ, Zhang S, Tang YS et al (2009a) Natural polyphenols as direct trapping agents of lipid peroxidation-derived acrolein and 4-hydroxy-trans-2-nonenal. Chem Res Toxicol 22(10):1721–1727
Zhu Q, Liang C-P, Cheng K-W, Peng X, Lo C-Y, Shahidi F et al (2009b) Trapping effects of green and black tea extracts on peroxidation-derived carbonyl substances of seal blubber oil. J Agric Food Chem 57(3):1065–1069

Effect of Lipophilic and Hydrophilic Thiols on the Lipid Oxidation

O. T. Kasaikina, K. M. Zinatullina, V. D. Kancheva, A. K. Slavova-Kasakova, and D. V. Loshadkin

1 Introduction

All living organisms on earth cannot exist without oxygen, yet oxygen is inherently dangerous to their existence. Respiratory chains in living nature and autoxidation of organic substances frequently generate different active radicals and peroxides $OH^{\bullet}$, $O_2^{\bullet-}$, $LO_2^{\bullet}$, H_2O_2, LOOH, so called reactive oxygen species (ROS), which appear to be responsible for oxygen toxicity.

Due to the crucial roles played by lipids for structural and signaling activities, the efficiency of the antioxidant network in controlling lipid reactivity and transformations is an interdisciplinary research field extended from chemistry to biology and medicine (Halliwell 2007). In this context, polyunsaturated fatty acid (PUFA) reactivity with free radicals is known to occur via two main processes: (1) the lipid peroxidation (Niki 2012), and (2) the cis–trans isomerization (Ferreri and Chatgilialoglu 2012).

For a long time, almost for all the last century, hydrocarbon and lipid (LH) oxidation was considered as a free radical chain branching process, which can be auto initiated or initiated by extern source (initiators, I, radiation, etc.) (Emanuel et al. 1965; Scott 1965; Emanuel and Gal 1978; Frankel 2005; Kamal-

O. T. Kasaikina (✉) · K. M. Zinatullina
N.N. Semenov Federal Research Center of Chemical Physics, Russian Academy of Sciences, Moscow, Russia
e-mail: okasai@yandex.ru; kasaikina@chph.ras.ru

V. D. Kancheva · A. K. Slavova-Kasakova
Institute of Organic Chemistry with Centre of Phytochemistry, Bulgarian Academy of Sciences, Sofia, Bulgaria
e-mail: vessela.kancheva@orgchm.bas.bg; adriana.kazakova@orgchm.bas.bg

D. V. Loshadkin
Yaroslavl State Technical University, Yaroslavl, Russia
e-mail: loshadkindv@ystu.ru

C. Bravo-Diaz (ed.), *Lipid Oxidation in Food and Biological Systems*,
https://doi.org/10.1007/978-3-030-87222-9_8

Eldin 2003; Denisov and Afanas'ev 2005). The rate of chain processes is equal to the product of chain initiation rate (W_i) and the length of the chain (ν):

$$W = W_i \cdot \nu \tag{1}$$

The chain initiation rate depends sufficiently on the hydroperoxide concentration:

$$W_i = w_0 + e\, k_d[LOOH] \tag{2}$$

Here, w_0 is the initiation rate without LOOH participation, k_d—apparent rate constant and e—the so called "radical escape" for LOOH decomposition.

So, the promotion of hydroperoxide decomposition into free radicals resulted in accelerated oxidation. On the contrary, the heterolytic nonradical reduction of hydroperoxides leads to the decrease of W_i and the oxidation rate.

The length of the chain is equal to the ratio of the rates of propagation and termination (W_t):

$$\nu = k_p[LH][LO_2^{\bullet}]/W_t \tag{3}$$

The most common way to protect products and materials against oxidation is to use chain breaking antioxidants (InH), which react with active radicals:

$$\begin{aligned} &\text{Chain termination}: InH + LO_2^{\bullet} \rightarrow In^{\bullet} + LOOH \\ &In^{\bullet} + In^{\bullet}, LO_2^{\bullet} \rightarrow \text{molecular products} \end{aligned} \tag{4}$$

Inhibitors increase the termination rate W_t, so, the chain length (ν) grows a decrease, and by this reason InH retard the oxidation as a whole (Halliwell 2007; Niki 2012; Ferreri and Chatgilialoglu 2012; Scott 1965; Emanuel and Gal 1978; Frankel 2005; Kamal-Eldin 2003; Denisov and Afanas'ev 2005).

Thio compounds have long been known as peroxide destroyers; they are traditionally used as synergistic antiperoxide additives in antioxidative compositions for lubricants (Hawkiks and Worthikgston 1963). Thiol antioxidants act through a variety of mechanisms, including (1) as components of the general thiol/disulfide redox buffer, (2) as metal chelators, (3) as radical quenchers, (4) as substrates for specific redox reactions (5) as specific reductants of individual protein disulfide bonds (thioredoxin) (Ulrich and Jakob 2019). So, antioxidant effect of thiols includes both reducing the chain initiating rate (R_i) and shortening the length of the chain (ν).

However, there are some circumstances that reduce the antioxidant effect of thiols, and sometimes even lead to an acceleration of lipid spoilage.

1. Thiyl radicals $RS^{\bullet}$ are known to catalyze the cis/trans isomerization of unsaturated fatty acids (LH) (Ferreri and Chatgilialoglu 2012; Chatgilialoglu et al. 2002; Chatgilialoglu and Ferreri 2005; Mengele et al. 2015; Chatgilialoglu and Bowry 2018). Thiyl radicals are generated from thiols under the radical stress in the

Scheme 1 *Cis-trans* isomerization of unsaturated lipids

$$r^{\bullet} + RSH \rightarrow RS^{\bullet} + rH \quad (i)$$

$$RS^{\bullet} + cisLH \Leftrightarrow \{RS\text{-}LH^{\bullet}\} \Leftrightarrow RS^{\bullet} + transLH \quad (ii)$$

Scheme 2 Thiol-ene reaction of thiols with unsaturated substances

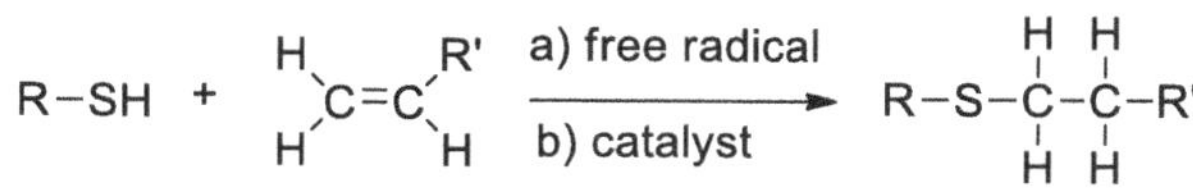

"radical repair reaction" as well as during the activity of some enzymes (Scheme 1).

In the free radical isomerization, the addition-elimination of a thiyl radical is enough to produce the mono-trans geometrical isomers (Chatgilialoglu and Ferreri 2005). Unsaturated fatty acid molecules present in the living organisms and high-quality natural oils adopt the *cis*-configuration (Afaf and Min 2008; Sebedio and Christie 1998). *Trans*-isomers appear usually in the course of hydrogenation and high-temperature treatment of natural oils. They are undesirable components, because *trans*-lipids incorporate into cell membranes and thus violate the balance of exchange processes.

2. Thiols can add to unsaturated compounds according to Scheme 2, and recently, these reactions are intensively discussed in connection with the so-called thiol-ene click chemistry (Kade et al. 2010; Turunc and Meier 2012; Koo et al. 2010; Vanslambrouck et al. 2021):

   The thiol-ene click reactions are mainly used to synthesize linear and branched heterochain polymers (Hoyle and Bowman 2010). The concept of click chemistry was first used by B. Sharpless in 2001 to refer to chemical reactions suitable for the rapid and reliable synthesis of potential drugs by combining individual small elements (Kolb et al. 2001).
3. Thiols can accelerate the oxidation of hydrocarbons and lipids owing to low yield of free radicals in the reactions with hydroperoxides. It has been shown (Mengele et al. 2015; Sies and Jones 2007), that thiols behave as inhibitors in the initial steps of the oxidation of substrates, thoroughly purified from peroxide impurities. However, real systems usually do contain hydroperoxides and/or hydrogen peroxide, so thiols, used as additives, can accelerate the oxidation.

   In this work taking 2-mercaptoethanol (RSH), the simplest thiol, as an example, we consider the kinetic features of oxidation and *cis-trans* isomerization of methyl linoleate in the presence of lipophilic thiol in hydrocarbon solution. The effects of hydrophilic thiol glutathione (GSH) on hydrogen peroxide decomposition, thiol-ene reaction of GSH with unsaturated phenol resveratrol (RVT) and the oxidation of sunflower oil and methyllinoleate in the micellar solution in the presence of GSH are discussed as well.

2-Mercaptoethanol (RSH) is known as an efficient radioprotector and antioxidant. It is widely used in the analysis of proteins and is added to components of enzymatic reactions to inhibit the oxidation of free sulfhydryl residues and maintain

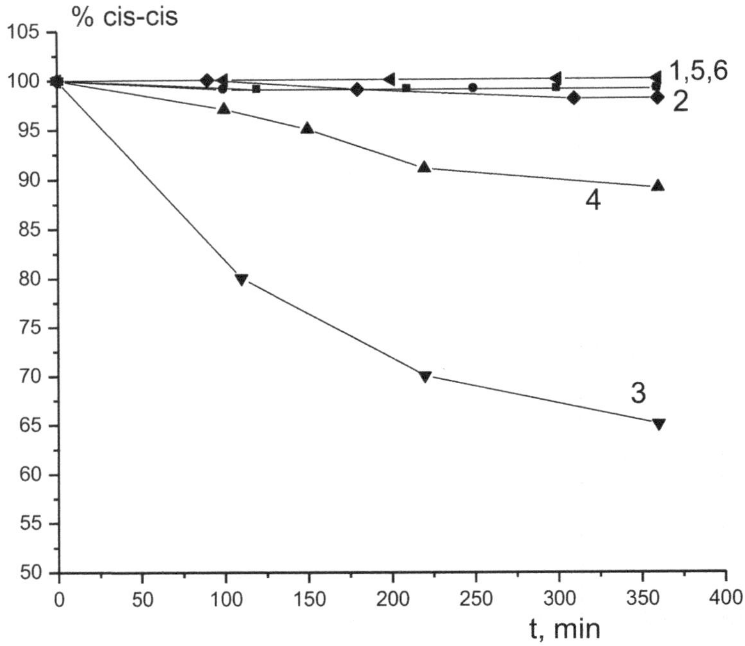

Fig. 1 The consumption of *cis/cis* LH, 5 mM, in cyclohexane solution at 50 °C, initiated by AIBN, 2.5 mM, (1–6), under N_2 (1,3,5) or O_2 (2,4,6) in the presence of RSH, 5 mM, (3–6); α-tocopherol, 2.5 mM, (5,6)

the protein activity, as well as to protect readily oxidizable compounds from oxygen (Roy 2005; Aitken et al. 2008).

Figures 1 and 2 illustrate the effect of mercaptoethanol (RSH) and thiyl radicals formed in exchange reactions between RSH and radicals generated during the thermal decomposition of the initiator azo-bis-isobutyronitrile (AIBN) in an atmosphere of N_2 and O_2 on the formation of *trans* isomers in non-chain (Fig. 1) consumption and oxygen uptake in the chain oxidation (Fig. 2) *cis-cis* methyllinoleate (LH). Figure 1 presents the kinetic curves of consumption of *cis, cis*-methyl linoleate (LH) in reactions with radicals produced in the decomposition of AIBN (cyclohexane was used as solvent) in the presence (curves 3 and 4) and in the absence of mercaptoethanol (RSH) (curves 1,2) in air(curves 4) and in nitrogen (curve 3) atmosphere. At low concentration of LH (5 mM) and at the initiation rate ($W_i = 3.8 \cdot 10^{-9}$ M/s), it is the non-chain regime ($\nu \leq 1$) and cisLH is practically not consumed. In the presence of mercaptoethanol (RSH), the starting *cis*LH is consumed and *trans*-isomers are formed. From comparison of the rates of *cis*LH consumption in the presence of RSH (curves 3 and 4: under N_2 $1.7 \cdot 10^{-7}$ M/s and under O_2 $3.7 \cdot 10^{-8}$ M/c) and W_i, it follows that thiyl radicals catalytically accelerate the *cis—trans*-isomerization and the number of catalytic cycles is of the order of a few dozen (~45). *Cis-trans*-isomerization occurs in accordance with Scheme 1,

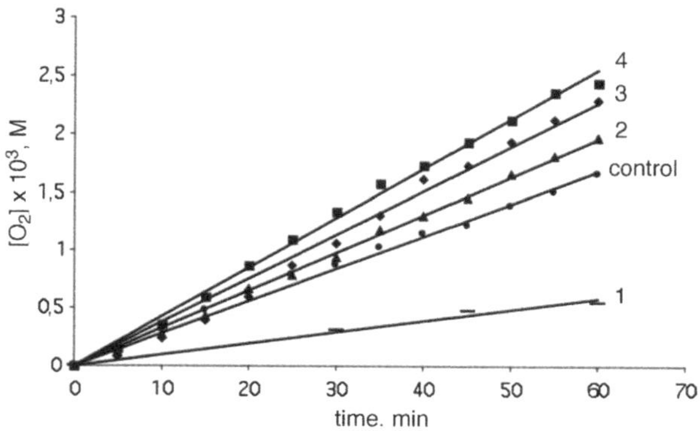

Fig. 2 Oxygen uptake during oxidation of *cis/cis* LH, 200 mM, initiated by AIBN (1–3), 4 mM, at 50 °C, by addition: (1) 50 mM AmH, (2) 50 mM RSH; (3) 50 mM RSH and 50 mM AmH; (4) the initiation by 1.6 mM ACh + 50 mM *t*-BuOOH +50 mM RSH

Table 1 Kinetics parameters of *cis/trans* isomerization and oxidation of methyllinoleate (200 mM), initiated by AIBN (4 mM) at 50 °C

No	Substratum LH, 200 mM	% *trans* in O_2	% *trans* in N_2	$W_{O2}\cdot10^7$, mol/(L·s)
Control	LH	0	0	6.76
1	LH + AmH (50 mM)	0.2	0.3	2.22
2	LH + RSH (50 mM)	7.1	14.2	7.94
3	LH + RSH (50 mM) + AmH (5 mM)	11.5	15.8	9.16
4	LH + RSH (50 mM) + ACh (1.6 mM) + LOOH (50 mM), without AIBN	2.7	2.5	10.2

supplemented by the reactions: $r^{\bullet} + LH \rightarrow rH + L^{\bullet}$; $L^{\bullet} + RSH \rightarrow LH + RS^{\bullet}$ (under N_2) and: $r^{\bullet} + O_2 \rightarrow rO_2^{\bullet}$; $rO_2^{\bullet} + LH \rightarrow rOOH + L^{\bullet}$; $L^{\bullet} + O_2 \rightarrow LO_2^{\bullet}$; $LO_2^{\bullet} + RSH \rightarrow RS^{\bullet}$ (under O_2).

RSH readily reacts with alkyl radicals ($k_{L\bullet} \sim 10^4\ M^{-1}\cdot s^{-1}$), being relatively inert toward peroxyl ones ($k_{LO2} \sim 10\ M^{-1}\cdot s^{-1}$). For this reason, the rate of isomerization in oxygen is lower than that in N_2 (see curves 3 and 4).

Figure 2 shows the kinetics of oxygen uptake during the oxidation of greater concentration of LH, initiated by AIBN, in the absence (control) or presence of additives. In the presence of 50 mM diphenylamine (AmH), the LH oxidation was inhibited confirming that the aromatic amines are antioxidants (Emanuel et al. 1965; Scott 1965; Emanuel and Gal 1978; Frankel 2005; Kamal-Eldin 2003; Denisov and Afanas'ev 2005) (curve 1). In contrast, 50 mM RSH added alone to the solution of LH induced its oxidation (curve 2) and with the mixture AmH+ RSH the oxidation further increased (curve 3).

These results are summarized in Table 1. The rates of O_2 uptake (W_{O2}) are reported along with the percentages of *trans*-isomers which were determined by GC analyses after work-up and in parallel experiments conducted with the same concentrations of reagents, but in a nitrogen atmosphere (columns 3 and 4). They show that during one hour, RSH catalyzes *trans*-isomer formation and slightly increases LH oxidation rate. On the other hand, AmH added alone decreased the oxidation rate and gave only traces of trans-isomers formation. However, in the joint presence of RSH and AmH, both LH oxidation rate and *cis-trans* isomerization rate are substantially increased under aerobic and anaerobic conditions. So, diphenylamine shows a synergistic effect with mercaptoethanol, increasing the rate of oxidation and *cis/trans* isomerization of LH in the joint presence.

The observed synergism is probably based on the high activity of diphenylaminyl radicals ($Am^{\bullet}$) in the abstraction of hydrogen. In the presence of oxygen, i.e. in the presence of peroxyl radicals in the system, AmH intercepts $rO_2^{\bullet}$ and replaces them with aminyl radicals that are more active in the reaction with RSH:

$$\mathrm{AmH} + \mathrm{rO_2}^{\bullet} \rightarrow \mathrm{Am}^{\bullet} + \mathrm{rO_2H}$$
$$\mathrm{Am}^{\bullet} + \mathrm{RSH} \rightarrow \mathrm{RS}^{\bullet} + \mathrm{AmH}$$

An increase in the rate of LH oxidation in the presence of mercaptoethanol (control and curve 2 in Fig. 2) may be due to the addition of O_2 to the adduct {RS-LH$^{\bullet}$} as well (see Scheme 1, (ii)):

$$\{\mathrm{RS\text{-}LH}^{\bullet}\} + \mathrm{O_2} \rightarrow \mathrm{L'O_2}^{\bullet}$$

This reaction affects both the rate of oxidation and the rate of *cis/trans* isomerization.

The curve 4 in Fig. 2 shows the O_2 uptake during LH oxidation initiated with a mixture of acetylcholine (ACh) and tert-butyl-hydroperoxide (*t*-BuOOH) which generate radicals with the rate equal to W_i initiated by AIBN in the control case. The increase of the oxidation rate in the presence of RSH and relative decrease of the yield of *trans*LH may be associated with interaction of RSH with *t*-BuOOH resulting in free radical formation.

Glutathione (GSH) is a water soluble thiol. This the most common cytosolic thiol belongs to endogenous biological antioxidants synthesized directly in living organisms. GSH interacts with hydroxyl radicals, reduces hydrogen peroxide, hydroperoxides, and –S–S– disulfide bonds, and prevents the oxidation of proteins (Aitken et al. 2008; Saito and Kawabata 2004; Winterbourn and Metodieva 1995; Sajewicz et al. 2015; Winterbourn 2016; Takashima et al. 2012). There are reports on significant changes in the GSH content during the development of many pathologies, in particular, Alzheimer's, Parkinson's, cardiovascular, and oncological diseases (Penninck 2000; Wu et al. 2004; Conway et al. 1987; Townsend et al. 2003; Estrela et al. 2006; Toyokuni 2014; Stavrovskaya 2000; Guo et al. 2018). The redox pair GSH/GSSG and H_2O_2 are central to the determination of redox homeostasis and intracellular information transmission-cellular signaling. In living organisms, hydroperoxides are reduced by glutathione peroxidases, enzymes specific for organs and tissues that use GSH as a substrate and efficiently reduce H_2O_2 and organic hydroperoxides, including the hydroperoxides of membrane polyunsaturated fatty acids. But with H_2O_2, the thiols can react directly. The reaction of reduction of H_2O_2 by thiols (TSH) is described by the stoichiometric equation (Abedinzadeh et al. 1989; Winterbourn and Hampton 2008; Winterbourn 2015, 2018; Zinatullina et al. 2019):

$$2TSH + H_2O_2 \rightarrow TSST + 2H_2O \quad (5)$$

However, the actual interaction of glutathione with H_2O_2 proceeds by a complex mechanism, including the formation of intermediate complexes GSH-GSH (Zinatullina et al. 2019, 2021), GSH-H_2O_2 (Abedinzadeh et al. 1989; Abedinzadeh 2001). It was found that the reaction between GSH and H_2O_2 is accompanied by the formation of radicals (Zinatullina et al. 2020, 2017a). Using the inhibitor method and employing an original radical acceptor (Zinatullina et al. 2016), it was shown that, in deionized water, the rate of radical formation in the reaction with H_2O_2 is increased in the following sequence: glutathione $\approx$ homocysteine $<$ cysteine. Using the spin trap method and employing 5,5-dimethyl-1-pyrroline-N-oxide, it was showed that the interaction between GSH and H_2O_2 really leads to the formation of thiyl radicals (Zinatullina et al. 2020). The radical yield is low ($<1\%$); however, it may be enough to initiate chain processes.

It was found (Zinatullina et al. 2017b, 2018, 2021) that, in the presence of H_2O_2 in aqueous solutions, thiol-ene chain reactions of GSH with unsaturated phenols resveratrol (RVT) and caffeic acid are initiated. Resveratrol and caffeic acid are plant polyphenols. They contain an unsaturated bond in the side substituents of the aromatic cycle. Recently, these phenols, in particular RVT (3,5,4-′-trihydroxystilbene), have attracted the attention of physicians and biochemists owing to the so-called "French paradox," i.e. an unusually low level of cardiovascular and oncological diseases despite a high-calorie diet with an abundance of fat that is observed in some regions of France against the background of regular consumption of red wine (Yu et al. 2012; Salehi et al. 2018).

Figure 3 shows that resveratrol (RVT) consumption is observed only in the joint presence of glutathione and H_2O_2. At the same concentrations of GSH and H_2O_2 the initial RVT consumption rate (W_{RVT}) increases linearly with an increase of RVT concentration (Fig. 4). It should be noted that the linear dependences in Fig. 4 cut off segments on the ordinate axis that are equal (within the error) to the rate of radical initiation (W_i), measured by the inhibitor method using a polymethine dye (a water-soluble radical acceptor (Zinatullina et al. 2016)). So, the rate of RVT consumption is satisfactorily described by Eq. (6) for chain reactions of oxidation and polymerization with quadratic chain termination on the leading chain radicals:

$$W_{RVT} = W_i + a[RVT]/W_i^{0.5} \quad (6)$$

Here, the parameter $a \cong 1.0\ (M{\cdot}s)^{-0.5}$ is similar to the ratio of the rate constants of the propagation (k_p) and chain termination reactions (k_t) $a = k_p/(2k_t)^{0.5}$.

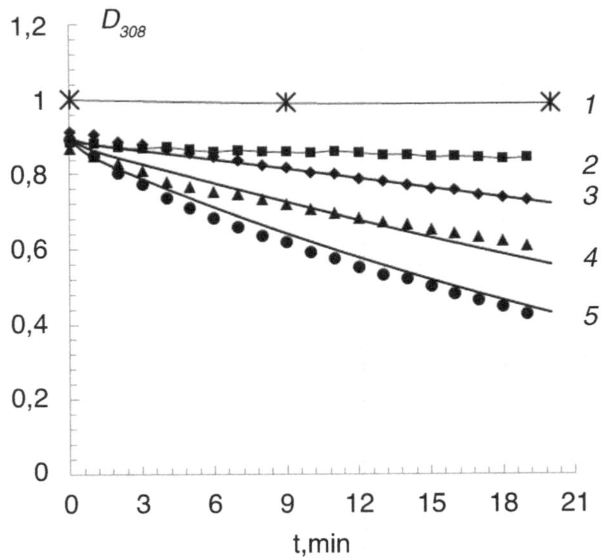

Fig. 3 Kinetic curves of consumption of 0.03 mM RVT in reaction with GSH (1) in the absence and (2–5) in the presence of 4.55 mM H_2O_2; the GSH concentration (mM): (1) 25 (RVT concentration of 0.033 mM), (2) 0, (3) 2.5, (4) 5, and (5) 10 (Zinatullina et al. 2021)

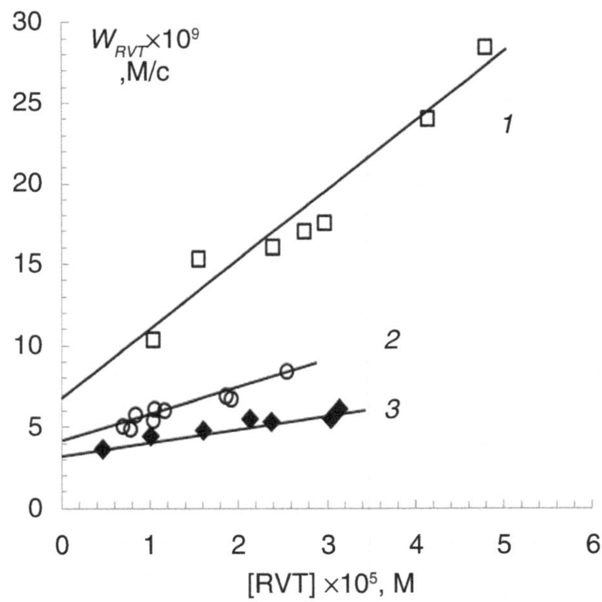

Fig. 4 Dependences of the RVT consumption rates (W_{RVT}) on the RVT concentration in the reaction mixture of 4.55 mM H_2O_2 with different initial GSH concentrations (mM): 1–10; 2–5; 3–2.5 (Zinatullina et al. 2021)

Analysis of the composition of products formed in the reactions of GSH with H_2O_2 and with RVT, made by electrospray ionization mass spectrometry in the positive-ion measuring mode, showed the following. (1) The initial GSH solution contains sufficiently stable dimers GSH-GSH (ions M'H$^+$ 615,17). (2) The main

Table 2 Effect of phosphate buffer solutions on the rates of GSH consumption and radical initiation (W_i) in the interaction of 10 mM GSH and 2 mM H_2O_2

Solvent	Solvent, pH	Reaction mixture, pH	W_{GSH} 10^6, M/s	W_i 10^9, M/s	Δ[GSH], consumed
Deionized water	7.0	3.25	0.8	5.3	<4 mM
PBS	7.4	5.2	1.1	0.45	5 mM
PB	7.2	7.0	7.1	0	6 mM

product of the GSH oxidation in the reaction with H_2O_2 in accordance with Eq. (5) is the corresponding disulfide GSSG (M"H^+ 613,16). (3) In a mixture of GSH, RVT and H_2O_2 in deionized water, the main products are GSSG disulfide and a product of M"'H^+ 568.16, the mass of which corresponds to hydroperoxide (PO_2H), which can be obtained as a result of the sequential addition of the thiyl radical $GS^•$ and oxygen to RVT:

$$GS^{\bullet} + RVT \Leftrightarrow P^{\bullet} \overset{+O_2}{\longrightarrow} PO_2{}^{\bullet} \overset{+GSH}{\longrightarrow} PO_2H + GS^{\bullet} \quad (7)$$

Using the experimental data on the kinetics, the product composition and the published data on reactions of GSH with H_2O_2 and thiyl radicals, in (Zinatullina et al. 2021) a kinetic model of the complex interaction between GSH and RVT in the presence of H_2O_2 in an aqueous medium at 37 °C is proposed. The model includes 19 quasi-elementary reactions with respective rate constants, in particular, the formation of intermediate GSH–H_2O_2 and GSH–GSH complexes, the formation of radicals, and their subsequent transformations into final products in reactions with RVT and GSH. A computer simulation based on the developed model adequately describes the kinetics and mechanism of this thiol-ene reaction.

In the last decade, much attention has been paid to the signaling role of glutathione, often in combination with H_2O_2, in oxidative stress regulating and the response of living organisms to external influences. The GSH molecule contains two carboxyl groups with pKa 2.5 and 3.7. Therefore, in water, GSH forms acidic solutions (pH <<7), and in alkaline solutions, it often shifts the pH to the acidic side.

From the comparison of the rates of GSH consumption and radical formation (W_i) in Table 2 it can be seen that in phosphate buffer solutions, the rate of thiol consumption increases and W_i decreases sharply. One of the reasons for the increase of W_{GSH} in PBS and especially in PB is the additional to H_2O_2 oxidation of GSH by air oxygen. At initial concentrations of 10 mM of thiol and 2 mM of H_2O_2, more thiols were consumed in buffer solutions than is required according to the stoichiometry of Eq. (5).

Figure 5 shows a nontrivial dependence of the GSH consumption rate in the reaction with H_2O_2 on the pH of the reaction mixture in phosphate buffer solutions. It can be seen that in areas close to the physiological pH values of 6.8–7.4, the rate increases exponentially. This means that at pH $\geq$ 7, GSH consumption occurs mainly in the reactions:

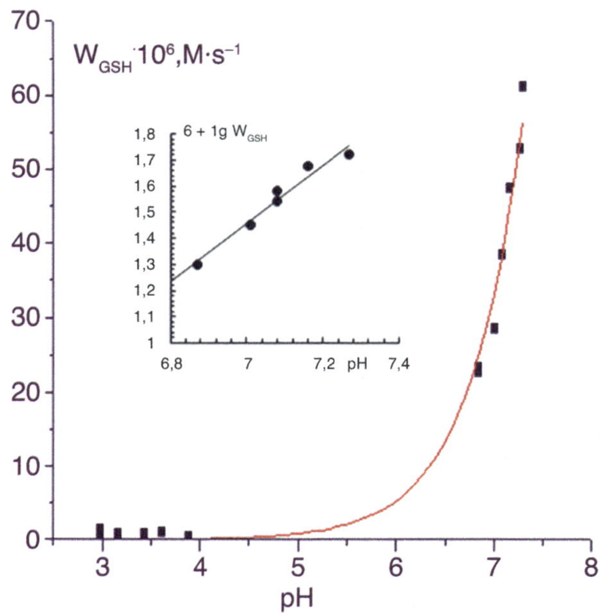

Fig. 5 (**a**) The effect of pH on the rate of consumption of 10 mM GSH in the reaction with 2 mM H_2O_2 in sodium-potassium phosphate buffer solutions; (**b**) The dependence of lgW_{GSH} on pH in the pH range of 6.8–7.4

$$GSH + OH^- \rightarrow GS^- + H_2O$$
$$GS^- + H_2O_2 \rightarrow GSOH + OH^-$$
$$GSOH + GSH \rightarrow GSSG + H_2O$$

In these reactions, as in an acidic medium, disulfide and water are formed, but the limiting stage is the thiolate-anion reaction.

2 Glutathione in Lipid Oxidation

The vast majority of studies on the biochemistry of GSH and other natural thiols are carried out under conditions close to physiological in animal organisms, i.e. in buffer solutions providing pH = 7.2–7.4. Under such conditions, the rate of reactions of thiols with ROS is largely determined by the contribution of the thiolate anion to these processes and depends on the pKa value of the SH bond. In a large review (Aldini et al. 2018) the data on the antioxidant and regenerative activity of natural thiols were analyzed and it was concluded that the antioxidant activity of thiol is due to the thiolate anion, the relative concentration of which is regulated by the acidity of the thiol.

Glutathione was tested under lipid autoxidation conditions at 80 °C in two concentrations (0.1 mM and 1.0 mM) (Fig. 6) in oxidation of triacylglicerols of sunflower oil (TGSO). It can be seen from this figure that GSH doesn't show chain-

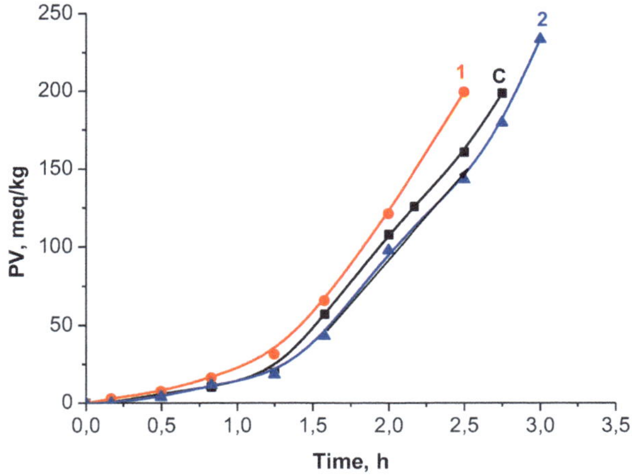

Fig. 6 Kinetic curves of lipid peroxide accumulation during TGSO (Triacylglicerols of sunflower oil) autoxidation at 80 °C in absence (control, C) and in presence of glutathione *GSH* in 0.1 mM and 1.0 mM

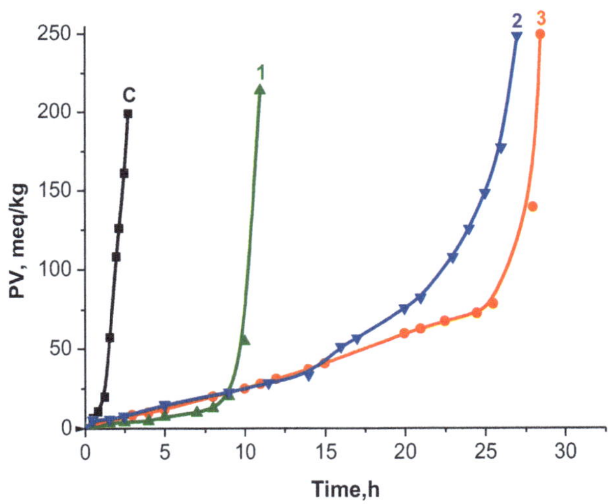

Fig. 7 Kinetic curves of lipid peroxide accumulation during TGSO autoxidation at 80 °C in absence (control, C) and in presence of adrenaline, tocopherol and epicatechine in 1.0 mM concentration

breaking antioxidant activity, i.e. its kinetic curves at both concentrations are almost the same as that for the control sample. Figure 7 demonstrates the effect of truly chain-breaking endogenous phenolic antioxidants adrenaline, tocopherol and epicatechine of different activities on TGSO oxidation under similar conditions. GSH is highly hydrophilic, water soluble thiol. Probably, in the oil medium, glutathione does not dissolve, but forms a transparent dispersion, the particles of which are shielded by TGSO ester groups that prevent the interaction of GSH with peroxyl radicals and hydroperoxides.

Oxidation of methyl linoleate (LH) is widely used as a model reaction for the oxidation of unsaturated lipids (Niki 2012; Frankel 2005; Loshadkin et al. 2020). For testing a variety of water soluble bioantioxidants and their mixtures, water micellar solutions LH are more convenient and successfully used as a kinetic model of the biological process of lipid peroxidation. Recently (Loshadkin et al. 2020), the kinetics of oxygen uptake (W_{O2}) during the chain oxidation of LH in micellar solutions of the nonionic surfactant Triton-100 (TX-100) initiated by the water-soluble initiator 2,2′-azobis (2-methylpropionamide) dihydrochloride (AAPH) has been studied up to the deep stages of LH oxidation. During the oxidation of a freshly prepared LH solution in the TX-100 micellar system, an initial non-stationary stage of increasing the oxygen absorption rate to the W_{st} value is observed, the duration of which decreases with an increase in the amount of added LH and the initiation rate. These features are due to structural changes over time in the micellar system, which, by the time W_{st} is reached, consists of mixed TX-100 micelles with formed hydroperoxides (~2% of TX-100), in the hydrophobic interior of which LH is solubilized. Such micelles provide complete interception of the radicals generated by the initiator, and chain oxidation of LH and TX-100 occurs with a quadratic chain termination.

The phosphate buffer was obtained by mixing 0.05 M solutions of NaH_2PO_4 (Merck) and Na_2HPO_4 (Merck), purified from traces of metals of variable valence using Chelex-100 resin (Bio-Rad). The rates of O_2 absorption during LH oxidation in a micellar solution of TX-100 were measured using a computerized biological oxygen monitor from Yellow Springs Instruments Co. Model 5300A (USA) with Clarke electrode as sensor. The oxidation rate was determined as the slope of the kinetic curves of the $[O_2]$ decrease in the reaction mixture. A comparison of the antioxidant effect of glutathione in LH oxidation in a micellar solution of TX-100 initiated by AAPH, in bidistillate (Fig. 8a) and in the phosphate buffer pH 7.4 (Fig. 8b) shows that in a phosphate buffer solution at the same gross concentrations of AAPH, TX-100 and LH, glutathione additives provide a deeper and longer inhibition than that in deionized water, i.e. the antiradical and antioxidant effects of thiols are more and brighter pronounced in the phosphate buffer. However, it

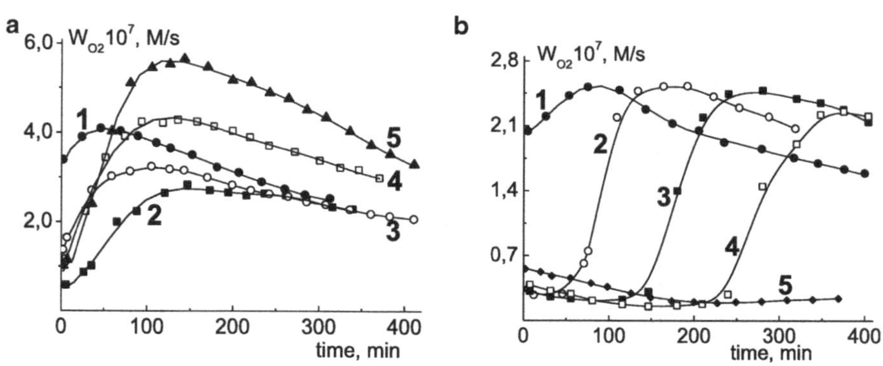

Fig. 8 The effect of GSH on the rate of oxygen uptake (W_{O2}) during the oxidation of AAPH-initiated (4 mM) methyl linoleate (10 mM) in a micellar solution of TX-100 (50 mM) in bidistillate (**a**) and the phosphate buffer pH 7.4 (**b**), 37 °C. Concentration of GSH (left) in bidistillate in mM: 1-0; 2-1; 3-2; 4-5; 5-10. Right, in the phosphate buffer pH 7.4 in mM: 1-0; 2-0,2; 3-0,5; 4-1; 5-2

should be noted that even in the phosphate buffer, the stoichiometric coefficient of inhibition for GSH is less than 1 and decreases with increasing concentration of GSH: 0.13; 0.085; 0.065 (Fig. 8b).

3 Conclusions and Perspective

For the first time, we detected the formation of radicals in the reaction of mercaptoethanol with hydroperoxides, which caused an acceleration of lipid oxidation in an organic medium (Mengele et al. 2015). To find out whether the formation of radicals accompanies the interaction of H_2O_2 with glutathione, known as the main endogenous bioantioxidant, the kinetics of this reaction was studied in pure deionized water to exclude the influence of transition metal impurities in buffer solutions. The radical formation was discovered and the rates of radical initiation were measured by the inhibitor method using the original radical acceptor. By the spin trap method, the formation of namely thiyl radicals was shown in the reaction of glutathione with H_2O_2. The radical yield is low ($<1\%$); however, it has been enough to initiate a chain thiol-ene reaction between GSH and resveratrol, plant phenol with unsaturated bond in a side substituent to aromatic ring, in the presence of H_2O_2 (Zinatullina et al. 2020, 2021). The rates of GSH consumption (W_{GSH}) and radical formation (W_i) in reaction of GSH with H_2O_2 are sensitive to the pH value and ionic composition of the medium. When the pH ≥ 7, W_{GSH} is increases exponentially, and W_i decreases sharply to 0. Therefore, animals and humans, whose physiological pH value is higher than 7.2, are protected from the formation of radicals, and for them glutathione demonstrates antioxidant properties in the best way. However, the detected reactions involving glutathione in neutral and acidic environments may be important, for example, in plants, or when using glutathione in winemaking, cosmetics or food additives.

Funding

This work was supported by the Russian Foundation for Basic Research (project no. 20-03-00753) and performed under a state task (project no. 0082-2018-0006, registration no. AAAA-A18-118020890097-1).

Conflict of Interest The authors declare that there is no conflict of interest regarding the publication of this manuscript.

References

Abedinzadeh Z (2001) Sulfur-centered reactive intermediates derived from the oxidation of sulfur compounds of biological interest. Can J Physiol Pharmacol 79(2):166–170. https://doi.org/10.1139/y00-085

Abedinzadeh Z, Gardes-Albert M, Ferradini C (1989) Kinetic study of the oxidation mechanism of glutathione by hydrogen peroxide in neutral aqueous medium. Can J Chem 67:1247

Afaf K-E, Min DB (eds) (2008) Lipid oxidation pathways. AOCS Press, Champain

Aitken CE, Marshall RA, Puglisi JD (2008) An oxygen scavenging system for improvement of dye stability in single-molecule fluorescence experiments. Biophys J 94(5):1826–1835. https://doi.org/10.1529/biophysj.107.117689

Aldini G, Altomare A, Baron G, Vistoli G, Carini M, Borsani L, Sergio F (2018) N-Acetylcysteine as an antioxidant and disulphide breaking agent: the reasons why. Free Radical Res 52(7): 751–762. https://doi.org/10.1080/10715762.2018.1468564

Chatgilialoglu C, Bowry VW (2018) Why not trans? Inhibited radical isomerization cycles and coupling chains of lipids and alkenes with alkane-thiols. J Org Chem 83(16):9178–9189. https://doi.org/10.1021/acs.joc.8b01216

Chatgilialoglu C, Ferreri C (2005) Trans lipids: the free radical path. Acc Chem Res 36:441. https://doi.org/10.1021/ar0400847

Chatgilialoglu C, Altieri A, Fischer H (2002) The kinetics of Thiyl radical-induced reactions of monounsaturated fatty acid esters. J Am Chem Soc 124:12816–12823

Conway JG, Neptun DA, Garvey LK, Popp JA (1987) Carcinogen treatment increases glutathione hydrolysis by gamma-glutamyltranspeptidase. Carcinogenesis 8:999–1004

Denisov ET, Afanas'ev IF (2005) Oxidation and antioxidants in organic chemistry and biology. CRC Press Taylor & Francis Group

Emanuel NM, Gal D (1978) Oxidation of ethylbenzene. A model reaction. Nauka, Moscow

Emanuel NM, Denisov ET, Maizus ZN (1965) Chain reactions of hydrocarbons oxidation in liquid phase. Nauka, Moscow

Estrela JM, Ortega A, Obrador E (2006) Glutathione critical reviews. Clin Lab Sci 43(2):143–181. https://doi.org/10.1080/10408360500523878

Ferreri C, Chatgilialoglu C (2012) Lipid isomerization in Encyclope-dia of radicals in chemistry. In: Chatgilialoglu C, Studer A (eds) Biology and materials. John Wiley & Sons Ltd, Chichester, UK, pp 1599–1622

Frankel EN (2005) Lipid oxidation. The Oily Press, Glasgow

Guo RW, Yang G, Feng ZJ, Zhu YJ et al (2018) Glutathione-induced amino-activatable micellar photosensitization platform for synergistic redox modulation and photodynamic therapy. Biomater Sci 6(5):1238–1249. https://doi.org/10.1039/c8bm00094h

Halliwell B (2007) Gutteridge, free radicals in biology and medicine, 4th edn. University Press, Oxford

Hawkiks WL, Worthikgston MA (1963) Synergistic antioxidant combinations. Carbon black substitutes. J Polymer Sci Part A 1:3489–3497

Hoyle CE, Bowman CN (2010) Thiol–ene click chemistry. Angew Chem Int Ed 49:1540. https://doi.org/10.1002/anie.200903924

Kade MJ, Burke DJ, Hawker CJ (2010) The power of thiol-ene chemistry. J Polym Sci A Polym Chem 48:743–750. https://doi.org/10.1002/pola.23824

Kamal-Eldin A (2003) Lipid oxidation pathway. AOCS Press, Champaign, IL

Kolb HC, Finn MG, Sharpless KB (2001) Click chemistry: diverse chemical function from a few good reactions. Angew Chem Int Ed 40(11):2004–2021. https://doi.org/10.1002/1521-3773(20010601)40:11<2004::aid-anie2004>3.3.co;2-x

Koo SPS, Stamenović MM, Prasath RA, Inglis AJ et al (2010) Limitations of radical thiol-ene reactions for polymer-polymer conjugation. J Polym Sci A Polym Chem 48:1699–1171

Loshadkin DV, Pliss EM, Kasaikina OT (2020) Features of methyl linoleate oxidation in triton X-100 micellar buffer solutions. Russ J Appl Chem 93(7):1083–1088. https://doi.org/10.1134/S1070427220070216

Mengele EA, Krugovov DA, Kasaikina OT (2015) Effect of mercaptoethanol on the hydrocarbon oxidation and cis—trans-isomerization of unsaturated lipid. Russ Chem Bull 64:846. https://doi.org/10.1007/s11172-015-0943-1

Niki E (2012) Lipid peroxidation in encyclopedia of radicals in chemistry. In: Chatgilialoglu C, Studer A (eds) Biology and materials. John Wiley & Sons Ltd, Chichester, UK, pp 1577–1597

Penninck MJ (2000) A short review on the role of glutathione in the response of yeasts to nutritional, environmental, and oxidative stresses. Enzym Microb Technol 26:737–742
Roy K-M (2005) Thiols and organic sulphides. In: Ullmann's encyclopedia of industrial chemistry. Wiley-VCH, Weinheim. https://doi.org/10.1002/14356007.a26_767
Saito S, Kawabata J (2004) Synergistic effects of thiols and amines on antiradical efficiency of protocatechuic acid. J Agric Food Chem 52:8163–8168
Sajewicz W, Zalewska M, Milnerowicz H (2015) Comparative study on thiol drugs' effect on tert-butyl hydroperoxide induced luminol chemiluminescence in human erythrocyte lysate and hemoglobin oxidation. Toxicol In Vitro 29:149
Salehi B, Mishra AP, Nigam M, Sener B, Kilic M, Sharifi-Rad M, Fokou PVT, Martins N, Sharifi-Rad J (2018) Resveratrol: a double-edged sword in health benefits. Biomedicines 6(3):91. https://doi.org/10.3390/biomedicines6030091
Scott G (1965) Atmospheric oxidation and Antioxidation. Elsevier, Amsterdam
Sebedio JL, Christie WW (1998) Trans fatty acids in human nutrition. The Oily Press, Dundee
Sies H, Jones DP (2007) Oxidative stress. In: Fink G (ed) Encyclopedia of stress, vol 3. Elsevier, San Diego, CA, pp 45–48
Stavrovskaya AA (2000) Cellular mechanisms of multidrug resistance of tumor cells. Biochemistry (Mosc) 65:95–106
Takashima M, Shichiri M, Hagihara Y, Yoshida Y, Niki E (2012) Reactivity toward oxygen radicals and antioxidant action of thiol compounds. Biofactors 38(3):240. https://doi.org/10.1002/biof.1014
Townsend DM, Tew KD, Tapiero H (2003) The importance of glutathione in human disease. Biomed Pharmacother 57:145–155
Toyokuni S (2014) Iron and thiols as two major players in carcinogenesis: friends or foes? Front Pharmacol 5:200
Turunc O, Meier AR (2012) The thiol-ene (click) reaction for the synthesis of plant oil derived polymers. Eur J Lipid Sci Technol 115:41–54. https://doi.org/10.1002/ejlt.201200148
Ulrich K, Jakob U (2019) The role of thiols in antioxidant systems. Free Radic Biol Med 140:14–27
Vanslambrouck S, Riva R, Ucakar B, Préat V, Gagliardi M, Molin DGB, Lecomte H, Jérôme C (2021) Thiol-ene reaction: an efficient tool to design lipophilic Polyphosphoesters for drug delivery systems. Molecules 26(6):1750. https://doi.org/10.3390/molecules26061750
Winterbourn CC (2015) Are free radicals involved in thiol-based redox signaling? Free Rad Biol Med 80:164–170
Winterbourn CC (2016) Revisiting the reactions of superoxide with glutathione and other thiols. Arch Biochemistry Biophysics 595:68
Winterbourn CC (2018 Aug 20) Biological production, detection and fate of hydrogen peroxide. Antioxid Redox Signal. 29(6):541–551. https://doi.org/10.1089/ars.2017.7425. Epub 2017 Dec 14
Winterbourn CC, Hampton MB (2008) Thiol chemistry and specificity in redox signaling. Free Radic Biol Med 45(5):549–461. https://doi.org/10.1016/j.freeradbiomed.2008.05.004
Winterbourn CC, Metodieva D (1995) Reaction of superoxide with glutathione and other thiols. Methods Enzymol 252:81–86
Wu G, Fang YZ, Yang S, Lupton JR, Turner ND (2004) Glutathione metabolism and its implications for health. J Nutr 134:489–492
Yu W, Fu Y-C, Wang W (2012) Cellular and molecular effects of resveratrol in health and disease. J Cell Biochem 11(3):752–759. https://doi.org/10.1002/jcb.23431
Zinatullina KM, Kasaikina OT, Kuzmin VA, Khrameeva NP, Shapiro BI (2016) Interaction of polymethine dyes with hydroperoxides and free radicals. Russ Chem Bull 65(12):2825. https://doi.org/10.1007/s11172-016-1663-x
Zinatullina KM, Kasaikina OT, Kuz'min VA, Khrameeva NP, Shapiro BI (2017a) Kinetic characteristics of the interaction of natural thiols with peroxyl radicals and hydrogen peroxide. Russ Chem Bull 66(7):1300. https://doi.org/10.1134/S0023158417050093

Zinatullina KM, Khrameeva NP, Kasaikina OT, Kuzmin VA, Shapiro BI (2017b) Kinetic characteristics of the interaction of resveratrol with peroxyl radicals and natural thiols in an aqueous medium. Russ Chem Bull 66(11):2145. https://doi.org/10.1007/s11172-017-1995-1

Zinatullina KM, Khrameeva NP, Kasaikina OT (2018) Interaction of natural thiols and catecholamines with reactive oxygen species. Bulg Chem Commun 50:25

Zinatullina KM, Kasaikina OT, Kuz'min VA, Khrameeva NP (2019) Interaction of glutathione with hydrogen peroxide. Kinetic Model Kinet Catal 60(3):266. https://doi.org/10.1134/S0023158419030169

Zinatullina KM, Kasaikina OT, Motyakin MV, Ionova IS, Degtyarev EN, Khrameeva NP (2020) Features of radical formation in the reactions of thiols with hydrogen peroxide. Russ Chem Bull 69(10):1865. https://doi.org/10.1007/s11172-020-2971-8

Zinatullina KM, Kasaikina OT, Khrameeva NP, Indeykina MI, Kononikhin AS (2021) Interaction between glutathione and resveratrol in the presence of hydrogen peroxide: a kinetic model. Kinet Catal 62(2):255. https://doi.org/10.1134/S0023158421020130

Control of Lipid Oxidation in Oil-in Water Emulsions: Effects of Antioxidant Partitioning and Surfactant Concentration

Marlene Costa, Fátima Paiva-Martins, Carlos Bravo-Díaz, and Sonia Losada-Barreiro

1 Introduction

Lipid emulsions are both an integral part of parenteral nutrition and carriers of bioactives (Losada-Barreiro et al. 2021; Costa et al. 2021; Calder et al. 2020; Bush et al. 2019). Different generations of lipid emulsions are available in the market, and they include combinations of different lipid components. Each lipid has its own fatty acid composition and biological effects, which can be more or less beneficial (Calder et al. 2020; Martindale et al. 2020). Soybean oils are the traditional lipid source in a variety of emulsions, (Calder et al. 2020) but in modern lipid emulsions fish oils became important components because of their biological effects, (Pawlik et al. 2011; Calder 2009) which are mainly attributed to the beneficial effects of the α-eicosapentaenoic acid (EPA, C20:5, ω-3) and docosahexaenoic acid (DHA, C22:6, ω-3) fatty acids, Scheme 1. Omega-3 fatty acids (FAs), play a critical role in supporting human health because DHA provides important brain and eye health benefits, while EPA and DHA together promote cardiovascular health, in fact, fish oils are commonly used as source of omega-3 FAs in dietary supplements (Calder et al. 2020; Martindale et al. 2020).

Most vegetable and animal oils are not consumed in their bulk form but are instead dispersed into foods as colloidal droplets in products such as dairy foods, processed meats, dressings, dips, beverages, desserts, yogurts, and sauces. Nevertheless, the use of some of these oils is a major challenge for the food and

M. Costa · F. Paiva-Martins · S. Losada-Barreiro (✉)
REQUIMTE/LAQV, Department of Chemistry and Biochemistry, Faculty of Science, University of Porto, Porto, Portugal
e-mail: sonia@uvigo.es

C. Bravo-Díaz
Departamento de Química Física, Facultad de Química, Universidade de Vigo, Vigo, Spain
e-mail: cbravo@uvigo.es

C. Bravo-Diaz (ed.), *Lipid Oxidation in Food and Biological Systems*,
https://doi.org/10.1007/978-3-030-87222-9_9

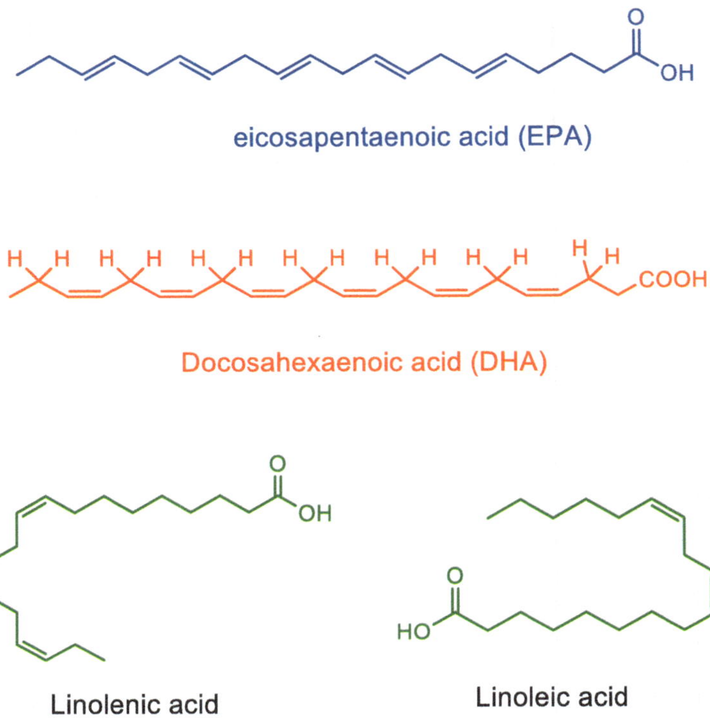

Scheme 1 Chemical structures of some unsaturated fatty acids

pharmaceutical industries due to their low hydrosolubility, unpleasant taste, low bioavailability and, mainly, because of its high susceptibility to oxidation, the higher the number of unsaturations, the higher their oxidation rate is (Foti 2007). Oxidation of unsaturated hydrocarbons usually starts at the allylic position *via* H-atom abstraction, (Foti 2007) although it has been reported that radical addition to the *p* bond is also possible (Ross et al. 2003). The C–H bond elongation and cleavage by radical attack is facilitated by the presence of double bonds that stabilize (by resonance) the intermediate C-centred radical. The propagation rates for stearic, oleic and linoleic acids (ratios 1: 1000: 10000) demonstrate the increasing activation of methylene groups for H-atom abstraction upon increasing the degree of unsaturation (Foti 2007). Polyunsaturated fatty acids are therefore subjected to fast autoxidation, the process being orders of magnitude faster than with the saturated counterparts (Foti 2007; Ross et al. 2003; Frankel 2005).

Emulsions have a high potential in delivering essential fatty acids and bioactives, since they can be designed to have good kinetic stability, oral bioavailability, mask undesirable flavors and can be used to fortify beverages, sauces and desserts, or to increase the bioactivity of omega-3 FAs formulations (McClements 2005; Salvia-Trujillo et al. 2017). On the contrary, the use of emulsions or emulsion-based formulations may have some drawbacks because the lipid components (e.g.,

$$\text{Propagation} \quad r_p \qquad LH + LOO^{\bullet} \xrightarrow[slow]{k_{pro}} LOOH + L^{\bullet} \qquad (1)$$

$$\text{Inhibition} \quad r_{inh} \qquad ArO\text{-}H + LOO^{\bullet} \xrightarrow{k_{inh}} ArO^{\bullet} + LOOH \qquad (2)$$

Scheme 2 Basic representation of the rate-limiting step of the propagation step of the lipid oxidation reaction (1) and of the competing inhibition reaction in the presence of antioxidants (2). *LH* any unsaturated lipid, $LOO^{\bullet}$ peroxyl radical, $L^{\bullet}$ lipid radical, *LOOH* lipid hydroperoxide, *ArO-H* antioxidant and $ArO^{\bullet}$ antioxidant radical

omega-3 FAs) are particularly susceptible to oxidation, consituing a serious economic burden because of the development of off-odors and harmful products, hindering their safe application in nutrition (Calder et al. 2020; Klek 2016; Roche 2012). Attempts to reduce the negative effects of the oxidation of lipids, and particularly omega-3 FAs, have increased in the last years by enhancing food processing, refrigeration and packaging, (Frankel 2005; Shahidi 2015) but, for economical and practical reasons, addition of antioxidants (AOs) is probably the best choice to retard the oxidation of bioactive lipids during the production, manipulation and storage of lipid-based formulations (Frankel 2005; Schaich et al. 2013). Scheme 2 shows the slow step of the propagation step and, when antioxidants are present, the competing inhibition reaction (Ross et al. 2003).

In this chapter we will focus on the control of the oxidative stability of lipid-based emulsions and on the efficiency of natural AOs in emulsions. As we will see, efficient antioxidants need to be located where the inhibition reaction takes place to maximize the rates of the inhibition reactions; otherwise they are uneficient and the lipid oxidation reaction proceeds (Losada-Barreiro et al. 2021; Costa et al. 2021; Ross et al. 2003; Frankel 2005). Some discussion on the effects of the hydrophobicity of antioxidants on their distributions in emulsions will be also included. Special emphasis will be given to the relationships between antioxidant distributions and efficiencies, which could eventually open new possibilities for employing natural AOs in emulsions or in functionalized lipid-based formulations.

2 Effects of Hydrophobicity on Antioxidant Efficiencies: Phenomenological Observations

Despite of extensive research, predicting the antioxidant efficiency in emulsified systems is still challenging (Costa et al. 2021; Decker et al. 2017; Laguerre et al. 2015; Berton-Carabin et al. 2014). The first systematic studies on the effects of hydrophobicity were summarized by Porter(Porter 1993) when he extracted patterns form a large number of confusing data, recognized that, other conditions kept constant, polar antioxidants were, in general, more effective in bulk oils (i.e., low surface to volume ratio systems) than in emulsifed systems (Porter 1993). On the contrary, Porter pointed out that lipophilic antioxidants were more effective in emulsions than in bulk oils. In other words, polar antioxidants apparently worked better in apolar fluids meanwhile apolar antioxidants work better in polar systems. This apparent contradiction is known as the "polar paradox" (Porter 1993; Shahidi and Zhong 2011). Porter was, most probably, aware of the difficulties in rationalizing this apparently anomalous behavior, recognizing that comparisons between antioxidants of different hydrophobicity should be taken with caution because, for instance, increasing the number of methyl substituents in antioxidants bearing different reactive moieties may change simulateneously their hydrophilic-lipophilic balance (HLB), their reduction potentials and the rates of reaction with radicals, suggesting that comparisons between homologous series of antioxidants bearing the same reactive moieties would provide better comparisons for testing the polar paradox (Nahas and Berdah 2013).

The practical value of the polar paradox came later when Frankel and coworkers (Frankel 2005; Schwarz et al. 2000) tested the polar paradox by systematically comparing new sets of antioxidants in emulsions and in bulk oils, hypothesizing the involvent of the interfacial layer (Nahas and Berdah 2013). Results demonstrated the versatility of the polar paradox as a predictive tool in some sets of antioxidants, recognizing the problem of interfacial modifications in the location of antioxidants. However, the lack of quantitative measurements of lipopilicity and hydrophilicity in different oils and the in the interfacial regions of emulsions, the hurdles in determining the effective concentrations of antioxidants in the oil, interfacial and aqueous regions, (Stöckman and Schwarz 1999; Oehlke et al. 2008) together with the difficulties found in selecting a propert method to monitor the oxidation reaction limited substantially their predictive power (Shahidi and Zhong 2011; Nahas and Berdah 2013).

Following Frankel's suggestions, other researchers started to investigate on interfacial effects more closely, measuring a variety of physical parameters such as surface tension (Lucas et al. 2010) and octanol-water partition coefficients, (Torres de Pinedo et al. 2007) but were not able to find acceptable correlations between the measured parameters and the antioxidant efficiencies. More recent studies reported on the non-linear effects observed in series of homologous antioxidants bearing the same reactive moieties but of different hydrophobicity. This non-linear effect is called the "cut-off" effect and reflects the experimental observation of an increase in the antioxidant efficiency upon increasing the hydrophobicity of the antioxidant up to a limit, after which the efficience decreases. Several putative hypotheses in attempting to provide a rationale to the anomalous observed behavior. Among them, limitations in the aqueous solubility of the AOs, changes in their physical properties, kinetic effects, changes in AO mobilities, self-aggregation, and changes in the location of the antioxidants and differences in the percentage o AO in the interfacial region of the emulsions were proposed (Laguerre et al. 2012, 2015; Panya et al. 2012). Unfortunately, no experimental evidence to prove, or discard, any of these hypotheses was provided (Laguerre et al. 2012, 2015). More recently, the same authors hypothesized that the "cut-off" effect governing antioxidant activity in lipid dispersions may be due to the fact that above a certain hydrophobicity, the transfer mechanism for antioxidants changes from diffusion to collision-exchange-separation (Laguerre et al. 2017).

Similar "cut-off" effects were observed when investigating the behavior a series of homologous 2,2,6,6-Tetramethylpiperidin-1-yl)oxyl or (2,2,6,6-tetramethylpiperidin-1-yl)oxidanyl (TEMPO) radicals (Lopez de Arbina et al. 2019; Aliaga et al. 2016) (see chapter by C. Aliaga in this book for further information) and in the various biological activities of series of homologous surfactants of different hydrophobicities (Balgavý and Devínsky 1996). Some of the most recent reports on the topic have demonstrated that the "cut-off" effect is a consequence of the subtle balance of the inter- and intra-molecular forces, which makes antioxidants to have a differential partitioning in the oil, aqueous and interfacial region of emulsions, modifying the effective concentrations of the antioxidants in those regions (Costa et al. 2016, 2020a, b, 2021; Meireles et al. 2019; Raimúndez-Rodríguez et al. 2019; Mitrus et al. 2019; Bravo-Díaz et al. 2015).

2.1 Determining the Distribution of AOs in Intact Emulsions

In emulsions, Scheme 3, antioxidants distribute between the oil, interfacial and aqueous regions and their distribution can be described by the partition constants: between oil and interfacial region, P_O^I, Eq. (1), and between aqueous and interfacial region, P_W^I, Eq. (2). The parenthesis in Eqs. (1–2) refer to the concentration of AOs in a particular region expressed in moles per liter of the volume of the particular region.

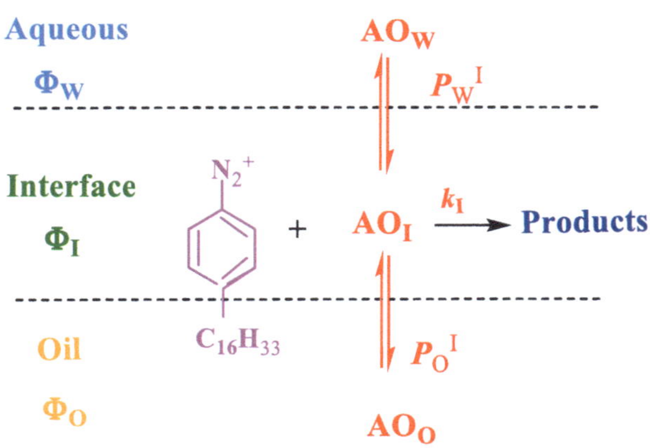

Scheme 3 Distribution of an antioxidant (AO) between the different regions of a model lipid-based emulsion. P_O^I = partition constant of the AO between oil and interface, P_W^I = partition constant of the AO between aqueous and interface and k_I = rate constant for the reaction between 4-hexadecylbenzenediazonium ion (16-ArN_2^+) and the antioxidant in the interfacial region. Φ_O, Φ_I and Φ_W stand for the volume fraction of the oil, surfactant (interfacial) and aqueous regions, respectively

Table 1 Main equations employed to determine the partition constants values, percentage of AOs between the different regions of emulsions and their effective interfacial molarities (AO_I)

Oil insoluble AOs $P_W{}^O <<< 1$	AOs of moderate hydrophobicity	Water insoluble AOs $P_W{}^O >>> 1$
$k_{obs} = \frac{k_1[AO]_T P_W^I}{\Phi_I P_W^I + \Phi_W}$	$k_{obs} = \frac{[AO_T] k_1 P_W^I P_O^I}{\Phi_O P_W^I + \Phi_I P_W^I P_O^I + \Phi_W P_O^I}$	$k_{obs} = \frac{k_1[AO]_T P_O^I}{\Phi_I P_O^I + \Phi_O}$
$\%AO_I = \frac{100 \Phi_I P_W^I}{\Phi_I P_W^I + \Phi_W}$	$\%AO_I = \frac{100 \Phi_I P_W^I P_O^I}{\Phi_O P_W^I + \Phi_I P_W^I P_O^I + \Phi_W P_O^I}$	$\%AO_I = \frac{100 \Phi_I P_O^I}{\Phi_I P_O^I + \Phi_O}$
$(AO_I) = \frac{[AO_I](\%AO_I)}{\Phi_I}$ (3)		

k_{obs} stands for the observed (or measured) rate constant between the AO and the chemical probe (see Sect. 2), $P_W{}^O$ stands for the partition constant of the antioxidant in binary oil-water sytems, $P_W{}^I$ and $P_O{}^I$ are the partition constants between the water-interfacial and oil-interfacial regions of emulsions, respectively. $\%AO_I$ is the percentage of AO in the interfacial region and (AO_I) is the effective concentration of the antioxidant in the interfacial region (expressed as moles per liter of interfacial volume). Details on their derivatization can be found in the chapter by Romsted et al. in the present book and elsewhere (Bravo-Díaz et al. 2015; Dar et al. 2017; Freiría-Gándara et al. 2018a; Romsted and Bravo-Díaz 2013; Gunaseelan et al. 2006)

To determine AO distributions in the intact emulsions, a kinetic method based on the reaction between an antioxidant and the chemical probe, 4-hexadecylbenzenediazonium ion ($16\text{-}ArN_2^+$) was developed. The chemical probe is exclusively located at the interface where reacts with the antioxidant, and the observed rate constant depends on the amount of antioxidant in the interfacial region, which in turn, depends on the partitioning. Only the relevant equations are given here, Table 1. Further details on the method and calculations can be found in the chapter by Romsted et al. in this book and elsewhere (Bravo-Díaz et al. 2015; Dar et al. 2017).

$$P_O^I = \frac{(AO_I)}{(AO_O)} \tag{1}$$

$$P_W^I = \frac{(AO_I)}{(AO_W)} \tag{2}$$

3 Control of Lipid Oxidation with Antioxidants

3.1 Effects of the Hydrophobicity of the Antioxidant and Emulsifier Concentration on the Distribution of Antioxidants

A careful choice of both the physicochemical properties of antioxidants and composition of emulsified systems are crucial because they strongly affect their effective concentrations in the oil, aqueous and interfacial regions, and thus, they are expected

Table 2 Values for the partition constant values for different series of homologous AOs in olive oil-in water emulsions and the estimated percentage of antioxidant in the interfacial region of the emulsions ($\Phi_I = 0.005$)

		O/W Emulsion		
AO structure	–R	P_O^I	P_W^I	%AO_I
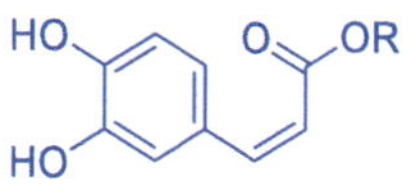 Caffeates	–H (CA)	–	349	75
	$-CH_2\,CH_3$ (C2)	405	3156	81
	$-(CH_2)_2\,CH_3$ (C3)	454	–	85
	$-(CH_2)_7\,CH_3$ (C8)	502	–	86
	$-(CH_2)_{15}\,CH_3$ (C16)	376	–	83
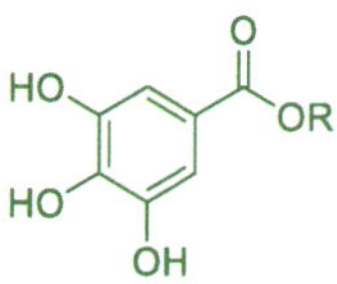 Gallates	–H (GA)	–	101	46
	$-(CH_2)_2\,CH_3$ (C3)	449	328	65
	$-(CH_2)_7\,CH_3$ (C8)	27	–	25
	$-(CH_2)_{11}\,CH_3$ (C12)	17	–	18
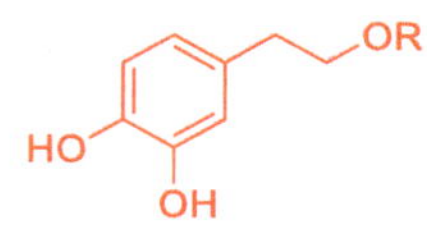 Hydroxytyrosol derivatives	–H (HT)	–	53	40
	$-CO-CH_2\,CH_3$ (C3)	171	373	56
	$-CO-(CH_2)_4\,CH_3$ (C6)	184	–	70
	$-CO-(CH_2)_6\,CH_3$ (C8)	296	–	79
	$-CO-(CH_2)_8\,CH_3$ (C10)	125	–	61
	$-CO-(CH_2)_{14}\,CH_3$ (C16)	52	–	40
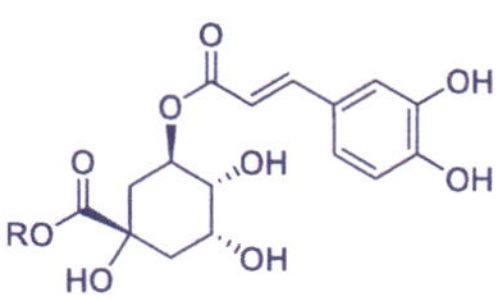 Chlorogenates	–H (CGA)	–	40	25
	$-CH_2\,CH_3$ (C2)	–	78	40
	$-(CH_2)_3\,CH_3$ (C4)	–	141	54
	$-(CH_2)_7\,CH_3$ (C8)	111	–	58
	$-(CH_2)_{11}\,CH_3$ (C12)	159	–	67

Data extracted from refs. (Meireles et al. 2019; Costa et al. 2015; Almeida et al. 2016; Losada-Barreiro et al. 2013)

to affect their antioxidant efficiency against free radicals. Several reports have been published in attempting to gain a better understanding on the factors that control the efficiency of antioxidants in emulsions (Meireles et al. 2019; Costa et al. 2015, 2020b; Almeida et al. 2016; Ferreira et al. 2018). Most of them point out the fact that the antioxidant efficiency depends on, among others, the interfacial concentrations of the antioxidants (Meireles et al. 2019; Costa et al. 2015, 2020a, b; Almeida et al. 2016).

To investigate the influence of the hydrophobicity of AOs on their distributions, series of AOs bearing the same radical scavenging moieties but of different hydrophobicity were used, Table 2. By grafting inert hydrocarbon chains of different length to the aromatic ring of antioxidants changes the hydrophilic-hydrophobic balance (HLB) of the AOs, and, consequently, their relative solubility in the different regions of the emulsion (Laguerre et al. 2012). Table 2 shows some of the reported

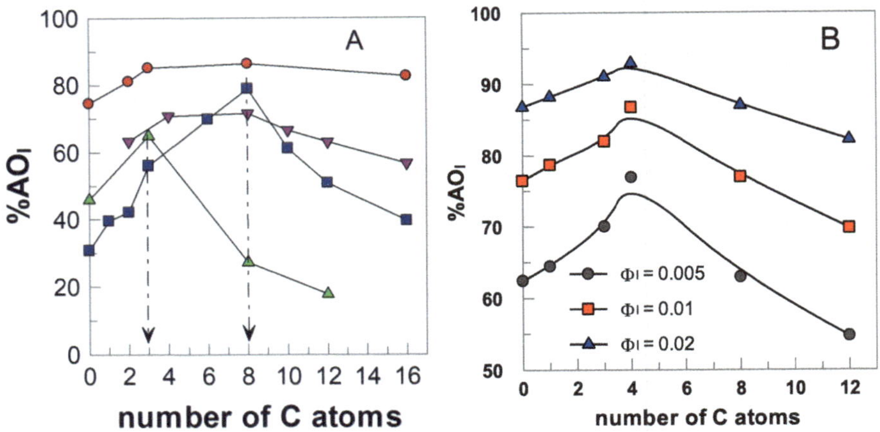

Fig. 1 (**a**) Variation of the percentage of antioxidants in the interfacial region, %AO_I, derived of caffeic acid (red circles), (Costa et al. 2015) hydroxytyrosol (blue squares) (Almeida et al. 2016), gallic acid (green triangles) (Losada-Barreiro et al. 2013) and protocatechuic acid (pink inverted triangles) (Silva et al. 2017) (**b**) Effects of surfactant volume fraction, Φ_I, on the percentage of gallic acid esters in soybean emulsions. Figure 1b is reprinted with permission from reference (Mitrus et al. 2019). Copyright (2019) American Chemical Society

values for the partition constants of homologous series of AOs in emulsions and the percentage of antioxidants in the interfacial region, which can be easily calculated by employing those values as indicated elsewhere (Freiría-Gándara et al. 2018a; Costa et al. 2015; Almeida et al. 2016; Freiría-Gándara et al. 2018b).

Partitioning experiments were carried out in the intact emulsions to avoid disruption of the existing equilibria (see Sect. 2). Results obtained for series of homologous antioxidants, Table 2, indicate that the percentage of AO in the interfacial region, % AO_I, does not increase linearly upon increasing the hydrophobicity of the antioxidant but shows a maximum at chain lengths of 3–12 C atoms, depending of particular AO chemical structure, Fig. 1a, after which %AO_I decreases. Similar results were obtained for different series of AOs in emulsions prepared with different oils, Fig. 1b, indicating that the parabolic-like variation is a common feature of the effects of hydrophobicity on the distribution of AOs in emulsions.

These results are in clear contradiction with the predictions of the polar paradox, indicating that the highest percentage in the interfacial region is not achieved for the most hydrophobic antioxidant but for that of medium-high hydrophobicity, highlighting a widely extended mistake in literature reports considering that AOs are incorporated into the interfacial region of the emulsion upon increasing the hydrophobicity of the antioxidant without taking into consideration that, in spite that the interfacial region is better solvent than water, it may be worst than the oil, so that AOs are solubilized in the oil region after reaching a certain hydrophobicity. The optimal alkyl chain length to reach the highest AO percentage in the interfacial region ranges 3–12 C atoms, (Costa et al. 2021; Laguerre et al. 2017) and it has been proposed that the optimal hydrophobicity is reached when the HLB of the AOs matches the HLB of the interfacial region (composed of a melange of oil, water and surfactant) (Losada-Barreiro et al. 2021; Costa et al. 2021; Mitrus et al. 2019; Bravo-Díaz et al. 2015; Losada-Barreiro et al. 2013).

The non-linear behavior observed in Fig. 1 is typical of the "*cut-off*" effect, discussed in Sect. 2, (Balgavý and Devínsky 1996) and it is, therefore, a natural consequence of the differential solubility of the antioxidants in the interfacial and oil regions as a result of the changes in the hydrophobicity of the antioxidants (Meireles et al. 2019; Costa et al. 2020a; Almeida et al. 2016; Ferreira et al. 2018; Freiría-Gándara et al. 2018b). The edge hydrophobicity that makes the antioxidant to be optimally located in the interfacial region of the emulsion depends on nature and position of the substuents in the aromatic ring of the phenolic antioxidant and on the composition of the interfacial region (where solvation forces may be relevant) but either an increase or decrease in this limiting hydrophobicity decreases the concentration on AO at the interfacial region.

3.2 Effects of the Surfactant Concentration

The surfactant concentration employed to stabilize kinetically the emulsions has also important effects on the effective concentrations of antioxidants (Raimúndez-Rodríguez et al. 2019; Mitrus et al. 2019; Ferreira et al. 2018; Freiría-Gándara et al. 2018b; Silva et al. 2017). The emulsifier volume fraction (Φ_I), which reflects the concentration of surfactant, affects the distribution of AOs in two opposite directions. For one side, an increase in Φ_I increases the percentage of AO at the interfacial region so that at $\Phi_I = 0.04$, more than 70% of all antioxidants are located in this region as illustrated in Fig. 2 for chlorogenic and gallic acid derivatives. This increase in $\%AO_I$ with Φ_I is usually less than two fold (for some antioxidants can be up to three fold, Fig. 2a). On the other hand, the increase in the interfacial volume is eight fold when Φ_I changes from 0.005 to 0.04. Therefore, the increase in the percentage of AOs in the interfacial region does not compensate the increase in the interfacial volume, and antioxidants are effectively diluted in the interfacial region upon increasing Φ_I. These effecs are illustrated in Figs. 2b for CGA and GA derivatives, and is in keeping with the predictions of Eq. 3 in Table 1.

It is worthwhile noting that, for any AO, its effective interfacial molarity is much higher (~20–250 fold, depending on Φ_I) than the stoichiometric concentration of the antioxidants ($[AO_T] = 0.2$–0.4 mM). These results indicate that AOs are effectively concentrated in the interfacial region, but the effective concentration depends on both Φ_I and the hydrophobicity of the antioxidant. Results, thus, suggest that the antioxidant efficiency of compounds can be increased by increasing in their scavenging ability (i.e., the rate constant, k_{inh}, of the inhibition process, reaction 2 in Scheme 2), but also by modulating its effective concentration at the interfacial region.

In contrast, reported values for the effective concentrations in the oil region indicate that the effective concentration of the AOs is only ~0.1–1.5 times higher

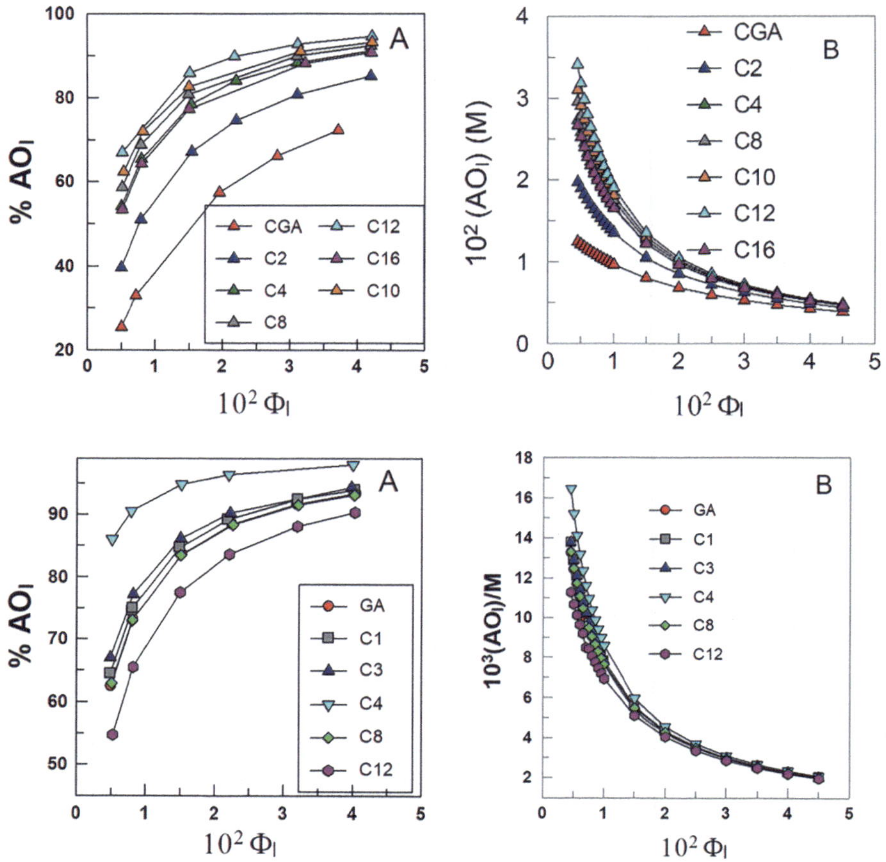

Fig. 2 Variations of the percentage of antioxidants in the interfacial region (**a**) and of the effective interfacial concentrations (**b**) of chlorogenic (CGA) and gallic acid (GA) derivatives in olive and soybean oil emulsions. Figures extracted from refs. (Meireles et al. 2019) (CGA) and (Mitrus et al. 2019) (GA)

than the stoichiometric concentration, meanwhile that in the aqueous region is much lower (Mitrus et al. 2019; Almeida et al. 2016; Ferreira et al. 2018).

3.3 Relationships Between the Antioxidant Efficiency and Interfacial Concentrations

Extensive research has been aimed at understanding the relationship between antioxidant hydrophobicity and its antioxidant efficiency in emulsions to control lipid oxidation (Costa et al. 2021; Laguerre et al. 2017; Mitrus et al. 2019; Ferreira et al. 2018; Freiría-Gándara et al. 2018b; Laguerre et al. 2020a, b). In spite of this, some

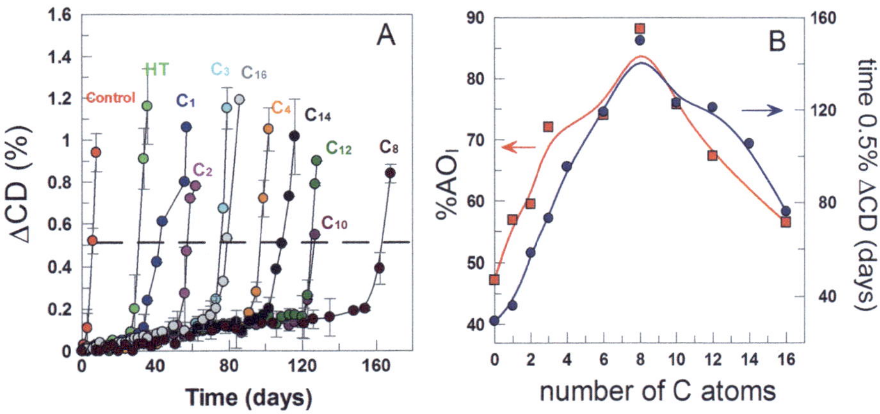

Fig. 3 (**a**) A typical oxidation kinetics curves obtained in 4:6 (O/W) olive oil emulsions in the presence of hydroxytyrosol ester derivatives (1% Tween 20, T = 60 °C). The control emulsion is prepared in the absence of AO. The dashed line is the time required to reach 0.5% conjugated dienes content, used as a marker of to evaluate the efficiency of the antioxidants. (**b**) Correlation between the percentage of hydroxytyrosol derivatives (red square), $\%AO_I$, with the antioxidant efficiency (blue circles). Reprinted with permission from reference (Almeida et al. 2016). Copyright (2016) American Chemical Society

aspects are still a matter of debate. The kinetic profiles of the inhibited peroxidation processes, Fig. 3a, are usually characterized by a slow buildup of primary oxidation products (e.g., conjugated dienes, CDs) in time followed by a faster production of CDs (propagation reaction, reaction 1 in Scheme 2). In the presence of antioxidants, the lipidic radicals produced in the system react with the antioxidant (reaction 2 in Scheme 2). During this relatively slow step (inhibition reaction 2), the reaction continues until the antioxidant is almost consumed so that when the antioxidant concentration is nearly depleted, the inhibition reaction becomes un-inhibited, and the rate of the overall oxidation reaction increases, Fig. 3a (Frankel 2005; Litwinienko 2005; Ingold and Pratt 2014; Valgimigli and Pratt 2012). The relative antioxidant efficiency can be determined either by measuring the length of the induction period τ_{ind} in Fig. 3a or by determining the time necessary to increase the formation of CDs in a given amount (for instance, $\Delta CD(\%) = 0.5$) as indicated by the dashed line in Fig. 3a).

The variations of both τ_{ind} or $\Delta CD(\%)$ with the length of the alkyl chain of the antoxidant derivatives show a parabolic-like dependence, increasing upon increasing the hydrophobicity of the AOs up to a maximum, a further increase in the hydrophibicty leads to a decrease in both τ_{ind} or $\Delta CD(\%)$, Fig. 3b.

Figure 3b shows that the variations of the efficiency of AOs (as asssed by the change in $\Delta CD(\%)$) and the percentage of AO in the interfacial region, $\%AO_I$, (or (AO_I)) parallel each other, providing a molecular explanation of the "cut-off" effect (Costa et al. 2015, 2021; Meireles et al. 2019; Raimúndez-Rodríguez et al. 2019; Mitrus et al. 2019; Ferreira et al. 2018; Freiría-Gándara et al. 2018b; Silva et al. 2017).

4 Conclusions

The early need to merely identify which natural materials had strong antioxidant efficiencies has been replaced now by the need to elucidate accurately how natural antioxidants act and how their efficiency can be improved. Both the polar paradox approach and the phenomenological "cut-off" effects are, in a broad context, widely used to interpret the effects of hydrophobicity on antioxidant efficiency, recognizing that antioxidants may be located in the lipid core, interfacial region and aqueous phase of oil-in-water emulsions (Lopez de Arbina et al. 2019; Balgavý and Devínsky 1996; Costa et al. 2015, 2017; Losada-Barreiro et al. 2013; Silva et al. 2017; Laguerre et al. 2009). However, both approaches fail in considering crucial parameters that have an enormous impact in the inhibition reaction such as the effective concentrations of the antioxidants in the various regions. This is a consequence, in part, of the lack of suitable methods to determine distributions in intact emulsions avoiding their rupture distorting the existing equilibria (Losada-Barreiro et al. 2021; Costa et al. 2021; Bravo-Díaz et al. 2015). In attempting to improve the polar paradox and to rationalize the "cut-off" effects, pseudophase approaches have been developed aiming to estimate the antioxidant location and their effective concentrations and linking them to the antioxidant efficiency. Estimations of the distributions of antioxidants and of their effective concentrations are made in the intact emulsions (in spite that emulsions are opaque) and under different experimental conditions, resulting in a large improvement in rationalizing the complex effects of antioxidants (Decker et al. 2017; Genot 2015). The pseudophase method is versatile enough to measure antioxidant interfacial location in a variety of model food emulsions, providing important information of how complex emulsion droplet interfaces with multiple emulsifiers impact on the efficiency of antioxidants (Losada-Barreiro et al. 2021; Costa et al. 2021). Application of pseudophase models to lipid-based emulsions allowed, for the first time, to obtain physical evidence of the interfacial region as the main reaction site between antioxidants and lipid radicals, (Ferreira et al. 2018; Freiría-Gándara et al. 2018b; Silva et al. 2017; Costa et al. 2017) allowing to show experimentally that there is a direct correlation between the antioxidant efficiencies and their effective interfacial concentrations. This is a crucial point in the development of new antioxidant strategies because, as pointed out by Genot, (Genot 2015) such correlations are crucial in rationalizing the complex effects of different properties (e.g. acidity, temperature, mixed emulsifiers, droplet sizes, etc.).

Evaluation of the effective concentrations of AOs in the interfacial region of an emulsion is, therefore, a crucial determinant of the success of an AO in inhibiting lipid oxidation. New antioxidants should be designed, therefore, to maximize both their free radical scavenging rate constants and their interfacial concentration to achieve optimal oxidative stabilities. Modulation of the hydrophilic/lipophilic properties of AOs using natural phenolics (e.g. chlorogenic and caffeic acids, hydroxytyrosol, etc.) to optimize its effective concentration in the interfacial region can be a significant contribution to control the lipid oxidation in emulsified systems. Expanding our knowledge on the factors that affect AO distributions will result in a better control of the oxidative degradation of lipid-based foods because addition of efficient antioxidants to stabilize omega-3 enriched emulsions (and other lipid-based systems) is a promising strategy that may enable a significant improvement in the oxidative stability and bioavailability of lipid-based formulations.

Acknowledgement This chapter was prepared during the sabbatical leave of C. B-D, supported by Universidad de Vigo. The authors thank FCT/MCTES (UIDB/QUI/50006/2020; PTDC/OCE-ETA/32492/2017–POCI-01-0145-FEDER-032492; doctoral grant SFRH/BD/100889/2014), REQUIMTE-LAQV (PTDC/OCE-ETA/32492/2017), and Xunta de Galicia (10TAL314003PR) for financial support.

References

Aliaga C, Bravo-Moraga F, Gonzalez-Nilo D, Márquez S, Lürh S, Mena G et al (2016) Location of TEMPO derivatives in micelles: subtle effect of the probe orientation. Food Chem 192:395–401

Almeida J, Losada-Barreiro S, Costa M, Paiva-Martins F, Bravo-Díaz C, Romsted LS (2016) Interfacial concentrations of Hydroxytyrosol and its lipophilic esters in intact olive oil-in-water emulsions: effects of antioxidant hydrophobicity, surfactant concentration, and the oil-to-water ratio on the oxidative stability of the emulsions. J Agric Food Chem 64:5274–5283

Balgavý P, Devínsky F (1996) Cut-off effects in biological activities of surfactants. Adv Colloid Interf Sci 66:23–63

Berton-Carabin CC, Ropers M-H, Genot C (2014) Lipid oxidation in oil-in-water emulsions: involvement of the interfacial layer. Compr Rev Food Sci Food Safety 13:945–977

Bravo-Díaz C, Romsted LS, Liu C, Losada-Barreiro S, Pastoriza-Gallego MJ, Gao X et al (2015) To model chemical reactivity in heterogeneous emulsions. Think Homogeneous Microemulsions Langmuir 31:8961–8979

Bush L, Stevenson L, Lane KE (2019) The oxidative stability of omega-3 oil-in-water nanoemulsion systems suitable for functional food enrichment: a systematic review of the literature. Crit Rev Food Sci Nutr 59(7):1154–1168. https://doi.org/10.1080/10408398.2017.1394268

Calder PC (2009) Rationale for using new lipid emulsions in parenteral nutrition and a review of the trials performed in adults: conference on 'malnutrition matters' symposium 4: hot topics in parenteral nutrition. Proc Nutr Soc 68(3):252–260

Calder PC, Waitzberg DL, Klek S, Martindale RG (2020) Lipids in parenteral nutrition: biological aspects. J Parenter Enter Nutr 44(S1):S21–SS7

Costa M, Losada-Barreiro S, Paiva-Martins F, Bravo-Díaz C, Romsted LS (2015) A direct correlation between the antioxidant efficiencies of caffeic acid and its alkyl esters and their concentrations in the interfacial region of olive oil emulsions. The pseudophase model interpretation of the "cut-off" effect. Food Chem 175:233–242

Costa M, Losada-Barreiro S, Paiva-Martins F, Bravo-Diaz C (2016) Optimizing the efficiency of antioxidants in emulsions by lipophilization: tuning interfacial concentrations. RSC Adv 6 (94):91483–91493

Costa M, Losada-Barreiro S, Paiva-Martins F, Bravo-Díaz C (2017) Physical evidence that the variations in the efficiency of homologous series of antioxidants in emulsions are a result of differences in their distribution. J Sci Food Agric 97(2):564–571

Costa M, Losada-Barreiro S, Bravo-Díaz C, Vicente AA, Monteiro LS, Paiva-Martins F (2020a) Influence of AO chain length, droplet size and oil to water ratio on the distribution and on the activity of gallates in fish oil-in-water emulsified systems: emulsion and nanoemulsion comparison. Food Chem 310:125716

Costa M, Losada-Barreiro S, Bravo-Díaz C, Vicente AA, Monteiro LS, Paiva-Martins F (2020b) Influence of AO chain length, droplet size and oil to water ratio on the distribution and on the activity of gallates in fish oil-in-water emulsified systems: emulsion and nanoemulsion comparison. Food Chem 310(1–9):125716

Costa M, Losada-Barreiro S, Paiva-Martins F, Bravo-Díaz C (2021) Polyphenolic antioxidants in lipid emulsions: partitioning effects and interfacial phenomena. Foods 10(3):539

Dar AA, Bravo-Diaz C, Nazir N, Romsted LS (2017) Chemical kinetic and chemical trapping methods: unique approaches for determining respectively the antioxidant distributions and interfacial molarities of water, counter-anions, and other weakly basic nucleophiles in association colloids. Curr Opin Colloid Interface Sci 32:84–93

Decker EA, DJ MC, Bourlieu-Lacanal C, Durand E, Figueroa-Espinoza MC, Lecomte J et al (2017) Hurdles in predicting antioxidant efficacy in oil-in-water emulsions. Tr Food Sci Technol 67 (Supplement C):183–194

Ferreira I, Costa M, Losada-Barreiro S, Paiva-Martins F, Bravo-Díaz C (2018) Modulating the interfacial concentration of gallates to improve the oxidative stability of fish oil-in-water emulsions. Food Res Int 112:192–198

Foti MC (2007) Antioxidant properties of phenols. J Pharm Pharmacol 59(12):1673–1685

Frankel EN (2005) Lipid oxidation. The Oily Press, PJ Barnes & Associates, Bridgwater, England

Freiría-Gándara J, Losada-Barreiro S, Paiva-Martins F, Bravo-Díaz C (2018a) Differential partitioning of bioantioxidants in edible oil–water and octanol–water systems: linear free energy relationships. J Chem Eng Data 63(8):2999–3007

Freiría-Gándara J, Losada-Barreiro S, Paiva-Martins F, Bravo-Díaz C (2018b) Enhancement of the antioxidant efficiency of gallic acid derivatives in intact fish oil-in-water emulsions through optimization of their interfacial concentrations. Food Funct 9(8):4429–4442

Genot C (2015) Distributions of phenolic acid antioxidants between the interfacial and aqueous regions of corn oil emulsions—a commentary. Eur J Lipid Sci Technol 117(11):1684–1686

Gunaseelan K, Romsted LS, Pastoriza-Gallego MJ, González-Romero E, Bravo-Díaz C (2006) Determining α-tocopherol distributions betweeen the oil, water and interfacial regions of macroemulsions: novel applications of electrocanalytical chemistry and a pseudophase kinetic model. Adv Colloid Interf Sci 123–126:303–311

Ingold KU, Pratt DA (2014) Advances in radical-trapping antioxidant chemistry in the 21st century: a kinetics and mechanisms perspective. Chem Rev 114(18):9022–9046

Klek S (2016) Omega-3 fatty acids in modern parenteral nutrition: a review of the current evidence. J Clin Med 5(3):34

Laguerre M, López-Giraldo LJ, Lecomte J, Figueroa-Espinoza MJ, Baréa B, Weiss J et al (2009) Chain length affects antioxidant properties of chlorogenate esters in emulsion: the cut-off theory behind the polar paradox. J Agric Food Chem 57:11335–11342

Laguerre M, Bayrasy C, Lecomte J, Chabi B, Decker EA, Wrutniak-Cabello C et al (2012) How to boost antioxidants by lipophilization? Biochimie 95:1–7

Laguerre M, Bayrasy C, Panya A, Weiss J, McClements DJ, Lecomte J et al (2015) What makes good antioxidants in lipid-based systems? The next theories beyond the polar paradox. Crit Rev Food Sci Nutr 55(2):183–201

Laguerre M, Bily A, Roller M, Birtić S (2017) Mass transport phenomena in lipid oxidation and Antioxidation. Annu Rev Food Sci Technol 8(1):391–411

Laguerre M, Tenon M, Bily A, Birtić S (2020a) Toward a spatiotemporal model of oxidation in lipid dispersions: a hypothesis-driven review. Eur J Lipid Sci Technol 122(3):1900209

Laguerre M, Bily A, Birtić S (2020b) Chapter 7 – Lipid oxidation in food. In: Galanakis CM (ed) Lipids and edible oils. Academic Press, Cambridge, MA, pp 243–287

Litwinienko G (2005) Chapter 7 – Analysis of lipid oxidation by differential scanning calorimetry. In: Kamal Elding A, Pokorny J (eds) Analysis of lipid oxidation. AOCS Press

Lopez de Arbina A, Losada-Barreiro S, Rezende MC, Vidal M, Aliaga C (2019) The location of amphiphobic antioxidants in micellar systems: the diving-swan analogy. Food Chem 279:288–293

Losada-Barreiro S, Bravo Díaz C, Paiva Martins F, Romsted LS (2013) Maxima in antioxidant distributions and efficiencies with increasing hydrophobicity of gallic acid and its alkyl esters. The pseudophase model interpretation of the "cut-off effect". J Agric Food Chem 61:6533–6543

Losada-Barreiro S, Bravo-Díaz C, Paiva-Martins F (2021) Why encapsulate antioxidants in emulsion-based systems, where they are located, and how location affects their efficiency. In: Aboudzadeh MA (ed) Emulsion-based encapsulation of antioxidants. Food bioactive ingredients. Cham, Switzerland, Springer Nature

Lucas L, Comelles F, Alcántara D, Maldonado O, Curcuroze M, Parra JL et al (2010) Surface-active properties of lipophilic antioxidants Tyrosol and Hydroxytyrosol fatty acid esters: a potential explanation for the nonlinear hypothesis of the antioxidant activity in oil-in-water emulsions. J Agric Food Chem 58:8021–8026

Martindale RG, Berlana D, Boullata JI, Cai W, Calder PC, Deshpande GH et al (2020) Summary of proceedings and expert consensus statements from the international summit "lipids in parenteral nutrition". J Parenter Enter Nutr 44(S1):S7–S20

McClements DJ (2005) Food emulsions. CRC Press, Boca Raton, USA

Meireles M, Losada-Barreiro S, Costa M, Paiva-Martins F, Bravo-Díaz C, Monteiro LS (2019) Control of antioxidant efficiency of chlorogenates in emulsions: modulation of antioxidant interfacial concentrations. J Sci Food Agric 99:3917–3925

Mitrus O, Żuraw M, Losada-Barreiro S, Bravo-Díaz C, Paiva-Martins F (2019) Targeting antioxidants to interfaces: control of the oxidative stability of lipid-based emulsions. J Agric Food Chem 67(11):3266–3274

Nahas R, Berdah D (2013) The polar paradox: how an imperfect conceptual framework accelerated our knowledge of antioxidant behavior. In: Logan A, Nienaber U, Pan X (eds) Lipid oxidation, challenges in food systems. AOCS Press, Urbana, IL, USA, pp 243–260

Oehlke K, Garamus V, Heins A, Stöckman H, Schwarz K (2008) The partitioning of emulsifiers in o/w emulsions: a comparative study of SANS, ultrafiltration and dialysis. J Colloid Interface Sci 322:294–303

Panya A, Laguerre M, Bayrasy C, Lecomte J, Villeneuve P, McClements D et al (2012) An investigation of the versatile antixoidant mechanism of action of rosmarinate alkyl esters in oil-in-water emulsions. J Agric Food Chem 60:2692–2700

Pawlik D, Lauterbach R, Turyk E (2011) Fish-oil fat emulsion supplementation may reduce the risk of severe retinopathy in VLBW infants. Pediatrics 127(2):223–228

Porter WL (1993) Paradoxical behavior of antioxidants in food and in biological systems. Priceton Scientific, Princeton, NJ, pp 93–122

Raimúndez-Rodríguez EA, Losada-Barreiro S, Bravo-Díaz C (2019) Enhancing the fraction of antioxidants at the interfaces of oil-in-water emulsions: a kinetic and thermodynamic analysis of their partitioning. J Colloid Interface Sci 555:224–233

Roche LD (2012) Oxidative stress: the dark side of soybean-oil-based emulsions used in parenteral nutrition. Oxid Antioxid Med Sci 1(1):11–14

Romsted LS, Bravo-Díaz C (2013) Modelling chemical reactivity in emulsions. Curr Opin Colloid Interface Sci 18:3–14

Ross L, Barclay C, Vinqvist MR (2003) Phenols as antioxidants. In: Rappoport Z (ed) The chemistry of phenols. Wiley, West Sussex, England

Salvia-Trujillo L, Soliva-Fortuny R, Rojas-Graü MA, McClements DJ, Martín-Belloso O (2017) Edible Nanoemulsions as carriers of active ingredients: a review. Annu Rev Food Sci Technol 8 (1):439–466

Schaich KM, Shahidi F, Zhong Y, NAM E (2013) Chapter 11 - Lipid oxidation. In: Biochemistry of foods, 3rd edn. Academic Press, San Diego, pp 419–478

Schwarz K, Huang SW, German JB, Tiersch B, Hartmann J, Frankel EN (2000) Activities of antioxidants are affected by colloidal properties of oil-in-water and water-in-oil emulsions and bulk oils. J Agric Food Chem 48:4874–4882

Shahidi F (2015) Handbook of antioxidants for food preservation, 1st edn. Woodhead Publiser, Sawston

Shahidi F, Zhong Y (2011) Revisiting the polar paradox theory: a critical overview. J Agric Food Chem 59:3499–3504

Silva R, Losada-Barreiro S, Paiva-Martins F, Bravo-Díaz C (2017) Partitioning and antioxidative effect of protocatechuates in soybean oil emulsions: relevance of emulsifier concentration. Euro J Lipid Sci Technol 119(6):1600274-n/a

Stöckman H, Schwarz K (1999) Partitioning of low molecular weight compounds in oil-in-water emulsions. Langmuir 15:6142–6149

Torres de Pinedo A, Peñalver P, Pérez-Victoria I, Rondón D, Morales JC (2007) Synthesis of new phenolic fatty acid esters and their evaluation as lipophilic antioxidants in an oil matrix. Food Chem 105:657–665

Valgimigli L, Pratt DA (2012) Antioxidants in chemistry and biology. In: encyclopedia of radicals in chemistry, biology and materials. John Wiley &Sons, USA. https://doi.org/10.1002/9781119953678.rad055

Effects of Emulsion Droplet Size on the Distribution and Efficiency of Antioxidants

Marlene Costa, Sonia Losada-Barreiro, Carlos Bravo-Díaz, and Fátima Paiva-Martins

1 Lipid-Based Emulsions

Emulsion technology can be employed to enrich a variety of different food and beverage products such as yogurt, milk and salad dressings, with bioactive lipids (e.g. polyunsaturated fatty acids, PUFAs) showing important advantages, by reducing and/or inhibiting their oxidation and, concomitantly, providing a range of health benefits (Zembyla et al. 2020).

An emulsion is a mixture of two or more immiscible liquids, typically oil and water, where oil droplets are dispersed in an aqueous phase (O/W emulsions) or aqueous droplets are dispersed in an oil (W/O emulsions). The preparation of emulsions that are kinetically stable over a required time period needs the addition of surface-active compounds (emulsifiers) that adsorb at the oil/water interface of emulsion droplets, weakening interfacial forces, and allowing the immiscible phases to be mixed (Goodarzi and Zendehboudi 2019). Many types of emulsifiers can be used for generating emulsions such as low molecular weight surface active compounds (Spans, Tweens, lecithins, etc), biopolymers (polysaccharides and proteins including starch, pectin, whey protein, soy protein and egg protein), and insoluble

M. Costa (✉) · F. Paiva-Martins
REQUIMTE-LAQV, Departamento de Química e Bioquímica, Faculdade de Ciências, Universidade do Porto, Porto, Portugal
e-mail: marlene.costa@fc.up.pt

S. Losada-Barreiro
REQUIMTE-LAQV, Departamento de Química e Bioquímica, Faculdade de Ciências, Universidade do Porto, Porto, Portugal

Departamento Química Física, Facultad de Química, Universidade de Vigo, Vigo, Spain

C. Bravo-Díaz
Departamento Química Física, Facultad de Química, Universidade de Vigo, Vigo, Spain

C. Bravo-Diaz (ed.), *Lipid Oxidation in Food and Biological Systems*,
https://doi.org/10.1007/978-3-030-87222-9_10

Table 1 Differences between macroemulsions, nanoemulsions and microemulsions related to size, stability, emulsification method among others. Adapted from references (Gupta et al. 2016; Nastiti et al. 2017; Mason et al. 2006)

	Macroemulsions	Nanoemulsions	Microemulsions
	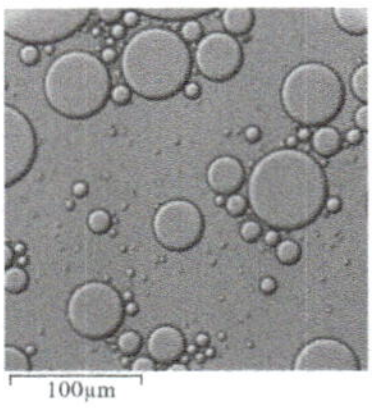	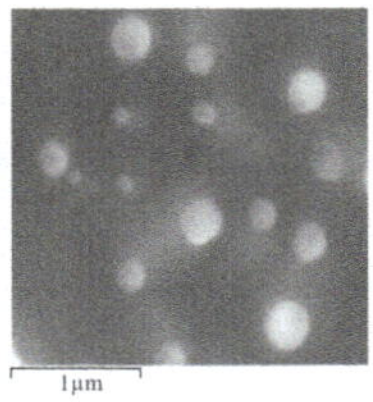	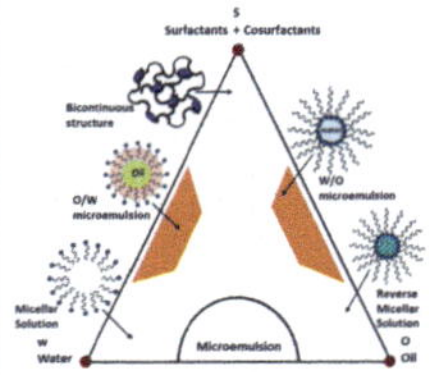
Size	1–100 μm	20–500 nm	10–100 nm
Shape	Spherical	Spherical	Spherical, lamellar
Physical appearance	Creamy	Transparent	Transparent
Stability	Thermodynamically unstable, weakly kinetically stable	Thermodynamically unstable, kinetically stable	Thermodynamically stable
Method of preparation	High energy methods	High & low energy methods	Low energy methods
Emulsifier concentration	Low	Moderate	High
Polydispersity	Often high (>40%)	Typically low (<20%)	Typically low (<10%)

particles in Pickering emulsions (hydrophobic silica particles, modified cellulose particles etc.) (Berton-Carabin et al. 2018).

Emulsions can be prepared by employing different high energy methods such as high-pressure homogenization and ultrasonication, or low energy methods, e.g., phase inversion temperature and emulsion inversion point (McClements et al. 2017). Description of the methods for the preparation of emulsions, their pros and cons, and of the use of the different emulsifiers will not be covered here, and the interested reader is referred to specialized reviews where they are elegantly discussed (Zembyla et al. 2020; McClements et al. 2017; McClements and Rao 2011).

The size of foods emulsion droplets can vary hugely and, according to the size of their droplets, can be classified in three types (McClements 2012): microemulsions (droplet sizes <100 nm), nanoemulsions (droplet sizes 100–400 nm) and macroemulsions (droplet sizes >400 nm).

The major differences between the different mentioned emulsified systems are the droplet sizes and their stability characteristics, as summarized in Table 1 (Gupta et al. 2016; Nastiti et al. 2017). Both macroemulsions and nanoemulsions are thermodynamically unstable, i.e. given sufficient time, phase separation will occur. Nanoemulsions can be kinetically stable over longer period of time than macroemulsions mainly due to their smaller average droplet diameter (Delmas et al. 2011). On the other hand, microemulsions are thermodynamically stable systems, what makes them sensitive to changes in temperature and composition, in contrast with nanoemulsions, which are very attractive for industrial application because they are quite insensitive to physical and chemical changes. Some studies have clarified the distinction between nanoemulsions and microemulsions detailing the differences between these two classes of liquid-in-liquid dispersions (McClements 2012; Anton and Vandamme 2011).

The distribution of the droplet diameter determines both the number of droplets and the total interfacial area. Therefore, the bigger the droplets, the smaller will be the interfacial area formed (Berton-Carabin et al. 2014). The nature and concentration of emulsifiers, the pH, composition and ionic strength of the aqueous phase are among the factors that will determine not only the electrostatic charge of the droplets but also the thickness of the interfacial layer, being both parameters involved in the physical and chemical stability of the emulsified systems (Tcholakova et al. 2008; Wilde et al. 2004; Dickinson 1992; Grigoriev and Miller 2009).

2 Physical Stability of Emulsions

Macro- and nanoemulsions are thermodynamically unstable systems that tend to break down over time through different mechanisms that can be linked to each other. These breakdown processes occur as a consequence of different factors such as the physicochemical properties of the emulsifier, pH, ionic strength, homogenization parameters, temperature, and interactions of dispersed phase with the continuous phase among others (Zembyla et al. 2020).

Details of their thermodynamic and kinetic stability and of their most common instability mechanisms are given elsewhere (Zembyla et al. 2020; McClements 2015). A brief summary is given below.

2.1 Thermodynamics and Kinetics

The trend of any colloidal systems will be always to reach the state with the lowest free energy, and thus the thermodynamic stability of an emulsified system can be assessed by the difference in Gibbs free energy between the emulsified and the breakdown states, ΔG, Scheme 1. Both macroemulsions and nanoemulsions need some energy input (mechanical stirring, high-pressure homogenization, etc.) to become metastable systems, which is is required to balance the increase in enthalpy caused by the contact between hydrophilic and hydrophobic molecules, which is not compensated by the increased entropy of mixing, that is, to overcome the Gibbs free energy needed to increase interfacial area between the two phases and to obtain emulsion droplets. Macroemulsions and nanoemulsions are thermodynamically unstable because the Gibbs free energy of emulsified system is higher than the free energy of the separate phases ($\Delta G_T<0$), Scheme 1. However, microemulsions are

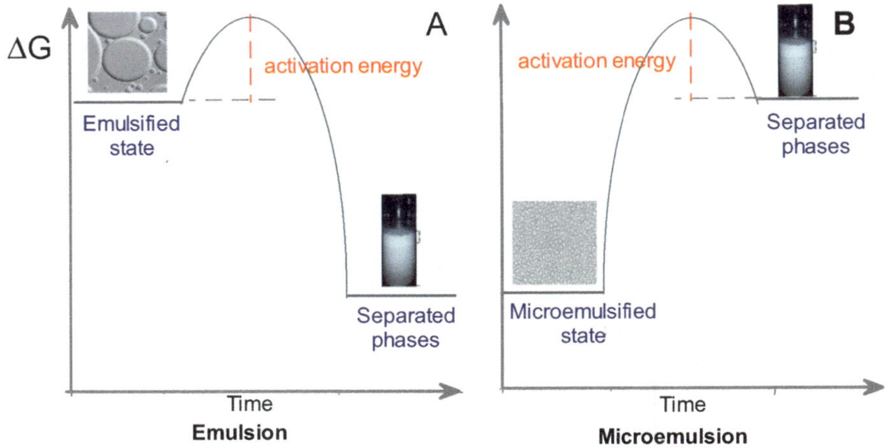

Scheme 1 Representation of the energy difference between the emulsified (**A**) or microemulsified (**B**) state and non-emulsified state. Emulsified systems have a higher free energy than the separated state whereas microemulsified systems have a lower free energy. Figure adapted from reference (McClements 2012)

thermodynamically stable due to the free energy of the emulsified system is lower than that of the separate phases ($\Delta G_T > 0$), remaining stable indefinitely if the formation conditions remain unaltered.

To stabilize kinetically the emulsions, amphiphilic molecules having both hydrophilic, water soluble, and lipophilic, oil soluble, residues capable of interact simultaneously with the oil and water need to be added to the binary oil-water system. Surfactants locate at the droplet surface with their hydrophilic part of the emulsifier pointing to the water and its lipophilic part pointing to the oil, forming a three-dimensional boundary surrounding the surface of the droplets, and this arrangement makes emulsions to be stabilized kinetically for periods ranging minutes to years. The choice of the emulsifier (cationic, anionic, zwitterionic and nonionic) is crucial in the formation of the emulsion and its long-term stability (McClements 2015). Some details on the factors that affect the stability of emulsions and their mechanisms are given below and the interested reader is referred to specialized literature reports (McClements 2005).

2.2 *Physical Destabilization Mechanisms*

Several mechanisms have been proposed to rationalize the instability of emulsions:

– Creaming and sedimentation occur as a consequence of gravitational separation when the density of the dispersed phase is lower or higher than the density of the continuous phase respectively. Oils commonly have densities lower than the

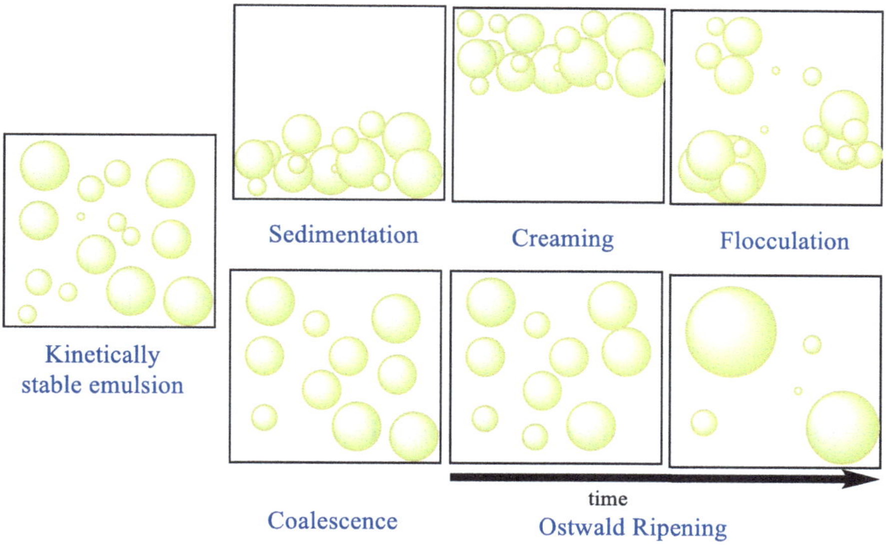

Scheme 2 Schematic representation of the different breakdown processes in emulsions: creaming or sedimentation, flocculation (aggregation), coalescence and Ostwald ripening (diffusion of dispersed molecules from small droplets to large ones across the continuous phase). Further details of the characteristics of each process and of methods to prevent or minimize them can be found elsewhere (Zembyla et al. 2020)

aqueous phase and so creaming is one of the most common forms of instability in O/W emulsions whereas sedimentation is one of the most common forms of instability in W/O emulsions. In both cases, this phenomenon usually does not change the droplet size. Creaming or sedimentation can be controlled in emulsified systems by decreasing the density differences between the dispersed and continuous phases, reducing the particle size or increasing the viscosity of the continuous phase.

- Flocculation is the process in which two or more droplets make contact but do not merge, maintaining their individual integrities (without changing droplet size). It is a consequence of a low droplet repulsion due to an unbalanced of the attractive and repulsive forces between droplets. Flocculation can be delayed in emulsified systems by increasing the viscosity of the continuous phase, droplet size and charges on the droplet surface, among others.
- Coalescence is the process by which two or more droplets merge together and form a larger droplet, which leads to a disruption of the interdroplet liquid film. In this sense, the driving force for coalescence is the interfacial film fluctuations. These fluctuations caused by the high mobility of molecules at the interfacial film breaks and thus, the droplets merge together and form a lager droplet causing coalescence. Coalescence is always irreversible while flocculation may be reversible (weak flocculation) or irreversible (strong flocculation). Coalescence is mostly controlled by the presence of emulsifiers adsorbed at the oil-water-

interface and it depends on their physicochemical properties. For example, coalescence occurs in O/W emulsions containing nonionic emulsifiers in which the electrical repulsive effect is negligible. The presence of charged emulsifiers at the droplet interface may produce electrostatic repulsion between the droplets, which tends to prevent droplet contact.

- Ostwald ripening is a process in which large droplets grow at the expense of smaller droplets due to molecular diffusion (mass transport) of the dispersed phase (oil) through the continuous phase (water). The driving force for Ostwald ripening is the difference in solubility between small and larger droplets. Ostwald ripening is retarded in emulsions with a narrow droplet size distribution.

3 Oxidative Stability of Emulsions: Lipid Oxidation

Lipid oxidation of unsaturated lipids has been, and still is, one of the most studied reactions in the food industry because it leads to food degradation and it is associated to the production of both toxic substances and undesirable of-odors. In general, lipid oxidation can be defined as the chemical reaction between unsaturated fatty acids (PUFAs) and reactive oxygen species (ROS), leading to the formation of free radicals and hydroperoxides that eventually decompose over time, Scheme 3.

The fate of the lipid oxidation reaction depends, among others, on the nature of the lipids involved and of the environment where the reaction takes place, It can be catalyzed and accelerated by the presence of metals and enzymes, the pH of emulsified systems, and, as in any other chemical reaction, its rate will depend on the nature of the reactants (e.g., the number of unsaturations in the lipids) and on environmental factors such as temperature, light,, the type of interface between lipids and oxygen (continuous lipid phase or dispersed in an emulsified system) and the presence or absence of antioxidants, which are commonly employed to halt or inhibit the reaction.

3.1 Brief Description of the Mechanism of Lipid Oxidation and Inhibition by Antioxidants

The fundamental processes that take place during lipid oxidation have been widely described by several authors. Lipid oxidation, particularly, autooxidation is a purely chemical and quite complex phenomenon in which $^3O_2$ reacts with lipid radicals, triggering radical reactions, capable of auto-propagation. Three steps or stages are usually identified: initiation, propagation, and termination (Scheme 3).

The initiation step takes place with the loss of a H-atom in the α-carbon of an unsaturated fatty acid. The reaction can be catalyzed by metals and accelerated by heating, UV-light, ionization radiations, food pigments such chlorophylls and

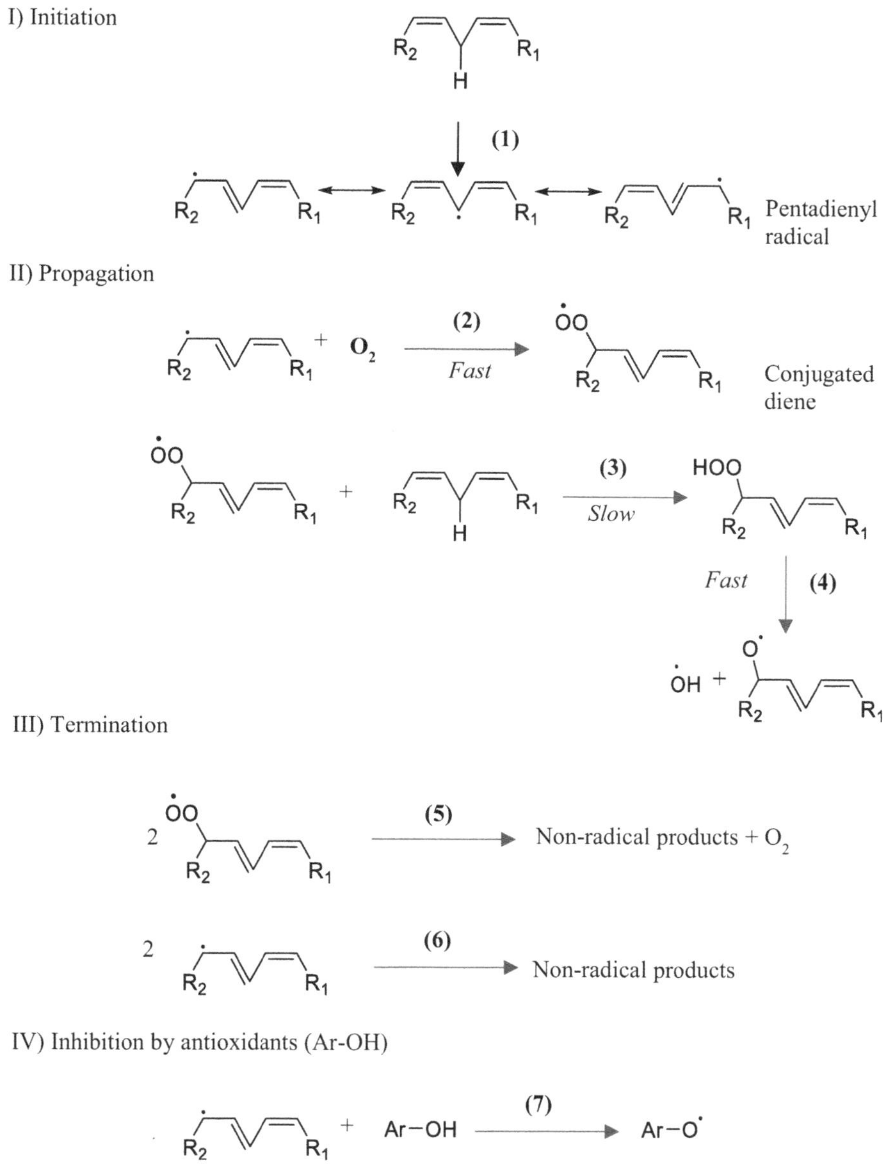

Scheme 3 The three main stages of the free radical oxidation of an unsaturated fatty acid (steps I, II and III) and their inhibition by radical scavenger antioxidants ArOH (step IV)

enzymes such as lipoxygenases. Due to the loss of the H-atom, lipid free radicals (Scheme 3, reaction 1) and H radicals ($H^{\bullet}$) are formed.

The difficulties in removing the H-atom (Scheme 3, reaction 1, 3 and 4) will dictate the rates at which the oxidation reactions occurs, and it is an extremely important factor controlling the oxidative stability of unsaturated fatty acids. The

C-H bond in a saturated alkyl chain normally has a dissociation energy of ~414 kJ/mol at 298 K. However, the C-H bond on an allyl carbon has a significantly lower dissociation energy, ~372 kJ/mol. On the other hand, being the C-H bond on a bis-allyl carbon, this dissociation energy is even lower, ~330 kJ/mol (Ma and Zaera 2006). Thus, due to their lower dissociation energy and greater stabilization of the radicals formed by resonance, bis-allylic hydrogens represent the most labile H in a fatty acid molecule. This explains the much higher oxidative instability of fish oils o when compared to other edible oils (e.g. olive oil). During the propagation step, the free radicals derived from the fatty acids react quickly with molecular oxygen to form peroxyl free radicals (Scheme 3, reaction 2). In turn, they can remove new H-atoms from other fatty acids (Scheme 3, reaction 3), slow step, producing more hydroperoxides and more free radicals, triggering the reaction. When more than one product is obtained, only one is represented as an example for the sake of simplicity.

The reaction terminates until the consumption of all unsaturated fatty acids or until free radicals react with each other, through various pathways, all of them leading to the formation of non-radical products. (Scheme 3, reaction 5 and 6).

To minimize the oxidation of lipids, exogenous antioxidants are added to the emulsions. Most of the antioxidants are phenolic in nature, working as radical scavenging radicals. They react with the peroxyl radicals, reaction 7 in Scheme 3, but they may also limit the formation of radicals in the initiation step (secondary antioxidants, reaction 1 in Scheme 3) (Frankel 2005; Ross et al. 2003). Efficient antioxidants are those molecules whose rate of trapping radicals (reaction 7 in Scheme 3) is equal to, or higher than, the rate of production of radicals (reaction 2–4 in Scheme 3) (Raimúndez-Rodríguez et al. 2019; Costa et al. 2017; Freiría-Gándara et al. 2018a, b; Ferreira et al. 2018). Details on the effects of antioxidants and their mechanisms of action can be found elsewhere (Frankel 1980; Shahidi et al. 1992; Foti et al. 2005).

3.2 *Effects of Droplet Sizes on the Oxidative Stability of Emulsions*

Relatively few reports can be found in the literature concerning the influence of the droplet sizes on the chemical stability of emulsified systems, in spite of its importance in the shelf-life of emulsified products (Berton-Carabin et al. 2014). This lack of studies may be due, in part, to misleading tendency to extrapolate the results obtained in bulk oils to emulsified systems without taking into consideration the kinetics of the processes involved, the effects of polarity of the different regions on the rates of the reactions, and, mainly, the lack of trustable values for the effective concentrations of antioxidants in each region of the emulsions.

Literature reports are frequently contradictory (Berton-Carabin et al. 2014; Kikuchi et al. 2014; Osborn and Akoh 2004; Neves et al. 2017); some reports conclude that a decrease in emulsion droplet sizes leads to a decrease in the oxidative stability of the emulsions, (Berton-Carabin et al. 2014; Uluata et al. 2016) meanwhile others reports presume that a decrease in the size of emulsion droplet results in a better oxidative stability of the emulsified system (Gohtani et al. 1999; Lethuaut et al. 2002; Azuma et al. 2009; Lee et al. 2011). In other studies, however, no significant effects of the droplet sizes on the oxidative stability are reported (Burakova et al. 2017; Romsted 2012). To further increase the controversial, one can find in the literature papers indicating that the oxidative stability of emulsions can vary widely with slight variations in the experimental conditions and when changing the nature of the oil used in the experiment,(Costa et al. 2020a) meanwhile some others attempt to rationalize the effects of droplet sizes on the basis of the opposite effects found between the physical and oxidative stabilities: small emulsion droplets improve the physical stability of the emulsions but decrease their oxidative stability (Berton-Carabin et al. 2014). The latter observation leads to the frequently argued hypothesis suggesting that the droplet size effects are related to the droplet surface areas, so that smaller emulsion droplets lead to higher interfacial areas, resulting in a greater "exposure" of the emulsion droplets to the oxygen and pro-oxidant compounds, leading to an increase in the lipid oxidation (Kabalnov 1994; Gruner et al. 2016).

Recently, Costa et al. argued, on kinetic grounds, that droplet sizes should not affect the oxidative stability of emulsions because the lipid oxidation and inhibition reactions take place, in different extents, in three-dimensional regions (oil, water and interfacial) and therefore the rates of the reactions should depend on concentrations (moles per liter of volume of the region) and not on surface areas (Costa et al. 2020b, c, 2021). To prove or discard the hypothesis, they prepared a series of fish oil emulsions and nanoemulsions of different oil to water ratio, determining some physical characteristics (among others their polydispersity degree and droplet sizes) and their oxidative stability, both in the presence and absence of antioxidants. The formulated emulsified systems were polydisperse and had quite different average droplet sizes, being their average values of 1282 nm (1:9 O/W emulsions), 4860 nm (4:6 O/W emulsions) and <300 nm (1:9 O/W nanoemulsions) when the emulsifier volume fraction employed to stabilize kinetically the emulsions was $\Phi = 0.005$ (Table 2).

Emulsions were allowed to oxidize spontaneously in the absence (control) and presence of antioxidants, and the relative antioxidant efficiency was assessed by employing Eq. (1) (Costa et al. 2020a)

$$\text{Relative increase in the oxidative stability} = \frac{\left(t_{(AO)} - t_{(C)}\right)}{t_{(C)}} \tag{1}$$

where $t_{(AO)}$ and $t_{(C)}$ are the time necessary for the samples containing each AO or the control, respectively, to increase by 0.5% the conjugated diene content. Figure 1 shows the effects of the droplet size on the chemical stability of emulsified systems in the absence and in the presence of gallates in 1:9 fish oil-in water emulsions and nanoemulsions and in 4:6 fish oil-in water emulsions, all of them containing a $\Phi_I = 0.005$. The chemical structures of the antioxidants are shown in Table 3.

Table 2 Characteristics of fish emulsions and nanoemulsions

	Nanoemulsions			Emulsions	
Φ_o	1.0	1.0	**1.0**	1.0	4.0
$10^2\ \Phi_I$	**0.5**	**1.0**	**2.0**	0.5	0.5
ς-Potential (mV)	−18.1	−14.2	−13.5	−22.6	Nd
10^6 d (m)	0.297	0.227	0.175	1.28	4.86
$10^{12}\ S_{droplet}\ (m^2)$	0.28	0.16	0.10	5.0	74.2
$10^{20}\ V_{droplet}\ (m^3)$	1.37	0.61	0.28	105	6007
$10^{-12}\ N_d$	73	163	357	9.5	0.67
$S_{total}\ (m^2)$	20.2	26.4	34.3	4.8	4.9
$10^2\ m_{T80\ available}/m^2$ of surface (g)	0.25	0.38	0.58	1.05	1.01
$10^2\ m_{T80,\ dropplet}$ (g)	0.53	0.69	0.90	0.12	0.13
$10^2\ m_{T80,\ excess}$ (g)	−0.03	0.31	1.10	0.38	0.37

d = droplet diameter; $V_{droplet} = \pi x d^3/6$, volume of one droplet; $S_d = \pi x d^2$, droplet surface; $N_d = V_{total\ oil}/V_{droplet}$, total number of droplets (V_{total}, oil is the total volume of oil per 10 g of emulsion); $S_{total} = 6xV_{total,\ oil}/d$, surface of all droplets; $m_{T80,\ droplet} = S_{total}\ x\ \Gamma_\infty$, mass of surfactant required for saturation per 100 g of emulsion (estimated using a interfacial coverage at saturation of $\Gamma_\infty = 2 \times 10^{-6}$ mol m^{-2}; $m_{T80,\ excess} = m_{T80-mT80,\ droplet}$, excess (g) of Tween 80 that may remain in the aqueous phase (Berton-Carabin et al. 2014; Costa et al. 2020a). Table 2 reprinted with permission from reference (Costa et al. 2020a)

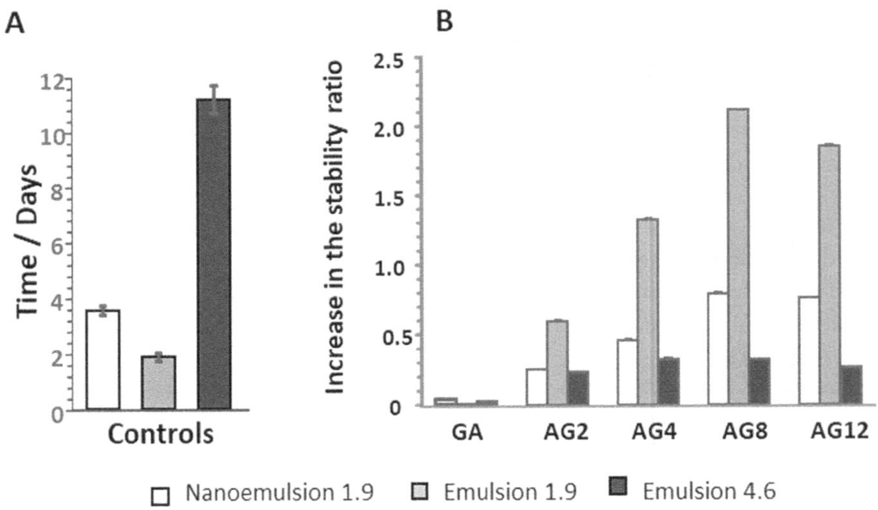

Fig. 1 (**A**) Changes in the time needed to reach a variation of the content of conjugated dienes of 0.5% in the emulsified systems prepared in the absence of AOs, T = 40 °C. (B) Influence of the hydrophobicity of gallic acid (GA) and gallates in their antioxidant efficiency ($\Phi_I = 0.005$) in the same emulsions studied in (**A**). Reprinted with permission from reference (Costa et al. 2020a)

Table 3 Partition constants values between the oil and interfacial region, P_O^I, and between the aqueous and interfacial region, P_W^I, obtained for derivatives of gallic acid (GA) and hydroxytyrosol (HT) in fish oil emulsions and in fish oil nanoemulsions

Compound	**R**	**GA**	**GA2**	**GA4**	**GA8**	**GA12**	
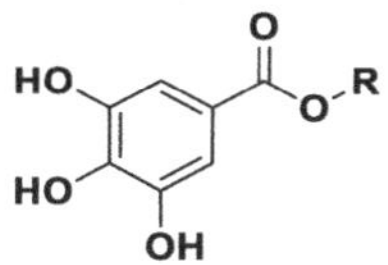		**−H**	**−(CH$_2$CH$_3$)**	**−(CH$_2$)$_5$CH$_3$**	**−(CH$_2$)$_7$CH3**	**−(CH$_2$)$_{11}$CH$_3$**	-
Emulsion O/W	P_W^I	118 ± 19	233 ± 32	559 ± 127	–	–	-
	P_O^I	–	706 ± 56	253 ± 56	183 ± 126	142 ± 37	-
Nanoemulsion O/W	P_W^I	109 ± 19	247 ± 32	578 ± 127	–	–	-
	P_O^I	–	750 ± 56	261 ± 56	189 ± 126	151 ± 37	-
Compound	**R**	**HT**	**HT2**	**HT6**	**HT8**	**HT12**	**HT16**
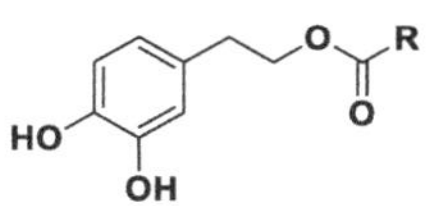		**−H**	**−CH$_3$**	**−(CH$_2$)$_4$CH$_3$**	**−(CH$_2$)$_6$CH$_3$**	**−(CH$_2$)$_{10}$CH$_3$**	**−(CH$_2$)$_{14}$CH$_3$**
Emulsion O/W	P_W^I	34 ± 1	207 ± 39	–	–	–	
	P_O^I	–	115 ± 22	89 ± 8	119 ± 45	97 ± 14	75 ± 12
Nanoemulsion O/W	P_W^I	30 ± 4	186 ± 14	–	–	–	–
	P_O^I	–	103 ± 8	79 ± 22	104 ± 13	89 ± 19	80 ± 13

Data extracted from (Costa et al. 2020a, b)

Figure 1A shows that, when carrying out oxidative experiments in the absence of antioxidants, under the same experimental conditions, the oxidative stability of 1:9 emulsions and nanoemulsions is different, as well as that of emulsions prepared with different oil-to-water ratio (o/w), the 4:6 emulsions being more stable than the 1:9 nanoemulsions. The observed increase in the oxidative stability of emulsions upon increasing the oil fraction is in accordance with reports in the literature (Osborn and Akoh 2004; Kargar et al. 2011; Sun and Gunasekaran 2009). Interestingly, emulsions and nanoemulsions of the same composition, Table 2, have different oxidative stability, the 1:9 nanoemulsions showing better oxidative stability than 1:9 (O/W) emulsions.

Upon addition of antioxidants to the emulsified systems, Fig. 1B, the same trends in the oxidative stabilities as in their absence were obtained. Figure 1B illustrates the effects of the hydrophobicity of the antioxidants on the oxidative stability, showing the expected "cut-off" effect (details on the cut-off effects can be found in the chapter by Losada-Barreiro in the present book). It is worthwhile noting that the same ranking of antioxidant efficiencies is obtained in emulsions and nanoemulsions independently of the sizes of the droplets.

Figure 2 further illustrates the effects of the droplet sizes and of the hydrophobicity of antioxidants (gallic acid and hydroxytyrosol derivatives, Table 3 in nanoemulsions prepared with different surfactant volume fractions Φ_I (Costa et al. 2020a, b). In 1:9 O/W nanoemulsions, an increase in Φ_I leads to a decrease in the droplet diameter (Table 2). In the presence of the most lipophilic AOs (AG4–AG12 and HT8–HT16 alkyl esters, Fig. 2A, and B respectively) a decrease in the droplet size decreases the oxidative stability of the system. However, for the most hydrophilic antioxidants (GA, GA2, HT, and HT2), an increase in the oxidative stability is observed upon increasing the surface area. Similar conclusions were reported for 4:6 O/W emulsions of the same composition (Costa et al. 2020a, b), where different trends in the oxidative stability with the increase in the droplet size were observed depending on the nature of the antioxidant used to stabilize the emulsion.

Results in Figs. 1 and 2 demonstrate, therefore, that no direct correlation can be established between the oxidative stability and the total interfacial surface areas, deserving a deeper analysis of the phenomena. It is worthwhile noting that, whatever are the effects the droplet sizes, the relative antioxidant efficiencies are not modified. For instance, in Fig. 2, the cut-off effects observed upon increasing the hydrophobicity of the antioxidants are evident, supporting the idea the oxidative stability of emulsions or nanoemulsions loaded with antioxidants cannot be predicted on the basis of droplet sizes but on the effective antioxidant concentrations (see chapter by Losada-Barreiro in this book for details). To prove, or discard, the idea, the partitioning and effective concentrations of antioxidants were determined.

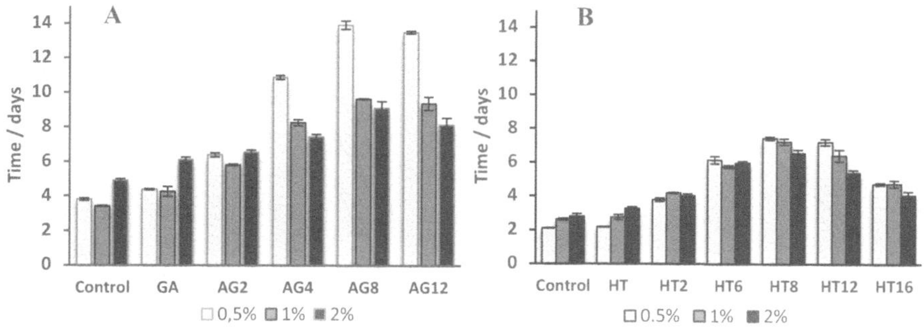

Fig. 2 Effects of the AO chain length at selected Φ_I ($\Phi_I = 0.005$, 0.01 and 0.02) on the oxidative stability of 1:9 (O/W) fish oil/Tween 80/citric buffer (pH 3.65) nanoemulsions. (**A**) Nanoemulsions containing gallic acid (GA) and its esters. (**B**) Nanoemulsions containing hydroxytyrosol (HT) and its esters. Reprinted with permission from reference (Costa et al. 2020a, b)

4 Effects of Droplet Size on the Partitioning of Antioxidants in Emulsions

The distribution and effective concentrations of AOs in emulsified systems has been determined during the last decade by the application of a kinetic method grounded on the pseudophase model (Bravo-Díaz et al. 2015). Details of the method can be found in the chapter by Bravo and Romsted in this book.

One of the main assumptions of the model is that the distribution of reactants in emulsions is independent of the droplet size of the emulsified system.(Costa et al. 2021; Bravo-Díaz et al. 2015; Meesters and Hooff 2013) AOs can be distributed between the different pseudophases according to their physicochemical properties. Table 3 is illustrative and shows the partition constant values obtained for gallates and hydroxytyrosol phenolipids in both emulsions and nanoemulsions composed of fish oil, acidic water and Tween 80 as surfactant. Results show that the distribution of AOs is not affected by different droplet sizes found in both emulsified systems, with differences lower than 6% for the determined partition constants, P_W^I and P_O^I, in the studied systems (Costa et al. 2020a, b). As a consequence, the distribution of these AOs in the emulsified systems prepared with the same ingredients and with the same concentrations is the same (Fig. 3).

5 Relationships Between the Antioxidant Efficiencies and the Interfacial Concentrations of AOs in Emulsions and Nanoemulsions

Antioxidants partition in different extents and their effective concentrations in the oil, aqueous and interfacial regions should be different, and this differential concentration should have an effect on the rate of the inhibition reaction as in any other

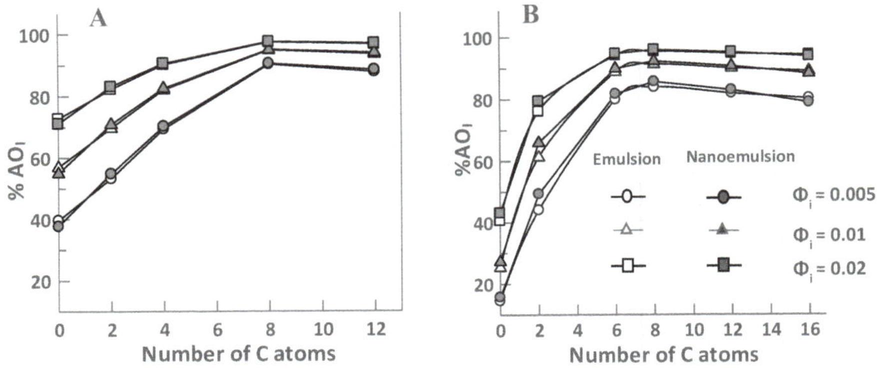

Fig. 3 Variation of the percentage of gallic acid (**A**) and hydoxytyrosol derivatives (**B**) in 1:9 O/W emulsions and nanoemulsions stabilized kinetically with different surfactant volume fractions. Figure 3A reprinted with permission from reference (Costa et al. 2020a). Figure 3B adapted from (Costa et al. 2020b)

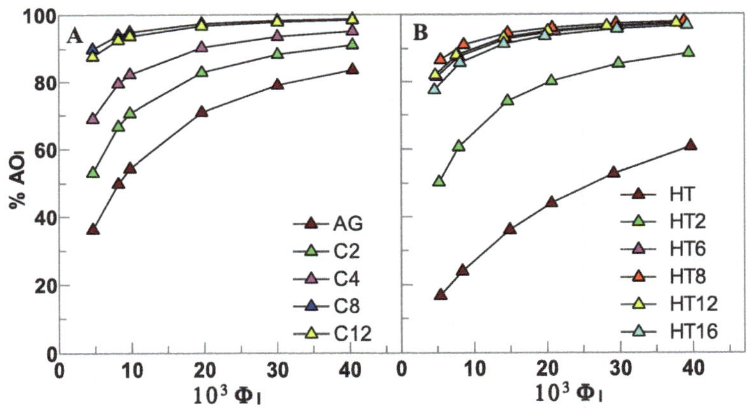

Fig. 4 Variation in the percentage of AG esters (**A**) and HT esters (**B**) with the emulsifier volume fraction in interfacial region of 1:9 fish oil nanoemulsions. Figure adapted from (Costa et al. 2020a, b)

chemical reaction taking place in a three-dimensional reaction site, in contrast with that when two-dimensional surfaces are involved (e.g., heterogeneous catalyses).

The effective concentrations of the antioxidants in the oil, water and interfacial regions can be easily determined from the partition constants (Ferreira et al. 2018; Costa et al. 2015, 2020a, b, c; Bravo-Díaz et al. 2015). Equation 2 shows the dependence of (AO_I) (expressed in moles per liter of interfacial volume) on both $\%AO_I$ and Φ_I values. Both parameters work in opposite directions. On one hand, the fraction of AO in the interfacial region, $\%AO_I$, increases upon increasing Φ_I, reaching a plateau when all antioxidants are incorporated into the interfacial region (Fig. 4). On the other hand, addition of surfactant, Φ_I, increases the interfacial

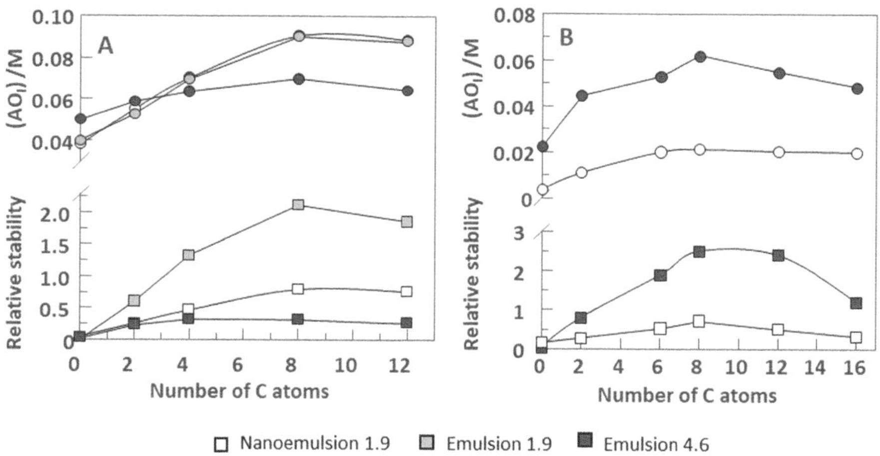

Fig. 5 Relationship between the AO interfacial concentrations of derivatives of gallic acid (**A**) and of hydroxytyrosol (**B**) and their antioxidant efficiency in the fish oil emulsion and nanoemulsions ($\Phi_I = 0.005$). Fig. 5A reprinted with permission from reference (Costa et al. 2020a). Figure 5B adapted from (Costa et al. 2020b)

volume, and the antioxidant is diluted (Costa et al. 2017, 2020a, b, c; Ferreira et al. 2018; Almeida et al. 2016).The increase in $\%AO_I$ does not compensate the increase in the interfacial volume and, as a consequence, the effective interfacial concentrations of GA decrease upon increasing Φ_I.

$$(AO_I) = \frac{[AO_T](\%AO_I)}{\Phi_I} \tag{2}$$

Figure 4 shows a comparison of the AO interfacial concentrations and the antioxidant efficiency of derivatives of gallic acid and hydroxytyrosol in emulsion and nanoemulsions prepared with the same composition. The antioxidant efficiency of AOs was given as the increase in the oxidative stability ratio, Eq. (1). As the droplet size has an insignificant effect on the distribution of each AO (Fig. 3) and the antioxidants employed had the same reactive moieties (Romsted 2012; Gruner et al. 2016), a parallel ranking of antioxidant efficiency was obtained in emulsions and nanoemulsions prepared with the same composition, (Fig. 5), (Costa et al. 2020a, b, c). In fact, a relationship between the AO efficiency and its interfacial concentration was found for the gallic acid and hydroxytyrosol esters (Fig. 5) in both emulsified systems (Romsted 2012; Gruner et al. 2016). Results are in keeping with literature reports demonstrating the direct relationship between the AO interfacial concentrations (AOI) and their efficiency in emulsions (Costa et al. 2015, 2020a, b, c; Ferreira et al. 2018; Meireles et al. 2019; Losada-Barreiro et al. 2013; Almeida et al. 2016). It is worth noting that the antioxidant efficiency of the more hydrophobic AOs in both series (C8, C12 and C16) decreased when the ratio oil/water increased when compared to the efficiency of the more hydrophilic homologous AOs (C2 and C4) because an increase in the oil fraction causes a shift of these more hydrophobic AOs from the interphase to the oil region. Consequently, in emulsified system with higher O/W ratio, only slightly differences between the efficiency of compounds were observed (Fig. 5).

6 Conclusions and Perspectives

Numerous researchers investigated the role of droplet sizes on the oxidative stability of emulsions and various hypothesis were proposed, but none of them provided experimental evidence supporting them. Even in current reports, it is frequently argued that the differential oxidative stability of emulsions is attributed to changes in surfaces areas. This is certainly misleading, because the oxidative stability of emulsions (typically measured as the rate of production of oxidation material), as well as the rate of any other chemical reaction, depends on the concentration of reactants at the reaction site and not on surface areas. Big droplets may contain more moles of a particular antioxidant than smaller ones, however, their effective concentrations in the aqueous, oil and interfacial regions is the same independently of the surface area of the droplets.

By employing physical methods in combination with the pseudophase model, sets of parameters that characterize physically the prepared emulsions and nanoemulsions and the determination of the effective concentrations of the antioxidants in the various regions allowed to establish correlations between droplet sizes and antioxidant efficiencies.

Any modification in the system or in the environmental conditions (e.g., oil to water ratio, temperature, etc.) that lead to changes in the effective concentrations of antioxidants in the various regions of emulsified systems have an impact on their oxidative stability. The work is not, however, finished as more experiments are needed to analyze the effects of various parameters, mainly the composition of the emulsions, on their physical and oxidative stability, which need to be balanced for practical applications.

References

Almeida J, Losada-Barreiro S, Costa M, Paiva-Martins F, Bravo-Díaz C, Romsted LS (2016) Interfacial concentrations of Hydroxytyrosol and its lipophilic esters in intact olive oil-in-water emulsions: effects of antioxidant hydrophobicity, surfactant concentration, and the oil-to-water ratio on the oxidative stability of the emulsions. J Agric Food Chem 64(25):5274–5283

Anton N, Vandamme TF (2011) Nano-emulsions and Micro-emulsions: clarifications of the critical differences. Pharm Res 28(5):978–985

Azuma G, Kimura N, Hosokawa M, Miyashita K (2009) Effect of droplet size on the oxidative stability of soybean oil TAG and fish oil TAG in oil-in-water emulsion. J Oleo Sci 58 (6):329–338

Berton-Carabin CC, Ropers M-H, Genot C (2014) Lipid oxidation in oil-in-water emulsions: involvement of the interfacial layer. Compr Rev Food Sci Food Saf 13(5):945–797
Berton-Carabin CC, Sagis L, Schroën K (2018 Mar 25) Formation, structure, and functionality of interfacial layers in food emulsions. Annu Rev Food Sci Technol 9(1):551–587
Bravo-Díaz C, Romsted LS, Liu C, Losada-Barreiro S, Pastoriza-Gallego MJ, Gao X et al (2015) To model chemical reactivity in heterogeneous emulsions. Think Homogeneous Microemulsions Langmuir 31(33):8961–8979
Burakova Y, Shi J, Schlup JR (2017) Impact of oil composition on formation and stability of emulsions produced by spontaneous emulsification. J Dispers Sci Technol 38(12):1749–1754
Costa M, Losada-Barreiro S, Paiva-Martins F, Bravo-Díaz C, Romsted LS (2015) A direct correlation between the antioxidant efficiencies of caffeic acid and its alkyl esters and their concentrations in the interfacial region of olive oil emulsions. The pseudophase model interpretation of the "cut-off" effect. Food Chem 175:233–242
Costa M, Losada-Barreiro S, Paiva-Martins F, Bravo-Díaz C (2017) Physical evidence that the variations in the efficiency of homologous series of antioxidants in emulsions are a result of differences in their distribution. J Sci Food Agric 97(2):564–571
Costa M, Losada-Barreiro S, Bravo-Díaz C, Vicente AA, Monteiro LS, Paiva-Martins F (2020a) Influence of AO chain length, droplet size and oil to water ratio on the distribution and on the activity of gallates in fish oil-in-water emulsified systems: emulsion and nanoemulsion comparison. Food Chem 310:125716
Costa M, Losada-Barreiro S, Bravo-Díaz C, Monteiro LS, Paiva-Martins F (2020b) Interfacial concentrations of Hydroxytyrosol derivatives in fish oil-in-water emulsions and Nanoemulsions and its influence on their lipid oxidation: droplet size effects. Foods. 9(12):1897
Costa M, Freiría-Gándara J, Losada-Barreiro S, Paiva-Martins F, Bravo-Díaz C (2020c) Effects of droplet size on the interfacial concentrations of antioxidants in fish and olive oil-in-water emulsions and nanoemulsions and on their oxidative stability. J Colloid Interface Sci 562:352–362
Costa M, Losada-Barreiro S, Paiva-Martins F, Bravo-Díaz C (2021) Polyphenolic antioxidants in lipid emulsions: partitioning effects and interfacial phenomena. Foods 10(3):539
Delmas T, Piraux H, Couffin A-C, Texier I, Vinet F, Poulin P et al (2011) How to prepare and stabilize very small nanoemulsions. Langmuir 27(5):1683–1692
Dickinson E (1992) Introduction to food colloids. Oxford university press, Oxford
Ferreira I, Costa M, Losada-Barreiro S, Paiva-Martins F, Bravo-Díaz C (2018) Modulating the interfacial concentration of gallates to improve the oxidative stability of fish oil-in-water emulsions. Food Res Int 112:192–198
Foti MC, Sortino S, Ingold KU (2005) New insights into solvent effects on the formal $HOO^{•}$ + $HOO^{•}$ reaction. ChemEur J 11:1942
Frankel EN (1980) Lipid oxidation. Prog Lipid Res 19(1):1–22
Frankel EN (2005) Lipid oxidation
Freiría-Gándara J, Losada-Barreiro S, Paiva-Martins F, Bravo-Díaz C (2018a) Differential partitioning of bioantioxidants in edible oil–water and octanol–water systems: linear free energy relationships. J Chem Eng Data 63(8):2999–3007
Freiría-Gándara J, Losada-Barreiro S, Paiva-Martins F, Bravo-Díaz C (2018b) Enhancement of the antioxidant efficiency of gallic acid derivatives in intact fish oil-in-water emulsions through optimization of their interfacial concentrations. Food Funct 9(8):4429–4442
Gohtani S, Sirendi M, Yamamoto N, Kajikawa K, Yamano Y (1999) Effect of droplet size on oxidation of docosahexaenoic acid in emulsion system. J Dispers Sci Technol 20(5):1319–1325
Goodarzi F, Zendehboudi S (2019) A comprehensive review on emulsions and emulsion stability in chemical and energy industries. Can J Chem Eng 97(1):281–309
Grigoriev DO, Miller R (2009) Mono- and multilayer covered drops as carriers. Curr Opin Colloid Interface Sci 14(1):48–59
Gruner P, Riechers B, Semin B, Lim J, Johnston A, Short K et al (2016) Controlling molecular transport in minimal emulsions. Nat Commun 7(1):10392

Gupta A, Eral HB, Hatton TA, Doyle PS (2016) Nanoemulsions: formation, properties and applications. Soft Matter 12(11):2826–2841

Kabalnov AS (1994) Can micelles mediate a mass transfer between oil droplets? Langmuir 10 (3):680–684

Kargar M, Spyropoulos F, Norton IT (2011) The effect of interfacial microstructure on the lipid oxidation stability of oil-in-water emulsions. J Colloid Interface Sci 357(2):527–533

Kikuchi K, Yamamoto S, Shiga H, Yoshii H, Adachi S (2014) Effect of reducing oil droplet size on lipid oxidation in an oil-in-water emulsion. Jpn J Food Eng 15:43–47

Lee SJ, Choi SJ, Li Y, Decker EA, McClements DJ (2011) Protein-stabilized nanoemulsions and emulsions: comparison of physicochemical stability, lipid oxidation, and lipase digestibility. J Agric Food Chem 59(1):415–427

Lethuaut L, Métro F, Genot C (2002) Effect of droplet size on lipid oxidation rates of oil-in-water emulsions stabilized by protein. J Am Oil Chem Soc 79(5):425–430

Losada Barreiro S, Bravo-Díaz C, Paiva-Martins F, Romsted LS (2013) Maxima in antioxidant distributions and efficiencies with increasing hydrophobicity of gallic acid and its alkyl esters. The pseudophase model interpretation of the "cutoff effect". J Agric Food Chem 61(26):6533–6543

Ma Z, Zaera F (2006) Organic chemistry on solid surfaces. Surf Sci Rep 61(5):229–281

Mason TG, Wilking JN, Meleson K, Chang CB, Graves SM (2006) Nanoemulsions: formation, structure, and physical properties. J Phys Condens Matter 18(41):R635–R666

McClements DJ (2005) Food emulsions: principles, practices, and techniques. CRC Press, Boca Raton

McClements DJ (2012) Nanoemulsions versus microemulsions: terminology, differences, and similarities. Soft Matter 8(6):1719–1729

McClements DJ (2015) Food emulsions: principles, practices, and techniques. CRC press, Boca Raton

McClements DJ, Rao J (2011) Food-grade Nanoemulsions: formulation, fabrication, properties, performance, biological fate, and potential toxicity. Crit Rev Food Sci Nutr 51(4):285–330

McClements DJ, Bai L, Chung C (2017) Recent advances in the utilization of natural emulsifiers to form and stabilize emulsions. Annu Rev Food Sci Technol 8(1):205–236

Meesters RJW, Hooff GP (2013) State-of-the-art dried blood spot analysis: an overview of recent advances and future trends. Bioanalysis 5(17):2187–2208

Meireles M, Losada-Barreiro S, Costa M, Paiva-Martins F, Bravo-Díaz C, Monteiro LS (2019) Control of antioxidant efficiency of chlorogenates in emulsions: modulation of antioxidant interfacial concentrations. J Sci Food Agric 99:3917–3925

Nastiti CMRR, Ponto T, Abd E, Grice JE, Benson HAE, Roberts MS (2017) Topical nano and microemulsions for skin delivery. Pharmaceutics 9(4):37

Neves MA, Wang Z, Kobayashi I, Nakajima M (2017) Assessment of oxidative stability in fish oil-in-water emulsions: effect of emulsification process, droplet size and storage temperature. J Food Process Eng 40(1):e12316

Osborn HT, Akoh CC (2004) Effect of emulsifier type, droplet size, and oil concentration on lipid oxidation in structured lipid-based oil-in-water emulsions. Food Chem 84(3):451–456

Raimúndez-Rodríguez EA, Losada-Barreiro S, Bravo-Díaz C (2019) Enhancing the fraction of antioxidants at the interfaces of oil-in-water emulsions: a kinetic and thermodynamic analysis of their partitioning. J Colloid Interface Sci 555:224–233

Romsted LS (2012) Introduction to surfactant self-assembly. In: Supramolecular chemistry: from molecules to nanomaterials. Wiley, Oxford

Ross L, Barclay C, Vinqvist MR (2003) Phenols as antioxidants. In: Rappoport Z (ed) The chemistry of phenols. J. Wiley & Sons, West Sussex, UK

Shahidi F, Janitha PK, Wanasundara PD (1992) Phenolic antioxidants. Crit Rev Food Sci Nutr 32 (1):67–103

Sun C, Gunasekaran S (2009) Effects of protein concentration and oil-phase volume fraction on the stability and rheology of menhaden oil-in-water emulsions stabilized by whey protein isolate with xanthan gum. Food Hydrocoll 23(1):165–174

Tcholakova S, Denkov ND, Lips A (2008) Comparison of solid particles, globular proteins and surfactants as emulsifiers. Phys Chem Chem Phys 10(12):1608–1627

Uluata S, McClements DJ, Decker EA (2016) Riboflavin-induced oxidation in fish oil-in-water emulsions: impact of particle size and optical transparency. Food Chem 213:457–461

Wilde P, Mackie A, Husband F, Gunning P, Morris V (2004) Proteins and emulsifiers at liquid interfaces. Adv Colloid Interf Sci 108:63–71

Zembyla M, Murray BS, Sarkar A (2020) Water-in-oil emulsions stabilized by surfactants, biopolymers and/or particles: a review. Trends Food Sci Technol 104:49–59

Nanoemulsions as Encapsulation System to Prevent Lipid Oxidation

Carla Arancibia and Natalia Riquelme

1 Introduction

In recent years, there has been a growing interest in designing structures for encapsulating, protecting, and delivering bioactive compounds to develop products that prevent diseases and promote health. Nanotechnology applied to food and nutraceutical products involves the design, creation, development, and use of material structures and systems in the nanometer scale (<200 nm) (Akhavan et al. 2018; Yu et al. 2018), where nanosized colloidal systems, such as nanoemulsions, have been regarded as useful tools for the protection of active ingredients, the dispersing hydrophobic compounds into aqueous systems, enhancing the stability and delivery of bioactive compounds (Kumar and Sarkar 2018; Liu et al. 2019a).

Oil-in-water nanoemulsions consist of small oil droplets (radii between 20 and 200 nm) that are dispersed into an aqueous phase, where each droplet is surrounded by emulsifier molecules (McClements 2015) (Fig. 1). They have considerable potential for encapsulating lipophilic compounds (Salvia-Trujillo et al. 2017) and can be designed to have different optical, rheological, and stability properties when controlling their composition and structure (McClements 2015). Nanoemulsions have several advantages concerning the use of conventional emulsions, since (1) they scatter light easily that tend to be more transparent and can be incorporated into a wide variety of products; (2) they have high physical stability against droplet aggregation and gravitational separation; and (3) they can increase the bioavailability of the lipid compounds incorporated in them, making them suitable for their application in functional foods or nutraceuticals products (Chung and McClements 2018; Zhang and McClements 2018; Zhang et al. 2020a).

C. Arancibia (✉) · N. Riquelme
Departamento de Ciencia y Tecnología de los Alimentos, Facultad Tecnológica, Universidad de Santiago de Chile (USACH), Estación Central, Chile
e-mail: carla.arancibia@usach.cl

C. Bravo-Diaz (ed.), *Lipid Oxidation in Food and Biological Systems*,
https://doi.org/10.1007/978-3-030-87222-9_11

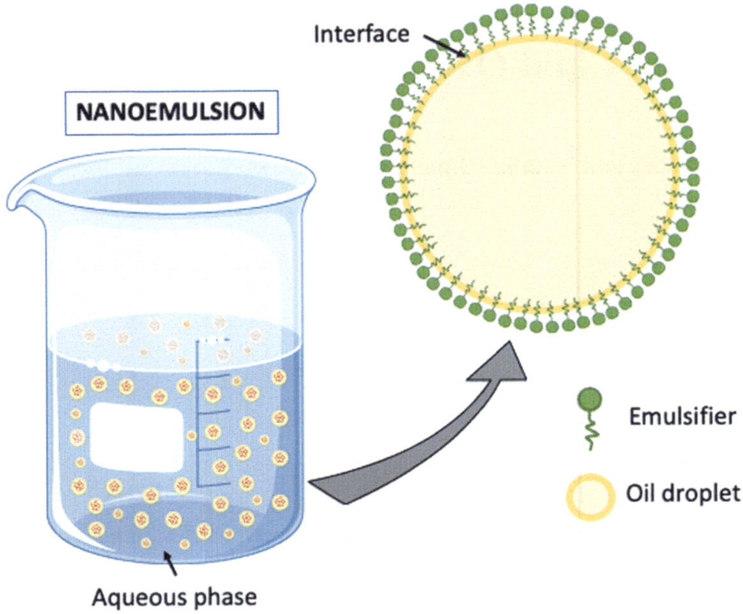

Fig. 1 Schematic representation of nanoemulsion structure and composition

Nanoemulsions can also improve the oxidative stability of bioactive compounds incorporated into them due to forming an interfacial layer around oil droplets that encapsulate lipid compounds, protecting them from degradation and hydrolysis (Kumar and Sarkar 2018; Sharma et al. 2019; Gasa-Falcon et al. 2020). Several studies have analyzed the use of nanoemulsions as an encapsulation system for different lipophilic bioactive compounds, such as omega 3 fatty acids, liposoluble vitamins, curcumin, carotenes, among others; obtaining that these colloidal systems can improve solubility and increase the chemical stability of bioactive compounds during storage and digestion process (Akhavan et al. 2018; Li et al. 2021; McClements and Öztürk 2021). However, this fact depends on nanoemulsions composition and interface's properties since these characteristics may potentially influence lipid oxidation rate (Waraho et al. 2011). Besides, the presence of prooxidants in the aqueous phase also affects the lipid oxidation of nanoemulsions, since they are able to interact with the lipid phase and promote their oxidation (Berton-Carabin et al. 2014). Hence, diverse strategies to prevent lipid oxidation of nanoemulsions have been applied, including incorporating antioxidant components, creating an interfacial steric barrier, and modifying the electric charge in order to repel transition metal ions (prooxidants) (Zhu et al. 2018; Wang et al. 2020. In that regard, this chapter will focus on nano-emulsification as a tool to improve the oxidative stability of lipid compounds incorporated into nanoemulsions.

2 Nanoemulsions

Nanoemulsions are colloidal systems that consist of small droplets dispersed in a continuous phase, where each droplet is surrounded by emulsifier molecules (McClements and Jafari 2018a; McClements 2021) (Fig. 1). These systems are similar to conventional emulsions, but their dispersed droplets are considerably smaller (20–200 nm of diameter) and are kinetically stable (Choi and McClements 2020; Pavoni et al. 2020a, b). They can be classified according to the relative spatial distribution of the aqueous and lipid phases, since water droplets dispersed into a lipid phase are water-in-oil (W/O) nanoemulsions, and the oil droplets dispersed into the aqueous phase correspond to oil-in-water (O/W) nanoemulsions (Zhang et al. 2020a), being the last ones the most used as encapsulating systems.

2.1 Nanoemulsion Preparation

O/W nanoemulsions can be fabricated using two approaches: low and high energy methods (Liu et al. 2019a). Low energy methods are mainly based on the spontaneous formation of small oil droplets in the aqueous phase when modifying lipid phase composition, emulsifier type, concentration, and temperature (McClements 2021). The most low-energy methods are phase inversion temperature, emulsion inversion, and spontaneous emulsification (Safaya and Rotliwala 2020; Vinh et al. 2020). On the other hand, high-energy methods have great potential for scaling up to industrial use. These methods are based on mechanical devices such as high-pressure homogenizer, microfluidizer, and ultrasonic homogenizer, which are capable of generating disruptive forces that break up the dispersed phase, thus allowing for small oil formation droplets (Espitia et al. 2019). High-energy methods can deliver enough energy to reduce the drops' dimensions, and as such, are considered an effective technique for elaborating oil-in-water nanoemulsions (Calligaris et al. 2016) (Table 1). The homogenization conditions play an essential role in obtaining physically stable nanoemulsions with a reduced droplet size (Håkansson and Rayner 2018). In this respect, various authors have studied how the homogenization process parameters affect the droplet size and stability of nanoemulsions, since the mean of the droplets' diameter decreases to nanoscale size when increasing the operation conditions of high energy methods (time of processing, energy, number of cycles, pressure, among others), giving rise to physically stable nanoemulsions, as observed in Table 1. Despite this, high-energy methods are extremely intense treatments for producing nanoemulsions with nano-size droplets (Qian and McClements 2011). Thus, it is necessary to be careful with "over-processing" behavior since larger droplet sizes can be generated due to these homogenization conditions that favor the coalescence of oil droplets, decreasing shelf life nanoemulsions (Páez-Hernández et al. 2019).

Table 1 Composition and preparation of bioactive nanoemulsions

Lipid phase	Emulsifier	Homogenization conditions	Droplet size	Goal	References
Algae oil rich in ω-3 fatty acids	Phytosterols structured in quillaja saponin	Ultrasound, 20 kHz, 70% amplitude, 5 min	152–165 nm	Enhance oxidative stability of algae oil-nanoemulsions using structured phytosterols	Chen et al. (2016)
ω-3 fatty acids	Tween 80 and alginate	Microfluidization, 15,000 psi, 3 cycles	135 nm (tween 80) 150–800 nm (tween 80 + alginate)	Inhibit lipid oxidation of ω-3 fatty acids-loaded nanoemulsions by anionic polysaccharide addition	Salvia-Trujillo et al. (2016)
Fish oil	Tween 80	Microfluidization, 20,000 psi, 5 cycles	82–162 nm	Prevent the lipid oxidation of fish oil-based nanoemulsion using different carrier oils	Walker et al. (2017)
Fish oil	Casein and modified starch (OSA)	Ultrasound, 20 kHz, 360 W, 7–22 min	~ 220 nm	Enhance oxidative stability of fish oil-nanoemulsions by casein incorporation	Yang et al. (2019)
Avocado oil	Tween 80 or lecithin	Ultrasound, 20 kHz, 90% amplitude, 20 min	103–249 nm	Compare the effect of different emulsifiers on the oxidative stability of avocado oil-based nanoemulsion	Arancibia et al. (2017)
Black pepper and cinnamon essential oils	Tween 20 and span 20	Ultrasound, 20 kHz, 750 W, 30% amplitude, 12 min High-pressure, 150 MPa, 5 cycle	32–320 nm	Study the effect of emulsification methods on the physicochemical, stability, and antimicrobial properties of nanoemulsions.	Jiménez et al. (2018)

2.2 Nanoemulsion Composition

Nanoemulsions are composed of three major components: emulsifiers, lipid and aqueous phase (Gasa-Falcon et al. 2020), which influence their physicochemical properties and play a role in the formation of small droplets during their elaboration.

2.2.1 Emulsifiers

Emulsifiers are small surfactant molecules or macromolecules whose function is to reduce the interfacial tension between the phases since they are rapidly adsorbed at the oil-water interface, which facilitates the formation of the nanoemulsions during the homogenization process, and the generation of repulsive interactions between the oil droplets, promoting a good physical stability during storage (Riquelme et al. 2019; Xu et al. 2019). A large number of emulsifiers have been used in the elaboration of nanoemulsions (Li et al. 2018a, b; Ryu et al. 2018; Acevedo-Estupiñan et al. 2019), where their nature and characteristics determine some of their properties and functionalities during the elaboration of nanoemulsions, due to chemical structure differences that can affect their arrangements and thickness on the interface (McClements and Jafari 2018b). For example, small molecular weight emulsifiers (surfactants) are more efficient to decrease the nanoemulsions' droplet size than high molecular weight polymers (polysaccharides or proteins) since the small ones rapidly are adsorbed onto the oil droplet surfaces during homogenization, favoring the formation of nanometric droplets (Ma et al. 2018; Flores-Andrade et al. 2021). In addition, in recent years, there has been a study of the use of a mixture of emulsifiers to elaborate nanoemulsions as a strategy to replace or reduce synthetic emulsifier concentration (Su and Zhong 2016; Li et al. 2018b; Riquelme et al. 2019), because of the increasing demand to provide consumers with more natural and eco-friendly products. Su and Zhong (2016) studied the mixture of sodium caseinate and Tween 20 on lemon oil-based nanoemulsions development, observing that the combination of emulsifiers significantly decreases the amount of synthetic emulsifier (Tween 20) from 1.2 and 0.2% w/v to prepare nanoemulsions with small droplets (67–120 mn). Riquelme et al. (2019) investigated the use of a binary and ternary emulsifier mixture (soy lecithin-Tween 80 and soy lecithin-Tween 80-quillaja saponin, respectively) to stabilize avocado oil-based nanoemulsions, founding that the binary emulsifier mixture (Tween 80-soy lecithin) allows obtaining nanoemulsions with a droplet size lower than 100 nm; and that the use a ternary emulsifier mixture (Tween 80- soy lecithin-quillaja saponin) improves their physical stability at different temperatures (30–90 °C) and ionic strengths (0–500 mM NaCl). Therefore, selecting the emulsifier or its mixtures is one of the most critical aspects of forming nanoemulsions due to their role in the emulsification process.

2.2.2 Lipid Phase

Different lipid ingredients have been used for nanoemulsion preparation, such as triacylglycerols, diacylglycerols, monoacylglycerols, edible oils, free fatty acids, non-polar essential oils, mineral oils, lipid substitutes, waxes, among others (Aswathanarayan and Vittal 2019). Besides, several hydrophobic compounds have also been incorporated into the lipid phase of nanoemulsions to improve their nutritional and functional properties, including bioactive lipids (lipid-soluble vitamins, omega-3 fatty acids, and conjugated linoleic acid), antimicrobials (essential oils, thymol, carvacrol, among others), antioxidants (tocopherols, carotenes, astaxanthin, and polyphenols), and nutraceuticals (curcumin, resveratrol, lutein, among others) (Walker et al. 2017; Sharif et al. 2017; McClements and Jafari 2018a; Doost et al. 2019; Park et al. 2019; Sharma et al. 2021; da Silveira et al. 2021). The lipid phase influences the physical and chemical stability of nanoemulsions, mainly for their composition (degree of unsaturation, prooxidant impurities, inherent antioxidants, added antioxidants), physical state (solid fat content and crystal properties, solubility, partitioning, and diffusion of antioxidants, and prooxidants) and physical properties (rheological characteristics and polarity) (Waraho et al. 2011).

Several studies using different oil types as carriers of bioactive compounds have been carried out, demonstrating that the stability of lipid compounds is correlated with density, viscosity, and unsaturation degree of oils used as carriers. For example, oils with lower viscosity, higher density, and lower unsaturation, such as sesame oil, maintain their stability during storage and increase the bioaccessibility of lycopene during *in vitro* digestion (Zhao et al. 2020). Additionally, the solubility of the carrier oils in the continuous phase has a significant impact on the physical stability of nanoemulsions since highly hydrophobic molecules, such as long-chain triglycerides-LCT, can improve the long-term stability of these systems (Zhao et al. 2020). Besides, low-soluble molecules (LCT) can act as a kinetic barrier between the lipid and aqueous phases, improving the oxidative stability of encapsulated bioactive compounds into nanoemulsions, making them less water-soluble and positively influencing their partitioning between the lipid droplets and aqueous phase (Liu et al. 2019a; Pavoni et al. 2020a, b). On the other hand, the fatty acid composition of carrier oils, mainly the chain length and unsaturation degree, impacts the nanoemulsions' stability during storage (Walker et al. 2017). For example, it has been demonstrated that the nanoemulsions elaborated with oils that contain polyunsaturated fatty acids (PUFAs) are more stable during storage than those with monounsaturated fatty acids-rich oils (Lozada et al. 2021). Therefore, the carrier oil selection not only facilitates the formation of nanoemulsions but also can improve their physical, chemical, and biological properties, since the degradation degree of bioactive compounds incorporated into nanoemulsion depends on the selected carriers' physicochemical properties, such as composition, polarity, viscosity, among other properties (Walker et al. 2017).

2.2.3 Aqueous Phase

The aqueous phase is typically composed of water and may contain other components, such as salts, co-solvents, and preservatives, impacting nanoemulsions' stability due to the changes in pH, ionic strength, and viscosity (McClements and Jafari 2018a; Wang et al. 2020). In particular, the presence of salts can modify the structural organization and electrical charge of the emulsifier's molecules at the oil-water interface and consequently affect the stability of nanoemulsions (Liu et al. 2019b). Different studies have reported a positive effect from the addition of salts in the nanoemulsions stability (Saberi et al. 2014; Xu et al. 2017), since the addition of NaCl (0–1 N) and $CaCl_2$ (0–0.5 N) can increase the dehydration of the surfactant headgroup, reducing the attractive van der Waals interactions among droplets, which decreases the coalescence phenomenon among droplets, improving the physical stability of nanoemulsions (Saberi et al. 2014). However, salts may also cause the destabilization of nanoemulsions due to the accumulation of salt ions around the surfaces of charged oil droplets, forming a double layer that neutralizes electric charge in the oil-water interface, thus decreasing the electrostatic repulsions between oil droplets and reducing their physical stability (Dickinson 2019). On the other hand, hydrocolloids are commonly used as a thickening or stabilizing agent since they can improve the physical stability of nanoemulsions by increasing aqueous phase viscosity (Salvia-Trujillo et al. 2017). The addition of thickeners into nanoemulsions boosts a higher resistance against flocculation, coalescence, and phase separation phenomena due to hydrocolloids forming an extended network in the aqueous phase that limits the movements of oil droplets (Arancibia et al. 2016). Also, some of them have emulsifying properties since they can be absorbed into the oil droplet surfaces, increasing the steric and electrostatic repulsion between the droplets, which improves nanoemulsion physical stability (Xu et al. 2017).

3 Nanoemulsions to Prevent Lipid Oxidation

The oxidative stability of nanoemulsions mainly depends on droplet size, interfacial characteristics, and composition, the presence of antioxidants and water-soluble prooxidants, among other factors (Waraho et al. 2011; Yi et al. 2019). Thus, it is possible in such a case to improve the oxidative stability of nanoemulsions by using antioxidants, eliminating the prooxidants present in the aqueous phase through the addition of chelating agents, and modulating the interfacial properties through the engineering of the oil-water interface (McClements 2015). In this sense, the oil-water interface plays a critical role in determining lipid oxidation for this type of colloidal system, since the lipid phase is separated from the aqueous phase through a protecting barrier that inhibits or minimizes the lipid oxidation rate (Wang et al. 2020; McClements and Decker 2018).

3.1 Engineering of the Oil-Water Interface

In recent years, the modulation of the oil-water interface has been considered to be one of the most promising strategies for preventing lipid oxidation of nanoemulsions, since oxidation reactions are often initiated by free radicals localized at the oil-water interface of the nanoemulsions, which can permeate across the interfacial barrier and induce lipid oxidation (Berton-Carabin et al. 2014; Zhang et al. 2015). Therefore, the barrier properties of the interfacial layer have a crucial role in limiting the permeation of these oxidative species, since the type of molecule adsorbed into the lipid droplets surface, their structure and location can minimize the interaction with the lipid phase and limit the permeation of oxidative species (water-soluble prooxidants) across the oil-water interface, improving the oxidative stability of nanoemulsions (Yi et al. 2019). In this respect, the increase of the interfacial thickness, the modification of the interface's electrical charge, and the modulation of the location of emulsifiers (interface or aqueous phase) have been studied as approaches to prevent lipid oxidation in nanoemulsions (Fig. 2).

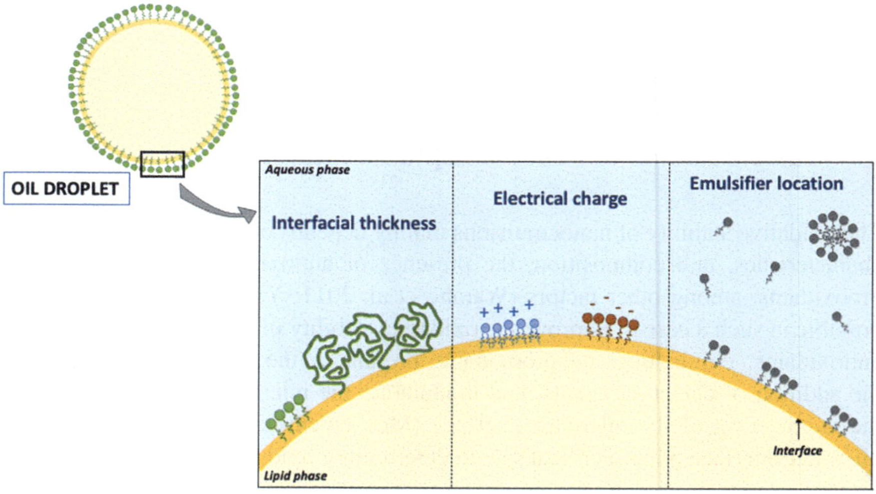

Fig. 2 Characteristics of oil-water interface that influence lipid oxidation of nanoemulsions

3.1.1 Interfacial Thickness

The thickness of the interfacial layer that surrounds the oil droplets plays an essential role in lipid oxidation since a thick interface can improve barrier properties between the lipid and aqueous phases, preventing or retarding the lipid oxidation of nanoemulsions (Zhu et al. 2018; Wang et al. 2020). High molecular weight emulsifiers, such as proteins and some polysaccharides, can form thick interfaces that improve nanoemulsions' oxidative stability by forming a steric barrier formation and electrostatic repulsion among oil droplets (McClements and Jafari 2018b; Li et al. 2020). Furthermore, different hydrocolloids can be dispersed in the aqueous phase by their ability to form a three-dimensional gel-like network, thereby hindering the movement of the prooxidants toward the lipid phase and, consequently, inhibiting lipid oxidation (Shao et al. 2020). On the other hand, it has been demonstrated that the complex formation among polymeric emulsifiers (protein-protein, protein-polysaccharide, and protein-surfactant) at the oil-water interface can generate viscoelastic interfacial films that improve the physicochemical stability of nanoemulsions due to their high elasticity and high mechanical strength, thus contributing to the resistance of oil droplets against coalescence (Schröder et al. 2017; Tamm and Drusch 2017; Zhu et al. 2018). To this effect, the interfacial structures among proteins and polysaccharides can be formed by noncovalent protein-polysaccharide complexes (mixed layer), electrostatic deposition between protein and polysaccharide (multilayer), and protein-polysaccharide conjugates (covalent layer), which affect the physicochemical properties and functional performance of emulsions (Cho and McClements 2009; Li et al. 2020). Liu et al. (2020) investigated how different ways of assembling faba bean protein isolate and chitosan affected the oil-in-water emulsion stability, where the formation of a complex between both hydrocolloids, before emulsification, decreased the physical and oxidative stability of O/W-emulsions. However, the addition of chitosan as a second layer (by electrostatic deposition) on faba-bean protein emulsified oil droplets increased the interfacial layer thickness/compactness (from 19–20 nm to 60–70 nm) and maintained the interfacial protein adsorption. This fact improved the oxidative stability of emulsions, observing a minor generation of lipid oxidation products (less conjugated dienes and secondary oxidation products) and protein oxidation biomarkers (less tryptophan loss and carbonyls formation). Zhang et al. (2020b) investigated the use of mixed plant-based emulsifiers (almond protein isolate-API and camellia saponin-CS) to inhibit lipid oxidation in walnut oil-in-water nanoemulsions. They found that mixed API/CS-emulsifiers inhibited lipid oxidation more effectively than the protein alone, where API/CS nanoemulsions showed a reduction of ~27% of hydroperoxides and ~33% of TBARs during six days of storage at pH 7, in comparison with sole use of API. This effect was attributed to the formation of complexes between emulsifiers, promoting a physical barrier that improved the steric inhibition of prooxidants. Thus, the modulation of the interface through thick film formation can be a helpful tool to prevent nanoemulsions lipid oxidation, where the strength and packing of this film play an essential role in the stability (García-Moreno et al. 2021).

3.1.2 Interfacial Charge

Another strategy used to prevent lipid oxidation is the promotion of electrostatic repulsion among oil droplets using the electrostatic property of emulsifiers since anionic-charged nanoemulsions are more susceptible to lipid oxidation than cationic ones because the negative charges promote the interaction with metals present in the aqueous phase (Waraho et al. 2011; Berton-Carabin et al. 2014). Choi et al. (2010) studied the influence of droplet charge on the chemical stability of citral-based nanoemulsions using cationic (lauryl alginate), non-ionic (polyoxyethylene (23) lauryl ether), and anionic (sodium dodecyl sulfate) emulsifiers. They concluded that positive and neutral charged oil droplets could attract charged metal ions from the aqueous phase, favoring the oxidative stability of citral incorporated into nanoemulsions and decreasing their degradation rates, where ~60% of the citral remained after one week of storage. In comparison, in the anionic nanoemulsions only 10% of the citral remained stable. In the same way, Hu et al. (2003) studied the effect of three protein emulsifiers (casein, whey protein, and soy protein isolate) on oxidative stability of corn oil-in-water nanoemulsions, observing a higher oxidative stability of the nanoemulsions when whey protein used as emulsifier at pH $<$ isoelectric point-PI (4.8). These results were attributed to pH $<$ PI producing cationic charge into oil droplets, which is able to repel prooxidative metals and decrease the lipid oxidation of nanoemulsions. Therefore, emulsifier electrical charged can have an impact on oxidative stability, especially when prooxidants compounds (e.g., transition metals) are in the aqueous phase, since metals can be attracted for the anionic emulsion droplet interface avoiding their interaction with the lipids in the emulsion droplet core (Fomuso et al. 2002).

3.1.3 Emulsifier Location

The promotion or inhibition of lipid oxidation also depends on the structure and location of emulsifiers (Waraho et al. 2011; Berton-Carabin et al. 2014; McClements and Decker 2018). In general terms, unabsorbed protein emulsifiers may interact with prooxidant components of the aqueous phase due to their ability to act as metal chelators or metal binders, free radical scavengers, or binding of secondary lipid products in dispersed systems, which greatly influences the chemical stability of nanoemulsions (Berton-Carabin et al. 2014). Gumus et al. (2017) determined the impact of the location (adsorbed versus non-adsorbed) of different legume proteins used as emulsifiers on fish oil-in-water nanoemulsions. They found that the rate of lipid oxidation was considerably less for the nanoemulsion with proteins located only at the interface, in comparison with those that contained proteins located both in the interface and at the aqueous phase (non-absorbed protein). This effect demonstrated that non-adsorbed proteins in the aqueous phase could inhibit lipid oxidation of nanoemulsions from the neutralization transition metals present, decreasing their interaction with the fatty acids of the lipid phase. Likewise, an excess of surfactant can form micelles in the aqueous phase, which encapsulate prooxidant agents or alter the properties of the interfacial layer, enhancing nanoemulsion oxidative stability, but emulsifier micelles can sometimes promote Ostwald ripening destabilization and thus impact negatively on their long-term stability (Zhang et al. 2020a; Chuesiang et al. 2018; Losada-Barreiro et al. 2020).

In summary, diverse strategies can be used to prevent nanoemulsions' lipid oxidation, where creating an interfacial barrier both steric and electrostatic formed by emulsifiers could be a helpful tool to improve the chemical stability of lipids incorporated into nanoemulsions.

3.2 *Addition of Antioxidants*

Antioxidants can be incorporated into nanoemulsions to extend their shelf-life due to their ability to inhibit or delay oxidation processes that occur under the influence of atmospheric oxygen or reactive oxygen species (Lourenço et al. 2019; Maqsoudlou et al. 2020). However, some antioxidants present several technological disadvantages related to their poor solubility in aqueous matrices and low chemical stability (Ozkan et al. 2019), making for a difficult application in aqueous product and thus posing a challenge for the food industry. Hence, their encapsulation into nanoemulsions has been investigated to be an efficient technology to disperse and protect the antioxidants into aqueous food systems. That regard, in recent years there have been extensive studios using nanoemulsions for the encapsulation of bioactive antioxidants such as phenolic compounds, vitamins, and carotenoids (Table 2), where the main focus has been on enhancing the water dispersion, chemical stability, effectiveness, and bioavailability of the antioxidants incorporated.

The effectiveness of antioxidants encapsulated in nanoemulsions is influenced by different factors, such as emulsifier composition and concentration, homogenization conditions, physical and chemical stability of lipid and aqueous phase. Thus, it is necessary to understand how antioxidants interact with the other ingredients of nanoemulsions and their effect on interfacial properties of nanoemulsions for modulating their antioxidant capacity. For example, the antioxidant efficiencies of caffeic acid and its alkyl esters in emulsions depend on their distributions between the different regions of emulsions (oil, interfacial and aqueous) and the efficiency with which each antioxidant molecule quenches the free radicals involved in the oxidation of polyunsaturated oils (Costa et al. 2016). Other studies have demonstrated that the polarity of antioxidants and, therefore, their location into the different emulsion regions affect the chemical stability of these colloidal systems. Cheng et al. (2019) reported that the use of flaxseed phenolics compounds (flaxseed lignan extract-FLE, secoisolariciresinol-SECO, secoisolariciresinol diglucoside-SDG and p-coumaric acid-CouA) in flaxseed oil-based nanoemulsions exerted both antioxidant effects, depending on its molecular characteristics and location. The incorporation of SECO and FLE into the nanoemulsions inhibited lipid oxidation, retarding the formation of hydroperoxides and TBARs in the sequence SECO<FLE<CouA<SDG, since the

Table 2 Use of nano-emulsified antioxidants to prevent lipid oxidation

Bioactive compound	Carrier oil	Emulsifier	Encapsulation efficiency	Droplet size	Principal outcome	References
Peppermint oil (PO)	Borage seed oil (BSO)	Modified starch (purity gum ultra)	70.3–96.5%	132–393 nm	The incorporation of PO improves the oxidative stability of BSO-based nanoemulsion due to the retention of their antioxidant capacity.	Rehman et al. (2020)
β-Carotene (BC) α-Tocopherol (VE)	Flaxseed oil (FSO) Medium-chain triglycerides (MCT)	Modified starch (purity gum ultra) and tween 80	–	124–207 nm	Carrier oils and emulsifiers influence the oxidative stability of nanoemulsions, where VE works as an antioxidant for the protection of BC in MCT-based nanoemulsion.	Sharif et al. (2017)
Gallic acid and alkyl gallates	Rapeseed oil	Sodium dodecyl sulfate (SDS)	–	80–110 nm	The medium and long-chain length gallate alkyl esters are more efficient to prevent lipid oxidation of nanoemulsions than the gallic acid.	da Silveira et al. (2021)
Thymol	High oleic sunflower oil (HOSO) Tricaprylin (TC) Cinnamaldehyde (CA)	Quillaja saponin	–	166–245 nm	Thymol-based nanoemulsions applied in chicken meat show a significant improvement of the antioxidant activity.	Doost et al. (2019)
α-Tocopherol (TOC)	Sunflower oil Sweet almond oil Olive oil Coconut oil	Tween 80 Poloxamer 188 Span 60 Span 80	79–98%	120–350 nm	The incorporation of TOC improves sunflower oil- nanoemulsions antioxidant activity.	Pinto et al. (2018)
Quercetin (QT)	Soybean oil Glycerol triacetate	Polyglyceryl-10 laurate (PGFE-10) Polyglycerol-6 monostearate (PGFE-6)	85%	152 nm	The quercetin nano-emulsified remains their antioxidant activity due to the high efficiency of encapsulation that keeps their complete functional activity after emulsification	Ni et al. (2017)

		Sucrose esters-11 (SE-11)				
Turmeric extract (TE)	Medium-chain triglycerides (MCT)	Lecithin Tween 80	–	84 nm	TE-nanoemulsions can protect the curcuminoids during gastric digestion and release them during intestinal digestion.	Park et al. (2019)
Curcumin (cur)	Curcumin root essential oil	Tween 20	64–94%	138–271 nm	Optimized curcumin-loaded nanoemulsion is more resistant to pepsin digestion, which increases the bioaccessibility of curcumin.	Sharma et al. (2021)

phenols presented a stronger molecular polarity and higher affinity for the oil droplets, whilst SDG and CouA, mainly located in the aqueous phase, can act as prooxidants. Kharat et al. (2018) analyzed the impact of the polarity and location of different antioxidants in protecting curcumin from degradation, studying the addition of four antioxidants (water-soluble: ascorbic acid and Trolox, amphiphilic: ascorbyl palmitate, and oil-soluble: α-tocopherol) into curcumin-loaded nanoemulsions. They observed that water soluble antioxidants were more effective in protecting curcumin from degradation than interfacial or oil soluble antioxidants, since an 82.2 and 82.6% of curcumin remained in the emulsions that contained ascorbic acid and Trolox, respectively. This effect on curcumin stability may be due to the fact that curcumin is highly resistant to degradation when dissolved in oil, but it degrades quickly when dissolved in aqueous phases (Kharat et al. 2017; Kharat and McClements 2019). Therefore, water-soluble antioxidants, mainly located within the aqueous phase, can protect the curcumin more effectively when nearing the aqueous phase. On the other hand, some phenolic compounds with interfacial antioxidant activity are capable of protecting food emulsions against lipid oxidation, acting as both chelating agents and free radical scavengers (Velderrain-Rodríguez et al. 2021). For example, the amphiphilic properties of certain phenolic compounds (flavonoids, such as catechin and quercetin) are responsible for their interactions with water-soluble pro-oxidants because of their tendency to serve as hydrogen donors, which contributes to the formation of metal coordination complexes with good stability (Zamora & Hidalgo, 2016). Therefore, the performance of antioxidants nano-emulsified will depend on their intrinsic factors, such as polarity, molecular characteristics, and surface activity, which determine their affinity by some of the three regions of nanoemulsions: interface, lipid and aqueous phase (Costa et al. 2016).

The efficiency of antioxidants also depends strongly on their concentration in the interface since the rate of any reaction depends on their effective concentration on the reaction site (Raimúndez-Rodriguez et al., 2019; Losada-Barreiro et al. 2020). Thus, an alternative for controlling the concentration of antioxidants on the oil-water interface is to modulate their hydrophilic-lipophilic balance (HLB) by modifying their chemical structure or environmental conditions (Silva et al. 2017). For example, a strategy to change the HLB value of antioxidants is to grafted non-reactive alkyl chains of different lengths into their structure, maintaining their radical scavenging properties (Silva et al. 2017; Ferreira et al. 2018). On the other hand, the effective interfacial concentration of antioxidants has been correlated directly with the antioxidant efficiency to prevent lipid oxidation of emulsions (Costa et al. 2016). Costa et al. (2020) studied the effects of the droplet size on the distribution and effective interfacial concentration of gallic acid in both coarse emulsions and nanoemulsions, and on their oxidative stability. They reported that the droplet size (80–1300 nm) has a negligible influence on the oxidative stability of the emulsified system. However, this stability decreases upon increasing surfactant concentration in the systems loaded with gallic acid, since the effective interfacial concentration of the antioxidant decreases. Therefore, the antioxidant efficiency of gallic acid in the nanoemulsions is always positively correlated with the concentration of antioxidants at the interfacial region.

On the other hand, it is possible to find antioxidants that form conjugates with emulsifying agents in the oil-water interface, which is considered a potential strategy for retarding or reducing lipid oxidation in nanoemulsions (McClements and Decker 2018). This type of antioxidant has a better affinity with the aqueous phase, where prooxidants agents are located, and can form an interfacial complex with emulsifiers that improve the oxidative stability of nanoemulsions (Cheng et al. 2019; Zhang et al. 2020a). For example, Artiga-Artigas et al. (2018) observed that lecithin (emulsifier) phosphate ions could interact with phenolic hydroxyl groups of the curcumin (antioxidant), getting an efficient entrapping of the lipophilic compound inside of phospholipids chains. This interaction generates an interfacial barrier with a high antioxidant capacity (23,801 mg TE/g and 3005 mg ET/g of DPPH and FRAP assays, respectively) that improves the oxidative stability of curcumin-loaded nanoemulsions. Similarly, Shehzad et al. (2021) indicated that curcumin and resveratrol could hold their optimal antioxidant capacity when they are in conjugation with polysaccharides (modified starches: Purity Gum 2000-PG2000 and Purity Gum Ultra-PGU). This conjugation improved the chemical stability of Tuna fish oil-based nanoemulsions, where a slight increment of peroxide value was observed after one month of storage at different temperatures (4°, 25° and 40 °C). Therefore, the formation of antioxidants and emulsifier complexes can be a helpful tool to increase the antioxidant capacity of nanoemulsions, where factors such as polarity and surface activity are essential for determining the efficacy of the antioxidant compounds against lipid oxidation.

4 Conclusions

Lipid oxidation control continues to pose a great challenge for the food industry, since it causes chemical and sensory deterioration and decreases the shelf-life of food products. In this sense, one of the current efforts to control lipid oxidation has focused on using nanoemulsions as encapsulation systems of bioactive compounds. However, it is still necessary to understand the oxidation mechanisms of these colloidal systems for developing new strategies to improve oxidative stability. Several studies mentioned in this chapter have demonstrated that the oil-water interface plays a crucial role in the lipid oxidation process, since emulsifier type and the presence of antioxidants can modify the oxidation degree of lipid components of nanoemulsions. In this regard, the use of emulsifiers with antioxidant activity or the incorporation of antioxidants that interact with emulsifiers into the oil-water interface could be an excellent alternative for preventing lipid oxidation of nanoemulsions. Thus, the development of antioxidant-emulsifiers complexes appears to be a new tool for improving the antioxidant effect of bioactive nanoemulsions.

References

Acevedo-Estupiñan MV, Gutierrez-Lopez GF, Cano-Sarmiento C, Parra-Escudero CO, Rodriguez-Estrada MT, Garcia-Varela R, García HS (2019) Stability and characterization of O/W free phytosterols nanoemulsions formulated with an enzymatically modified emulsifier. LWT-Food Sci Technol 107:151–157

Akhavan S, Assadpour E, Katouzian I, Jafari SM (2018) Lipid nano scale cargos for the protection and delivery of food bioactive ingredients and nutraceuticals. Trends Food Sci Technol 74:132–146

Arancibia C, Navarro-Lisboa R, Zúñiga RN, Matiacevich S (2016) Application of CMC as thickener on nanoemulsions based on olive oil: physical properties and stability. Int J Polymer Sci 2016:6280581

Arancibia C, Riquelme N, Zúñiga R, Matiacevich S (2017) Comparing the effectiveness of natural and synthetic emulsifiers on oxidative and physical stability of avocado oil-based nanoemulsions. Innovative Food Sci Emerg Technol 44:159–166

Artiga-Artigas M, Lanjari-Pérez Y, Martín-Belloso O (2018) Curcumin-loaded nanoemulsions stability as affected by the nature and concentration of surfactant. Food Chem 266:466–474

Aswathanarayan JB, Vittal RR (2019) Nanoemulsions and their potential applications in food industry. Front Sustain Food Syst 3:95

Berton-Carabin CC, Ropers MH, Genot C (2014) Lipid oxidation in oil-in-water emulsions: involvement of the interfacial layer. Compr Rev Food Sci Food Saf 13:945–977

Calligaris S, Plazzotta S, Bot F, Grasselli S, Malchiodi A, Anese M (2016) Nanoemulsion preparation by combining high-pressure homogenization and high-power ultrasound at low energy densities. Food Res Int 83:25–30

Chen XW, Chen YJ, Wang JM, Guo J, Yin SW, Yang XQ (2016) Phytosterol structured algae oil nanoemulsions and powders: improving antioxidant and flavor properties. Food Funct 7:3694–3702

Cheng C, Yu X, McClements DJ, Huang Q, Tang H, Yu K, Xiang X, Chen P, Wang X, Deng Q (2019) Effect of flaxseed polyphenols on physical stability and oxidative stability of flaxseed oil-in-water nanoemulsions. Food Chem 301:125207

Cho YH, McClements DJ (2009) Theoretical stability maps for guiding preparation of emulsions stabilized by protein-polysaccharide interfacial complexes. Langmuir 25:6649–6657

Choi SJ, Decker EA, Henson L, Popplewell LM, McClements DJ (2010) Influence of droplet charge on the chemical stability of citral in oil-in-water emulsions. J Food Sci 75:C536–C540

Choi SJ, McClements DJ (2020) Nanoemulsions as delivery systems for lipophilic nutraceuticals: strategies for improving their formulation, stability, functionality and bioavailability. Food Sci Biotechnol 29:149–168

Chuesiang P, Siripatrawan U, Sanguandeekul R, McLandsborough L, Julian McClements D (2018) Optimization of cinnamon oil nanoemulsions using phase inversion temperature method: impact of oil phase composition and surfactant concentration. J Colloid Interface Sci 514:208–216

Chung C, McClements DJ (2018) Characterization of physicochemical properties of nanoemulsions: appearance, stability and rheology. In: Nanoemulsions. Academic Press, San Diego, CA, pp 543–572

Costa M, Losada-Barreiro S, Paiva-Martins F, Bravo-Díaz C (2016) Physical evidence that the variation in the efficiency of homologous series of antioxidants in emulsions are a result of differences in their distribution. J Sci Food Agr 97:564–571

Costa M, Freiría-Gándara J, Losada-Barreiro S, Paiva-Martins F, Bravo-Díaz C (2020) Effects of droplet size on the interfacial concentration of antioxidants in fish and olive oil-in-water emulsions and nanoemulsions and on their oxidative stability. J Colloid Interface Sci 562:352–362

da Silveira TFF, Laguerre M, Bourlieu-Lacanal C, Lecomte J, Durand E, Figueroa-Espinoza MC, Baréa B, Barouh N, Castro IA, Villeneuve P (2021) Impact of surfactant concentration and antioxidant mode of incorporation on the oxidative stability of oil-in-water nanoemulsions. LWT-Food Sci Technol 141:110892

Dickinson E (2019) Strategies to control and inhibit the flocculation of protein-stabilized oil-in-water emulsions. Food Hydrocoll 96:209–223

Doost AS, Van Camp J, Dewettinck K, Van der Meeren P (2019) Production of thymol nanoemulsions stabilized using Quillaja saponin as a biosurfactant: antioxidant activity enhancement. Food Chem 293:134–143

Espitia PJP, Fuenmayor CA, Otoni CG (2019) Nanoemulsions: synthesis, characterization, and application in bio-based active food packaging. Compr Rev Food Sci Food Saf 18:264–285

Ferreira I, Costa M, Losada-Barreiro S, Paiva-Martins F, Bravo-Díaz C (2018) Modulating the interfacial concentration of gallates to improve the oxidative stability of fish oil-in-water emulsions. Food Res Int 112:192–198

Flores-Andrade E, Allende-Baltazar Z, Sandoval-González PE, Jiménez-Fernández M, Beristain CI, Pascual-Pineda LA (2021) Carotenoid nanoemulsions stabilized by natural emulsifiers: whey protein, gum Arabic, and soy lecithin. J Food Eng 290:110208

Fomuso LB, Corredig M, Akoh CC (2002) Effect of emulsifier on oxidation properties of fish oil-based structured lipid emulsions. J Agric Food Chem 50:2957–2961

García-Moreno PJ, Yang J, Gregersen S, Jones NC, Berton-Carabin CC, Sagis LM, Hoffman SV, Marcatili P, Overgaard MT, Hansen EB, Jacobsen C (2021) The structure, viscoelasticity and charge of potato peptides adsorbed at the oil-water interface determine the physicochemical stability of fish oil-in-water emulsions. Food Hydrocoll 115:106605

Gasa-Falcon A, Odriozola-Serrano I, Oms-Oliu G, Martín-Belloso O (2020) Nanostructured lipid-based delivery systems as a strategy to increase functionality of bioactive compounds. Foods 9:325

Gumus CE, Decker EA, McClements DJ (2017) Impact of legume protein type and location on lipid oxidation in fish oil-in-water emulsions: lentil, pea, and faba bean proteins. Food Res Int 100:175–185

Håkansson A, Rayner M (2018) General principles of nanoemulsion formation by high-energy mechanical methods. In: Nanoemulsions. Academic Press, San Diego, CA, pp 103–139

Hu M, McClements DJ, Decker EA (2003) Lipid oxidation in corn oil-in-water emulsions stabilized by casein, whey protein isolate, and soy protein isolate. J Agric Food Chem 51:1696–1700

Jiménez M, Domínguez JA, Pascual-Pineda LA, Azuara E, Beristain CI (2018) Elaboration and characterization of O/W cinnamon (*Cinnamomum zeylanicum*) and black pepper (*Piper nigrum*) emulsions. Food Hydrocoll 77:902–910

Kharat M, Du Z, Zhang G, McClements DJ (2017) Physical and chemical stability of curcumin in aqueous solutions and emulsions: impact of pH, temperature, and molecular environment. J Agric Food Chem 65:1525–1532

Kharat M, Zhang G, McClements DJ (2018) Stability of curcumin in oil-in-water emulsions: impact of emulsifier type and concentration on chemical degradation. Food Res Int 111:178–186

Kharat M, McClements DJ (2019) Recent advances in colloidal delivery systems for nutraceuticals: a case study-delivery by design of curcumin. J Colloid Interface Sci 557:506–518

Kumar DL, Sarkar P (2018) Encapsulation of bioactive compounds using nanoemulsions. Environ Chem Lett 16:59–70

Li Z, Dai L, Wang D, Mao L, Gao Y (2018a) Stabilization and rheology of concentrated emulsions using the natural emulsifiers quillaja saponins and rhamnolipids. J Agric Food Chem 66:3922–3929

Li Y, Wu CL, Liu J, Zhu Y, Zhang XY, Jiang LZ, Qi BK, Zhang XN, Wang ZJ, Teng F (2018b) Soy protein isolate-phosphatidylcholine nanoemulsions prepared using high-pressure homogenization. Nano 8:307

Li M, McClements DJ, Liu X, Liu F (2020) Design principles of oil-in-water emulsions with functionalized interfaces: mixed, multilayer, and covalent complex structures. Compr Rev Food Sci Food Saf 19:3159–3190

Li G, Zhang Z, Liu H, Hu L (2021) Nanoemulsion-based delivery approaches for nutraceuticals: fabrication, application, characterization, biological fate, potential toxicity and future trends. Food Funct 12:1933–1953
Liu Q, Huang H, Chen H, Lin J, Wang Q (2019a) Food-grade nanoemulsions: preparation, stability and application in encapsulation of bioactive compounds. Molecules 24:4242
Liu J, Guo Y, Li X, Si T, McClements DJ, Ma C (2019b) Effects of chelating agents and salts on interfacial properties and lipid oxidation in oil-in-water emulsions. J Agric Food Chem 67:13718–13727
Liu C, Pei R, Peltonen L, Heinonen M (2020) Assembling of the interfacial layer affects the physical and oxidative stability of faba bean protein-stabilized oil-in-water emulsions with chitosan. Food Hydrocoll 102:105614
Lourenço SC, Moldão-Martins M, Alves VD (2019) Antioxidants of natural plant origins: from sources to food industry applications. Molecules 24:4132
Losada-Barreiro S, Bravo-Díaz C, Paiva-Martins F (2020) Why encapsulate antioxidants in emulsion-based systems, where they are located, and how location affects their efficiency. In: Emulsion-based Encapsulation of Antioxidants: Design and Performance. Springer International Publishing, Cham, pp 1–39
Lozada MIO, Maldonade IR, Rodrigues DB, Santos D, Sanchez BAO, de Souza PEN, Longo JP, Amaro GB, de Oliveira LD (2021) Physicochemical characterization and nano-emulsification of three species of pumpkin seed oils with focus on their physical stability. Food Chem 343:128512
Ma P, Zeng Q, Tai K, He X, Yao Y, Hong X, Yuan F (2018) Development of stable curcumin nanoemulsions: effects of emulsifier type and surfactant-to-oil ratios. J Food Sci Technol 55:3485–3497
Maqsoudlou A, Assadpour E, Mohebodini H, Jafari SM (2020) Improving the efficiency of natural antioxidant compounds via different nanocarriers. Adv Colloid Interf Sci 278:102122
McClements DJ (2015) Appearance. In: Food emulsions: principles, practices, and techniques, 3rd edn. CRC Press and Taylor & Francis Group, Boca Raton
McClements DJ, Decker E (2018) Interfacial antioxidants: a review of natural and synthetic emulsifiers and co-emulsifiers that can inhibit lipid oxidation. J Agric Food Chem 66:20–35
McClements DJ, Jafari SM (2018a) General aspects of nanoemulsions and their formulation. In: Nanoemulsions. Academic Press, Cambridge, MA, pp 3–20
McClements DJ, Jafari SM (2018b) Improving emulsion formation, stability and performance using mixed emulsifiers: a review. Adv Colloid Interf Sci 251:55–79
McClements DJ (2021) Advances in edible nanoemulsions: digestion, bioavailability, and potential toxicity. Prog Lipid Res 81:101081
McClements DJ, Öztürk B (2021) Utilization of nanotechnology to improve the handling, storage and biocompatibility of bioactive lipids in food applications. Foods 10:365
Ni S, Hu C, Sun R, Zhao G, Xia Q (2017) Nanoemulsions-based delivery systems for encapsulation of quercetin: preparation, characterization, and cytotoxicity studies. J Food Process Eng 40: e12374
Ozkan G, Franco P, De Marco I, Xiao J, Capanoglu E (2019) A review of microencapsulation methods for food antioxidants: principles, advantages, drawbacks and applications. Food Chem 272:494–506
Páez-Hernández G, Mondragón-Cortez P, Espinosa-Andrews H (2019) Developing curcumin nanoemulsions by high-intensity methods: impact of ultrasonication and microfluidization parameters. LWT-Food Sci Technol 111:291–300
Park SJ, Hong SJ, Garcia CV, Lee SB, Shin GH, Kim JT (2019) Stability evaluation of turmeric extract nanoemulsion powder after application in milk as a food model. J Food Eng 259:12–20
Pavoni L, Perinelli DR, Ciacciarelli A, Quassinti L, Bramucci M, Miano A, Casettari L, Cespi M, Bonacucina G, Palmieri GF (2020b) Properties and stability of nanoemulsions: how relevant is the type of surfactant? J Drug Delivery Sci Technol 58:101772

Pavoni L, Perinelli DR, Bonacucina G, Cespi M, Palmieri GF (2020a) An overview of micro- and nanoemulsions as vehicles for essential oils: formulation, preparation and stability. Nano 10:135
Pinto F, de Barros DP, Fonseca LP (2018) Design of multifunctional nanostructured lipid carriers enriched with α-tocopherol using vegetable oils. Ind Crop Prod 118:149–159
Qian C, McClements DJ (2011) Formation of nanoemulsions stabilized by model food-grade emulsifiers using high-pressure homogenization: factors affecting particle size. Food Hydrocoll 25:1000–1008
Raimúndez-Rodríguez EA, Losada-Barreiro S, Bravo-Díaz C (2019) Enhancing the fraction of antioxidants at the interfaces of oil-in-water emulsions: a kinetic and thermodynamic analysis of their partitioning. J Colloid Interface Sci 555:224–233
Rehman A, Jafari SM, Tong Q, Karim A, Mahdi AA, Iqbal MW, Aadil RM, Ali A, Manzoor MF (2020) Role of peppermint oil in improving the oxidative stability and antioxidant capacity of borage seed oil-loaded nanoemulsions fabricated by modified starch. Int J Biol Macromol 153:697–707
Riquelme N, Zúñiga RN, Arancibia C (2019) Physical stability of nanoemulsions with emulsifier mixtures: replacement of tween 80 with quillaja saponin. LWT- Food Sci Technol 111:760–766
Ryu V, McClements DJ, Corradini MG, Yang JS, McLandsborough L (2018) Natural antimicrobial delivery systems: formulation, antimicrobial activity, and mechanism of action of quillaja saponin-stabilized carvacrol nanoemulsions. Food Hydrocoll 82:442–450
Saberi AH, Fang Y, McClements DJ (2014) Effects of salts on formation and stability of vitamin E-enriched mini-emulsions produced by spontaneous emulsification. J Agric Food Chem 62:11246–11253
Safaya M, Rotliwala YC (2020) Nanoemulsions: a review on low energy formulation methods, characterization, applications and optimization technique. Mater Today Proc 27:454–459
Salvia-Trujillo L, Decker EA, McClements DJ (2016) Influence of an anionic polysaccharide on the physical and oxidative stability of omega-3 nanoemulsions: antioxidant effects of alginate. Food Hydrocoll 52:690–698
Salvia-Trujillo L, Soliva-Fortuny R, Rojas-Graü MA, McClements DJ, Martín-Belloso O (2017) Edible nanoemulsions as carriers of active ingredients: a review. Annual Rev Food Sci Technol 8:439–466
Schröder A, Berton-Carabin C, Venema P, Cornacchia L (2017) Interfacial properties of whey protein and whey protein hydrolysates and their influence on O/W emulsion stability. Food Hydrocoll 73:129–140
Sharif HR, Goff HD, Majeed H, Liu F, Nsor-Atindana J, Haider J, Liang R, Zhong F (2017) Physicochemical stability of β-carotene and α-tocopherol enriched nanoemulsions: influence of carrier oil, emulsifier and antioxidant. Colloids Surf A Physicochem Eng Asp 529:550–559
Sharma S, Cheng SF, Bhattacharya B, Chakkaravarthi S (2019) Efficacy of free and encapsulated natural antioxidants in oxidative stability of edible oils: special emphasis on nanoemulsion-based encapsulation. Trends Food Sci Technol 91:305–318
Sharma N, Kaur G, Khatkar SK (2021) Optimization of emulsification conditions for designing ultrasound assisted curcumin loaded nanoemulsion: characterization, antioxidant assay and release kinetics. LWT-Food Sci Technol 141:110962
Shao P, Feng J, Sun P, Xiang N, Lu B, Qiu D (2020) Recent advances in improving stability of food emulsion by plant polysaccharides. Food Res Int 137:109376
Shehzad Q, Rehman A, Jafari SM, Zuo M, Khan MA, Ali A, Khan S, Karim A, Usman M, Hussain A, Xia W (2021) Improving the oxidative stability of fish oil nanoemulsions by co-encapsulation with curcumin and resveratrol. Colloids Surf B: Biointerfaces 199:111481
Silva R, Losada-Barreiro S, Paiva-Martins F, Bravo-Díaz C (2017) Partitioning and antioxidative effect of protocatechuates in soybean oil emulsions: relevance of emulsifier concentration. Eur J Lipid Sci Technol 119:1600274
Su D, Zhong Q (2016) Lemon oil nanoemulsions fabricated with sodium caseinate and tween 20 using phase inversion temperature method. J Food Eng 171:214–221

Tamm F, Drusch S (2017) Impact of enzymatic hydrolysis on the interfacial rheology of whey protein/pectin interfacial layers at the oil/water-interface. Food Hydrocoll 63:8–18

Velderrain-Rodríguez GR, Salvia-Trujillo L, González-Aguilar GA, Martín-Belloso O (2021) Interfacial activity of phenolic-rich extracts from avocado fruit waste: influence on the colloidal and oxidative stability of emulsions and nanoemulsions. Innovative Food Sci Emerg Technol 69:102665

Vinh TDT, Hien LTM, Dao DTA (2020) Formulation of black pepper (*Piper nigrum* L.) essential oil nanoemulsion via phase inversion temperature method. Food Sci Nutr 8:1741–1752

Walker RM, Gumus CE, Decker EA, McClements DJ (2017) Improvements in the formation and stability of fish oil-in-water nanoemulsions using carrier oils: MCT, thyme oil, & lemon oil. J Food Eng 211:60–68

Wang C, Sun C, Lu W, Gul K, Mata A, Fang Y (2020) Emulsion structure design for improving the oxidative stability of polyunsaturated fatty acids. Compr Rev Food Sci Food Saf 19:2955–2971

Waraho T, McClements DJ, Decker EA (2011) Mechanisms of lipid oxidation in food dispersions. Trends Food Sci Technol 22:3–13

Xu X, Luo L, Liu C, McClements DJ (2017) Utilization of anionic polysaccharides to improve the stability of rice glutelin emulsions: impact of polysaccharide type, pH, salt, and temperature. Food Hydrocoll 64:112–122

Xu X, Sun Q, McClements DJ (2019) Enhancing the formation and stability of emulsions using mixed natural emulsifiers: hydrolyzed rice glutelin and quillaja saponins. Food Hydrocoll 89:396–405

Yang L, Qin X, Kan J, Liu X, Zhong J (2019) Improving the physical and oxidative stability of emulsions using mixed emulsifiers: casein-octenyl succinic anhydride modified starch combinations. Nano 9:1018

Yi J, Ning J, Zhu Z, Cui L, Decker EA, McClements DJ (2019) Impact of interfacial composition on co-oxidation of lipids and proteins in oil-in-water emulsions: competitive displacement of casein by surfactants. Food Hydrocoll 87:20–28

Yu H, Park JY, Kwon CW, Hong SC, Park KM, Chang PS (2018) An overview of nanotechnology in food science: preparative methods, practical applications, and safety. J Chem 2018:5427978

Zamora R, Hidalgo FJ (2016) The triple defensive barrier of phenolic compounds against the lipid oxidation-induced damage in food products. Trends Food Sci Technol 54:165–174

Zhang Y, Tan C, Abbas S, Eric K, Xia S, Zhang X (2015) Modified SPI improves the emulsion properties and oxidative stability of fish oil microcapsules. Food Hydrocoll 51:108–117

Zhang Z, McClements DJ (2018) Overview of nanoemulsion properties: stability, rheology, and appearance. In: Nanoemulsions. Academic Press, Cambridge, pp 21–49

Zhang R, Zhang Z, McClements DJ (2020a) Nanoemulsions: an emerging platform for increasing the efficacy of nutraceuticals in foods. Colloids Surf B: Biointerfaces 194:111202

Zhang S, Tian L, Yi J, Zhu Z, Decker EA, McClements DJ (2020b) Mixed plant-based emulsifiers inhibit the oxidation of proteins and lipids in walnut oil-in-water emulsions: almond protein isolate-camellia saponin. Food Hydrocoll 109:106136

Zhao C, Wei L, Yin B, Liu F, Li J, Liu X, Wang J, Wang Y (2020) Encapsulation of lycopene within oil-in-water nanoemulsions using lactoferrin: impact of carrier oils on physicochemical stability and bioaccessibility. Int J Biol Macromol 153:912–920

Zhu Z, Zhao C, Yi J, Liu N, Cao Y, Decker EA, McClements DJ (2018) Impact of interfacial composition on lipid and protein co-oxidation in oil-in-water emulsions containing mixed emulsifiers. J Agric Food Chem 66:4458–4468

The Effects of Association Colloids on Lipid Autoxidation in Bulk Oils

Krzysztof Dwiecki and Ewa Bąkowska

1 Introduction

1.1 Minor Components of Vegetable Oils

Vegetable oils are food products obtained from raw materials of plant origin, usually from various oilseeds and fruits, such as the fruits of the olive tree *Olea europacea* and of the oil palm *Elaeis guineensis* (Belitz et al. 2009). The three most commonly produced vegetable oils in the world are palm oil, soybean oil, and rapeseed oil (Siger et al. 2015).

The main component of oils are triacylglycerols, also called triacylglycerides (TAG). The TAG content of most vegetable oils is in the 95–98% (w/w) range (Xenakis et al. 2010). In general, the TAG percentage is lower in crude oils and cold-pressed oils, and increases as a result of the refining process. For example, the content of triacylglycerols in crude soybean oil is 95–97%, while in refined, bleached, deodorized oil it increases to over 99% (Xenakis et al. 2010). According to Fine et al. (Fine et al. 2016), TAG concentration in refined rapeseed and sunflower oils is 98%. In turn, Xenakis et al. (2010) indicated that refined cottonseed, canola, palm, and sunflower oils are more than 99% TAG, while virgin olive oil is over 98.5% TAG.

In addition to TAG, oils also contain so-called minor oil components, such as monoacylglycerols and diacylglycerols, free fatty acids, phospholipids, sterols, tocopherols, tocotrienols, chlorophyll, carotenoids, phenolic compounds, ubiquinone, proteins, trace metals, products of lipid autoxidation, and water (Xenakis et al. 2010), (Fine et al. 2016), (Ghazani and Marangoni 2013). There is a greater

K. Dwiecki (✉) · E. Bąkowska
Department of Food Biochemistry and Analysis, Poznań University of Life Sciences, Poznań, Poland
e-mail: krzysztof.dwiecki@up.poznan.pl

C. Bravo-Diaz (ed.), *Lipid Oxidation in Food and Biological Systems*,
https://doi.org/10.1007/978-3-030-87222-9_12

percentage of minor components in crude oils and cold-pressed oils, and this decreases as a result of the refining process.

Monoacylglycerols are present in oils in smaller quantities (less than 0.2% w/w) than diacylglycerols (0.8–5.8% (w/w)). Crude palm and olive oils are the richest in diacylglycerols (Xenakis et al. 2010).

Free fatty acids are commonly considered to reduce oil quality and are therefore removed during neutralization. According to Ghazani and Marangoni (2013), refining, bleaching, and deodorization can decrease the free fatty acid content of canola oil from 0.3–1.2% to 0.03%.

As reported by Xenakis et al. (2010), the phospholipids content of crude oils falls within the range of 1.5–2.5% for soybean oil, 0.7–0.9% for cottonseed oil, 2.7–3.5% for canola oil, 0.006–0.013% for palm oil, 0.5–1.0% for sunflower oil, and 0.004–0.014% for virgin olive oil. As a result of refining the levels of these compounds decrease to 0.003–0.045% in the case of soybean oil and 0.012% in the case of palm oil. Phospholipids are removed mainly during the degumming process.

Sterols are one group of compounds present in oils that are beneficial from the nutritional point of view, but some of them are removed during refining. The richest sources of sterols are crude corn oil (924.3 mg/100 g) and crude rapeseed oil (823.8 mg/100 g). For rapeseed oil, it has been found that the total sterol content decreases from 823.8 mg/100 g in the crude oil to 767.1 mg/100 g after refining. It should be noted that free sterols are mostly removed—being reduced from 336.2 mg/100 g to 278.3 mg/100 g—while esters remain practically unchanged at 475.4 mg/100 g in crude oil and 484.7 mg/100 g in the refined oil (Verleyen et al. 2002).

Other important nutritional components include tocopherols and tocotrienols. As these possess antioxidant properties, they play an important role in ensuring oxidative stability of oil. The concentration of total tocopherols usually differs in the crude and refined oils. In the case of soybean oil, which is rich in these compounds, the total tocopherol level in the crude oil falls in the range 1094–2484 mg/kg, while in the refined product the range is 200–3327 mg/kg. In turn, sunflower oil contains 725–1892 mg/kg tocopherols in the crude product and 451–1289 mg/kg in the refined oil. Rapeseed oil has 464–1458 mg/kg tocopherols in the crude product and 227–1234 mg/kg in the refined product (Fine et al. 2016).

Other antioxidants present in vegetable oils include phenolic compounds (polyphenols). Rapeseeds contains about ten times greater levels of phenolic compounds (mainly sinapic acid and its derivatives) than other oilseeds (Siger et al. 2015). However, due to their relative hydrophilicity, only a small fraction of the phenolic compounds present in rapeseed is transferred to the oil during the pressing and extraction processes, and a significant part is also removed during refining. Rapeseeds contain as much as 2514–17,693 mg/kg of polyphenols, whereas the crude oil has only 113–629 mg/kg, and refined rapeseed oil contains only 2.1 mg/kg (Fine et al. 2016). Olive oil has a relatively high concentration of polyphenols in the range of 48–145 mg/kg for extra virgin olive oil, 11 mg/kg for extra light olive oil, and 10 mg/kg for cold-pressed olive oil (Xenakis et al. 2010).

The minor oil components also include pigments such as chlorophyll pigments and carotenoids. Although carotenoids are antioxidants, chlorophylls may act pro-oxidatively through photosensitized oxidation. For this reason, chlorophyll dyes are removed during refining in the bleaching process. In the case of crude canola oil the concentration of these pigments lies in the range 4–30 mg/kg, but after refining, bleaching, and deodorization this decreases to less than 0.025 mg/kg (Ghazani and Marangoni 2013). Crude canola oil contains about 130 ppm of carotenoids, mostly xanthophylls (90%) and carotenes (10%) (Ghazani and Marangoni 2013). In turn, in commercially available refined rapeseed oil, the content of β-carotene has been measured at 2.02 ppm (Rokosik et al. 2019).

Vegetable oils contain traces of metal ions (iron, copper, zinc, lead), which act as prooxidants; as such they should be present in lower concentration in the refined oil. In crude canola oil, the quantities of iron and copper present are 0.5–1.5 mg/kg and less than 0.2 mg/kg, respectively, whereas the refined oil contains less than 0.2 mg/kg of iron and less than 0.02 mg/kg of copper (Ghazani and Marangoni 2013).

Despite their hydrophobicity, oils also contain traces of water. Water in oil originates from the extraction and refining processes. The concentration of water in oils can change during storage, once the container has been opened, via absorption from the environment or loss from the oil. In commercially available refined vegetable oils, the water content usually ranges from 0.02% to 0.03% (200–300 ppm) (Xenakis et al. 2010; Budilarto and Kamal-Eldin 2015a). In unfiltered Greek olive oil samples, a higher concentration of 0.09–0.31% has been noted (Xenakis et al. 2010). Higher water content is observed predominantly in cold-pressed, unrefined oils. According to Siger et al. (2017), the concentration of water in cold-pressed common beech, chia, milk thistle, black cumin, white poppy, and black poppy oils was 911 ppm, 437 ppm, 779 ppm, 358 ppm, 583 ppm, and 831 ppm, respectively.

2 Amphiphilic Properties of Minor Oil Components

The minor components of vegetable oils may be hydrophobic (as with carotenoids), hydrophilic (such as phenolic compounds and proteins), or amphiphilic (monoacylglycerols, diacylglycerols, free fatty acids, phospholipids, sterols, and products of lipid autoxidation such as hydroperoxides, aldehydes, ketones, and epoxides) (Xenakis et al. 2010; Budilarto and Kamal-Eldin 2015a). Amphiphilic molecules possess both hydrophilic and hydrophobic parts and when dispersed in water, their hydrophilic parts preferably interact with water, whereas their hydrophobic parts come into contact with air or a nonpolar solvent. Amphiphiles thus undergo self-assembly and aggregate to form various structures. These assemblies are based on the repellant and coordinating forces between the hydrophobic and hydrophilic moieties of the molecules that constitute them and the surrounding medium (Wang et al. 2012). The type of structures that are formed depends on the hydrophilic–lipophilic balance (HLB) of the molecules. The HLB is calculated from the weight percentage of the molecule's hydrophilic to hydrophobic groups, and ranges from 1 to 20 (Kralova and Sjöblom 2009). Amphiphilic molecules with a relatively low HLB (3.0–6.0) preferably form water-in-oil emulsions, while a high HLB value (8.0–18.0) facilitates the formation of oil-in-water emulsions. Those minor oil components that have low HLBs—such as free fatty acids, monoacylglycerols, and diacylglycerols, with HLB values of 1.0, 3.4–3.8, and 1.8 respectively—favor the formation and stabilization of reverse micelles in water-in-oil emulsions. Phospholipids, which have an intermediate HLB of about 8.0 are able to form various structures: spherical reverse micelles in bulk oil containing a small amount of water (less than 300 ppm), and lamellar structures, combined with other surface-active compounds (Budilarto and Kamal-Eldin 2015a).

3 Formation of Association Colloids in Bulk Oils

Amphiphilic minor oil components are surface-active compounds (surfactants) and are therefore able to accumulate at the oil–water interface and lower surface tension. At concentrations above their critical micelle concentration (CMC), they self-aggregate and form association colloids (Kittipongpittaya et al. 2014). Association colloids are physical structures formed by amphiphilic compounds, which self-aggregate in a nonpolar environment (bulk oil) with a low water content. Because of the presence of small amounts of water in bulk oil, minor amphiphilic components may concentrate at the oil–water interface and form micellar structures. When association colloids are formed, they can change the physical and chemical properties of the bulk oil (Kittipongpittaya et al. 2014; Rokosik et al. 2020a). As mentioned earlier, vegetable bulk oils contain a lower concentration of water. They can thus be considered a type of water-in-oil nanoemulsion, and the formation of association colloids is in this case possible.

The two main types of association colloids found in vegetable oils are reverse micelles and lamellar structures (Fig. 1). Reverse micelles are micelles in which the polar groups are directed towards the center—a nanoscale hydrophilic core called the "water pool", which is stabilized by the monolayer of surface-active molecules. It was observed that the size of water droplets increases with the increasing concentration of water in the oil. Reverse micelles are dynamic structures that move by Brownian motion and which exchange water between each other. They have a large interface between their oil and water phases, which ensures that there is contact between the polar and nonpolar compounds. For this reason, reverse micelles are considered to be effective nanoreactors. Reverse micelles are mainly formed by amphiphiles with low HLB values, such as free fatty acids, monoacylglycerols, diacylglycerols, and phospholipids (Xenakis et al. 2010).

Lamellar structures consist of alternating layers of lamellae. In crude vegetable oils, layers of water and oil are separated by layers of appropriately oriented surface-active compounds. The HLB value means that such lamellar structures are mainly formed by phospholipids, which can be accompanied by sterols. These structures in

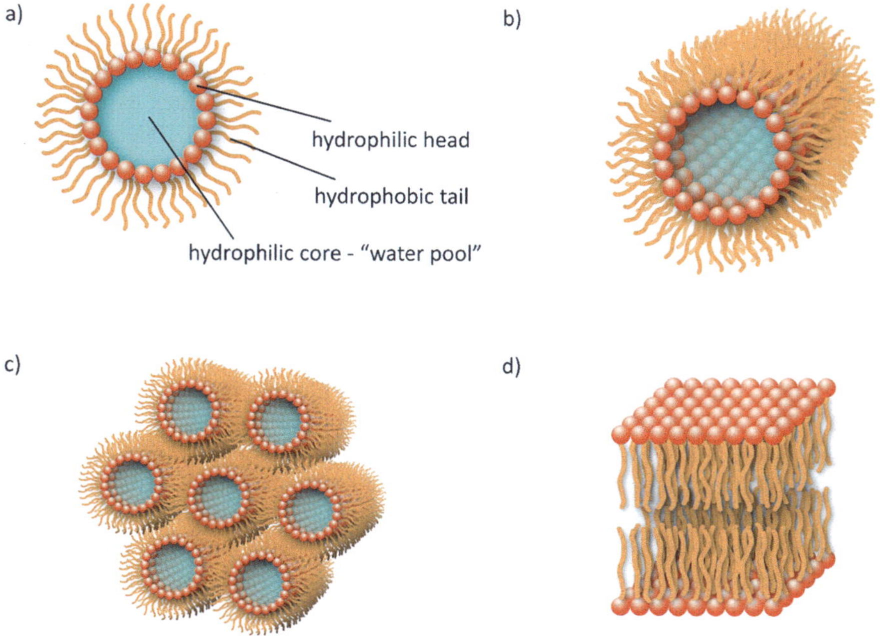

Fig. 1 Schematic representation of structures formed in bulk oil: (**a**) spherical reverse micelles, (**b**) cylindrical structures, (**c**) hexagonal structures, (**d**) lamellar structures

vegetable oils are so far relatively poorly studied and understood (Xenakis et al. 2010).

A necessary condition for the formation of association colloids is that the critical micelle concentrations (CMC) of amphiphilic compounds is exceeded. The CMC values of many minor oil components have already been established, though the values can vary in different oils. The CMC of a specific amphiphile may also be affected by other amphiphiles present in the oil, whether derived from raw materials or from products of lipid oxidation. For this reason, it is significant whether crude oil, refined oil, or refined oil additionally purified of minor components (stripped oil) is used.

In the case of phospholipids, the CMC is affected by the chain length of fatty acyl residues and by the polarity of the head group. It has been demonstrated that DOPC is able to form physical structures at much lower concentrations than phosphatidylcholine with short-chain fatty acyl residues (1,2-dibutyryl-sn-glycero-3-phosphocholine), on account of its lower critical micelle concentration (Chen et al. 2011a). The CMCs of phospholipid 1,2-dioleoyl-sn-glycero-3-phosphocholine (DOPC) is in the range 40–51.13 μmol/kg of oil (Rokosik et al. 2020a; Cui et al. 2014), whereas DOPE (1,2-dioleoyl-sn-glycero-3-phosphoethanolamine) had 200 μmol/kg (Kittipongpittaya et al. 2014). These differences have been explained by the different chemical structures of the compounds. The choline head group of

DOPC has a greater polarity than the ethanolamine moiety found in DOPE. As a result, DOPC has a stronger tendency to accumulate at the oil–water interface. Additionally, the critical micelle concentration of phospholipids depends on their hydration index—the extent to which the molecule has imbibed water, expressed on an arbitrary scale from 0 to 100. DOPC, with a hydration index of 100, has a lower CMC than DOPE with its hydration index of 16.

Commercial bulk oils contain a mixture of different minor components, which reduces the CMC in comparison to the CMC values that individual compounds would hold. Some of these components, such as phosphatidylcholine, are necessary for the formation of colloids, while others (like DOPE, stigmasterol, oleic acid, diacylglycerols, and other minor oil compounds) act as cosurfactants, reducing the CMC of phospholipids (Kittipongpittaya et al. 2014; Kim et al. 2019). However, according to Cui et al. (2014) the addition of DOPE alone increases the CMC of DOPC. The amount of water is a critical factor in the formation of association colloids in bulk oil. However, the minimum concentration of water needed to form colloids is difficult to determine, as it depends on the type of amphiphilic substance forming the micelles. For example, aggregation of stigmasterol in stripped corn oil was not observed with 400 ppm of water (Chen et al. 2010), while Cui et al. (2014) reported formation of DOPC micelles at 200–300 ppm of water. In refined commercially available vegetable oils, the water content usually ranges from 200 to 300 ppm, and these oils also contain various amphiphiles. It may be concluded that association colloids are present in practically all types of vegetable oil.

According to Chen et al. (2010), CMC differences in various media (oils) are due to differences in the fatty acid chain lengths and the presence of other minor components, which are surface active and were not completely removed during stripping. Critical micelle concentrations of amphiphilic minor oil components derived from the raw material (such as phospholipids) may be affected by lipid oxidation products. Many of these products (such as aldehydes) are amphiphilic. For this reason, CMC may change during oil storage, due to the lipid autoxidation process (Jo and Lee 2021).

Crude and commercial vegetable oils usually contain small amounts of water, which produces an oil–water interface in bulk oils. In such an environment, amphiphilic minor components migrate and preferably concentrate on the interface, forming the energetically favored association colloids (Xenakis et al. 2010). Although association colloids arise spontaneously in vegetable oils, in most studies concerning their properties, they are formed by stripping the oil of minor components or using a mixture of triacylglycerols and adding selected surface-active compounds (amphiphilic minor components) in a controlled manner.

The structure of aggregates formed in bulk oils from phospholipids depends on their type and concentration. In stripped soybean oil, 1,2-dibutyryl-sn-glycero-3-phosphocholine (DC_4PC) is able to form cylindrical structures, whereas DOPC there forms spherical structures. Both phospholipids have the same hydrophilic headgroups, but DOPC has cis-oleic fatty acids on the sn-1 and sn-2 glycerol positions, whereas DC_4PC possess two butyl fatty acids. Amphiphilic compounds such as phospholipids have a large tail area and a small headgroup area, and for this reason they assemble in oils into spherical reverse micelles, because of the critical packing parameter (CPP) being greater than unity. The CPP of DC_4PC in stripped soybean oil is approximately less than half lower than that of DOPC, assuming no difference in the effective area of the headgroup. In this case, CPP depends mainly on hydrocarbon tail length. A low CPP value often results in the formation of cylindrical structures (Chen et al. 2010). The self-assembly of DOPC into association colloids has also been confirmed by Rokosik et al. (2020a, b). Cui et al. (2014) observed that DOPE forms reverse micelles in stripped soybean oil, but DHPE (1,2-dihexanoyl-sn-glycero-3-phosphoethanolamine) does not aggregate into association colloids at concentrations up to 10,000 μmol. This indicates that the length of the phospholipid fatty acid residue that affects HLB has a significant effect on the ability of specific compounds to form association colloids. The differences between DOPE and DOPC CMC probably result from phosphatidylcholine interacting more strongly with water than does DOPE. Phosphatidylethanolamine is less hydratable than phosphatidylcholine due to the DOPE amine group, which tends to form hydrogen bonds directly with adjacent phosphate molecules. For this reason, DOPE's lipid–lipid interactions are stronger than its lipid–water interactions (Cui et al. 2014; Kanamoto et al. 1981; Franks 2009).

Subramanian et al. (2001) observed the formation of mixed micelles from phospholipids present in crude soybean oil. The micellar structure included phosphatidylcholine (PC), phosphatidylinositol (PI), phosphatidylethanolamine (PE), phosphatidic acid (PA), and phytosphingolipids. The CMC of micelles decreased with increasing degree of phospholipid hydration. Because the hydration indices of PC, PI, PE, and PA were 100, 44, 16, and 8.5 (in arbitrary units with 100 as the maximum), Subramanian et al. believe that the CMC of mixed micelles is greater than that formed from PC. The qualitative and quantitative composition of phospholipids in the oil thus affects the formation of mixed micelles. As this results from the hydration index, PA can significantly affect the assembly of these structures.

Stigmasterol has been found to be capable of forming aggregates in stripped corn oil (Kittipongpittaya et al. 2016). Oleic acid, in turn, does not assemble, probably because of the repulsion of its charged head groups (Kittipongpittaya et al. 2016).

According to Lehtinen et al. (2017), temperature, water, and free fatty acid content strongly affect the self-assembly of lecithin in rapeseed oil. Those authors noted the formation of lecithin cylindrical reverse micelles. An increase in temperature from room temperature to 70 °C resulted in a decrease in the CMC value. CMC decreasing with increasing temperature is a general tendency observed in many experiments. Because of the assembly of oleic acid molecules around polar groups of lecithin, suppression of reverse micelle formation was observed. This resulted from the increased solubility of lecithin in oil. This phenomenon has been observed at a very low concentration of water in oil. At slightly higher water levels, oleic acid acts as a cosurfactant, stabilizing reverse micelles. At the concentration of lecithin above the CMC, the addition of the critical amount of water causes the formation of lamellar structures. Assembly of these structures leads to phase separation, with the consequent removal of the reverse micelles from the oil phase. Phase separation begins at a significantly higher water level in the presence of oleic acid. The behavior of phospholipids in oil may be substantially altered by a very small change in water concentration or the presence of other surface-active compounds (Lehtinen et al. 2017).

4 Interactions of Antioxidants with Association Colloids

Antioxidants in vegetable oils may interact with association colloids, altering their structure and properties, as well as the kinetics of the lipid autoxidation process and the effectiveness of the antioxidants themselves. Rokosik et al. investigated interactions of phenolic compounds (sinapic and ferulic acids) with DOPC association colloids in stripped rapeseed oil (Rokosik et al. 2020a). Measurement of fluorescent probe emission demonstrated the incorporation of phenolic acids into the micelle structure. Another possibility is that the antioxidants were present at the water–oil interface, in the vicinity of the fluorescence probe. The observation of the static mechanism of probe fluorescence quenching confirmed the formation of complexes from phenolics and NBD-PE embedded in the reverse micelles. NBD-PE anisotropy measurements showed that interactions of sinapic and ferulic acid with DOPC reverse micelles resulted in changes in their structure (in particular reduced rigidity). In another study, Rokosik et al. (2020b) found that canolol (common in the rapeseed oil product of sinapic acid decarboxylation) may be incorporated into the structure of DOPC reverse micelles, but is not present in their hydrophilic part. In the case of α-tocopherol, a hydrophobic antioxidant, no interactions with DOPC association colloids were recorded. However, in stripped rapeseed oil without the addition of DOPC, the formation of micellar structures consisting of α-tocopherol was observed. A mixture of canolol and α-tocopherol at a relatively high concentration (500 μmol of each antioxidant/kg of oil) led to changes in the hydrophilic part of the reverse micelles (Rokosik et al. 2020b). Fadel et al. (2017) produced micellar aggregates composed of polyglyceryl-3-diisostearate in vegetable oil. The resulting reverse micelles allow efficient solubilization of polar antioxidants in the oil medium.

5 The Effects of Association Colloids on Lipid Autoxidation in Bulk Oils

The association colloids found in vegetable oils may affect the lipid autoxidation process and the effectiveness of antioxidant action. Brimberg (1993a, b) postulated that the transition from the initiation to the propagation phase of lipid oxidation is governed by the CMC of hydroperoxides and its modification by other amphiphilic compounds. According to Brimberg and Kamal-Eldin (2003), oxidation of lipids in bulk oils begins as a pseudo-first-order reaction, in which hydroperoxides are formed. When the hydroperoxides exceed their CMC, they begin to aggregate into reverse micelles. At this point, the reaction rate changes to second order and autoxidation passes to the propagation phase. Consequently, the efficiency of pro-oxidants and antioxidants depends on how they modulate the hydroperoxide CMC. However, the presence of both water from the raw material or the refining process and of amphiphilic compounds derived from seeds and fruits means that association colloids are already present at the initiation stage. Certainly, the amphiphilic oxidation products (such as hydroperoxides) affect the structure of reversed micelles.

Budilarto and Kamal-Eldin (2015b) showed that micelle size in sunflower and canola oils increases during and slightly before the end of the induction phase; their size then reaches a maximum and they collapse. Association colloids, though the formation of the oil–water interface, can physically affect the location of lipids and the oxidation process. Both hydroperoxides that are surface-active compounds and metal ions migrate to the oil–water interface. This facilitates the decomposition of hydroperoxides, catalyzed by the presence of transition metal ions, resulting in the formation of free radical products and acceleration of the autoxidation processes (Rokosik et al. 2020a). Free fatty acids and monoacylglycerols also act as prooxidants by concentrating at the oil–water interface and speeding up lipid hydroperoxide decomposition—the micellar effect. In this way, association colloids provide a reaction site for oxidation to take place—a "nanoreactor" (Budilarto and Kamal-Eldin 2015a).

As in oil–water emulsions, oxidation also occurs in interfacial regions. The smaller interface in bulk oil (compared to the emulsion) means the autoxidation is slower and the effective concentration of interfacial antioxidants is higher. As a result of the smaller interface surface, the effect of amphiphilic compounds derived from raw material and of amphiphilic oxidation products on the properties of the interfacial region and the structure of reverse micelles is greater.

According to the hypothesis of Budilarto and Kamal-Eldin (2015a), formation of a nanoemulsion and molecular organization are the most important factors for lipid autoxidation during the initial stage (induction period), with free radical reactions becoming relevant in the propagation and termination phases. Primary antioxidants and synergists stabilize micelles during the initial stage, whereas during propagation they mainly scavenge free radicals (Budilarto and Kamal-Eldin 2015a).

Chen et al. (2010) investigated the influence of DOPC and DC_4PC on the oxidation kinetics of soybean oil lipids. These two phospholipids have the same choline hydrophilic group. For this reason, their effect on lipid autoxidation by chemical pathways would be expected to be similar. However, it turned out that the spherical structures formed in stripped soybean oil by DOPC were prooxidative, whereas the cylindrical structures formed by DC_4PC had no effect on autoxidation rate. It must therefore be concluded that association colloids formed from phospholipids in bulk oil affect the autoxidation rate of lipids and their impact depends on the form/shape of structures. The prooxidative effect of DOPC reverse micelles has been confirmed by Chen et al. (2011a) and by Rokosik et al. (2020a, b).

The antioxidant properties of phospholipids have however been demonstrated: these include an ability to chelate metal ions, decompose hydroperoxides, scavenge free radicals, and increase the antioxidant potential of tocopherols. Physical structures assembled from phospholipids may increase the antioxidant activity of tocopherols by allowing them to concentrate at the oil–water interface, where lipid oxidation is most intense (Chen et al. 2010; Koga and Terao 1995). In this way a physical proximity of antioxidants and prooxidants is created. According to Chen et al. (2011a), DOPC in stripped soybean oil improves the activity of α-tocopherol and its water-soluble analogue at low concentrations (10 μmol), while at high concentrations of both antioxidants (100 μmol), a decrease in their activity was recorded. Those authors believe that DOPC reverse micelles form a negatively charged interface that attracts prooxidative metal ions. The presence of both antioxidants in the same location as the transition metals allows them to reduce metals into a more prooxidative state, thereby increasing the rate of lipid oxidation. This phenomenon is seen as a decrease in antioxidant efficiency. In this mechanism, we can also observe the embedding of surface-active hydroperoxides into reverse micelles. These substances are the substrate in metal-promoted lipid autoxidation, accelerating the process.

It has also been demonstrated that the hydrophilic Trolox has a higher antioxidant potential than the hydrophobic α-tocopherol. On the basis of NBD-PE fluorescence, it was found that emission of this probe was affected by Trolox much more strongly than in the case of α-tocopherol. This indicates that both antioxidants have different locations in DOPC reverse micelles. This may be due to the better solubility of Trolox in water, which leads it to be located in the aqueous phase, where α-tocopherol is practically insoluble in water, so at best it could be present in the oil–water interface. Such differences in the location of Trolox and α-tocopherol may affect their antioxidant and prooxidant activities.

Kittipongpittaya et al. confirmed that Trolox and α- tocopherol take up different locations in mixed micelles made of DOPC, DOPE, stigmasterol, oleic acid, and diacylglycerols in stripped corn oil (Kittipongpittaya et al. 2014). Those authors noted that α-tocopherol was unlikely to concentrate at the oil–water interface, while Trolox could partition at the same location as NBD-PE at the association colloids' oil–water interface. Nevertheless, the association structures were not found to have an effect on the antioxidant efficiency of either antioxidant at concentrations of 10 or 50 μmol/kg.

It has however been shown that association colloids assembled from mixed components substantially decrease the oxidative stability of oil. Cui et al. observed the prooxidant action of micelles obtained from a mixture of DOPC and DOPE in stripped soybean oil (Cui et al. 2014). In another study, Cui et al. (2015) explained how DOPC and DOPE reverse micelles affect the activity of antioxidants: it was found that DOPC association colloids decreased the activity of 100 μmol α-tocopherol and Trolox. DOPE reverse micelles increase the antioxidant activity of α-tocopherol by regenerating oxidized α-tocopherol (quinone) by the phosphatidylethanolamine primary amine group. However, no effect of DOPE on the physical location of α-tocopherol was found.

Homa et al. (2015) recorded that hydroperoxide type is related to prooxidant activity. Hydroperoxides with DHA, EPA, and α-linolenic acids have high surface activities, which facilitates their ability to form associates and to accelerate oxidation in fish and soybean oils. In turn, linoleic acid and oleic acid hydroperoxides are less surface active, and micelles formed with them in high linoleic safflower and high oleic safflower oils are not pro-oxidative.

Rokosik et al. (2020a) have suggested that the presence of DOPC reverse micelles in stripped rapeseed oil accelerates the decomposition of hydroperoxides to hexanal. They indicate that the antioxidant effect of sinapic acid in bulk rapeseed oil is affected by DOPC reverse micelles. The formation of association colloids from amphiphilic lipid autoxidation products (hydroperoxides, hexanal and others) reduces the effectiveness of sinapic acid in oil without added DOPC. In another study, Rokosik et al. (2020b) noted that DOPC association colloids affect the antioxidant efficiency of canolol and α-tocopherol. Simultaneously, at concentrations of 100 μmol for both substances, a decrease of their antioxidant action effectiveness over time was observed. This effect probably results from the increased concentration of amphiphilic autoxidation products, which affects the structure of DOPC micelles or the formation of mixed micelles with antioxidants, modifying their effectiveness. It has been shown that α-tocopherol affects levels of both lipid hydroperoxides and hexanal, though in canolol only hexanal is affected. These differences may come from the different location of the antioxidants in the association colloids. It was also demonstrated that the antioxidant synergism of canolol and α-tocopherol only occurs in DOPC reverse micelles and is not observed in oil lacking these structures. This indicates the important role of association colloids in the antioxidant synergism phenomenon.

Homma et al. (2016) observed that the charge of association colloids affects lipid oxidation in ethyl oleate. Structures formed from anionic, cationic, and nonionic surfactants respectively retard, accelerate or have no effect on oxidation rates.

Changes in antioxidants efficiency at the interface in bulk oil may also be related to the "cut-off" effect, which is described as the parabolic dependence of antioxidant efficiency on molecular chain length. It is explained by changes in reactant diffusivity, antioxidants' ability to self-aggregate, and their solubility in the interfacial region. Molecule chain length may also affect reactivity orientation at the interface (Costa et al. 2021). In this context, it should be stated that the presence of amphiphilic minor compounds in bulk oil, the formation of association colloids, changes in their structure and in the ability of antioxidants to aggregate at the interface may affect the cut-off effect.

The effectiveness of antioxidants in the presence of association colloids should also be considered in the context of the polar paradox, according to which polar antioxidants are more efficient than nonpolar antioxidants in bulk oils (Costa et al. 2015). Lipid oxidation in oil-in-water emulsions occur in the interfacial region. Similarly, in bulk oil—which can be considered a nanoemulsion—the oil–water interface plays an important role in autoxidation. Surface-active hydroperoxides and metal ions accumulate at the interface in association colloids and, as a result of the metal-catalyzed decomposition of hydroperoxides, free radicals are formed, accelerating autoxidation. In this sense, association colloids play a catalytic role. The presence of polar antioxidants in colloids can effectively inhibit this process, in line with the polar paradox. The formation of reverse micelles, changes in their structure and composition (such as due to the appearance of amphiphilic oxidation products) may alter the effective interfacial concentration of antioxidants. This decrease in antioxidants' effective interfacial concentration in an oil–water emulsion was observed as a result of an increase in the molar volume of the surfactant to volume of emulsion ratio (Costa et al. 2020).

The rate and mechanism of lipid oxidation in association colloids depends on many factors. The concentration of the substrates, antioxidants, and prooxidants in association colloids depends on the quantity of surface-active compounds available to form such structures (free fatty acids, phospholipids, antioxidants). The location and orientation of reactants in colloids, as well as changes in their structure at different temperatures, affect lipid oxidation rate. For example, the anionic interface of micelles attracts metals catalyzing the decomposition of hydroperoxides (Chaiyasit et al. 2007).

6 The Role of Water in the Formation of Association Colloids and its Effects on Lipid Autoxidation in Bulk Oil

The water content of an oil changes during prolonged storage after the package has been opening. This happens due to the absorption of water from the environment and its evaporation from the oil, but also due to oxidation (Chen et al. 2011b; Park et al. 2014). Water content negatively affects the quality of oils: it is involved in triacylglycerol hydrolysis to free fatty acids in thermal processes and in the presence of lipase. Water is a solvent for many amphiphilic and hydrophilic substances, which affects their reactivity. This concerns both compounds with antioxidant properties, such as phenolic compounds or ascorbic acid, as well as prooxidative substances, such as lipid hydroperoxides or transition metals (Kittipongpittaya et al. 2016). Additionally, water is one of the basic requirements for the formation of association colloids in the oil. Water activity (which may affect the oxidation rate) may be influenced by amphiphilic compounds and by the presence of association colloids (Chaiyasit et al. 2007).

Due to its hydrophilic structure, water is practically immiscible with oil, with a solubility ranging from 0.05 to 0.3%. Water may have an effect on the critical micelle concentration of minor components in oil. Changes in the CMC of medium chain triacylglycerols and of lecithin in corn oil with water content were described by Kim et al. (2018), who argued that such changes are associated with the available water, which creates sites or areas in bulk oil for packaging amphiphilic compounds, such as phospholipids (PL). The size and number of association colloids are affected by the water-to-phospholipid ratio. If the concentration of PL is greater than the water content, the PL forms a large number of small-sized micelles. Demand for PL is thus high, which is equivalent to high CMC. An increase in humidity will lead to an increase in the area available for binding with the PL. More PL is needed to pack the available space in the water molecules, thus increasing CMC. At a humidity higher than 900 ppm, water molecules have a greater affinity for each other than for PL molecules. Small association colloids merge and form larger structures, reducing the interface and thus decreasing CMC (Kim et al. 2018).

Most studies have focused on the formation of association colloids from phospholipids. The appearance of additional surfactants, and even minor changes in water content, may alter the structure of phospholipid micelles (Rokosik et al. 2020a; Cui et al. 2014; Xu et al. 2019). According to Lehtinen et al. (2017), oleic acid in the presence of a low water concentration will accumulate around lecithin polar groups in rapeseed oil, increasing its solubility in oil and thus increasing the CMC value—that is, the concentration of lecithin necessary to create association colloids. The interaction of oleic acid and lecithin, however, is strictly dependent on the water content in the system: greater water content causes the precipitation of lecithin in the form of laminar structures, while added oleic acid acts as a cosurfactant and decreases the rate of phase separation, stabilizing reverse micelles (Lehtinen et al. 2017).

Although water is only present in trace amounts in oil, it has a significant impact on oxidation and other processes, thanks to its ability to dissolve antioxidants and pro-oxidants. However, water, apart from its interaction with other amphiphilic and hydrophilic substances, has no effect on the rate of autoxidation (Kittipongpittaya et al. 2016; Chen et al. 2011b). In the absence of surfactants (minor components) association colloids, considered as oxidation centers, are not formed, hence the rate of oxidation is similar to that in oils with different moisture contents (Budilarto and Kamal-Eldin 2015a).

Aldehydes such as propanal, hexanal, and nonanal accelerate the lipid oxidation rate in soybean oil and alter its water content (in comparison to soybean oil without aldehydes), but the changes depended on the type of aldehyde and on the initial relative humidity of the oil. Nonanal and hexanal contributed to the removal of water from soybean oil, while propanal stabilized the moisture (Jo and Lee 2021).

In turn, Kim et al. (2014) confirmed that water may function as a substrate for the formation of volatile oxidation products during the autoxidation of bulk oil. Park et al. (2014) found that changes in water content in corn oil and the rate of oxidation depend on the availability of oxygen and on storage temperature.

Water content and mobility could affect the size and number of association colloids, as well as the activity of reactants at the interface or in the water core. The dimensions of inner water core in association colloids may determine the type of molecules incorporated into the colloid structure. Those factors may affect lipid oxidation rate in bulk oil (Chaiyasit et al. 2007).

7 Methods Used in Association Colloids Studies

Studies of the properties of association colloids in vegetable oils and the effects they have on autoxidation usually require the formation of colloidal structures under controlled conditions. It is thus necessary to remove amphiphilic minor components before assembling the micelles or lamellar structures from compounds added to the oil in carefully measured amounts. The removal of minor compounds from oil is generally performed using column chromatography with n-hexane as an eluent. The most common types of chromatographic bed used include silica gel, silicic acid, activated charcoal, and aluminum oxide, which have undergone the appropriate activation. Beds are used individually or in combination. Aluminum oxide mainly removes antioxidants from vegetable oils (Nyström et al. 2007; Khuwijitjaru et al. 2009; Romero et al. 2007). Chromatographic beds containing silica gel are used to purify oil of tocopherols and sterols (Hrádková et al. 2013; Fang et al. 2017). Verleyen et al. (2002) separated free and esterified sterols from vegetable oils using silica gel. Most researchers use two chromatographic methods together: silicic acid and activated charcoal (Kittipongpittaya et al. 2016; Homma et al. 2015; Boon et al. 2008). Activated charcoal is used mainly to remove pigments, especially chlorophyll and carotenoids. It is often employed due to its low cost and strong adsorption properties, but it is not very selective of the compounds it removes, therefore it is often used alongside additional adsorbents (Rajczykowski and Loska 2016). Rokosik et al. (2019) in their research into appropriate chromatographic beds for purifying rapeseed oil from its minor components considered all the types just mentioned. They determined effectiveness of each bed type by measuring the concentration of the residual dyes (chlorophyll, carotenoids), sterols, and tocopherols in the purified oil. As a result, a three-stage optimized method for removing minor components from rapeseed oil was developed. The oil was dissolved in n-hexane (1: 1 v/v) and passed sequentially through the column with silicic acid, activated charcoal, and a second layer of silicic acid; the partially purified oil was then directed to a column with activated charcoal and aluminum oxide. In the last step, a column containing silica gel was used (Rokosik et al. 2019).

Association colloids are formed in bulk oils in the presence of small amounts of water, the concentration of which influences the properties of the micellar structures. It is thus often necessary to determine water content. For this purpose, the thermogravimetric method and the Karl Fischer titration are used. Drying the oil in the oven is straightforward and inexpensive, but is time-consuming and can lead to many errors; for example, volatile substances (volatile fatty acids) are lost at the drying temperature, which may falsify results (Xie et al. 2017). For this reason, the most common methods are the Karl Fischer volumetric (Leito and Jalukse 2019) and coulometric methods, which are fast, accurate, and reproducible. Due to the trace amounts of water in oils, the coulometric method is recommended (Felgner et al. 2008).

Spectroscopy is generally used in determining the CMC of amphiphilic minor oil components, with TCNQ spectroscopy being most common. The absorbance of 7,7,8,8-tetracyanoquinodimethane clearly increases as a result of micelle formation, due to the charge transfer of TCNQ in the presence of aggregates. The critical micelle concentration is determined as an inflexion point in the semilogarithmic plot showing the dependence of TCNQ absorbance (at 480 nm) on the amphiphilic substance concentration (Kittipongpittaya et al. 2014; Cui et al. 2014; Chen et al. 2010; Subramanian et al. 2001). The CMC may be also determined on the basis of NBD-PE fluorescence probe emission intensity, which increases upon formation of reverse micelles in oil as a result of changes in the microenvironment of the probe. As with the TCNQ technique, CMC is read from the plot of fluorescence intensity versus surfactant concentration (Rokosik et al. 2020a, b).

Fluorescence probes may also be used to detect association colloids and to find changes in the environment, structure, and interactions of micelles with antioxidants. DAF (5-dodecanoylaminofluorescein) probes can be used to observe how amphiphilic components like phosphatidylcholine can increase the exposure of surfactants to the aqueous phase of association colloids. DAF may also record changes in pH resulting from the addition of oleic acid to bulk oil. In turn, the ability of DAF to detect free radicals may be employed to determine how amphiphilic compounds impact oxidative reactions in oil (Chaiyasit et al. 2008). Amphiphilic NBD-PE fluorescence probes ((N-(7-nitrobenz-2-oxa-1,3-diazol-4-yl)-1,2-dihexadecanoyl-sn-glycero-3-phosphoethanolamine, triethylammonium salt) may detect the formation of association colloids: exposure of the NBD-PE head group to a polar environment leads to a decrease in fluorescence intensity. This probe is preferentially located at the oil–water interface and orients its hydrophilic moiety toward the water core. The imino group or the oxygen molecule on the NBD probe can hydrogen bond with water molecules, leading to fluorescence quenching. The increased NBD-PE fluorescence intensity in the presence of reverse micelles formed from phospholipids may be due to the ability of amphiphilic compounds to compete for the oil–water interface, thus decreasing interactions between NBD-PE and water, and leading to an increase in fluorescence intensity (Kittipongpittaya et al. 2014; Rokosik et al. 2020a).

NBD-PE may be used to investigate the physical location of antioxidants in association colloids. For example, Trolox can partition in the same location as NBD-PE, at the oil–water interface of association colloids. This leads to a measurable decrease in probe fluorescence intensity (Kittipongpittaya et al. 2014). Measurements of NBD-PE fluorescence quenching in combination with fluorescence decay (lifetime) allow it to be determined whether the mechanism of fluorescence quenching is static or dynamic. The static quenching mechanism implies that the ground-state nonfluorescent complex between NBD-PE and an antioxidant has been formed. In this way, the presence of sinapic and ferulic acids in the DOPC reverse micelles structure or at the oil–water interface in the vicinity of fluorescence probe could be determined (Rokosik et al. 2020a). Fluorescence anisotropy measurements are used to study changes in the structure of association colloids. This value reflects the rotational freedom of the fluorescent molecule. The increase in fluorescence anisotropy of NBD-PE molecules embedded in micelles shows the formation of structures with enhanced rigidity. Changes in rigidity were observed as a result of the incorporation of sinapic and ferulic acid into DOPC reverse micelles, and after formation of canolol molecular associations in stripped rapeseed oil (Rokosik et al. 2020a, b).

Small-angle X-ray scattering (SAXS) and cryogenic transmission electron microscopy (cryo-TEM) can determine the structural properties and morphology of reverse micelles; some of the structures detected in this way in vegetable oil include spherical, cylindrical, hexagonal, and lamellar shapes (Fig. 1), (Chen et al. 2010; Lehtinen et al. 2017).

An association of two triacylglycerol molecules forming a backbone (2.5 nm) has been detected in olive oil using dynamic light scattering (DLS) (Xenakis et al. 2010). To measure the hydrodynamic radius and polydispersity index of colloidal structures in oil, modified thin-layer cell dynamic light-scattering instruments with 3D optics can be also applied. The use of this variant of the DLS method avoids the problem of multiple scattering in the undiluted sample and the fluorescence of chlorophylls present in the oil. Static small angle light scattering (SALSA) allows the detection of colloids up to 15 μm. This is a good technique for turbid samples, such as virgin olive oil (Papadimitriou et al. 2013).

The surface activity of amphiphilic compounds can be determined by measuring its effect on the interfacial tension of oil using interfacial tensiometry (with a drop-shape analyzer) (Kittipongpittaya et al. 2014, 2016).

8 Conclusions

Vegetable oils are often thought of as homogeneous products. However, the presence of small amounts of water mean they are better understood as nanoemulsions. Amphiphilic compounds derived from the raw material and amphiphilic oxidation products accumulate at the interface, forming association colloids (spherical reverse micelles, as well as lamellar, cylindrical, and hexagonal structures). Lipid oxidation takes place mainly in the interfacial region. The accumulation of pro-oxidants (such as hydroperoxides decomposed to free radicals in the presence of metal ions) and amphiphilic or hydrophilic antioxidants at the interface can significantly affect this process. In addition, amphiphilic compounds that do not participate in oxidation reactions, due to their presence in the interfacial region, can modify the physical properties of the association colloids. They can thus indirectly affect the effectiveness of pro-oxidants and antioxidants by changing the location and orientation of reactants in the colloid structure. Association colloids hence play an important role in the autoxidation of vegetable oils. Controlling the processes taking place in them—for example by altering the composition of the amphiphilic compounds—may contribute to better ways of managing this unfavorable autoxidation.

Acknowledgements The authors very much appreciate the financial support of the National Science Centre, Poland, under grant NCN 2016/23/B/NZ9/02793.

References

Belitz H-D, Grosch W, Schieberle P (2009) Food chemistry. Springer, Berlin, Heidelberg. http://link.springer.com/10.1007/978-3-540-69934-7

Boon CS, Xu Z, Yue X, McClements DJ, Weiss J, Decker EA (2008) Factors affecting lycopene oxidation in oil-in-water emulsions. J Agric Food Chem 4(56):1408–1414

Brimberg UI (1993a) On the kinetics of the autoxidation of fats. J Am Oil Chem Soc 70(3):249–254

Brimberg UI (1993b) On the kinetics of the autoxidation of fats. II. Monounsaturated substrates. J Am Oil Chem Soc 70(11):1063–1067

Brimberg UI, Kamal-Eldin A (2003) On the kinetics of the autoxidation of fats: influence of pro-oxidants, antioxidants and synergists. Eur J Lipid Sci Technol 105(2):83–91

Budilarto ES, Kamal-Eldin A (2015a) The supramolecular chemistry of lipid oxidation and antioxidation in bulk oils: supramolecular chemistry of lipid oxidation. Eur J Lipid Sci Technol 117(8):1095–1137

Budilarto ES, Kamal-Eldin A (2015b) Water content and micelle size change during oxidation of sunflower and canola oils: lipid oxidation: water content & micelle size. Eur J Lipid Sci Technol 117(12):1971–1977

Chaiyasit W, Elias RJ, McClements DJ, Decker EA (2007) Role of physical structures in bulk oils on lipid oxidation. Crit Rev Food Sci Nutr 47(3):299–317

Chaiyasit W, McClements DJ, Weiss J, Decker EA (2008) Impact of surface-active compounds on physicochemical and oxidative properties of edible oil. J Agric Food Chem 56(2):550–556

Chen B, Han A, McClements DJ, Decker EA (2010) Physical structures in soybean oil and their impact on lipid oxidation. J Agric Food Chem 58(22):11993–11999

Chen B, Han A, Laguerre M, McClements DJ, Decker EA (2011a) Role of reverse micelles on lipid oxidation in bulk oils: impact of phospholipids on antioxidant activity of α-tocopherol and Trolox. Food Funct 2(6):302–309

Chen B, McClements DJ, Decker EA (2011b) Minor components in food oils: a critical review of their roles on lipid oxidation Chemistry in bulk oils and emulsions. Crit Rev Food Sci Nutr 51 (10):901–916

Costa M, Losada-Barreiro S, Paiva-Martins F, Bravo-Díaz C, Romsted LS (2015) A direct correlation between the antioxidant efficiencies of caffeic acid and its alkyl esters and their concentrations in the interfacial region of olive oil emulsions. The pseudophase model interpretation of the "cut-off" effect. Food Chem 175:233–242

Costa M, Freiría-Gándara J, Losada-Barreiro S, Paiva-Martins F, Bravo-Díaz C (2020) Effects of droplet size on the interfacial concentrations of antioxidants in fish and olive oil-in-water emulsions and nanoemulsions and on their oxidative stability. J Colloid Interface Sci 562:352–362

Costa M, Losada-Barreiro S, Paiva-Martins F, Bravo-Díaz C (2021) Polyphenolic antioxidants in lipid emulsions: partitioning effects and interfacial phenomena. Foods 10(3):539

Cui L, Kittipongpittaya K, McClements DJ, Decker EA (2014) Impact of Phosphoethanolamine reverse micelles on lipid oxidation in bulk oils. J Am Oil Chem Soc 91(11):1931–1937

Cui L, McClements DJ, Decker EA (2015) Impact of phosphatidylethanolamine on the antioxidant activity of α-tocopherol and Trolox in bulk oil. J Agric Food Chem 63(12):3288–3294

Fadel O, Girard L, Gomes Rodrigues D, Bauduin P, Le Goff X, Rossignol-Castera A (2017) I in. Micellization in vegetable oils: a structural characterisation. Colloids Surf B Biointerfaces 154:279–286

Fang B, Zhang M, Shen YM (2017) Importance of the higher retention of tocopherols and sterols for the oxidative stability of soybean and rapeseed oils. J Food Sci Technol 54(7):1938–1944

Felgner A, Schlink R, Kirschenbühler P, Faas B, Isengard H-D (2008) Automated Karl Fischer titration for liquid samples – water determination in edible oils. Food Chem 106(4):1379–1384

Fine F, Brochet C, Gaud M, Carre P, Simon N, Ramli F (2016) I in. Micronutrients in vegetable oils: the impact of crushing and refining processes on vitamins and antioxidants in sunflower, rapeseed, and soybean oils: micronutrients in vegetable oils. Eur J Lipid Sci Technol 118 (5):680–697

Franks F (2009) Water science reviews. Cambridge University Press, Cambridge

Ghazani SM, Marangoni AG (2013) Minor components in canola oil and effects of refining on these constituents: a review. J Am Oil Chem Soc 90(7):923–932

Homma R, Suzuki K, Cui L, McClements DJ, Decker EA (2015) Impact of association colloids on lipid oxidation in Triacylglycerols and fatty acid ethyl esters. J Agric Food Chem 63 (46):10161–10169

Homma R, Johnson DR, McClements DJ, Decker EA (2016) Influence of iron solubility and charged surface-active compounds on lipid oxidation in fatty acid ethyl esters containing association colloids. Food Chem 199:862–869

Hrádková I, Merkl R, Šmidrkal J, Kyselka J, Filip V (2013) Antioxidant effect of mono- and dihydroxyphenols in sunflower oil with different levels of naturally present tocopherols. Eur J Lipid Sci Technol 115(7):747–755

Jo S, Lee J (2021) Evaluation of the effects of aldehydes on association colloid properties and oxidative stability in bulk oils. Food Chem 338:127778

Kanamoto R, Wada Y, Miyajima G, Kito M (1981) Phospholipid- phospholipid interaction in soybean oil. J Am Oil Chem Soc 58(12):1050–1053

Khuwijitjaru P, Yuenyong T, Pongsawatmanit R, Adachi S (2009) Degradation kinetics of gamma-Oryzanol in antioxidant-stripped Rice bran oil during thermal oxidation. J Oleo Sci 58 (10):491–497

Kim JY, Kim M-J, Lee J (2014) Role of moisture on the lipid oxidation determined by D2O in a linoleic acid model system. Food Chem 146:134–140

Kim J, Kim M-J, Lee J (2018) The critical micelle concentration of lecithin in bulk oils and medium chain triacylglycerol is influenced by moisture content and total polar materials. Food Chem 261:194–200

Kim J, Woo Y, Ryu J, Kim M-J, Lee J (2019) Lecithin near its critical micelle concentration increases oxidative stability of non-stripped corn oil but not stripped corn oil. Eur J Lipid Sci Technol 121:1800219

Kittipongpittaya K, Panya A, Cui L, McClements DJ, Decker EA (2014) Association colloids formed by multiple surface active minor components and their effect on lipid oxidation in bulk oil. J Am Oil Chem Soc 91(11):1955–1965

Kittipongpittaya K, Panya A, Decker EA (2016) Role of water and selected minor components on association colloid formation and lipid oxidation in bulk oil. J Am Oil Chem Soc 93(1):83–91

Koga T, Terao J (1995) Phospholipids increase radical-scavenging activity of vitamin E in a bulk oil model system. J Agric Food Chem 43(6):1450–1454

Kralova I, Sjöblom J (2009) Surfactants used in food industry: a review. J Dispers Sci Technol 30 (9):1363–1383

Lehtinen O-P, Nugroho RWN, Lehtimaa T, Vierros S, Hiekkataipale P, Ruokolainen J (2017) I in. Effect of temperature, water content and free fatty acid on reverse micelle formation of phospholipids in vegetable oil. Colloids Surf B Biointerfaces 1(160):355–363

Leito I, Jalukse L (2019) Traceability, validation and measurement uncertainty in chemistry, vol 3 practical examples. https://doi.org/10.1007/978-3-030-20347-4

Nyström L, Achrenius T, Lampi A-M, Moreau RA, Piironen V (2007) A comparison of the antioxidant properties of steryl ferulates with tocopherol at high temperatures. Food Chem 101(3):947–954

Papadimitriou V, Dulle M, Wachter W, Sotiroudis TG, Glatter O, Xenakis A (2013) Structure and dynamics of veiled virgin olive oil: influence of production conditions and relation to its antioxidant capacity. Food Biophys 8(2):112–121

Park JW, Kim JY, Kim M-J, Lee J (2014) Evaluation of oxygen-limitation on lipid oxidation and moisture content in corn oil at elevated temperature. J Am Oil Chem Soc 91(3):439–444

Rajczykowski K, Loska K (2016) Heavy metals adsorption on aluminium oxide and stabilization of spent adsorbent using a SULSTAR technology. Arch Waste Manag Environ Prot 18(2):67–72

Rokosik E, Dwiecki K, Rudzińska M, Siger A, Polewski K (2019) Column chromatography as a method for minor components removal from rapeseed oil. Grasas Aceites 70(3):316

Rokosik E, Siger A, Rudzińska M, Siejak P, Dwiecki K (2020a) Formation of phospholipid association colloids in rapeseed oil and their effect on lipid autoxidation in the presence of Sinapic and Ferulic acid. Eur J Lipid Sci Technol 122(2):1900243

Rokosik E, Siger A, Rudzińska M, Dwiecki K (2020b) Antioxidant activity and synergism of canolol and α-tocopherol in rapeseed oil is affected by the presence of phospholipid association colloids. LWT 133:110095

Romero N, Robert P, Masson L, Ortiz J, González K, Tapia K (2007) Effect of α-tocopherol, α-tocotrienol and Rosa mosqueta shell extract on the performance of antioxidant-stripped canola oil (brassica sp.) at high temperature. Food Chem 104(1):383–389

Siger A, Kaczmarek A, Rudzińska M (2015) Antioxidant activity and phytochemical content of cold-pressed rapeseed oil obtained from roasted seeds: antioxidant activity and phytochemical content. Eur J Lipid Sci Technol 117(8):1225–1237

Siger A, Dwiecki K, Borzyszkowski W, Turski M, Rudzińska M, Nogala-Kałucka M (2017) Physicochemical characteristics of the cold-pressed oil obtained from seeds of Fagus sylvatica L. Food Chem 225:239–245

Subramanian R, Ichikawa S, Nakajima M, Kimura T, Maekawa T (2001) Characterization of phospholipid reverse micelles in relation to membrane processing of vegetable oils. Eur J Lipid Sci Technol 103:93–97

Verleyen T, Forcades M, Verhe R, Dewettinck K, Huyghebaert A, De Greyt W (2002) Analysis of free and esterified sterols in vegetable oils. J Am Oil Chem Soc 79(2):117–122

Wang C, Wang Z, Zhang X (2012) Amphiphilic building blocks for self-assembly: from Amphiphiles to supra-amphiphiles. Acc Chem Res 45(4):608–618

Xenakis A, Papadimitriou V, Sotiroudis TG (2010) Colloidal structures in natural oils. Curr Opin Colloid Interface Sci 15(1–2):55–60

Xie W-Q, Gong Y-X, Yu K-X (2017) Efficient quantification of water content in edible oils by headspace gas chromatography with vapour phase calibration: quantification of water in edible oils. J Sci Food Agric 98(8):3208–3212

Xu N, Shanbhag AG, Li B, Angkuratipakorn T, Decker EA (2019) Impact of phospholipid–tocopherol combinations and enzyme-modified lecithin on the oxidative stability of bulk oil. J Agric Food Chem 67(28):7954–7960

Antioxidant Activity of Edible Isothiocyanates

Jakub Cedrowski, Jacek Grebowski, and Grzegorz Litwinienko

1 From Mustard Oils to Isothiocyanates

Isothiocyanates (ITCs)[1] are organic compounds with the –N=C=S functional group. They are sulfur analogues of isocyanates (R-NCO) and isomers of thiocyanates (R-SCN). ITCs were formerly named mustard oils because of specific odour and taste that those compounds give to numerous higher plants, particularly those belonging to *Brassicaceae* family (older name *Cruciferae*) (Kjær 1961). The history of exploration of mustard oils before 1913 was presented by Gildmeister (1913). Presumably, the first tangible evidences of the occurrence of natural ITCs appeared in 17th century, when the volatile oil formation during distillation of mustard seed was observed (Fenwick et al. 1983). In 1831 Robiquet and Boutron isolated isothiocyanate precursor, sinalbin (Fig. 1a), from the seeds of the white mustard (*Sinapis alba*) (Robiquet and Boutron 1831).

Few years later, the next glucosinolate, sinigrin (Fig. 1b) was isolated by Bussy from the seeds of black mustard (*Brassica nigra*) (Bussy 1840). Long before the idea

[1]Main abbreviations and acronyms: **ABTS/ABTS$^{+•}$**, radical cation formed from 2,2′-azino-bis (3-ethylbenzothiazoline-6-sulfonic) acid**;** **DPPH$^{•}$**, 2,2-diphenyl-1-picrylhydrazyl radical; **ERN**, erucin; **ERS**, erysolin; **FRAP**, ferric reducing antioxidant power; **GL**, glucosinolate(s); **GSL-ERN**, glucoerucin; **GSL-RPS**, glucoraphasatin; **GSL-SFE**, glucoraphenin; **GSL-SFN**, glucoraphanin; **HAT**: hydrogen atom transfer; **ITC**, isothiocyanate(s), **ORAC**, oxygen radical absorbance capacity; **ROS**, reactive oxygen species; **RPS**, raphasatin; **SFE**, sulforaphene; **SFN**, sulforaphane; **SOD**, superoxide dismutase; **TEAC**, Trolox Equivalent Antioxidant Capacity; **TOSC**, total oxidant scavenging capacity.

J. Cedrowski · G. Litwinienko (✉)
Faculty of Chemistry, University of Warsaw, Pasteura 1, Warsaw, Poland
e-mail: litwin@chem.uw.edu.pl

J. Grebowski
Department of Molecular Biophysics, Division of Radiobiology, Faculty of Biology and Environmental Protection, University of Lodz, Pomorska 141/143, Lodz, Poland

C. Bravo-Diaz (ed.), *Lipid Oxidation in Food and Biological Systems*,
https://doi.org/10.1007/978-3-030-87222-9_13

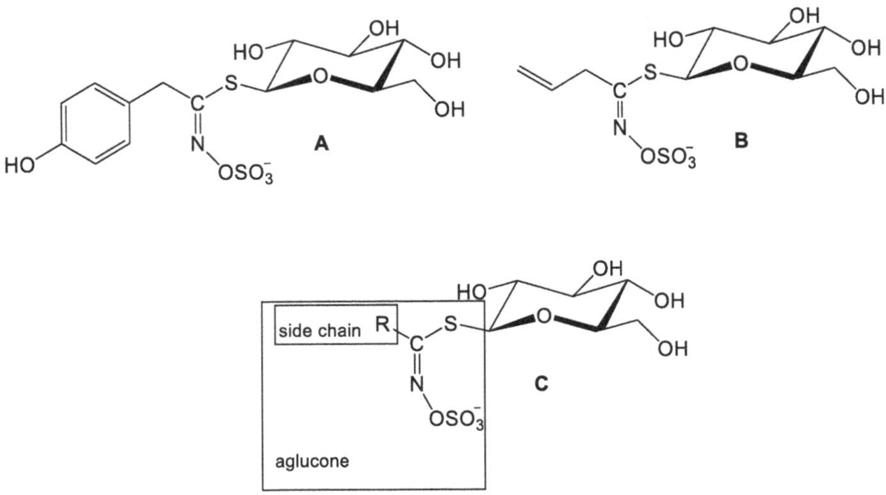

Fig. 1 Structures of sinalbin (**a**) and sinirgin (**b**), historically first glucosinolates isolated from plant material, and general structure of glucosinolates (anionic form)

of enzymes were elaborated, it had been noticed that in water and in the presence of some plant material (now we know that this plant material contained enzyme myrosinase, EC 3.2.1.147) both isolated glucosinolates decomposed to the corresponding mustard oils (Boutron and Fremy 1840). Empirical formula (C_4H_5NS) for mustard oil isolated from black mustard seeds (*Brassica nigra*) was proposed in 1844 independently by Will and Wertheim (Will 1844; Wertheim 1844). In that time this mustard oil was considered as an allyl ester of thiocyanic acid ($H_2C{=}CH{-}CH_2{-}SCN$). The erroneous interpretation was corrected to allyl isothiocyanate in 1868, when Hoffman proposed that the carbon of the allyl group is directly attached to nitrogen instead of sulfur and suggested that thiocyanates R-SCN have isocompounds (R-NCS) called isothiocyanates, ITCs (Hofmann 1868). Such structure of mustard oils was later confirmed by Billeter and Gerlich (Billeter 1875; Gerlich 1875).

Linear structure of –N=C=S group was proposed as early as in 1929 by Perschke (1929). On the other hand, Dadieu and Kohlrausch proposed a cyclic structure for –NCS group, after investigation the Raman effect related to allyl isothiocyanate (Dadieu and Kohlrausch 1930; Dadieu 1931). However, studies on dipole moment of phenyl ITC (Bergmann and Tschudnowsky 1932) and of Raman spectra of methyl and ethyl ITCs (Goubeau and Gott 1940) confirmed the linearity of –NCS group, rejecting its cyclic structure. Microwave spectral analysis of isothiocyanic acid (H–NCS) and of methyl isothiocyanate (CH_3–NCS) supported this conclusion and allowed to describe the distances between the atoms: 1.22 Å and 1.56 Å for N–C and C–S respectively, with the ideal linearity of N–C–S (180° for non isotopically enriched group) (Dousmanis et al. 1953; Beard and Dailey 1949) in agreement with the results of crystallographic analysis of isothiocyanate complexes with transition metals and isothiocyanato-posphine complex (Kam-Wing Lo et al. 2001; Chang

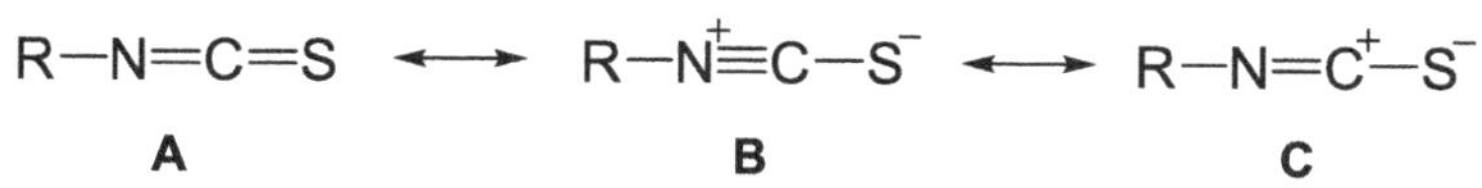

Fig. 2 Mesomeric structures proposed for –NCS group

et al. 2005; Nuzzo et al. 2016). Isothiocyanate group can be represented by canonical (mesomeric) structures A–C shown in Fig. 2, containing cumulated double bonds (A), and ionic structures with either a triple (B) or double N–C (C) bond and a negative charge localized at sulfur atom.

Analysis of dipole moments for mesomeric structures bonded to the aromatic ring, indicated that the contribution of structure B is insignificant, thus, A and C express the structure of –NCS group in a best way (Drobnica et al. 1977).

2 Sources of Natural Isothiocyanates

ITCs could be divided into two basic groups, synthetic and naturally occurring ITCs. In organic synthesis the ITCs are conventionally prepared from primary amines in direct reactions with thionyl compounds (e.g. thiophosgene $CSCl_2$ and its derivatives) or through dithiocarbamates obtained after a reaction of primary amine with carbon disulfide CS_2. Dithiocarbamates could be further converted into corresponding ITCs by using various desulfurization agents, e.g. heavy-metal salts, hydrogen peroxide, carbodiimides, thrimethylsilyl chloride and others (Drobnica et al. 1977; Sharma 1989). Alternative methods for synthesis of ITCs have been also developed, for example the reaction of elemental sulfur (S_8) with organic isocyanides (R–NC) and thermal decomposition of thioureas (designed for synthesis of aryl ITCs) should be mentioned herein (Drobnica et al. 1977; Sharma 1989). High toxicity of majority of chemicals (reagents and catalysts as well as solvents) used for common syntheses of ITCs causes that there is still a need to develop less harmful and more sustainable synthetic procedures for obtaining ITCs (Nickisch et al. 2021).

Natural ITCs are products of decomposition of corresponding glucosinolates (GSLs) - secondary metabolites in plants, mainly in *Brassicales* (Fenwick et al. 1983; Fahey et al. 2001; Hanschen et al. 2014; Spencer and Daxenbichler 1980; Blažević et al. 2020). The general structure of glucosinolate (GSL) is presented in Fig. 1c, with S-β-D-glucopyrano unit anomerically connected to an aglucone moiety. The aglucone is O-sulfated (Z)-thiohydroximate function with a side chain R, which usually is the same as a side chain of the amino acid used by plant for biosynthesis of GSL (Fahey et al. 2001; Blažević et al. 2020). Majority of all isolated GSLs are bioproduced from Ala, Glu, Val, Leu, Ile, Trp, Phe/Tyr and Met while some have uncertain precursor (Blažević et al. 2020).

During enzymatic hydrolysis the whole R is incorporated into the evolved isothiocyanate R-NCS, as presented in Fig. 3. The hydrolysis is catalyzed

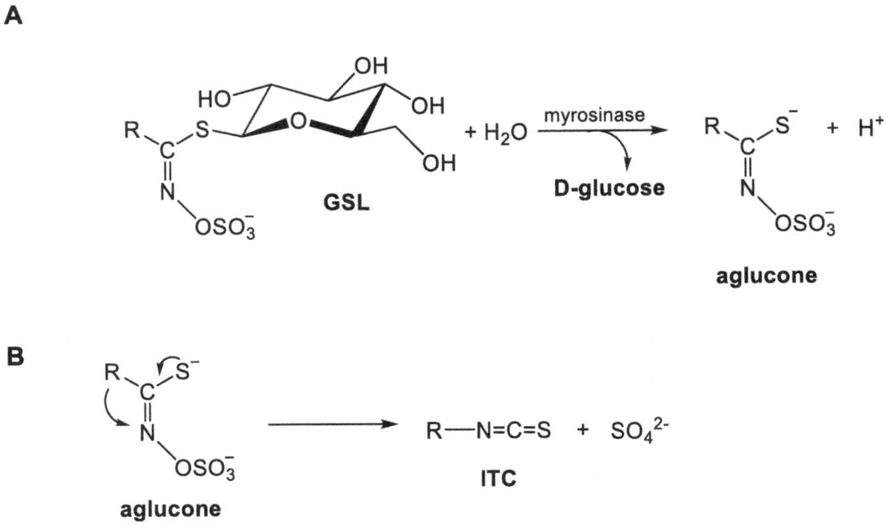

Fig. 3 Myrosinase-catalyzed hydrolysis of GSL to aglucone (**A**), and a subsequent "Lossen-like" rearrangement of the aglucone into ITC (**B**)

enzymatically by plant enzymes, myrosinases (thioglucoside glucohydrolases, EC 3.2.1.147), under physiological conditions and proceeds via unstable aglucone undergoing rearrangement into ITC (Blažević et al. 2020; Ettlinger and Lundeen 1956; Nguyen et al. 2020).

Up to date more than 130 different GSLs have been identified and isolated (Blažević et al. 2020; Nguyen et al. 2020). Among several ways of their classification (Blažević et al. 2020) probably the most practical is the one with respect a structure of the side chain R. For instance, side chain could be aliphatic (straight or branched, containing hydroxyl group, carbonyl group or double bond), "aromatic" (containing phenyl, benzoate or indole moiety), polyglycolysated or sulfur-containing (Fahey et al. 2001). The group of GSLs with the R containing sulfur (in particular methylsulfanylalkyl, methylsulfinylalkyl, methylsulfonylalkyl and their analogues) and their corresponding ITCs are the most studied due to their health-promoting effects (Valgimigli and Iori 2009; de Figueiredo et al. 2013) and abundant occurrence in commonly consumed vegetables belonging to *Brassicales* order of plants, mainly from *Brassicaceae* family such as broccoli, cabbages, Brussels sprouts, radish, and others (Fahey et al. 2001; Blažević et al. 2020; Kjær 1960). Some examples of sulfur containing ITCs are presented in Fig. 4.

Among described ITCs, erysolin (ERS) was discovered first, isolated from the seed of *Erysimum Perofskianum* (Schneider and Kaufmann 1912). Interestingly, sulforaphane (SFN) was found and isolated from the leaves of *Lepidium draba* L. (hoary cress) by Prochazka in 1959 (Prochazka 1959), although the compound was synthesized by Schmid and Karrer ten years earlier (Schmid and Karrer 1948). Erucin (ERN) also was synthesized by Schmid and Karrer in 1948, 7 years before Kjaer and Gmelin identified ERN in the extract from the seed of *Eruca sativa* Miller

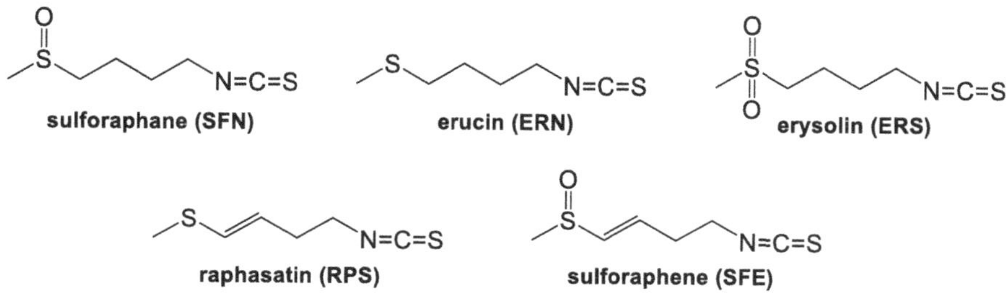

Fig. 4 Examples of ITCs containing sulfur atom incorporated into carbon chain and their trivial names

(rocket, a worldwide common alimentary plant) (Kjær and Gmelin 1955). SFN and ERN as well as several other ITCs could be obtained in diet by consuming common *Brassica* vegetables (Fenwick et al. 1983; Fahey et al. 2001; International Agency for Research on Cancer, World Health Organization, Centre international de recherche sur le cancer (Lyon) and Kongress 2004). For instance, rich sources of SFN are broccoli, some cultivars of cabbage and Brussels sprouts while ERN predominantly could be obtained from rocket (International Agency for Research on Cancer, World Health Organization, Centre international de recherche sur le cancer (Lyon) and Kongress 2004). ERS is less widespread in edible plants comparing to SFN and ERN, but certain amounts of ERS, along with SFN and ERN, could be found in cabbage and red cabbage, rutabaga or turnip (Fenwick et al. 1983). Moreover, radish is abundant source of raphasatin (RPS) as well as of sulforaphene (SFE) (Visentin et al. 1992). Some GSLs (gluconapin, glucobrassicanapin, progoitrin, napoleiferin, indolyl and 4-OH-glucobrassicin) occur in important agricultural crops like in *Brassica napus* L. (rapeseed), and after extraction of the oil those GSLs remain with the rapeseed meal (a protein source for animals) but their concentration is relatively high, giving a "hot" and pungent taste, thus, together with erucic acid, tannins, and synapine, GSLs are considered as anti-nutritional components. Negative effects of GSLs on animals are relative to their concentration in diet.

3 Bioactivity

Clinical observations as well as experimental models carried out in recent years have shown that carcinogenesis is a multi-stage process that may begin with an oncogenic DNA mutation in a single cell (Klaunig and Kamendulis 1999; Vineis et al. 2010; Nam et al. 2021). In this process, normal cells acquire the features of neoplastic cells, such as: uncontrolled growth, avoidance of suppressor growth signals, failure to undergo the process of apoptosis, the ability to invade distant tissues and change metabolism (Hanahan and Weinberg 2011). One of the most promising actions to reduce the risk of neoplastic changes is a chemoprevention (Greenwald 2001; Kelloff et al. 2000), which has been defined as the use of natural or synthetic chemical compounds to inhibit, delay or reverse the process of carcinogenesis (Kelloff et al. 1994). Chemopreventive compounds are characterized by low toxicity and ability to neutralize carcinogens, therefore they show little side effects (Levi et al. 2001; Baek et al. 2009; Crooker et al. 2018). Due to their anti-carcinogenesis strategies, chemopreventive agents can be divided into two groups: suppressive compounds affecting already initiated neoplastic cells and anti-initiating agents, blocking the process of neoplasticity (Landis-Piwowar and Iyer 2014). And so, suppressors mainly affect the expression of genes responsible for the proliferation and division of neoplastic cells, reducing their number and restoring the correct level and course of apoptosis (Manson et al. 2000). Anti-initiating agents may act by scavenging free radicals, stimulating the repair of damaged DNA, or influencing the level of Phase-2 enzymes (Valgimigli and Iori 2009; Wu et al. 2004).

ITCs may act in both ways but they exert their antioxidant activity mostly (often entirely) through transcriptional activation of multiple antioxidant proteins, such as NQO1, glutathione-S-transferase (GST), epoxide hydrolase, ferritin, glutamate cysteine ligase (GCL), glutathione peroxidase, glutathione reductase, heme oxygenase, thioredoxin and thioredoxin reductase, and UDP-glucuronosyl transferase (Valgimigli and Iori 2009; Zhang et al. 2005; Ye and Zhang 2001; Thimmulappa et al. 2002). In vitro studies have shown that some ITCs, especially SFN, selectively induce Phase-2 enzymes and they reduce the activity of Phase-1 enzymes involved in the conversion of procarcinogens to more cytotoxic derivatives (Bonnesen et al. 2001; Jakubíková et al. 2005; Svehlíková et al. 2004). It was found that the protective effect of the *Brassicaceae* extracts may come as a result of increased activity of Phase-2 enzymes in the large intestine (Kassie et al. 2003). Interestingly, ITCs also possess properties that may favor their use as therapeutic agents for large intestine cancer development, since numerous in vitro studies have shown that the primary mechanism of action of ITCs in the carcinogenesis promotion stage is to induce apoptosis and inhibit the colorectal neoplastic cell cycle (Gamet-Payrastre et al. 2000; Parnaud et al. 2004; Smith et al. 2004; Vanduchova et al. 2019; Hudlikar et al. 2020; Nandini et al. 2020). The pro-apoptotic effect of ITCs obtained from broccoli sprouts has been also demonstrated in the animal models with carcinogenesis process induced by 1,2-dimethylhydrazine (Smith et al. 2003).

The mechanism of ITCs interaction with Phase-2 enzymes is mainly based on reactions with electrophilic –N=C=S groups, especially a carbon atom surrounded by more electro-negative sulfur and nitrogen atoms. Therefore, -NCS moiety can react readily with –SH residues under physiological conditions and the isothiocynanate molecule is converted to dithiocarbamate, which is equivalent to the reduction of Nrf2-Keap1 binding and thus the induction of the Phase-2 enzymes genes dependent on Nrf2 (Talalay et al. 2003; Magesh et al. 2012), see Fig. 5. Through the induction of this signaling pathway some toxic substances may be inactivated, including free radicals, capable of damaging DNA (Giudice and Montella 2006). Moreover, Nrf2 reduces the level of pro-inflammatory factors, such as: cyclooxygenase 2 (COX-2), nitric oxide synthase (iNOS), IL-1β, IL-6 and TNF-α (Khor TO et al. 2006). At the same time, inflammation is accompanied by excessive formation of Reactive Oxygen Species (ROS), which in turn directly

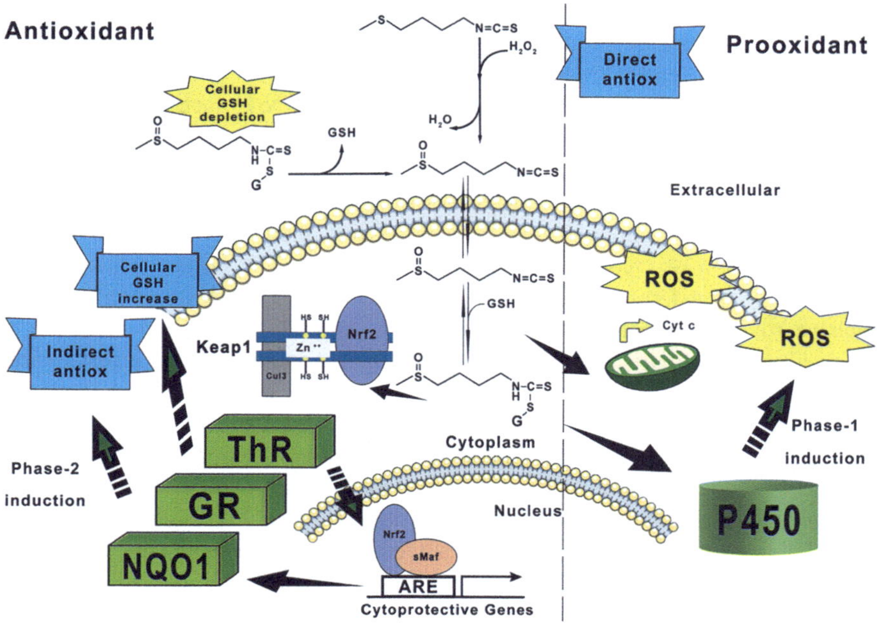

Fig. 5 The illustration of the cellular antioxidant machinery presented in review by Valgimigli and Iori, (Valgimigli and Iori 2009) with ERN as initial ITC responsible for removal of peroxide (here H_2O_2) with formation of SFN which enters the cell as an adduct with glutathione (diffusion transport). Both forms (adduct or free form) can interact with Keap 1 proteins and trigger a release of NRf2 factor, activating transcriptional overexpression of antioxidant Phase-2 enzymes. The scheme presents also interactions with cytochrome P450 enzymes, increasing the productions of ROS (see the text). The figure reprinted with permission and slightly modified

interact with mitogen-activated protein kinases (MAPK) and thus affect pathways dependent on both Nrf2 and the NF-κB transcription factor, playing the major role in the whole process.

There is increasing evidence of the protective effect of ITCs toward oxidation. For example, studies on animals showed that feeding the rats for 5 days/week for 14 weeks with 200 mg/day of dried broccoli sprouts that contained glucoraphanin (GSL-SFN, SFN precursor), caused a significant decrease of oxidative stress in cardiovascular and kidney tissues, as shown by increased glutathione (GSH) content and decreased the level of oxidized GSH, decreased protein nitrosylation, as well as increased GSH reductase and GSH peroxidase in comparison to control groups. Increased levels of GSH was also reported in correlation with dietary intake of ITCs (Valgimigli and Iori 2009). In another investigation, in the ischemia reperfusion as a model for oxidative stress-mediated injury, SFN improved changes in the lipid hydroperoxides, glutathione, creatinine clearance, kidney weight, and histologic abnormalities (Yoon et al. 2008). Through their ability to modulate oxidative stress and inflammatory processes, ITCs could be used as potential neuroprotecting agents in the course of the Parkinson's disease (Sita et al. 2016).

The chemopreventive properties of ITCs are not only limited to the direct induction of detoxification enzymes by activating the Nrf2 factor. Some studies have shown that ITCs have ability to regulate gene expression through epigenetic mechanisms such as DNA methylation, chromatin modification and miRNA activation (Li and Tollefsbol 2010). It has been shown that SFN may promote demethylation of the Nrf2 promoter region to increase activation of Nrf2, which induces chemoprevention of colon cancer (Zhou et al. 2019). The polymorphism of the glutathione S-transferase isoenzymes GSTM1 and GSTT1 genes probably plays a decisive role in the beneficial effect of ITCs. Anti-Glutathione-S-Transferase is involved in the detoxification of numerous chemical carcinogens and is responsible for the metabolism of ITCs. This was confirmed by Seow et al. (2002), who reported a 57% reduction in the risk of colorectal cancer development in the group of individuals lacking the GSTM1 and GSTT1 genes consuming large amounts of ITCs, compared to individuals having a diet poor in these compounds. The analysis of the results taking into consideration only the presence of ITCs in the diet without comorbidity of genetic conditions of the subjects, showed no significant differences between the consumption of ITC and a risk of cancer. This relationship between the GST gene polymorphism and a diet may be important in further epidemiological studies.

In vitro studies have shown that some ITCs, especially SFN, selectively induce Phase-2 enzymes and at the same time reduce the activity of Phase-1 enzymes involved in the conversion of procarcinogens to more cytotoxic derivatives (Jakubíková et al. 2005). Kassie et al. (2003) in their animal studies showed that the protective effect of 2-amino-3-methylimidazo quinoline (derived from the *Brassicaceae*) against aberrant crypt foci was due to their direct effect on Phase-2 detoxification enzyme, UDP-glucuronyl transferases, present in liver. It was also found that the protective effect of the *Brassicaceae* vegetables extracts may come as a result of increased activity of Phase-2 enzymes in the large intestine (Olejnik et al. 2010). Interestingly, ITCs possess properties that may favor their use as therapeutic agents for large intestine cancer development, since numerous in vitro studies have shown that the primary mechanism of action of ITCs in the carcinogenesis promotion stage is to induce apoptosis and inhibit the colorectal neoplastic cell cycle (Olejnik et al. 2010). The pro-apoptotic effect of ITCs obtained from broccoli sprouts has been also demonstrated in animal models in which the carcinogenesis process was induced by 1,2-dimethylhydrazine (Smith et al. 2003; Olejnik et al. 2010).

3.1 Prooxidant Action

On one hand ITCs are considered as promising protective compounds, but on the other, they can be rapidly accumulated in the cell, even up to 200-fold over extracellular levels, and what drives that accumulation is the reaction with -SH groups of cellular thiols, particularly glutathione, see Fig. 5. In turn GSH-conjugated ITCs are rapidly removed from the cell to undergo dissociation and ITC molecule diffuse into the cell again and undergo further conjugation, establishing a cycle consuming the cellular pool of thiols. Apart from cellular thiol depletion observed in vitro (Zhang et al. 2005), ITCs can increase oxidative stress also by increasing the cellular concentration of ROS (Valgimigli and Iori 2009). For that instance, the –N=C=S moiety of some ITCs could undergo a redox reaction with Cu(I) and hydrogen peroxide, leading to the formation of superoxide (Murata et al. 2000). Moreover, ITCs were reported to damage mitochondria, thus promoting oxidation (Nakamura et al. 2002). It has been also shown that ITCs might undergo oxidative desulfurization to the corresponding isocyanate by cytochrome P-450 (Lee 1996), see Fig. 5. The overproduction of ROS might not be a transient phenomenon preceding the long-lasting antioxidant activity (Valgimigli and Iori 2009). Synthetic SFN caused a 50–70% increase in oxidative DNA damage in vitro through a Fenton-like reaction (Srinivasan et al. 2002), while in vivo experiments performed on male Wistar rats indicated that long-term treatment with 0.1 or 0.3 mg/kg of phenethyl ITC profoundly affected metabolic parameters and resulted in increased levels of malondialdehyde, as a marker of oxidative stress, in liver (Okulicz et al. 2005).

To conclude, it should be emphasized that ITCs are of great interest in terms of their antioxidant, anti-inflammatory and chemopreventive capacities as the result of the multiple induction of Phase-1 and Phase-2 enzymes, however, their pro-oxidant behavior calls for better understanding of their dualistic nature depending on the experimental design (in vivo vs in vitro), dosage and a variety of ITCs structures.

4 Direct Reactions of Isothiocyanate (N=C=S) Moiety with Radicals

Distribution of the π electrons densities increases the polar, electron-accepting character of –NCS, with the negative charge on nitrogen and sulfur atoms and positive charge on carbon atom (Drobnica et al. 1977). Because of highly electrophilic central carbon atom, the –NCS group can react with many nucleophiles (Sharma 1989). The typical reactions of –NCS group are nucleophilic addition reaction on carbon atom and cycloaddition reactions on the cumulated N=C and C=S bonds. When –NCS group is bonded to the aromatic ring it acts as electron-withdrawing substituent, with a negative inductive effect and a positive mesomeric effect, similar to that exhibiting by halogens. However, it was suggested that there is no possibility to perform electrophilic substitution reaction to the aromatic ring (in phenyl isothiocyanate) because the electrophile would react with nucleophilic moiety of –NCS group firstly/spontaneously (with nitrogen or even more likely, with sulfur atom) (Drobnica et al. 1977).

Isothiocyanate functional group attached to an alkyl chain is rather unreactive towards radicals. Stannyl radicals form with R-NCS unstable C-centered radical adduct $Bu_3Sn\text{-}S\text{-}^{\bullet}C{=}N\text{-}R$ that decomposes with generation of alkyl radical and tributyltin thiocyanate (Bu_3SnSCN)(Crich and Quintero 1989). Other examples are limited to synthetically useful reactions of aromatic isothiocyanates (attacked by aromatic radicals) (Leardini et al. 1997; Benati et al. 2000) or sulfonyl isothiocyanates reacting with alkyl radicals (Barton et al. 1992) with formation of alkyl cyanates. Another example is the generation of imidoyl radicals from isothiocyanates (sulfur atom from NCS group is attacked by C, Si or Sn-centered radicals) (Matteo et al. 2007).

In contrast to rather limited literature on the direct radical attack on –NCS group, there are several reports on the antiradical activity of isothiocyanates containing other functionalities.

5 Reactions of ITCs and GSLs with ROS

5.1 Reaction with Peroxides

Decomposition of peroxides and hydroperoxides is not connected with breaking the chain of peroxidation, however, such kind of antioxidant activity is considered as a direct antioxidant action because the elimination of ROS prevents the initiation of peroxidation. Such direct antioxidant activity can be important not only for cancer prevention, described previously and connected with the indirect antioxidant action, but also for effective reduction of oxidative stress and reduction of cardiovascular problems. Clinical studies indicated that the supplementation of extracts from the *Brassicaceae* and legume sprouts (100 g for 2 weeks) reduces H_2O_2-induced DNA damage in peripheral blood lymphocytes (Gill et al. 2004). In vitro experiments confirmed significant antigenotoxic activity against H_2O_2 damage in HT29 human colon cells (Gill et al. 2004), thus, the ability of ITCs to decompose hydrogen peroxide was studied in a few works. Barillari and coworkers (Barillari et al. 2005a) measured the rate constants for reactions of hydrogen peroxide and *t*-BuOOH with glucoerucin (GSL-ERN, major glucosinolate found in rocket seeds and sprouts) and its corresponding ITC, erucin, ERN. In water, at 37 °C GSL-ERN decomposed $H_2O_2$with the rate constant $6.9 \times 10^{-2}\ M^{-1}s^{-1}$ and 15 times smaller for *t*-BuOOH ($k = 4.5 \times 10^{-3}\ M^{-1}s^{-1}$). The same peroxides do not react with GSL-SFN, therefore, after the HPLC-MS analysis of the oxidation products the authors put hypothesis that the process involves sacrificial oxidation of methylsulfanyl to methylsulfinyl group (but further oxidation to sulfone does not occur), see Fig. 6.

Peroxide – decomposing properties were also observed for glucoraphasatin (GSL-RPS, a secondary metabolite isolated from *Raphanus sativus* L.): reaction of GSL-RPS with an excess of H_2O_2 within 60 minutes at 37 °C quantitatively converted GSL-RPS (2 mM solution) into glucoraphenin (GSL-SFE), Fig. 7

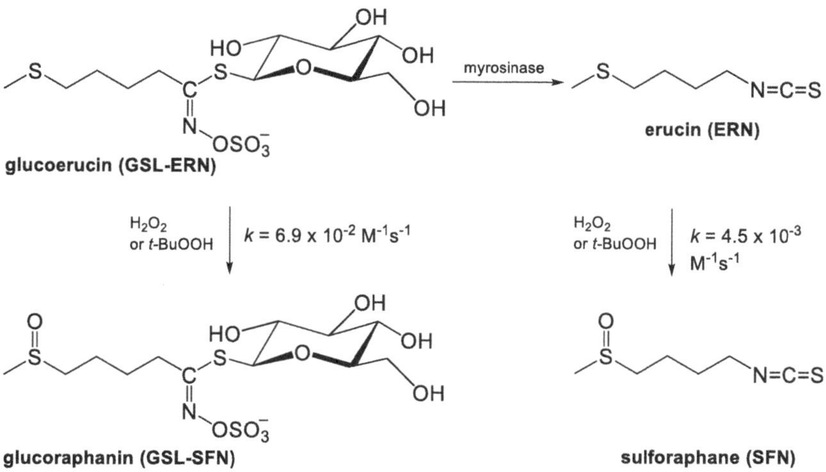

Fig. 6 Conversion of methylsulfanyl to methylsulfinyl group by hydroperoxides during oxidation of GSL-ERN/ERN to GSL-SFN/SFN (Barillari et al. 2005a)

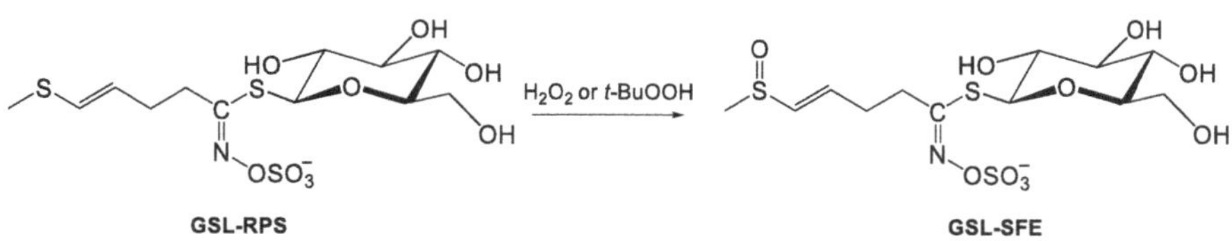

Fig. 7 Oxidation of glucoraphasatin (GSL-RPS) to glucoraphenin (GSL-SFE) by hydroperoxides

(Barillari et al. 2005b). More detailed kinetic measurements for GSL-RPS reactions with H_2O_2 and *t*-BuOOH (Papi et al. 2008) showed that rate constants are ca. 5-fold smaller (1.9×10^{-2} $M^{-1}s^{-1}$ and 9.5×10^{-4} $M^{-1}s^{-1}$, respectively) than for analogous reaction GSL-ERN→GSL-SFN. As the only difference between GSL-ERN and GSL-RPS is double C=C bond in GSL-RPS, its slower oxidation by hydroperoxides was assigned to small deactivating effect of C=C group.

5.2 Reactions of ITCs with Artificial Model Radicals

Several assays employ artificial radicals (or metal ions) reacting with individual compounds or multi-component plant extracts containing potential antioxidants. Among the proposed methods are the oxygen radical absorbance capacity (ORAC), ferric reducing antioxidant power (FRAP), total oxidant scavenging capacity (TOSC), 2,2-diphenyl-1-picrylhydrazyl ($DPPH^•$) assay, the Trolox Equivalent Antioxidant Capacity (TEAC) and other assays. Frequently (and erroneously) those methods are considered as ways of determination of the antioxidant activity,

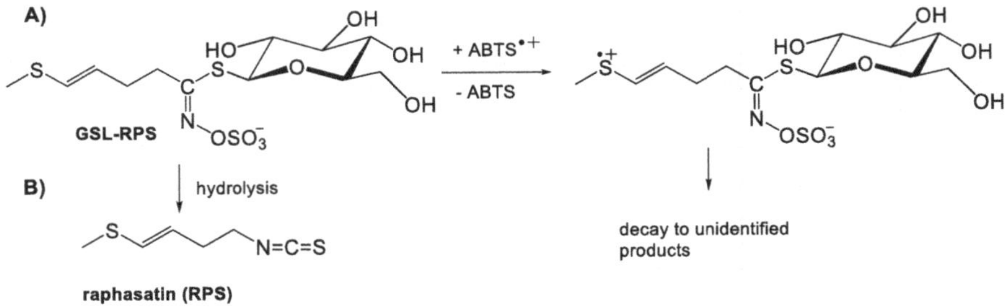

Fig. 8 (**a**) Reaction of glucoraphasatin (GSL-RPS) with $ABTS^{•+}$, (**b**) RPS, the corresponding isothiocyanate, is evolved during the hydrolysis of GSL-RPS

Table 1 Bimolecular rate constant *k* for reaction of DPPH radical with GSL-RPS, RPS and some phenols measured in methanol, 298K (Papi et al. 2008)

Compound	k [$M^{-1}s^{-1}$]	Solvent
GSL-RPS	14	MeOH/water 99:1
RPS	43.1	MeOH
α-tocopherol	420	MeOH
2,6-di-t-butyl-6-methoxyphenol (BHA)	3.9	EtOH
2,6-di-t-butyl-6-methylphenol (BHT)	0.86	EtOH

however, the results are not relevant to the rate of the radical-scavenging process because the reducing capacity is measured at given time (several minutes) after mixing the solutions of reactants, and results are provided as percent of scavenged radicals (Amorati and Valgimigli 2015).

In TEAC method the decay of ABTS radical cation ($ABTS^{•+}$) is monitored at 734 nm and compared to the results obtained for Trolox (standard water soluble antioxidant, analog of α-tocopherol with phytyl chain replaced with carboxyl group). Barillari and coworkers reported that at 30 °C and pH 7.4 GSL-RPS reacts with $ABTS^{•+}$ (Fig. 8) but its reducing capacity was ca. 8-fold smaller than for Trolox (Barillari et al. 2005b). They also noticed that oxidation of methylsulfanyl (CH_3S) group to methylsufinyl ($CH_3S{=}O$) group, deactivated the formed glucoraphenin GSL-SFE towards $ABTS^{•+}$. The same lack of reactivity was observed for 3-butenyl glucosinolate (gluconapin). Thus, the authors concluded that the initial step of reaction with $ABTS^{•+}$ includes an electron transfer and formation of transient radical cation $RS^{•+}$. Similar step was described for enzymatic oxidation of sulfides to sulfoxides by hemoproteins such as cytochrome P-450 and peroxidases (Peñéñory et al. 2005).

GSL-RPS and corresponding 4-methylthio-3-butenyl isothiocyanate (RPS) react also with DPPH radical, with surprisingly high rate constants, see Table 1 (Papi et al. 2008): RPS reacts ten times slower than α-tocopherol and both, GSL-RPS and RPS are much more reactive toward $DPPH^•$ than hindered phenolic antioxidants BHT and BHA.

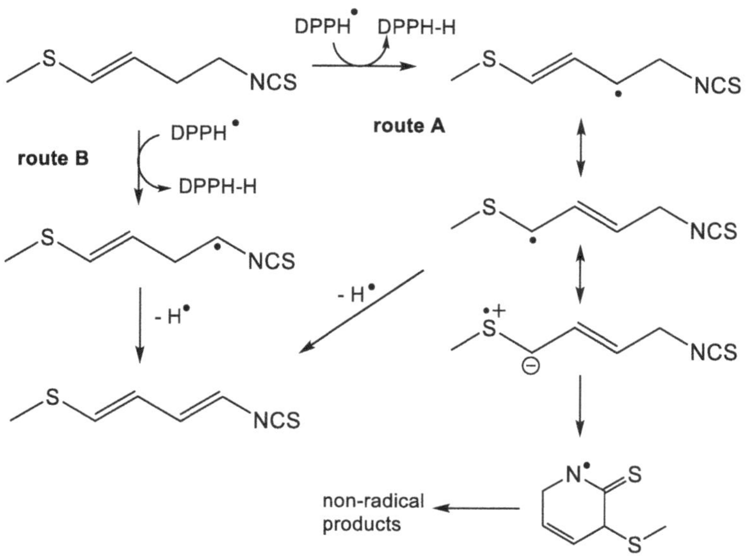

Fig. 9 Two mechanisms of reaction of RPS with DPPH radical proposed by Papi et al. (2008)

The high rate of reaction of GSL-RPS and RPS was a matter of mechanistic considerations and the authors proposed that allylic hydrogen atom is abstracted by the radical (Fig. 9, route A). However, such explanation seems to be improbable because hydrogen atom abstractions from C–H bonds by DPPH• are very slow, with the upper limit of $k \sim 10^{-3}\ M^{-1}s^{-1}$ (bis-allyl C–H) for DPPH• reacting with 1,4-cyclohexadiene (Valgimigli et al. 1996).

An alternative explanation for such fast reaction with DPPH• was based on the observation that RPS reacts three times faster than GSL-RPS. As the main difference is a presence of –NCS group in RPS, the authors suggested that H atom is abstracted from C–H adjacent to NCS. The effectiveness of scavenging of DPPH• by RPS and GSL-RPS isolated from Tunisian *Raphanus sativus* has been recently confirmed (Salah-Abbès et al. 2010).

Definitely, the presence of double bond and terminal methylsulfanyl group is important because six other GSLs presented in Fig. 10a do not efficiently react with DPPH• (and with ABTS•+) (Cabello-Hurtado et al. 2012), however, ORAC test carried out for all six GSLs resulted in moderate activity, comparable to ascorbic acid and glutathione, and the ORAC results for the corresponding isothiocyanates (obtained by enzymatic hydrolysis) were surprisingly high. The best activity (trolox equivalent was 3) was recorded for glucobrassicin and the products of its enzymatic decomposition, which authors considered as indol-3-ylmethyl isothiocyanate (Fig. 10b), but such high reducing power should be rather assigned to intrinsic ability of indole to reduce radicals, as many indole derivatives including melatonin, serotonin and tryptophan (Herraiz and Galisteo 2004). Moreover, indol-3-ylmethyl

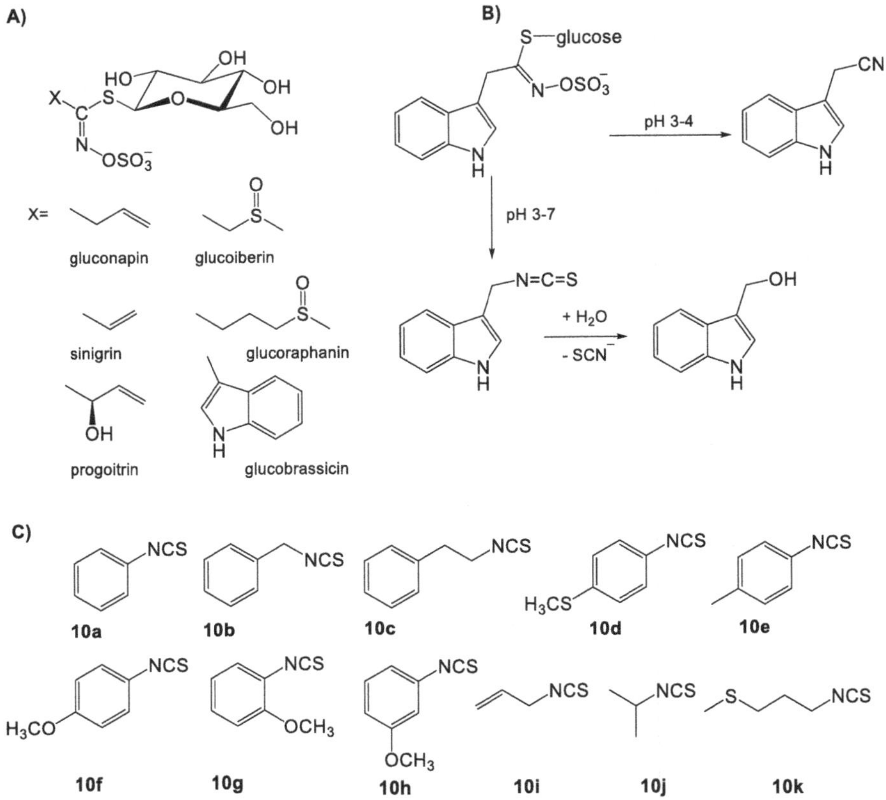

Fig. 10 (**a**) Six glucosinolates studied by Cabello-Hurtado et al. (2012), (**b**) Decomposition of glucobrassicin to indole-3-carbinol, (**c**) Eleven ITCs studied by Burčul et al. (2018)

isothiocyanate is unstable and undergoes fast decomposition to indole-3-carbinol (Blažević et al. 2020; Agerbirk et al. 2008), see Fig. 10b.

Another example of isothiocyanate that obviously will indicate the radical scavenging ability is 4-hydroxybenzyl isothiocyanate (4-HBITC) (Montaut et al. 2017). The EC_{50} parameter (the effective concentration needed for a 50% decrease the radicals concentration during 30 minutes of reaction) measured in ethanol, with EPR detection of DPPH$^\bullet$, determined for 4-HBITC was 7 times smaller than IC_{50} = 57.1 mM for ERN. This result is not surprising, because 4-HBITC is a phenol and behaves as typical phenolic scavenger of radicals.

The reaction of SFN with DPPH$^\bullet$ was also monitored by Yuan et al. (2010) who proposed mechanism that H atom is abstracted from α C–H bond with respect to methylsulfinyl group, that is rather not possible because S=O (and C=O) groups are electron withdrawing, and their presence should cause a significant increase of the bond strength of the adjacent C–H. The same authors studied reaction of DPPH$^\bullet$ with benzyl isothiocyanate and concluded that isothiocyanate group did not participate in

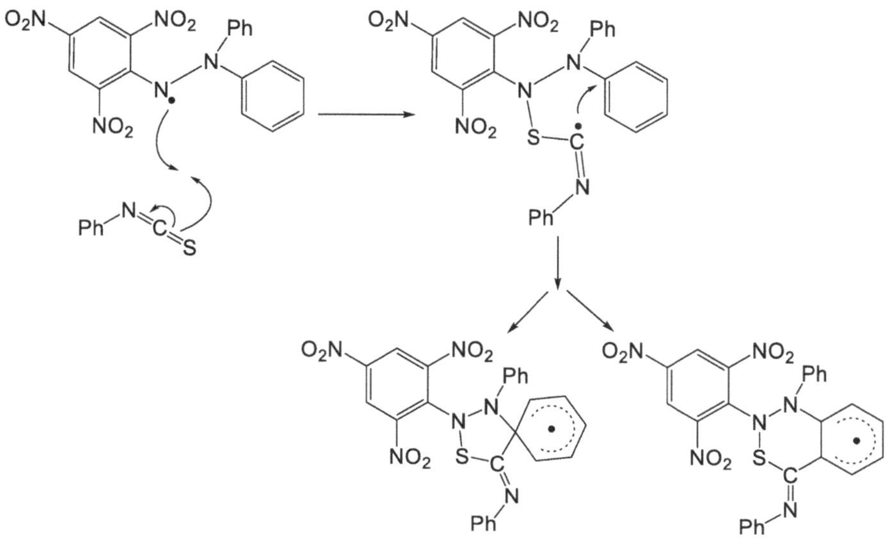

Fig. 11 Proposition of the mechanism of reaction of DPPH$^•$ with phenyl isothiocyanate (Burčul et al. 2018)

the reaction, because -NCS functionality was be still identified in the reaction products.

The DPPH$^•$ scavenging ability of eleven aromatic and aliphatic ITCs presented in Fig. 10c was studied by Burcul et al. (2018). Within the series Ph-NCS, $PhCH_2$-NCS, $PhCH_2CH_2$-NCS (compounds **10a–c**) the first one was the only active toward DPPH$^•$. No scavenging activity was also observed for X-Ph-NCS with X=methyl or methoxy group attached to phenyl ring, compounds **10e–h**. Three aliphatic ITCs, namely allyl, isopropyl and 3-(methylsulfanyl)propyl ITC (trivial name ibervirin), see compounds **10i–k** in Fig. 10c, did not scavenge DPPH$^•$, therefore, the authors proposed the mechanism in which aromatic isothiocyanate reacts with DPPH radical with formation of α-(arylsulfanyl)imidoyl radical, see Fig. 11. The proposed mechanism, though speculative, was inspired by works of Zanardi and coworkers (Leardini et al. 1997; Benati et al. 2000, 2003) focused on the mechanisms of attack of aryl radicals on aromatic isothiocyanates, the reaction of practical importance in organic synthesis.

In conclusion to this section, several reports describe the results of reaction of SFN and other ITCs or extracts from broccoli/broccoli seeds with artificial model radicals like ABTS$^{•+}$ or DPPH$^•$. Unfortunately, the results are very scattered, from the ones demonstrating rather strong anti-radical activity of unsaturated isothiocyanate (raphasatin, RPS) (Salah-Abbès et al. 2010; Ligen et al. 2017) to very moderate or no activity of saturated ITCs (Ligen et al. 2017; Farag and Motaal 2010). Direct comparison of the results reported in different publications might be confusing because some authors define IC_{50}/EC_{50}/SC_{50} parameters expressed as concentration of extract, not the concentration of active compound in the extract. Thus, the information that SFN is better scavenger than 4-hydroxybenzyl isothiocyanate (4-HBITC) (Montaut et al. 2017) basing on their EC_{50} parameters (DPPH$^•$ measurements in methanol) taken from two separate works is misleading. It should also be noted that in several papers the results obtained for extracts indicated strong activity which disappeared when purified ITCs were used. Taking into account that the extracts might contain some endogeneous fitoantioxidants, the reports about their increased antiradical activity should be read with caution.

5.3 Reactions with Other Radicals

Regardless the confirmed activity as peroxide decomposers, SFN and ERN (and their glucosinolates) do not scavenge peroxyl radicals fast enough to efficiently break the peroxidation chain. Peroxidation of styrene in chlorobenzene initiated by azo-initiator at 37 °C in the presence of SFN or ERN at concentrations 10-500 μM showed neither inhibition nor retardation effect (Barillari et al. 2005a). No inhibition/retardation effect was also observed for unsaturated RPS during peroxidation of styrene in chlorobenzene and peroxidation of methyl linoleate in SDS micelles (Papi et al. 2008). The lack of antioxidant activity of GSL-RPS and RPS was surprising because both compounds exhibited high activity toward DPPH$^•$ (between reactivity of α-tocopherol and BHA/BHT), therefore, another explanation was proposed that in aerobic environment C-centered alkyl radicals (Fig. 9) immediately react with molecular oxygen and are converted into peroxyl radicals (chain propagating species). The authors suggested that antioxidant activity of those compounds could be manifested in oxygen poor biological systems (Papi et al. 2008).

Yuan et al. studied broccoli extracts and two ITCs, SFN and benzyl isothiocyanate with several assays, including mentioned above DPPH$^•$ test but also superoxide scavenging assay (generated from pyrogallol at pH 8.2), lipid peroxidation with the detection of thiobarbituric acid adducts (TBARS assay) and hydroxyl radical scavenging (Yuan et al. 2010). Comparing to SFN, benzyl isothiocyanate was not active and authors concluded that NCS group is not essential for antiradical activity of ITCs.

Superoxide assay (generated from pyrogallol and detected by luminol chemiluminescence, pH=10.2) was also employed by Ligen et al. (2017) and they observed that SFN was less active than the broccoli seed extracts. Furthermore, after the SFN solution was incubated in water bath (80 °C), its free radical scavenging activity decreased, and Ligen and coworkers concluded that the products of thermal degradation of SFN are not free radical scavengers because do not contain a methylsulfinyl group (Ligen et al. 2017).

Burcul and coworkers employed ORAC assay for the same series of ITCs as for DPPH$^•$ (described above, see Fig. 10c). However, reactivity toward alkylperoxyl radicals (generated from ABAP) gave different range of activity than for DPPH$^•$, with the best activity determined for 4-methylsulfanylphenyl ITC, comp. **10d** (Burčul et al. 2018). The Rancimat test was carried out for the same series of eleven

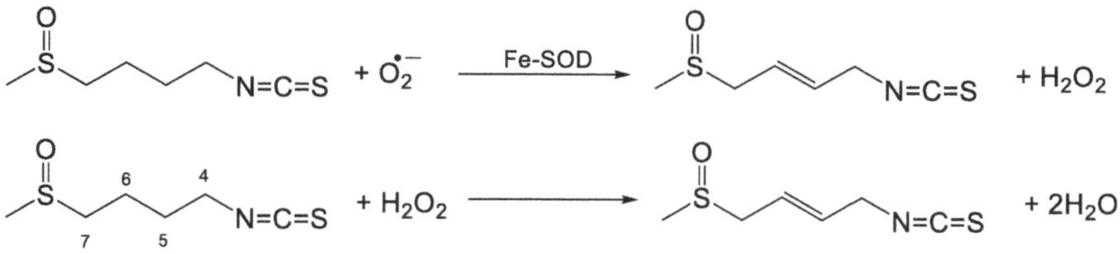

Fig. 12 Theoretical model for two step reduction of superoxide radical anion to water by two molecules of SFN, with hydrogen peroxide as intermediate. First reaction is catalysed by Fe^{3+} complex with three histidine units

ITCs indicated that eight of them protected the lipid against autoxidation at 120 °C, with phenyl ITC as the most active inhibitor of oxidation, the activity of other compounds was not correlated with their structures. Interestingly, 3-(methylsulfanyl)propyl isothiocyanate (compound no. **10k**) was more effective than 4-methylsulfanylphenyl ITC (comp. **10d**) (Burčul et al. 2018).

Other reports, not relevant to kinetic methods, compare the scavenging activity for short series of isothiocyanates with ORAC, hydroxyl radical scavenging, superoxide scavenging and DPPH$^{\bullet}$ assays with the results expressed as IC_{50} or percent of scavenged radicals compared to control sample (Salah-Abbès et al. 2010; Manesh and Kuttan 2003; Montaut et al. 2012).

The experimental results have been recently supplemented with two theoretical works. The calculations performed by Prasad et al. (Prasad and Mishra 2015) indicate that SFN alone cannot efficiently scavenge superoxide radical anion in gas phase or aqueous media because the corresponding reaction barriers are very high. However, the reaction with $O_2^{\bullet-}$ becomes almost barrierless in the presence of SOD-mimicking complexes (Fe^{3+} ion with three histidine rings), see Fig. 12. The catalytic role of Fe-SOD was interpreted as due to strong attraction of $O_2^{\bullet-}$ which destabilizes the local polarization of the aqueous medium. The formed H_2O_2 is subsequently converted into two molecules of water during reaction with another molecule of SFN (no catalyst is needed in this step). The scavenging of superoxide radical anion and hydrogen peroxide was found to involve the unusual mechanism of double hydrogen transfer.

Figure 12 presents the abstraction of two H atoms from position C5–C6 of SFN by H_2O_2, however, the calculated Gibbs energy barriers for double H atom abstraction from all three vicinal carbons C4–C5, or C5–C6, or C6–C7 are very small, and the rate constants for such concerted processes in water are 10^{11} or 10^{12} $M^{-1}s^{-1}$ (calculated at the B3LYP/6-311+G(d) and M06-2X/6-311+G(d) levels of theory). These results are interesting and put new light on the preventive character of antioxidant action of SFN, with some extension on the biological systems containing SOD or iron complexes that could catalyse the reaction of SFN with superoxide.

In another theoretical work (Akbari and Namazian 2020) three mechanisms of reaction of SFN with hydroxyl, methoxyl, hydroperoxyl radicals and superoxide radical anion are considered: Single Electron Transfer (SET), Radical Addition, and HAT in gas phase, water and octanol. The products are, respectively: radical cation

$SFN^{\bullet+}$, radical adduct $SFN\text{-}R^{\bullet}$, and radical $[SFN_{(-H)}]^{\bullet}$. The authors concluded that SFN can scavenge $^{\bullet}OH$ and $^{\bullet}OCH_3$ radicals via HAT mechanism with H atom abstracted from methylene at position α to isothiocyanate group.

5.4 Activity at Elevated Temperatures

Increasing interest in health benefits from dietary intake of ITCs present in *Brassica/Brassicaceae* vegetables resulted in some works about the stability of ITCs and their glucosinolates during processing of vegetables in food industry and preparing food in house kitchens. The studies included pressure and thermal treatment as well as pH and presence of other food additives. Some of those works included also the effects of pH and temperature on the activity of myrosinase that during cutting, slicing or grinding is released from the cellular compartments and catalyses the transformation of GSLs into ITCs. Reactivity and stability of glucosinolates has been reviewed by Hanschen et al. (2014) and the authors extended their review on the information on the reactivity of several isothiocyanates (including allyl and aromatic ITCs). For purpose of this section the effect of temperature on the stability of sulforaphane, sulforaphene and erucin (SFN, SFE, ERN) in context of potential antiradical activity is described.

First and the most comprehensive analysis of products of thermal degradation of SFN in aqueous solution at 50 and 100 °C and the mechanism of degradation (Fig. 13a) was presented by Jin and coworkers (Jin et al. 1999), basing on the detected products (GC-MS detection): 4-methylsulfanyl-3-butenyl isothiocyanate (**13a**), dimethyldisulfide (**13b**), methyl(methylsulfanyl)methyl disulfide (**13c**), S-methyl methylsulfanylsulfinate (**13d**), and the corresponding sulfonate (**13e**).

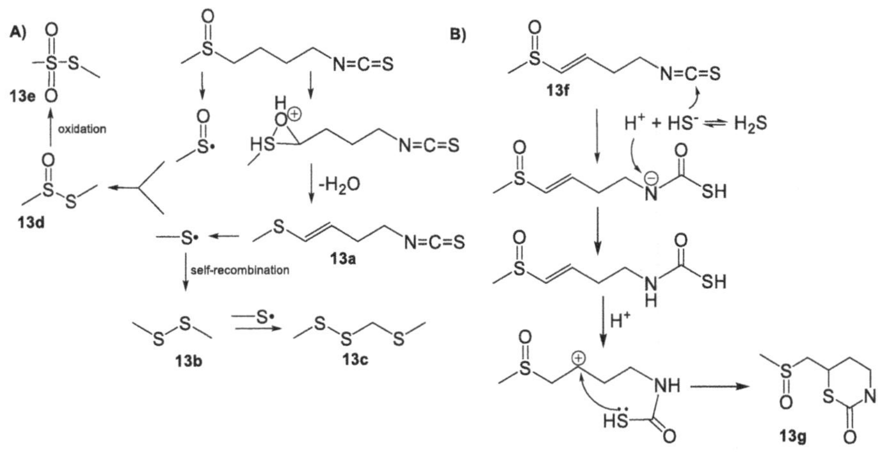

Fig. 13 Decomposition pathways of SFN (**a**) and SFE (**b**)

Compound **13a** is not stable and, apart from decomposition shown in Fig. 13a, can be oxidized to SFE (**13f**). The mechanism of hydrolytic decomposition of this ITC (present in radish seeds as glucoraphenin, GSL-SFE) was studied by Song and coworkers by means of HPLC, ESI-MS and NMR (Song et al. 2013). The process carried out at 25 °C and pH 7.0 during 12h for 300 mg of **13f** and 2g of ground radish seeds resulted in identification of 6-[(methylsulfinyl)methyl]-1,3-thiazinan-2-thione (compound **13g**) as the main product, with the yield dependent on the concentration of H_2S and on pH (at low pH decomposition is very slow). Hydrogen sulfide is necessary for the first step of decomposition and the authors suggested that the ground radish seeds are a source of H_2S because a disruption of plant tissues produces various kinds of volatile sulfur compounds including hydrogen sulfide and methanethiol-related compounds.

The rate of decomposition of SFN and SFE at elevated temperatures (but without identification of the products) is reported in a few works. Van Eylen et al. studied the degree of thermal decomposition of methanol solutions of synthetic SFN and phenylethyl isothiocyanate (mixed with the broccoli juice) heated for 20 min at temperatures 30–90 °C in 250 μL glass capillaries (anaerobic conditions). The ITCs were stable until 60 °C and were degraded for more than 90% after a 20 min treatment at 90 °C. (Van Eylen et al. 2007). However, the authors did not discuss the role of methanol as a medium reacting with isothiocyanates, a reaction resulting in formation of thiocarbamates R–N=C=S+R′OH→RNH–(C=S)–OR′ (Walter and Bode 1967), with the mechanism similar to attack of H_2S on isothiocyanate group in SFE (Fig. 13b). The reaction is accelerated by strong H–bond acceptors like DMSO (Satchell et al. 1990), perhaps due to faster nucleophilic attack of ROH–O=SMe_2 on electrophilic carbon in the isothiocyanate group thus, the methylsulfinyl functional group in SFN or other carbonyl or sulfinyl compounds present in broccoli juice might support the reaction with methanol.

Wu and coworkers (2010, 2013) reported that decomposition of SFN in aqueous, buffered system at temperatures 50–90 °C is a slow, pH dependent, first order process. The slowest decomposition was at pH 2.2 and temp. 60 °C (k_{decomp}=0.01 h^{-1}, the half-life time, $t_{1/2}$, was 69 h) and much faster decomposition was observed at pH 6 and temp. 90 °C, with k_{decomp}=0.8 h^{-1} and $t_{1/2}$ =0.9 h. Moreover, they noticed that SFN in broccoli extracts was more stable than isolated and purified SFN (Wu et al. 2014). Thermal behavior of SFN at 90 °C is comparable to stability of allyl isothiocyanate in nonaqueous environment (paraffin, 80 °C), with k_{decomp} = 0.71 h^{-1} and $t_{1/2}$ = 0.97 h measured by Neoh et al. (2012). For the same allyl ITC at 120 °C the parameters were 1.57 h^{-1} and 1.1 h, respectively.

Described in previous sections antiradical properties of ITCs were deduced on the results of experiments carried out at ambient temperatures (or 30–37 °C). Two attempts to estimate antioxidant activity of SFN and other ITCs at elevated temperatures gave non-conclusive results. The Rancimat test at 120 °C demonstrated the protecting activity of several ITCs (Burčul et al. 2018) whereas the superoxide assay done at ambient temperature with SFN samples previously incubated at 80 °C indicated a decrease of activity, thus, the authors concluded that products of thermal degradation of SFN are not active towards radicals (Ligen et al. 2017).

The effect of two ITCs: ERN and SFN on the oxidative stability of sunflower oil and linolenic acid (LNA, $C_{18:3}$) was explored by Cedrowski et al. (2020). The authors monitored thermal effects of non-isothermal oxidation by differential scanning calorimetry (DSC) and determined the temperatures of start of oxidation (relevant to oxidation induction time in isothermal oxidation) and calculated kinetic parameters using methodology elaborated earlier for non-isothermal oxidation of bulk lipids (Litwinienko et al. 2000; Litwinienko and Kasprzycka-Guttman 2000; Ulkowski et al. 2005; Czochara et al. 2016). Neither ERN nor SFN inhibited the oxidation of linolenic acid, which is easily oxidizable polyunsaturated lipid (its oxidation starts at 90 °C), whereas both ITCs were good inhibitors when added to sunflower oil, which is more stable, its oxidation starts at 140 °C. The apparent activation energy calculated for thermal oxidation of pure sunflower oil was 103 $\pm$ 4 kJ/mol and increased to 119 $\pm$ 8 kJ/mol in the presence of 10mM ERN or 5mM of SFN. Similarly, the overall first order rate constant for oxidation of pure oil calculated at 100 °C (lag phase) was 5.8×10^{-3} min^{-1} but in the presence of 10 mM ERN decreased to 2.9×10^{-3} min^{-1}. The inhibitory effects of ERN and SFN were also noticed for oxidation of soy lecithin, another lipid undergoing oxidation when the system is heated to temperature about 40–50 °C higher than temperature of oxidation of linolenic acid (Cedrowski et al. 2021). The kinetic neutrality of both ITCs at temperatures below 100 °C and their inhibiting effect during oxidations at temperatures above 120 °C in bulk lipid environment suggest that at first step methylsulfanyl group of ERN is oxidized to methylsulfinyl group, then SFN undergoes decomposition as described by Jin et al. (1999) and presented in Fig. 13a, with formation of methylsulfanyl and methylsulfinyl radicals that are good trapping agents for lipidperoxyl radicals (Lynett et al. 2011). Additional explanation for inhibiting activity of SFN at temperatures above 120 °C was also proposed as due to Cope-like elimination and formation transient methanesulfenic acid, see Fig. 14.

Sulfenic acids can react with alkylperoxyl radicals with a near diffusion-controlled rate and the hypothesis about formation of sulfenic acids can be supported by some experimental evidences of antioxidant action of dialkyl sulfoxides during oxidation of hydrocarbons at elevated temperatures (Koelewijn and Berger 1972; Block 1992). Direct detection of transient sulfenic acids is not possible but GC-MS analysis of thermal decomposition of SFN at 100 and 160 °C indicated the products shown in Fig. 14, that could be assigned to both mechanisms, therefore peroxyl radicals might be trapped by sulfenic acids or by methylsulfanyl/methylsulfinyl radicals evolved from SFN (Cedrowski et al. 2020, 2021). Those two ITCs are examples of biocompounds which become chain-breaking antioxidants above 120 °C (thermal release of radical trapping products), in contrast to phenolic antioxidants that lose their antioxidant activity at elevated temperatures.

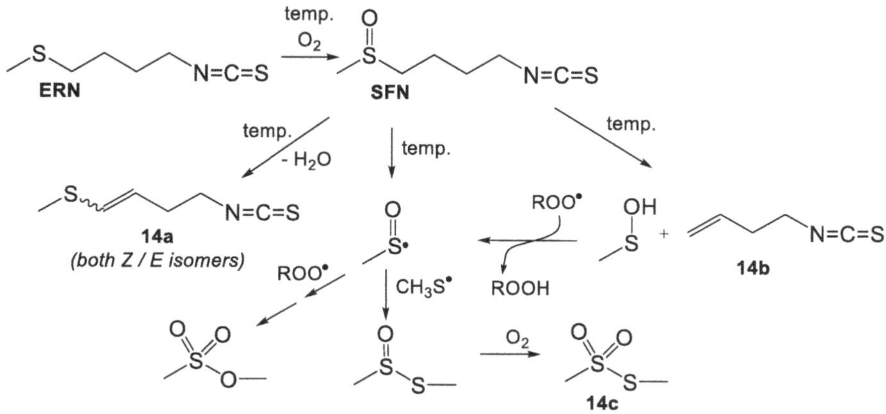

Fig. 14 Oxidation of ERN to SFN and its decomposition transient methanesulfenic acid and methylsulfinyl radicals able to trap lipid peroxyl radicals. The products **14a–c** were detected by GC-MS as stable products of oxidation of ERN and SFN at 100–160 °C (Cedrowski et al. 2020, 2021)

6 Conclusions

Mustard oils (isothiocyanates, ITC) have been studied for over two hundred years as those compounds are components of many vegetables giving them specific taste and flavour. In last decade of XX century the series of pioneering publications indicating many health benefits and chemopreventive properties of sulforaphane triggered the intensive research on the bioactivity and antiradical activity of ITCs and their precursors (glucosinolates). The knowledge cumulated during the last 30 years allows to confirm that bioactivity of ITCs (mainly sulforaphane and its analogues) is connected with their "indirect antioxidant action", that is, these ITCs activate a multi-component (genetic and enzymatic) antioxidant machinery within the cell.

Many attempts to prove "direct antioxidant action" of ITCs, i.e., their anti-radical behaviour at ambient temperature, gave negative results (with exception of •OH which is the strongest oxidant) or, as for the assays employing artificial radicals, are not conclusive because of scattered data coming from not standardized methodologies. On the other hand, there are experimental evidences that ITCs bearing methylsulfanyl moiety are able to reduce hydroperoxides, thus, this group of compounds would prevent the cell against non-radical ROS generated during the oxidative stress.

In contrast to the studies of reactivity at ambient temperatures, the accessible data about thermal stability of ITCs confronted with the experimental results indicating inhibitory effect of sulforaphane and erucin during oxidation of lipids in bulk-phase at temperature above 100 °C suggest that the observed inhibitory effect is due to thermal decomposition of ITCs releasing sulfenic acids and methylsulfinyl radicals, the species being good trapping agents for lipidperoxyl radicals.

Acknowledgements Financial support from the National Science Centre, Poland (NCN grant OPUS No. 2018/31/B/ST4/02354 and NCN grant PRELUDIUM No. 2014/15/N/ST5/02939) is gratefully acknowledged.

References

Agerbirk N, De Vos M, Kim JH, Jander G (2008) Indole glucosinolate breakdown and its biological effects. Phytochem Rev 8(1):101

Akbari E, Namazian M (2020) Sulforaphane: a natural product against reactive oxygen species. Comput Theor Chem 1183:112850

Amorati R, Valgimigli L (2015) Advantages and limitations of common testing methods for antioxidants. Free Radic Res 49(5):633–649

Baek SJ, McEntee MF, Legendre AM (2009) Review paper: cancer chemopreventive compounds and canine cancer. Vet Pathol 46(4):576–588

Barillari J, Canistro D, Paolini M, Ferroni F, Pedulli GF, Iori R, Valgimigli L (2005a) Direct antioxidant activity of purified glucoerucin, the dietary secondary metabolite contained in rocket (*Eruca sativa* Mill.) seeds and sprouts. J Agric Food Chem 53(7):2475–2482

Barillari J, Cervellati R, Paolini M, Tatibouët A, Rollin P, Iori R (2005b) Isolation of 4-methylthio-3-butenyl glucosinolate from *Raphanus sativus* sprouts (kaiware daikon) and its redox properties. J Agric Food Chem 53(26):9890–9896

Barton DHR, Jaszberenyi JC, Theodorakis EA (1992) The invention of radical reactions. Part XXIII new reactions: Nitrile and thiocyanate transfer to carbon radicals from sulfonyl cyanides and sulfonyl isothiocyanates. Tetrahedron 48(13):2613–2626

Beard CI, Dailey BP (1949) The microwave spectra of CH_3NCS and CH_3SCN. J Am Chem Soc 71 (3):929–936

Benati L, Leardini R, Minozzi M, Nanni D, Spagnolo P, Zanardi G (2000) Cascade radical reaction of 2-alkynyl-substituted aryl radicals with aryl isothiocyanates: A novel entry to benzothieno [2,3-b]quinolines through α-(arylsulfanyl)imidoyl radicals. J Org Chem 65(25):8669–8674

Benati L, Calestani G, Leardini R, Minozzi M, Nanni D, Spagnolo P, Strazzari S, Zanardi G (2003) Cascade radical reactions via α-(arylsulfanyl)imidoyl radicals: Competitive [4 + 2] and [4 + 1] radical annulations of alkynyl isothiocyanates with aryl radicals. J Org Chem 68(9):3454–3464

Bergmann E, Tschudnowsky M (1932) Die Dipolmente einiger organischer Stickstoffverbindungen. Z Phys Chem 17B(1):100–106

Billeter O (1875) Ueber organische Sulfocyanverbindungen. Ber Dtsch Chem Ges 8(1):462–466

Blažević I, Montaut S, Burčul F, Olsen CE, Burow M, Rollin P, Agerbirk N (2020) Glucosinolate structural diversity, identification, chemical synthesis and metabolism in plants. Phytochemistry 169:112100

Block E (1992) The organosulfur chemistry of the genus *Allium* – Implications for the organic chemistry of sulfur. Angew Chem Int Ed Eng 31(9):1135–1178

Bonnesen C, Eggleston IM, Hayes JD (2001) Dietary indoles and isothiocyanates that are generated from cruciferous vegetables can both stimulate apoptosis and confer protection against DNA damage in human colon cell lines. Cancer Res 61(16):6120–6130

Boutron F, Fremy E (1840) Untersuchung des schwarzen und weissen Senssamens. Justus Liebigs Ann Chem 34(2):230–232

Burčul F, Generalić Mekinić I, Radan M, Rollin P, Blažević I (2018) Isothiocyanates: cholinesterase inhibiting, antioxidant, and anti-inflammatory activity. J Enzyme Inhib Med Chem 33 (1):577–582

Bussy A (1840) Untersuchungen über die Bildung des ätherischen Senföls. Justus Liebigs Ann Chem 34(2):223–230

Cabello-Hurtado F, Gicquel M, Esnault M-A (2012) Evaluation of the antioxidant potential of cauliflower (Brassica oleracea) from a glucosinolate content perspective. Food Chem 132 (2):1003–1009
Cedrowski J, Dąbrowa K, Krogul-Sobczak A, Litwinienko G (2020) A lesson learnt from food chemistry—elevated temperature triggers the antioxidant action of two edible isothiocyanates: erucin and sulforaphane. Antioxidants 9(11):1090
Cedrowski J, Dąbrowa K, Przybylski P, Krogul-Sobczak A, Litwinienko G (2021) Antioxidant activity of two edible isothiocyanates: Sulforaphane and erucin is due to their thermal decomposition to sulfenic acids and methylsulfinyl radicals. Food Chem 353:129213
Chang X, Lee K-E, Il Jeon S, Kim Y-J, Lee HK, Lee SW (2005) Bis(isothiocyanato)bis(phosphine) complexes of group 10 metals: reactivity toward organic isocyanides. Dalton Trans 23:3722–3731
Crich D, Quintero L (1989) Radical chemistry associated with the thiocarbonyl group. Chem Rev 89(7):1413–1432
Crooker K, Aliani R, Ananth M, Arnold L, Anant S, Thomas SM (2018) A review of promising natural chemopreventive agents for head and neck cancer. Cancer Prev Res (Philadelphia, PA) 11(8):441–450
Czochara R, Kusio J, Symonowicz M, Litwinienko G (2016) Fullerene C_{60} derivatives as high-temperature inhibitors of oxidative degradation of saturated hydrocarbons. Ind Eng Chem Res 55(37):9887–9894
Dadieu A (1931) Raman-Effekt und Konstitutions-Probleme, II. Mitteil.: Cyanverbindungen. Ber Dtsch Chem Ges (A and B Series) 64(2):358–361
Dadieu A, Kohlrausch KWF (1930) Raman-Effekt und Chemie. Ber Dtsch Chem Ges (A and B Series) 63(2):251–282
de Figueiredo SM, Filho SA, Nogueira-Machado JA, Caligiorne RB (2013) The anti-oxidant properties of isothiocyanates: a review. Recent Pat Endocr Metab Immune Drug Discov 7 (3):213–225
Dousmanis GC, Sanders TM Jr, Townes CH, Zeiger HJ (1953) Structure of HNCS from microwave spectra. J Chem Phys 21(8):1416–1417
Drobnica L, Kristián P, Augustín J (1977) Chapter - The chemistry of the —NCS group. In: Patai S (ed) Cyanates and their thio derivatives. PATAI'S chemistry of functional groups, vol 2. John Wiley & Sons Ltd, Chichester, pp 1003–1221
Ettlinger MG, Lundeen AJ (1956) The structures of sinigrin and sinalbin; an enzymatic rearrangement. J Am Chem Soc 78(16):4172–4173
Fahey JW, Zalcmann AT, Talalay P (2001) The chemical diversity and distribution of glucosinolates and isothiocyanates among plants. Phytochemistry 56(1):5–51
Farag MA, Motaal AAA (2010) Sulforaphane composition, cytotoxic and antioxidant activity of crucifer vegetables. J Adv Res 1(1):65–70
Fenwick GR, Heaney RK, Mullin WJ, Van Etten CH (1983) Glucosinolates and their breakdown products in food and food plants. CRC Crit Rev Food Sci Nutr 18(2):123–201
Gamet-Payrastre L, Li P, Lumeau S, Cassar G, Dupont MA, Chevolleau S, Gasc N, Tulliez J, Tercé F (2000) Sulforaphane, a naturally occurring isothiocyanate, induces cell cycle arrest and apoptosis in HT29 human colon cancer cells. Cancer Res 60(5):1426–1433
Gerlich G (1875) Ueber Pseudopropyl- und Allyrhodanür. Justus Liebigs Ann Chem 178(1):80–91
Gildmeister E (1913) The volatile oils, vol 1. John Wiley & Sons, New York, pp 1–432
Gill CI, Haldar S, Porter S, Matthews S, Sullivan S, Coulter J, McGlynn H, Rowland I (2004) The effect of cruciferous and leguminous sprouts on genotoxicity, in vitro and in vivo. Cancer Epidemiol Biomark Prev 13(7):1199–1205
Giudice A, Montella M (2006) Activation of the Nrf2-ARE signaling pathway: a promising strategy in cancer prevention. BioEssays 28(2):169–181
Goubeau J, Gott O (1940) Die Raman-Spektren einiger Rhodanverbindungen und die Struktur der Rhodanid-Gruppe. Ber Dtsch Chem Ges (A and B Series) 73(2):127–133

Greenwald P (2001) From carcinogenesis to clinical interventions for cancer prevention. Toxicology 166(1-2):37–45

Hanahan D, Weinberg RA (2011) Hallmarks of cancer: the next generation. Cell 144(5):646–674

Hanschen FS, Lamy E, Schreiner M, Rohn S (2014) Reactivity and stability of glucosinolates and their breakdown products in foods. Angew Chem Int Ed 53(43):11430–11450

Herraiz T, Galisteo J (2004) Endogenous and dietary indoles: A class of antioxidants and radical scavengers in the ABTS assay. Free Radic Res 38(3):323–331

Hofmann AW (1868) Ueber die dem Senföl entsprechenden Isomeren der Schwefelcyanwasserstoffäther. Ber Dtsch Chem Ges 1(1):25–28

Hudlikar R, Wang L, Wu R, Li S, Peter R, Shannar A, Chou PJ, Liu X, Liu Z, Kuo HD, Kong AN (2020) Epigenetics/epigenomics and prevention of early stages of cancer by isothiocyanates. Cancer Prev Res (Philadelphia, PA) 14(2):151–164

International Agency for Research on Cancer, World Health Organization, Centre international de recherche sur le cancer (Lyon), & Kongress (2004) Chapter -- Glucosinolates, isothiocyanates and indoles. Cruciferous vegetables, isothiocyanates and indoles. In: IARC handbooks of cancer prevention, vol 9. IARC Press, Lyon, pp 13–23

Jakubíková J, Sedlák J, Mithen R, Bao Y (2005) Role of PI3K/Akt and MEK/ERK signaling pathways in sulforaphane- and erucin-induced phase II enzymes and MRP2 transcription, G2/M arrest and cell death in Caco-2 cells. Biochem Pharmacol 69(11):1543–1552

Jin Y, Wang M, Rosen RT, Ho C-T (1999) Thermal degradation of sulforaphane in aqueous solution. J Agric Food Chem 47(8):3121–3123

Kam-Wing Lo K, Chun-Ming Ng D, Hui W-K, Cheung K-K (2001) Luminescent rhenium (I) polypyridine complexes with an isothiocyanate moiety–versatile labelling reagents for biomolecules. J Chem Soc Dalton Trans 18:2634–2640

Kassie F, Uhl M, Rabot S, Grasl-Kraupp B, Verkerk R, Kundi M, Chabicovsky M, Schulte-Hermann R, Knasmüller S (2003) Chemoprevention of 2-amino-3-methylimidazo[4,5-f]quinoline (IQ)-induced colonic and hepatic preneoplastic lesions in the F344 rat by cruciferous vegetables administered simultaneously with the carcinogen. Carcinogenesis 24(2):255–261

Kelloff GJ, Boone CW, Steele VE, Fay JR, Lubet RA, Crowell JA, Sigman CC (1994) Mechanistic considerations in chemopreventive drug development. J Cell Biochem Suppl 20:1–24

Kelloff GJ, Crowell JA, Steele VE, Lubet RA, Malone WA, Boone CW, Kopelovich L, Hawk ET, Lieberman R, Lawrence JA, Ali I, Viner JL, Sigman CC (2000) Progress in cancer chemoprevention: development of diet-derived chemopreventive agents. J Nutr 130(2S Suppl):467s–471s

Khor TO, Huang MT, Kwon KH, Chan JY, Reddy BS, Kong AN (2006) Nrf2-deficient mice have an increased susceptibility to dextran sulfate sodium-induced colitis. Cancer Res 66 (24):11580–11584

Kjær A (1960) Chapter - Naturally derived isothiocyanates (mustard oils) and their parent glucosides. In: Zechmeister L (ed) Fortschritte der Chemie organischer Naturstoffe/Progress in the Chemistry of Organic Natural Products/Progrés Dans la Chimie des Substances Organiques Naturelles, vol 18. Springer, Vienna, pp 122–176

Kjær A (1961) Chapter - Naturally occurring isothiocyanates and their parent glycosides. In: Kharasch N (ed) Organic sulfur compounds. Pergamon, New York, pp 409–420

Kjær A, Gmelin R (1955) Isothiocyanates XI. 4-methylthiobutyl isothiocyanate, a new naturally occurring mustard oil. Acta Chem Scand 9(3):542–544

Klaunig JE, Kamendulis LM (1999) Mechanisms of cancer chemoprevention in hepatic carcinogenesis: modulation of focal lesion growth in mice. Toxicol Sci 52(2 Suppl):101–106

Koelewijn P, Berger H (1972) Mechanism of the antioxidant action of dialkyl sulfoxides. Recl Trav Chim Pays-Bas 91(11):1275–1286

Landis-Piwowar KR, Iyer NR (2014) Cancer chemoprevention: current state of the art. Cancer Growth Metastasis 7:19–25

Leardini R, Nanni D, Pareschi P, Tundo A, Zanardi G (1997) α-(Arylthio)imidoyl radicals: [3 + 2] radical annulation of aryl isothiocyanates with 2-cyano-substituted aryl radicals. J Org Chem 62 (24):8394–8399

Lee M-S (1996) Enzyme induction and comparative oxidative desulfuration of isothiocyanates to isocyanates. Chem Res Toxicol 9(7):1072–1078

Levi MS, Borne RF, Williamson JS (2001) A review of cancer chemopreventive agents. Curr Med Chem 8(11):1349–1362

Li Y, Tollefsbol TO (2010) Impact on DNA methylation in cancer prevention and therapy by bioactive dietary components. Curr Med Chem 17(20):2141–2151

Ligen Z, Yuanfeng W, Yuke S, Lei Z, Mupunga J, Jianwei M, Shiwang L (2017) Broccoli seed extracts but not sulforaphane have strong free radical scavenging activities. Int J Food Sci Technol 52(11):2374–2381

Litwinienko G, Kasprzycka-Guttman T (2000) Study on the autoxidation kinetics of fat components by differential scanning calorimetry. 2. Unsaturated fatty acids and their esters. Ind Eng Chem Res 39(1):13–17

Litwinienko G, Daniluk A, Kasprzycka-Guttman T (2000) Study on autoxidation kinetics of fats by differential scanning calorimetry. 1. Saturated C_{12} - C_{18}fatty acids and their esters. Ind Eng Chem Res 39(1):7–12

Lynett PT, Butts K, Vaidya V, Garrett GE, Pratt DA (2011) The mechanism of radical-trapping antioxidant activity of plant-derived thiosulfinates. Org Biomol Chem 9(9):3320–3330

Magesh S, Chen Y, Hu L (2012) Small molecule modulators of Keap1-Nrf2-ARE pathway as potential preventive and therapeutic agents. Med Res Rev 32(4):687–726

Manesh C, Kuttan G (2003) Anti-tumour and anti-oxidant activity of naturally occurring isothiocyanates. J Exp Clin Cancer Res 22(2):193–199

Manson MM, Gescher A, Hudson EA, Plummer SM, Squires MS, Prigent SA (2000) Blocking and suppressing mechanisms of chemoprevention by dietary constituents. Toxicol Lett 112–113:499–505

Matteo M, Daniele N, Piero S (2007) Imidoyl radicals in organic synthesis. Curr Org Chem 11 (15):1366–1384

Montaut S, Grandbois J, Rossi LS, Kamal S, Khouri J, Ménard MG, Joly HA (2012) Composition of *Dithyrea wislizenii* fruit extract and free-radical scavenging activity of its constituents. Can J Chem 90(8):652–659

Montaut S, Benson HJ, Kay M, Guido BS, Mahboob SS, Chénier J, Gasparetto J-L, Joly HA (2017) Probing the free-radical scavenging activity of the extract, the major glucosinolate and isothiocyanate of *Eruca sativa* Mill. and *Lepidium densiflorum* Schrad. seeds. J Food Compos Anal 61:52–58

Murata M, Yamashita N, Inoue S, Kawanishi S (2000) Mechanism of oxidative DNA damage induced by carcinogenic allyl isothiocyanate. Free Radic Biol Med 28(5):797–805

Nakamura Y, Kawakami M, Yoshihiro A, Miyoshi N, Ohigashi H, Kawai K, Osawa T, Uchida K (2002) Involvement of the mitochondrial death pathway in chemopreventive benzyl isothiocyanate-induced apoptosis. J Biol Chem 277(10):8492–8499

Nam AS, Chaligne R, Landau DA (2021) Integrating genetic and non-genetic determinants of cancer evolution by single-cell multi-omics. Nat Rev Genet 22(1):3–18

Nandini DB, Rao RS, Deepak BS, Reddy PB (2020) Sulforaphane in broccoli: The green chemoprevention!! Role in cancer prevention and therapy. J Oral Maxillofac Pathol 24(2):405

Neoh TL, Yamamoto C, Ikefuji S, Furuta T, Yoshii H (2012) Heat stability of allyl isothiocyanate and phenyl isothiocyanate complexed with randomly methylated β-cyclodextrin. Food Chem 131(4):1123–1131

Nguyen VPT, Stewart J, Lopez M, Ioannou I, Allais F (2020) Glucosinolates: natural occurrence, biosynthesis, accessibility, isolation, structures, and biological activities. Molecules 25 (19):4537

Nickisch R, Conen P, Gabrielsen SM, Meier MAR (2021) A more sustainable isothiocyanate synthesis by amine catalyzed sulfurization of isocyanides with elemental sulfur. RSC Adv 11 (5):3134–3142

Nuzzo S, Twamley B, Platts JA, Baker RJ (2016) Characterisation of isothiocyanic acid, HNCS, in the solid state: trapped by hydrogen bonding. Chem Commun 52(90):13296–13298

Okulicz M, Bialik I, Chichłowska J (2005) The time-dependent effect of gluconasturtiin and phenethyl isothiocyanate on metabolic and antioxidative parameters in rats. J Anim Physiol Anim Nutr 89(11–12):367–372

Olejnik A, Tomczyk J, Kowalska K, Grajek W (2010) The role of natural dietary compounds in colorectal cancer chemoprevention. Postep Hig Med Dosw 64:175–187

Papi A, Orlandi M, Bartolini G, Barillari J, Iori R, Paolini M, Ferroni F, Fumo MG, Pedulli GF, Valgimigli L (2008) Cytotoxic and antioxidant activity of 4-methylthio-3-butenyl isothiocyanate from *Raphanus sativus* L. (kaiware daikon) sprouts. J Agric Food Chem 56(3):875–883

Parnaud G, Li P, Cassar G, Rouimi P, Tulliez J, Combaret L, Gamet-Payrastre L (2004) Mechanism of sulforaphane-induced cell cycle arrest and apoptosis in human colon cancer cells. Nutr Cancer 48(2):198–206

Peñéñory AB, Argüello JE, Puiatti M (2005) Novel model sulfur compounds as mechanistic probes for enzymatic and biomimetic oxidations. Eur J Org Chem 2005(1):114–122

Perschke W (1929) Über die Struktur der dreiatomigen Radikale der Rhodanwasserstoff- und Stickstoffwasserstoffsäure. Ber Dtsch Chem Ges (A and B Series) 62(11):3054–3056

Prasad AK, Mishra PC (2015) Mechanism of action of sulforaphane as a superoxide radical anion and hydrogen peroxide scavenger by double hydrogen transfer: a model for iron superoxide dismutase. J Phys Chem B 119(25):7825–7836

Prochazka Z (1959) Isolation of sulforaphane from hoary cress (*Lepidium draba* L.). Collect Czechoslov Chem Commun 24:2429–2430

Robiquet PJ, Boutron F (1831) Sur la sememce de moutarde. J Pharm Chim 17:279–282

Salah-Abbès JB, Abbès S, Abdel-Wahhab MA, Oueslati R (2010) In-vitro free radical scavenging, antiproliferative and anti-zearalenone cytotoxic effects of 4-(methylthio)-3-butenyl isothiocyanate from Tunisian *Raphanus sativus*. J Pharm Pharmacol 62(2):231–239

Satchell DPN, Satchell RS, Wassef WN (1990) The kinetics and mechanism of addition of water and alcohols to *p*-nitrophenyl isothiocyanate. The effects of added dimethyl sulphoxide. Z Naturforsch B 45(7):1032–1036

Schmid H, Karrer P (1948) Synthese der racemischen und der optisch aktiven Formen des Sulforaphans. Helv Chim Acta 31(6):1497–1505

Schneider W, Kaufmann H (1912) Untersuchungen über Senföle. II. Erysolin, ein Sulfonsenföl aus *Erysimum perowskianum*. Justus Liebigs Ann Chem 392(1):1–15

Seow A, Yuan JM, Sun CL, Van Den Berg D, Lee HP, Yu MC (2002) Dietary isothiocyanates, glutathione S-transferase polymorphisms and colorectal cancer risk in the Singapore Chinese Health Study. Carcinogenesis 23(12):2055–2061

Sharma S (1989) Isothiocyanates in heterocyclic synthesis. Sulfur Rep 8(5):327–454

Sita G, Hrelia P, Tarozzi A, Morroni F (2016) Isothiocyanates are promising compounds against oxidative stress, neuroinflammation and cell death that may benefit neurodegeneration in Parkinson's disease. Int J Mol Sci 17(9):1454

Smith TK, Mithen R, Johnson IT (2003) Effects of Brassica vegetable juice on the induction of apoptosis and aberrant crypt foci in rat colonic mucosal crypts *in vivo*. Carcinogenesis 24 (3):491–495

Smith TK, Lund EK, Parker ML, Clarke RG, Johnson IT (2004) Allyl-isothiocyanate causes mitotic block, loss of cell adhesion and disrupted cytoskeletal structure in HT29 cells. Carcinogenesis 25(8):1409–1415

Song D, Liang H, Kuang P, Tang P, Hu G, Yuan Q (2013) Instability and structural change of 4-methylsulfinyl-3-butenyl isothiocyanate in the hydrolytic process. J Agric Food Chem 61 (21):5097–5102

Spencer GF, Daxenbichler E (1980) Gas chromatography−mass spectrometry of nitriles, isothiocyanates and oxazolidinethiones derived from cruciferous glucosinolates. J Sci Food Agric 31:359–367

Srinivasan P, Vadhanam MV, Arif JM, Gupta RC (2002) A rapid screening assay for antioxidant potential of natural and synthetic agents *in vitro*. Int J Oncol 20(5):983–986

Svehlíková V, Wang S, Jakubíková J, Williamson G, Mithen R, Bao Y (2004) Interactions between sulforaphane and apigenin in the induction of UGT1A1 and GSTA1 in CaCo-2 cells. Carcinogenesis 25(9):1629–1637

Talalay P, Dinkova-Kostova AT, Holtzclaw WD (2003) Importance of phase 2 gene regulation in protection against electrophile and reactive oxygen toxicity and carcinogenesis. Adv Enzym Regul 43(1):121–134

Thimmulappa RK, Mai KH, Srisuma S, Kensler TW, Yamamoto M, Biswal S (2002) Identification of Nrf2-regulated genes induced by the chemopreventive agent sulforaphane by oligonucleotide microarray. Cancer Res 62(18):5196–5203

Ulkowski M, Musialik M, Litwinienko G (2005) Use of differential scanning calorimetry to study lipid oxidation. 1. Oxidative stability of lecithin and linolenic Acid. J Agric Food Chem 53 (23):9073–9077

Valgimigli L, Iori R (2009) Antioxidant and pro-oxidant capacities of ITCs. Environ Mol Mutagen 50(3):222–237

Valgimigli L, Ingold KU, Lusztyk J (1996) Solvent effects on the reactivity and free spin distribution of 2,2-diphenyl-1-picrylhydrazyl radicals 1. J Org Chem 61(22):7947–7950

Van Eylen D, Oey I, Hendrickx M, Van Loey A (2007) Kinetics of the stability of broccoli (*Brassica oleracea* Cv. Italica) myrosinase and isothiocyanates in broccoli juice during pressure/temperature treatments. J Agric Food Chem 55(6):2163–2170

Vanduchova A, Anzenbacher P, Anzenbacherova E (2019) Isothiocyanate from broccoli, sulforaphane, and its properties. J Med Food 22(2):121–126

Vineis P, Schatzkin A, Potter JD (2010) Models of carcinogenesis: an overview. Carcinogenesis 31 (10):1703–1709

Visentin M, Tava A, Iori R, Palmieri S (1992) Isolation and identification for trans-4-(methylthio)-3-butenyl glucosinolate from radish roots (*Raphanus sativus* L.). J Agric Food Chem 40 (9):1687–1691

Walter W, Bode K-D (1967) Syntheses of thiocarbamates. Angew Chem Int Ed Eng 6(4):281–293

Wertheim T (1844) Ueber das flüchtige Oel der *Alliaria officinalis*. Justus Liebigs Ann Chem 52 (1):52–55

Will H (1844) Untersuchungen über die Constitution des ätherischen Oels des schwarzen Senfs. Justus Liebigs Ann Chem 52(1):1–51

Wu L, Noyan Ashraf MH, Facci M, Wang R, Paterson PG, Ferrie A, Juurlink BHJ (2004) Dietary approach to attenuate oxidative stress, hypertension, and inflammation in the cardiovascular system. Proc Natl Acad Sci U S A 101(18):7094–7099

Wu H, Liang H, Yuan Q, Wang T, Yan X (2010) Preparation and stability investigation of the inclusion complex of sulforaphane with hydroxypropyl-β-cyclodextrin. Carbohydr Polym 82 (3):613–617

Wu Y, Mao J, Mei L, Liu S (2013) Kinetic studies of the thermal degradation of sulforaphane and its hydroxypropyl-β-cyclodextrin inclusion complex. Food Res Int 53(1):529–533

Wu Y, Mao J, You Y, Liu S (2014) Study on degradation kinetics of sulforaphane in broccoli extract. Food Chem 155:235–239

Ye L, Zhang Y (2001) Total intracellular accumulation levels of dietary isothiocyanates determine their activity in elevation of cellular glutathione and induction of Phase 2 detoxification enzymes. Carcinogenesis 22(12):1987–1992

Yoon HY, Kang NI, Lee HK, Jang KY, Park JW, Park BH (2008) Sulforaphane protects kidneys against ischemia-reperfusion injury through induction of the Nrf2-dependent phase 2 enzyme. Biochem Pharmacol 75(11):2214–2223

Yuan H, Yao S, You Y, Xiao G, You Q (2010) Antioxidant activity of isothiocyanate extracts from broccoli. Chin J Chem Eng 18(2):312–321

Zhang Y, Li J, Tang L (2005) Cancer-preventive isothiocyanates: dichotomous modulators of oxidative stress. Free Radic Biol Med 38(1):70–77

Zhou JW, Wang M, Sun NX, Qing Y, Yin TF, Li C, Wu D (2019) Sulforaphane-induced epigenetic regulation of Nrf2 expression by DNA methyltransferase in human Caco-2 cells. Oncol Lett 18 (3):2639–2647

Photosensitized Lipid Oxidation: Mechanisms and Consequences to Health Sciences

Isabel O. L. Bacellar, Rosangela Itri, Daniel R. Rodrigues, and Mauricio S. Baptista

1 Introduction

Quite expectedly, the main difference between photoinduced lipid oxidation and traditional lipid oxidation mechanisms relies on its light-dependent initiation. Lacking extended conjugated double-bonds, most lipids do not absorb visible light on their own, therefore requiring an intermediate species to trigger light-dependent initiation mechanisms. This role is fulfilled by photosensitizers, which are organic or inorganic compounds that absorb visible light and inflict photochemical or photophysical alterations on other molecular entities. The key characteristic of photosensitizers is that light absorption promotes them to reactive excited states, which trigger processes that would otherwise be unavailable to their ground states. Many definitions limit the term 'photosensitizer' to cases where the original light-absorbing compound is not consumed in the ensuing reactions (i.e., is regenerated, similarly to a catalyst) (Braslavsky 2007); however, there is mounting evidence that key biological processes elicited by molecules classically regarded as photosensitizers actually involve their consumption (Bacellar et al. 2018; Tasso et al. 2019). Being this observation especially true for lipid oxidation, we herein apply the term

I. O. L. Bacellar · M. S. Baptista (✉)
Departament of Biochemistry, University of São Paulo, Institute of Chemistry, São Paulo, Brazil
e-mail: isabel@iq.usp.br; baptista@iq.usp.br

R. Itri
Departament of Experimental Physics, University of São Paulo, Institute of Physics, São Paulo, Brazil
e-mail: itri@if.usp.br

D. R. Rodrigues
Department of Chemistry and Molecular Physics, University of São Paulo, São Carlos Institute of Chemistry, São Paulo, Brazil
e-mail: drcardoso@usp.br

C. Bravo-Diaz (ed.), *Lipid Oxidation in Food and Biological Systems*,
https://doi.org/10.1007/978-3-030-87222-9_14

more broadly, also to include compounds that undergo permanent or temporary inactivation as a result of photoinduced processes.

Photosensitized oxidations start with light absorption, which promotes photosensitizers from their ground state to an excited state—in the case of most organic photosensitizers, an excited singlet state. This excited state typically has the same spin multiplicity as the ground state, since changes in spin multiplicity violate the selection rules for electronic transitions ($\Delta S = 0$ and $\Delta L = \pm 1$). However, most rules have exceptions and excited singlet states may yield excited triplet states upon undergoing intersystem crossing (ISC), a forbidden, non-radiative transition resulting in a change of spin multiplicity (Braslavsky 2007). For many light-absorbing molecules (e.g., dyes used as fluorophores), ISC has a low probability of occurrence. Yet, photosensitizers often bear heavy atoms or extended conjugated systems that favor this otherwise spin-forbidden photophysical process, accounting for their high excited triplet state quantum yields. Excited triplet states are key to the activity of photosensitizers: because decaying back to the ground state entails yet another spin-forbidden transition, excited triplet states of organic photosensitizers generally live longer (usually nanosecond to microsecond range) than their singlet counterparts (usually picosecond to nanosecond range). Their longer lifetimes raise the probability of effective collisions with molecules nearby, allowing for significant diffusion in fluids and tissues and for electron or energy transfer to occur. Many of these events result in chemical or physical quenching of the photosensitizer, without triggering any chemically or biologically significant event; however, in other cases, they result in electron or hydrogen atom transfer reactions (e.g., forming lipid radicals) or in energy transfer processes (e.g., to molecular oxygen forming a reactive excited state known as 'singlet oxygen', $O_2(^1\Delta_g)$ or $^1O_2$ for simplicity) (Girotti 2001). Although through different routes, both hydrogen atom transfer and energy transfer processes are potential initial steps of lipid photooxidation (Fig. 1). These photooxidation mechanisms are commonly referred to as "Type I" and "Type II" pathways, even though different authors often define Type I and Type II differently (Baptista et al. 2017).

In the next subsections, we explore how each of these pathways triggers lipid oxidation, looking at possible oxidation products and delineating the conditions in which one pathway or the other may be expected to occur. We also look at the impact of lipid photooxidation in lipid membranes, providing insight into the biophysical and biological effects of light-induced lipid oxidation. We then switch gears to consequences and applications of lipid photooxidation, covering its role in photodamage of skin and photodynamic therapy (PDT), an anticancer and antimicrobial clinical modality based on photoactive drugs. We note that we will not cover in detail the post-irradiation (light-independent) lipid peroxidation reactions that can be more damaging than the primary Type I or Type II mechanisms. Post-irradiation lipid peroxidation reactions are not different from those that take place in purely chemical oxidation processes, being driven by radicals, reductants and redox-active

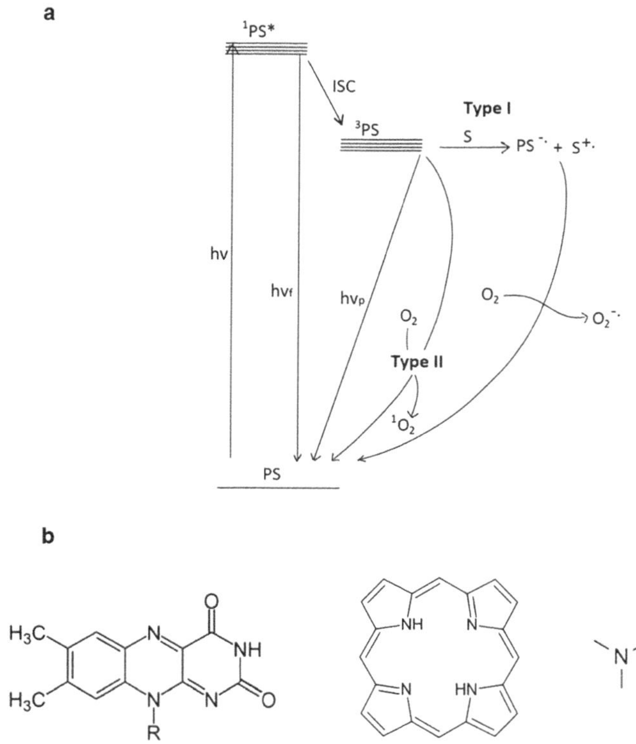

Fig. 1 (**a**) Modified Jablonski diagram illustrating Type I and Type II photosensitization mechanisms. hv = photon absorption, hv_f = fluorescence, hv_p = phosphorescence, ISC = intersystem crossing, O_2 = ground state molecular oxygen, $^1O_2$ = singlet-excited oxygen, $O_2^{-\bullet}$ = superoxide ion, PS = ground state photosensitizer, $^1PS^*$ = singlet-excited photosensitizer, $^3PS^*$ = triplet-excited photosensitizer, S = reducing substrate. (**b**) Molecular structure of common photosensitizers. On the left, a core structure of a flavin molecule, which not only is the main endogenous skin UVA-absorbing photosensitizer but also provides a lead for synthetic photosensitizers (Wei et al. 2021; Silva et al. 2015). In the center, a generic structure of porphyrin, which is also an endogenous skin photosensitizer and provides the base structure for a series of synthetic photosensitizers used in Photodynamic Therapy (PDT), including porphyrin derivatives, chlorins, chlorophyllin and bacteriochlorin (Uchoa et al. 2015). On the right, the structure of methylene blue, a synthetic, positively charged photosensitizer highly used in PDT, particularly in anti-microbial applications (Garcez et al. 2011)

metal ions (e.g., iron and copper) that promote one-electron reduction of phospholipid and cholesterol hydroperoxides (Girotti 2001). We additionally refer readers interested in understanding photosensitizer photochemical and photophysical properties to a recent review on this subject (Quina 2021).

1.1 Singlet Oxygen as an Initiator

Singlet oxygen formation is a hallmark of photoinduced lipid oxidation. For a photosensitizer to generate singlet oxygen, its excited triplet state needs to lay at least 94 kJ mol^{-1} above the ground state (Wilkinson et al. 1995; Schweitzer and Schmidt 2003). From a spatial and kinetic perspective, singlet oxygen generation additionally depends on excited triplet states of photosensitizers being generated in an oxygen-rich microenvironment, given that steep variations in oxygen concentration are observed even within lipid bilayers (Bacellar et al. 2019). If both conditions are met, excited triplet states may transfer energy to molecular oxygen, yielding singlet oxygen. Once formed, singlet oxygen diffuses from its generation site. However, being an excited state, there is only a limited distance along which it can diffuse before decaying to its ground state. The average diffusion distance of singlet oxygen in one dimension is given by $(D\tau)^{1/2}$, where D is its diffusion coefficient and τ is its lifetime. The latter parameter is mainly measured through singlet oxygen's signature near-infrared phosphorescence, and is strongly dependent on the media: while singlet oxygen has a lifetime of 3.7 μs in water, in deuterium oxide its lifetime increases to 60 μs (Wilkinson et al. 1995; Schweitzer and Schmidt 2003). As a consequence, diffusion distances also vary significantly with the environment: while in water singlet oxygen has an average diffusion distance of ~100 nm, in cells this distance is estimated to be 4 times shorter due to the presence of quenchers and to a reduced diffusion coefficient (Hackbarth et al. 2016). Although direct measurements of singlet oxygen lifetimes in lipid membranes still lack and are challenging to obtain, measurements in hydrated lipid films place the value in the 5–20 μs range (Bacellar et al. 2019). However, it should be noted that singlet oxygen molecules generated within lipid bilayers are typically able to escape the membrane before being quenched—even though they may reencounter lipids along their diffusion route. Conversely, singlet oxygen molecules generated in water at a distance within the average diffusion distance of singlet oxygen may reach lipids membranes within their lifetime (Hackbarth and Röder 2015). As a result, water-soluble photosensitizers can extensively promote singlet-oxygen mediated lipid oxidation, in sharp contrast to electron or hydrogen transfer mechanisms, which require photosensitizer excited states to come in contact with lipid molecules.

Differently from ground state molecular oxygen, singlet oxygen has an unoccupied π^*_{2p} orbital. This distinct electronic configuration allows singlet oxygen to participate in cycloaddition reactions without the spin-forbidden constraints faced by its ground state, granting its high reactivity towards electron-rich compounds. As a result, in the biological environment, singlet oxygen reacts with amino acids, nucleic acids and lipids, not to mention antioxidants (Bacellar et al. 2015). Reactions between singlet oxygen and lipids typically require lipids to bear allylic hydrogens, and proceed through a mechanism known as 'singlet oxygen *ene* reaction'. As a result, only unsaturated lipids (sterols or fatty acid-derived) are oxidized by singlet oxygen, even though both saturated and unsaturated lipids may physically quench it (Krasnovsky et al. 1983; Frankel 1984; Alberti and Orfanopoulos 2010). The singlet

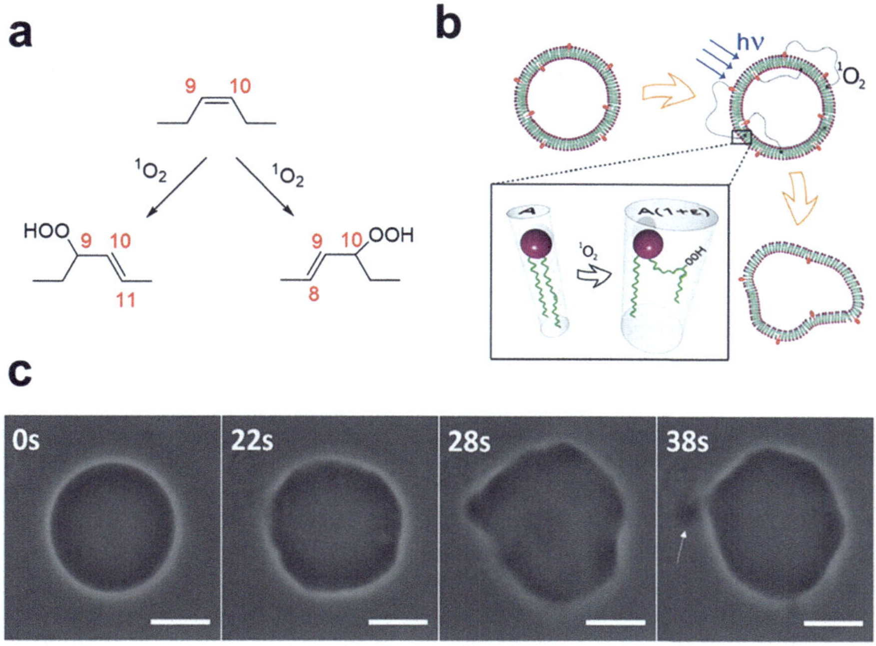

Fig. 2 (**a**) The reaction between singlet oxygen and monounsaturated lipids yields two isomeric allylic hydroperoxides, differing in terms of hydroperoxide group and double bond position. (**b**) Representation of the migration of a hydroperoxide group towards the membrane interface, imposing an increase in membrane surface area. (**c**) Optical microscopy images in phase-contrast mode of 1,2-dioleoyl-*sn*-glycero-3-phospho-(1′-rac-glycerol) (DOPG) giant unilamellar vesicle dispersed in a porphyrin-based photosensitizer ($TPPS_{2a}$) solution under photooxidation. The membrane displays increasing fluctuations with irradiation time (in seconds, indicated in the upper corner of each snapshot) due to an increase in surface area. This photoinduced area excess is released through the emission of buds, as indicated by the white arrow in the last snapshot. Scale bars represent 10 μm. Panel (**b**) was reproduced from Weber et al. 2014 with permission of The Royal Society of Chemistry. Panel (**c**) reproduced from Tsubone et al. 2019, with permission from Elsevier

oxygen *ene* reaction results in lipid hydroperoxides with the unsaturation shifted to the allylic position, in the *E (trans)* configuration. For example, oxidation of oleic acid (18:1, Δ9), which has an unsaturation between carbons 9 and 10, yields two positional isomers (Fig. 2a): one bearing a -OOH group attached to carbon 9 and having an unsaturation between carbons 10 and 11, and the other bearing a -OOH group attached to carbon 10 and having an unsaturation between carbons 8 and 9 (Frankel 1984; Alberti and Orfanopoulos 2010).

In the absence of high temperatures, radicals (including lipid radicals) and redox-active metal ions, lipid hydroperoxides may largely accumulate (Girotti 2001; Halliwell and Gutteridge 2015). Indeed, studies in model membranes demonstrated that photosensitizers operating solely through singlet oxygen generation can convert unsaturated lipids into their hydroperoxide counterparts to a large extent (Weber et al. 2014). The resulting lipid membranes undergo a surface area increase due to the migration of hydroperoxide groups towards the aqueous interface (Fig. 2b) (Wong-Ekkabut et al. 2007; De Rosa et al. 2018). This conformational change may result in membrane remodeling (Fig. 2c) yet without a significant change in membrane permeability (Bacellar et al. 2018; Weber et al. 2014; Riske et al. 2009; Tsubone and Baptista 2019a). Apart from the inherent stability of hydroperoxides, their accumulation is further facilitated by photosensitizers being able to keep generating singlet oxygen as long as molecular oxygen and light irradiation last. Differently from electron or hydrogen-atom transfer reactions, the energy transfer mechanism involved in singlet oxygen production immediately recovers photosensitizers into their ground state (Fig. 1). Therefore, singlet oxygen production may continue unless photosensitizers are bleached by directly reacting with singlet oxygen or by any competing processes (e.g., electron transfer reactions).

In conditions where singlet oxygen is the only oxidizing agent, hydroperoxides are the main redox-active intermediates that may accumulate in the membranes. For polyunsaturated lipids, the reaction of a pre-formed hydroperoxide with a second singlet oxygen molecule may occur, yielding lipids bearing multiple -OOH groups (Neff and Frankel 1982; Neff and Frankel 1984). However, for monounsaturated lipids, there is no evidence of additional *ene* reaction events occurring at the newly-formed double bond. Hydroperoxides may as well yield further products by one-electron reduction; yet, most of them are redox-inactive (Bacellar et al. 2018; Girotti 2001). The formation of stable, non-radical products that may accumulate is a marked difference of photoinduced lipid oxidation if compared with light-independent lipid oxidation or even to photoinduced lipid oxidation driven by electron or hydrogen-atom transfer: singlet oxygen mediated-lipid oxidation does inflict lipid oxidation, yet in a radical-independent pathway that deviates from the typical initiation-propagation-termination paradigm.

1.2 Photosensitizer Excited States as Initiators

Excited states are better oxidants and reducers than their corresponding ground states (Turro et al. 2009). As a result, photosensitizer excited states are often more prone than their ground-state counterparts to engage in direct reactions with biomolecules, including lipids. This probability increases for excited triplet states, which typically live longer than excited singlet states and are thus more likely to encounter their chemical targets within their lifetime. Also, despite singlet n,π^* states being more reactive than triplet-excited n,π^* states in electron transfer and hydrogen-atom abstraction reactions, the latter show higher efficiency since back-hydrogen-atom-abstraction reaction is not allowed due to the different spin multiplicities. For example, deactivation of singlet-excited vitamin B2 by plant phenols leads to a non-productive chemical quenching, recovering singlet-excited vitamin B2 into the ground state after diffusion-controlled electron transfer from phenol and fast charge recombination (Cardoso et al. 2012; Cardoso et al. 2006), Fig. 3.

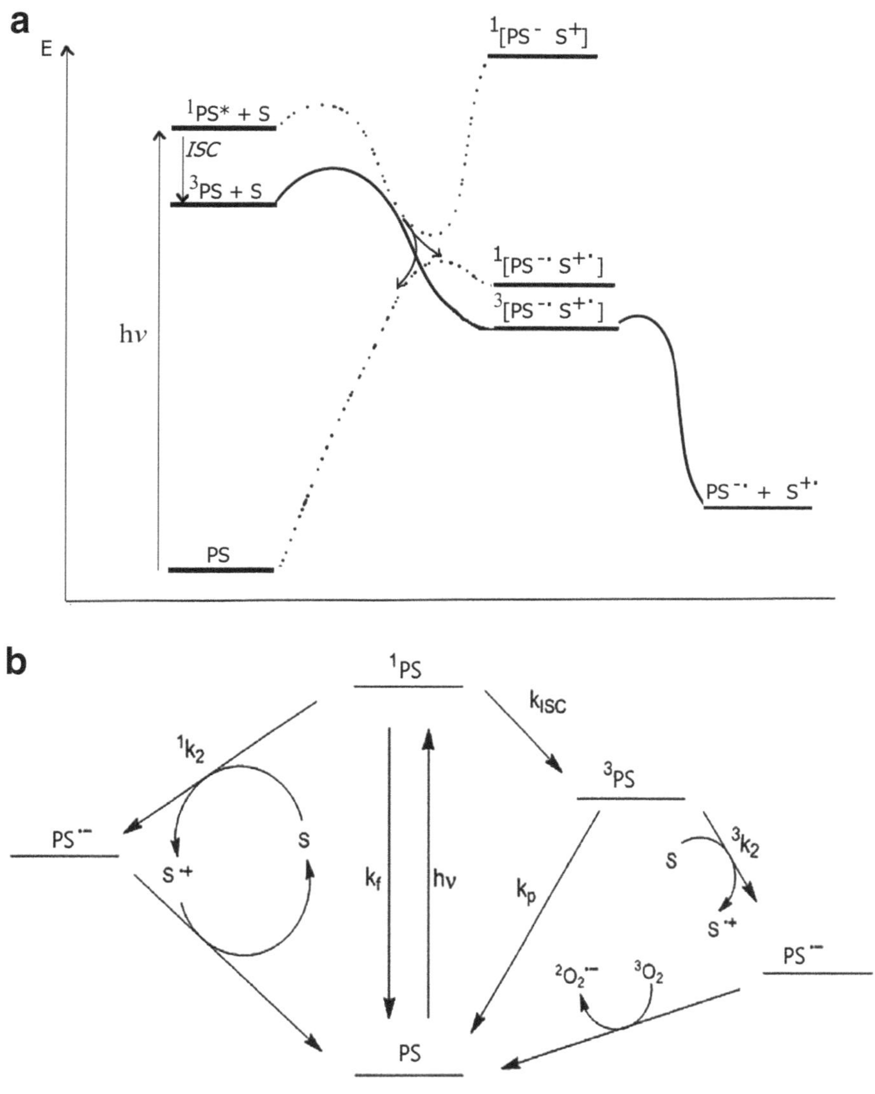

Fig. 3 (**a**) Generalized correlation diagram for electron transfer from a reducing substrate (S) to singlet- or triplet-excited photosensitizer (PS), illustrating the lower activation energy of the singlet reaction channel leading to a non-productive chemical quenching ($^1k_2$). (**b**) Illustrative diagram of photoinduced oxidation of reducing substrates (S) by direct reactions with excited photosensitizers (PS). The non-productive singlet channel chemical quenching recovers the PS in the ground state, due to back-electron-transfer reaction from the radical pair stage. The triplet channel can lead to substrate oxidation through electron-transfer, hydrogen atom transfer, or proton-coupled electron-transfer mechanism ($^3k_2$). *hν* indicates light absorption by PS and k_f and k_p are rate constants for fluorescence and phosphorescence decay pathways, respectively

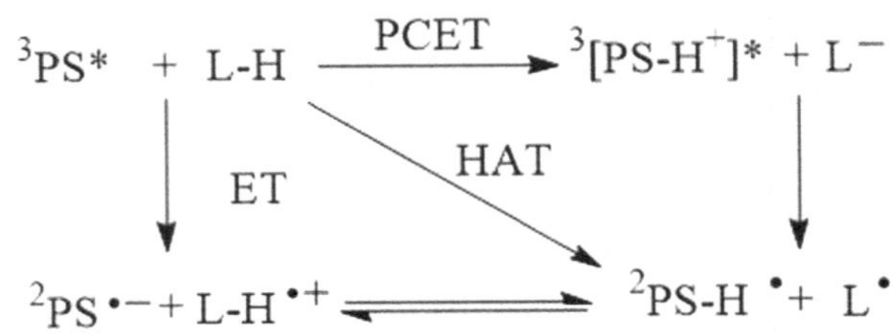

Fig. 4 Thermodynamic cycle for electron-transfer (ET), hydrogen atom transfer (HAT), and proton-coupled electron-transfer mechanism (PCET) for deactivation of triplet photosensitizer (PS) by lipids (LH)

Whether an excited state of a photosensitizer will or not be able to react with a lipid substrate depends both on thermodynamics and, especially in biological systems, also on kinetics and spatial distribution. Here, the requirements for photosensitizers to be close to lipids are significantly more stringent than for singlet-oxygen mediated pathways. This is because excited triplet states may have shorter average diffusion distances than singlet oxygen, due to a combination of smaller diffusion coefficients and often also shorter lifetimes. For the photosensitizer methylene blue (MB), for example, the average diffusion distance of its excited triplet state was estimated to be only 25 nm (i.e., 3.4 times shorter than the average diffusion distance of singlet oxygen) (Bacellar et al. 2018). Considering that excited states—especially singlet ones—may live have shorter lifetimes compared with MB's excited triplet state (1.5 μs in water) (Junqueira et al. 2002), proximity between photosensitizers and lipids becomes a prerequisite for extensive initiation of photooxidation by direct reactions with excited states. This effect is even more pronounced in biological media, where membranes and other biological structures constitute physical barriers for photosensitizer diffusion. Indeed, higher partition in lipid membranes was shown to predict photocytotoxicity of photosensitizers, as well as their capability of permeabilizing membranes (Bacellar et al. 2018; Pavani et al. 2009; Bacellar et al. 2014). For longer-lived excited states, the importance of photosensitizer and targets is decreased and the ratio between the concentration of oxygen and reducing substrates and oxygen will be the main determinant for the occurrence of direct reactions. This is the case of triplet states of biologically relevant photosensitizers like flavins (vitamin B2), which have an average diffusion distance estimated to be around 100 nm (McQuaid et al. 2016).

From a reactivity point of view, the reduction potentials or hydrogen atom affinity of excited photosensitizers need to be compatible with oxidation potentials around 2 V *vs.* NHE for unsaturated lipids or bond dissociation enthalpies for hydrogen abstraction of lipids (ranging from 294 kJ mol^{-1} for methyl linoleate to 276 kJ mol^{-1} for methyl docosahexanoate), as illustrated in the thermodynamic cycle for electron-transfer, hydrogen-atom transfer, and proton-coupled electron-transfer processes for deactivation of triplet excited photosensitizers by polyunsaturated lipids, Fig. 4. Based on the thermodynamic cycle depicted in Fig. 4, and considering the E' for oxidizing unsaturated lipids to be around 2 V *vs.* NHE (de Almeida et al. 2014;

Huvaere et al. 2010) and reduction potential of 1.48 V and 1.77 V *vs.* NHE for triplet excited methylene blue (^{3}MB) and triplet excited riboflavin (^{3}Rib), the electron-transfer mechanism is not operational ($\Delta G^o > 0$). In these cases, the chemical quenching of the excited triplet state is based on the transfer of hydrogen atoms (HAT). These reactions lead to carbon-centered lipid radicals, which in turn react with molecular oxygen to form peroxyl radicals (Yin et al. 2011). Once formed, peroxyl radicals propagate lipid peroxidation identically to light-independent lipid peroxidation, until radical reactions are terminated by antioxidants (including enzymes, in the biological environment) or by radical disproportionation.

Peroxidation initiation reactions also convert photosensitizers into radical species, meaning that further reactions are needed to return these chemically modified photosensitizers to their original state (Foote 1968). In the absence of adequate conditions or reagents for proper regeneration of photosensitizers, direct reactions by excited states contribute to the bleaching of photosensitizers. For many years, the search for new photosensitizers involved discarding those more susceptible to photobleaching. Even though reverting photobleaching is still seen as advantageous, we now understand that bleaching may not only be a necessary evil but also an indication that photosensitizers are engaging in direct reactions that have key biological repercussions and may be highly desirable in applications such as PDT (Bacellar et al. 2018; Tasso et al. 2019).

Given that, especially for polyunsaturated lipids, peroxyl radicals may abstract hydrogen atoms from non-oxidized lipids in the propagation phase, direct reactions between lipids and photosensitizer excited states indirectly yield lipid hydroperoxides. This observation means that, for a photosensitizer whose excited state directly reacts with lipids and also generates singlet oxygen (e.g., chlorophyll a and flavins), there are two possible sources of lipid hydroperoxides (Zhou et al. 2021). Oftentimes, there are differences in terms of positional and stereoisomerism for hydroperoxides generated by each of these pathways (e.g., oleic acid (18:1, Δ9) forms only two isomers upon singlet oxygen oxidation, but forms eight isomers upon radical-mediated oxidation) (Frankel 1984). Cholesterol-derived hydroperoxides should also be mentioned, given that cholesterol comprises a considerable portion of mammalian membrane lipids. Since cholesterol hydroperoxides are relatively stable and easy to separate chromatographically, the position at which the hydroperoxide addition occurs can be used to infer the type of reaction responsible for its formation. Free-radical reactions favor the attack on the carbon 7 of cholesterol, forming mainly 3β-hydroxycholest-5-ene-7α-hydroperoxide (7α-ChOOH) and 3β-hydroxycholest-5-ene-7β-hydroperoxide (7β-ChOOH), whereas singlet oxygen forms mainly 3-β-hydroxy-5α-cholest-6-ene-5-hydroperoxide (5α-ChOOH), 3β-hydroxycholest-4-ene-6α-hydroperoxide (6α-ChOOH), and 3β-hydroxycholest-4-ene-6-β-hydroperoxide (6β-ChOOH), Fig. 5 (Doleiden et al. 1974; Girotti 1992).

Hydroperoxides generated through either pathway also serve as substrates for additional reactions by excited photosensitizers and their radicals, directly or indirectly forming peroxyl or alkoxyl radicals that may further propagate lipid peroxidation and undergo other types of reactions. The formation of alkoxyl radicals was shown to have important consequences to the biophysical impacts of lipid

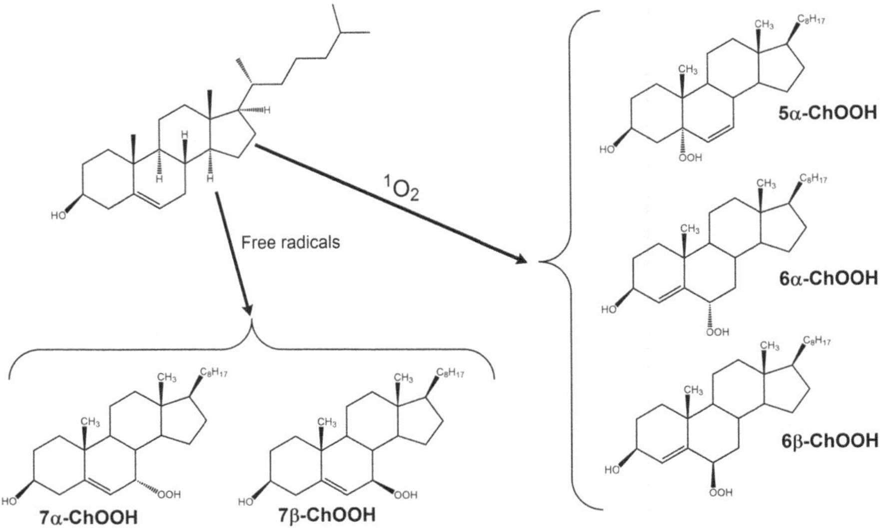

Fig. 5 Hydroperoxide regioisomers for oxidation mediated by free radicals (7α-ChOOH and 7β-ChOOH) and by singlet oxygen (5α-ChOOH, 6α-ChOOH and 6β-ChOOH)

photooxidation on membranes, namely membrane permeabilization. It was previously believed that lipid hydroperoxides could undergo Hock cleavage, a specific acid-catalysis mechanism that could hypothetically break a C-C bond in the vicinity of the hydroperoxide group, forming truncated lipid aldehydes directly from lipid hydroperoxides (Bacellar et al. 2018). However, the Hock cleavage mechanism was recently refuted for phospholipid hydroperoxides, as membranes composed solely of phospholipid hydroperoxides continue to sustain sugar gradients upon exposure to acids down at least to pH 3.5. Yet, the formation of truncated lipid aldehydes was shown to be compatible with the β-scission of alkoxyl radicals (not to be confused with the peroxyl radical β-scission) (Bacellar et al. 2018; Gardner 1989). In this mechanism, the C-C bond adjacent to the carbon bearing the alkoxyl group suffers homolytic cleavage, yielding a lipid aldehyde and a short-chain carbon-centered radical. This mechanism explained the formation of truncated lipid aldehydes in amounts as low as 1 mol%, which yet correlated with high increases in membrane permeability (Bacellar et al. 2018) (Fig. 6). Lipid alkoxyl radicals can be formed from peroxides or hydroperoxides via one-electron reduction. In the absence of redox-active metals, they can only be formed when photosensitizers initiate radical-mediated lipid oxidation. Therefore, direct reactions with excited photosensitizers may be considered essential to the pathways leading to membrane permeabilization. This notion only recently became clear and represents a paradigm shift from a long-lasting belief that singlet-oxygen mediated oxidation was the main source of photoinduced membrane damage and permeabilization. As a result, we foresee a shift from optimizing photosensitizers based on singlet oxygen production to optimizing these molecules based on the reactivity of their excited states.

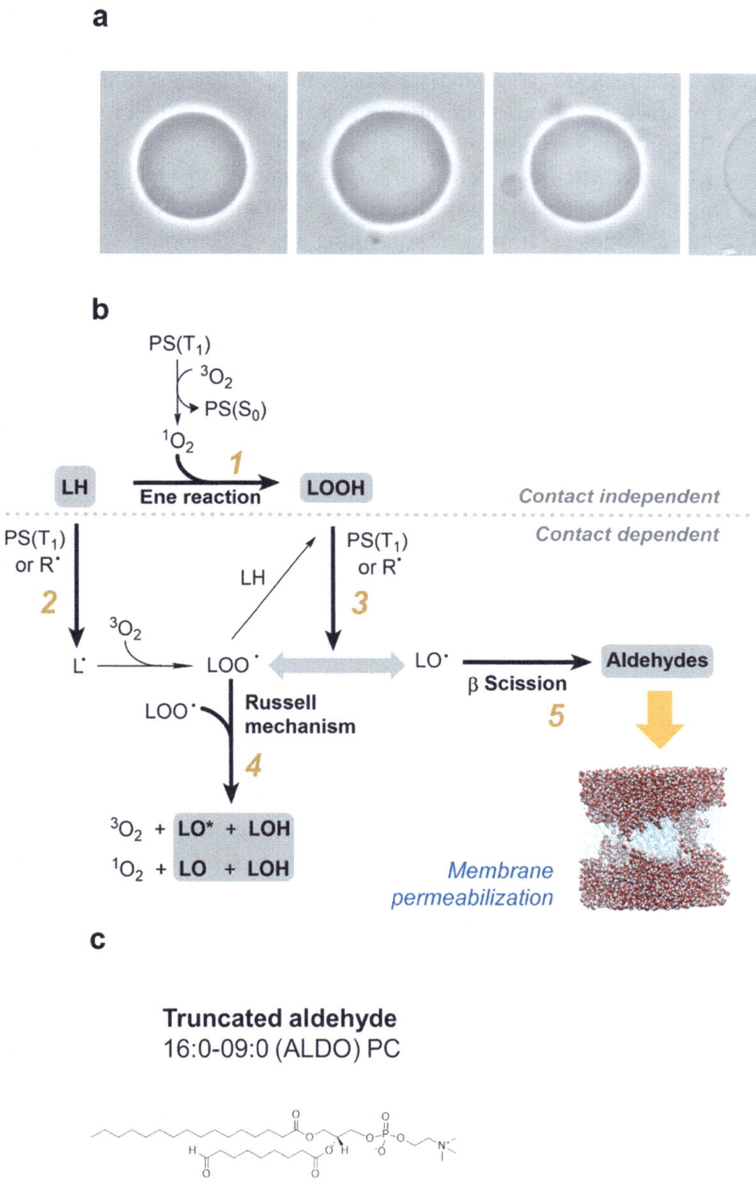

Fig. 6 (**a**) Time sequence of a 1,2-dioleoyl-*sn*-glycero-3-phosphocholine (DOPC) giant unilamellar vesicle dispersed in a methylene blue solution under photooxidation, as observed by optical microscopy in phase-contrast mode. The fading of the optical contrast in the last snapshot evidences an increase in membrane permeability, with mixing of the external and internal solutions. (**b**) Chemical pathways for photoinduced membrane permeabilization. The map distinguishes between contact-independent and contact-dependent processes, which rely on singlet oxygen or on direct reactions between PSs and lipids, respectively. PS(S_0), PS(T_1): PS ground and excited triplet states; $^3O_2$,$^1O_2$: ground and excited singlet states of oxygen; $R^•$: generic radical species; LH: non-oxidized lipid; $L^•$, $LOO^•$, $LO^•$: lipid carbon-centered, peroxyl and alkoxyl radicals; LOOH, LOH, LO, LO^*: lipid hydroperoxide, alcohol, ketone and excited state ketone. A snapshot of a

Initiation reactions continue while light, photosensitizers and substrates last, while propagation reactions may persist even after irradiation halts. The later stages of photoinduced oxidations are increasingly challenging to study, given the variety of reactions and products involved, a complexity that is only magnified for polyunsaturated lipids. The intricacy of possible outcomes is well illustrated by the Russell mechanism, which starts when two peroxyl radicals combine to yield a tetroxide intermediate. This intermediate promptly decomposes forming a lipid alcohol, an excited lipid ketone and molecular oxygen or forming a lipid alcohol, a lipid ketone and singlet oxygen. Even though the Rusell mechanism would be a termination step by definition, it is often referred as a form of "photochemistry in the dark", for it generates excited states that can trigger novel initiation events (Russell 1957; Howard and Ingold 1968). Besides further fueling peroxidation, this mechanism was suggested to be the main source of lipid alcohols and ketones during photosensitized oxidation of lipid membranes. As a hallmark of the Russell mechanism, alcohols and ketones are produced in a 1:1 ratio, a proportion that was observed when the chemical composition of liposomes was monitored during photoinduced oxidation (Bacellar et al. 2018).

1.3 Determinants of Photooxidation Mechanisms and Their Outcomes

Photosensitizers may act through singlet-oxygen generation alone, direct reactions by excited states alone, or by a combination of both mechanisms. Thermodynamic and kinetic factors shift this balance in one direction or another. First, different photosensitizers have different photochemical and photophysical properties (Quina 2021). As a consequence, excited states may be more or less energetic, live shorter or longer, have higher or lower singlet oxygen generation quantum yields, and be more or less prone to electron or hydrogen atom transfer reactions. Compare for example the energy of MB's excited triplet state (*ca.* 142 kJ mol^{-1}) to erythrosine and rose Bengal's, which lay around *ca.* 30 kJ mol^{-1} higher (Gollnick et al. 1970).

Second, the physical-chemical properties of photosensitizers vary widely: while some photosensitizers are highly hydrophilic and consequently do not interact efficiently with the membranes, others may be fully embedded in lipid bilayers and ready to engage in electron or energy transfer reactions. As a consequence, water-soluble photosensitizers may have a higher contribution of singlet-oxygen

---

Fig. 6 (continued) simulated aldehyde membrane, showing pore opening, is also provided. Reprinted with permission from Bacellar et al. 2018, with permission from the American Chemical Society. (**c**) Structure of a truncated lipid aldehyde, 1-palmitoyl-2-(9′-oxo-nonanoyl)-*sn*-glycero-3-phosphocholine (16:0–09:0 (ALDO) PC), implicated in membrane permeabilization. The membrane shown in panel b is exclusively composed of 16:0–09:0 (ALDO) PC, with lipids in cyan and water molecules in white (hydrogens) and red (oxygens)

mediated lipid oxidation than membrane-bound analogs (Bacellar et al. 2018). Not only the proximity to lipids modulates oxidation pathways, but also the nature of lipids (Bacellar et al. 2018; Hill et al. 2012) and their concentration relative to oxygen. Fluid phase lipid membranes are richer in oxygen than the surrounding solution, with oxygen concentration varying steeply depending on membrane depth. Yet, gel-phase lipid membranes can have oxygen concentrations that are even smaller than in bulk water (Bacellar et al. 2019). At the same time, the number of unsaturations in the phospholipid fatty acids favor contact-dependent reactions, as demonstrated by the increase in photobleaching rates of the photosensitizer with the increase in the number of allylic and bis-allylic hydrogens (Bacellar et al. 2018). As membrane oxidation progresses and new substrates (e.g., hydroperoxides) become available or depleted, the balance between both pathways is expected to further change.

Finally, we note that action mechanisms of photosensitizers are also affected by their chemical environment beyond the composition of lipid membranes (Uchoa et al. 2015; Bacellar et al. 2015; Pavani et al. 2009). Phenothiazinium photosensitizers such as MB, for example, aggregate at high concentrations or high ionic strengths. Dimerization triggers electron transfer reactions between the excited triplet state and the remaining ground state photosensitizer in the dimer, forming semi-reduced and semi-oxidized radical species that may react with biomolecules and start radical-mediated pathways (Junqueira et al. 2002). While negatively charged surfaces were shown to induce the dimerization of MB, zwitterionic lipid membranes reduce the aggregation of phenothiazinium photosensitizers (Bacellar et al. 2014; Severino et al. 2003).

Even though activation by light is the common element for both mechanisms of photosensitizer-induced lipid oxidation, the outcomes of photooxidation processes are unique to each situation. They not only depend on the original lipid composition and the particular properties of each photosensitizer, but also on the microenvironment where the excited state is generated. In cells, where the complexity of microenvironments and chemical composition raises dramatically, predicting photoinduced oxidation mechanisms is still a real challenge. Recently, lipidomic approaches have started to investigate changes in lipid composition as a result of photooxidation (Anthonymuthu et al. 2017; Hu et al. 2017), while cell biology studies delineate mechanisms leading to phototoxicity and the role of membrane damage (Libardo et al. 2017; Leonarduzzi et al. 2000; Volinsky and Kinnunen 2013). Yet, connecting changes in lipid composition to biological outcomes remains a challenge. In the next section, we explore how lipid bilayer models attempt to bridge this gap and highlight that different oxidized lipids affect bilayers in completely distinct ways.

2 Photooxidation Affects the Structure and Properties of Membranes

It is well known that a controlled amount of oxidized lipids is required for cell signaling, maturation and differentiation, as well as apoptosis. However, the products of lipid oxidation, if uncontrolled, may have deleterious effects on the functioning of the cell and be involved in a variety of diseases (Nicolson and Ash 2014; Bochkov et al. 2016). These changes have a major impact not only on the physical properties of membranes (due to hydrophilic groups, see Fig. 2 and Fig. 6) but also on cell physiology (Volinsky and Kinnunen 2013; Catalá 2009) and probably on the membrane response to extracellular signals. In terms of oxidized lipid concentration, there is a vague knowledge of oxidized lipid basal values in fluids and cells, which can be in the nM concentration range in healthy cells and higher (sub-μM) in pathological conditions (Reis 2017).

So far, the comprehensive understanding of how oxidized lipids impact cell membranes has been facilitated by work carried out on model systems with predominant phospholipid classes such as phosphatidylcholines. Interestingly, experimental results from photodamaged liposomes revealed membrane permeabilization associated with the formation of truncated lipid aldehydes in amounts as low as 1 mol% (Bacellar et al. 2018), similarly to studies in which membranes were pre-formed with truncated lipid aldehydes (Ytzhak and Ehrenberg 2014; Runas et al. 2016; Runas and Malmstadt 2015). In contrast, molecular dynamic simulations indicate that phospholipid bilayer poration requires a high amount of lipids-bearing oxidatively truncated acyl chains (Boonnoy et al. 2015), suggesting that in practice permeabilization of lipid bilayers may depend on out-of-equilibrium conditions or aggregation of oxidized lipids resulting in higher local concentrations.

Among hydroperoxide lipids, 1-palmitoyl-2-oleoyl-*sn*-glycero-3-phosphocholine hydroperoxide (POPC-OOH) has been experimentally and theoretically investigated in a variety of model systems. Due to the more hydrophilic character of the hydroperoxide group in respect to the lipid acyl chain milieu, this polar group tends to migrate to the bilayer surface (De Rosa et al. 2018; Siani et al. 2016; Garrec et al. 2014; Sankhagowit et al. 2014) (Fig. 2). As a consequence, an excess of membrane surface area results, in association with lipid molecular area increase (De Rosa et al. 2018; Riske et al. 2009; Siani et al. 2016) and a reduction of both the membrane elastic modulus (Weber et al. 2014) and thickness values (De Rosa et al. 2018), yet without compromising membrane permeability (Weber et al. 2014). Furthermore, the presence of hydroperoxide lipids in membranes also lowers the energy barrier for membrane electroporation (Rems et al. 2019; Yusupov et al. 2017; Corvalán et al. 2020).

It is well recognized that ordered lipid domains in plasma membranes are involved in signaling, protein traffic, endo- and exocytosis (Simons and Ikonen 1997; Lingwood and Simons 2010; Sezgin et al. 2017) and that oxidized phospholipids can induce phase separation in model lipid vesicles. Lipid-protein domains presented in the cytoplasmic membranes, also known as rafts, are enriched in sphingolipids and cholesterol. Any external stimuli changing rafts' features may compromise cellular homeostasis. Although membrane composition/bioactivity relationships remain elusive, it was shown that the existence of rafts in plasma membranes correlates well with the observation of liquid-ordered domains (Lo) in model membranes (Baumgart et al. 2007) of ternary mixtures composed of a high-melting transition temperature lipid, a low-melting transition temperature lipid and cholesterol. Remarkably, in-situ photo-oxidation of 1-palmitoyl-2-oleoyl-*sn*-glycero-3-phosphocholine (POPC)/1,2-dipalmitoyl-*sn*-glycero-3-phosphocholine (DPPC)/cholesterol giant unilamellar vesicles resulted in Lo—liquid disordered (Ld) phase coexistence (Haluska et al. 2012). It has also been recently demonstrated that the replacement of POPC by its hydroperoxide counterpart (POPC-OOH) in these vesicles drives Lo-Ld phase separation, whereas the gradual replacement of POPC-OOH by a lipid with an oxidized shortened chain (mimicking the progress of a photochemical reaction) promotes lipid mixing followed by an increase in membrane permeability (Tsubone et al. 2019b) (Fig. 7). The initial Lo-Ld phase segregation attributed to POPC hydroperoxidation can be related to (1) its larger mean molecular area occupied in the bilayer compared with that of the non-oxidized species (POPC-OOH ~78 Å > POPC ~65 Å) (De Rosa et al. 2018) and (2) the shorter tail lengths, rendering them 20% thinner (De Rosa et al. 2018). Therefore, the lipid packing favors lipid demixing and Ld-Lo phase coexistence. In the case of truncated lipids, the chains' lower packing parameter is in agreement with domain disruption and contributes to membrane permeabilization. Indeed, results from molecular dynamics simulation showed that truncated lipids pack well with cholesterol in model membranes, favoring lipid mixing (Khandelia et al. 2014).

3 The Effect of Light-Induced Oxidation on Skin Lipids

The consequences of light exposure to the lipids present in the dermis and epidermis have been studied by many authors since the 1970s. Contradictory results are the rule rather than the exception, which is a consequence of different irradiation regimes, varied biological responses of individuals and the use of different experimental techniques. Here we provide an overview of this extensive literature.

The skin performs several fundamental functions in the human organism, protecting internal tissues against several processes such as water loss, invasion by organisms (microbiota), reaction with pollutants and the effects of solar radiation. The outermost protective layer of the skin, called stratum corneum (SC), is similar to a brick wall. The "bricks" are proteinaceous corneocytes, which essentially are cross-linked supramolecular arrangements of keratin fibrillar macro-proteins. These structures are inserted and covalently linked in a "cement" made of specialized lipids, which self-organize in lamellar phases. The esterification of hydroxylated ceramides to involucrin and other corneocyte proteins gives rigidity to the lipid envelope and the whole SC, providing great strength and chemical resistance to the skin (Madison et al. 1987; Elias 1991) (Fig. 8).

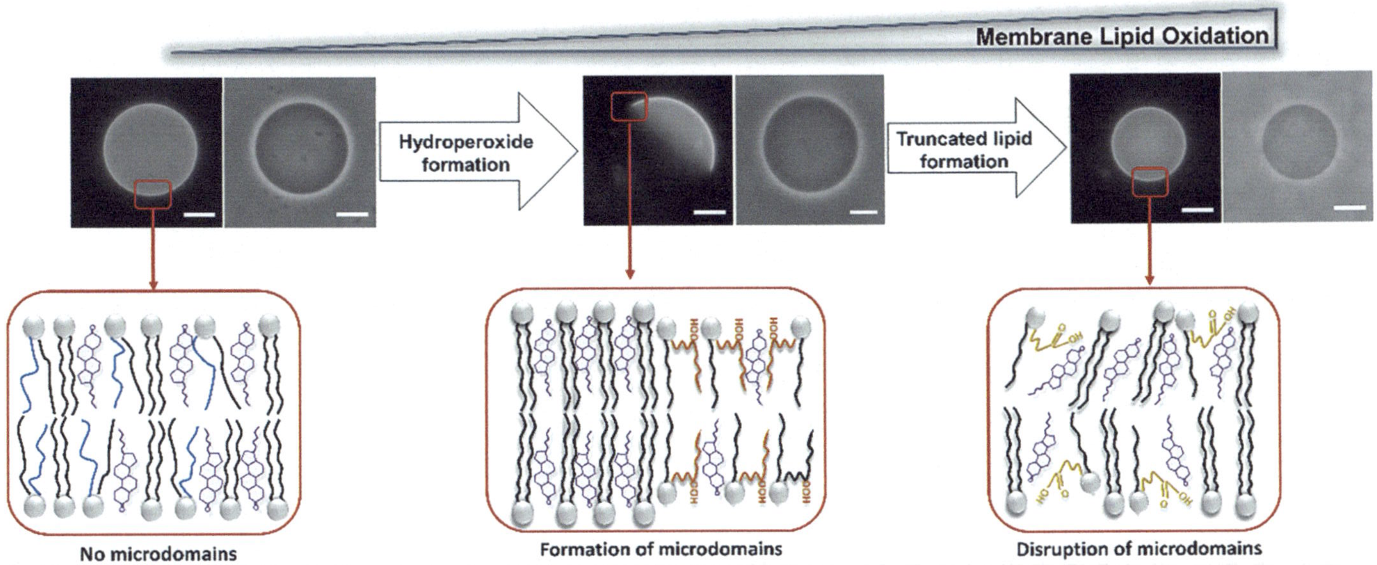

Fig. 7 Formation and disruption of microdomains under photooxidation of 1-palmitoyl-2-oleoyl-*sn*-glycero-3-phosphocholine (POPC)/1,2-dipalmitoyl-*sn*-glycero-3-phosphocholine (DPPC)/cholesterol model membranes. From left to right: initially, the membrane composed of mixed non-oxidized lipids displays a homogeneous phase. During the oxidation process, hydroperoxide lipids are formed and favor micro-scaled liquid-ordered (Lo)—liquid disordered (Ld) phase coexistence. Hydroperoxides lipids may subsequently be transformed into oxidized lipids with shortened chains, leading to lipid rearrangement/mixing that results in a single homogeneous phase. Scale bar represents 10 μm. This figure was reproduced from Tsubone et al. 2019, with permission from Elsevier

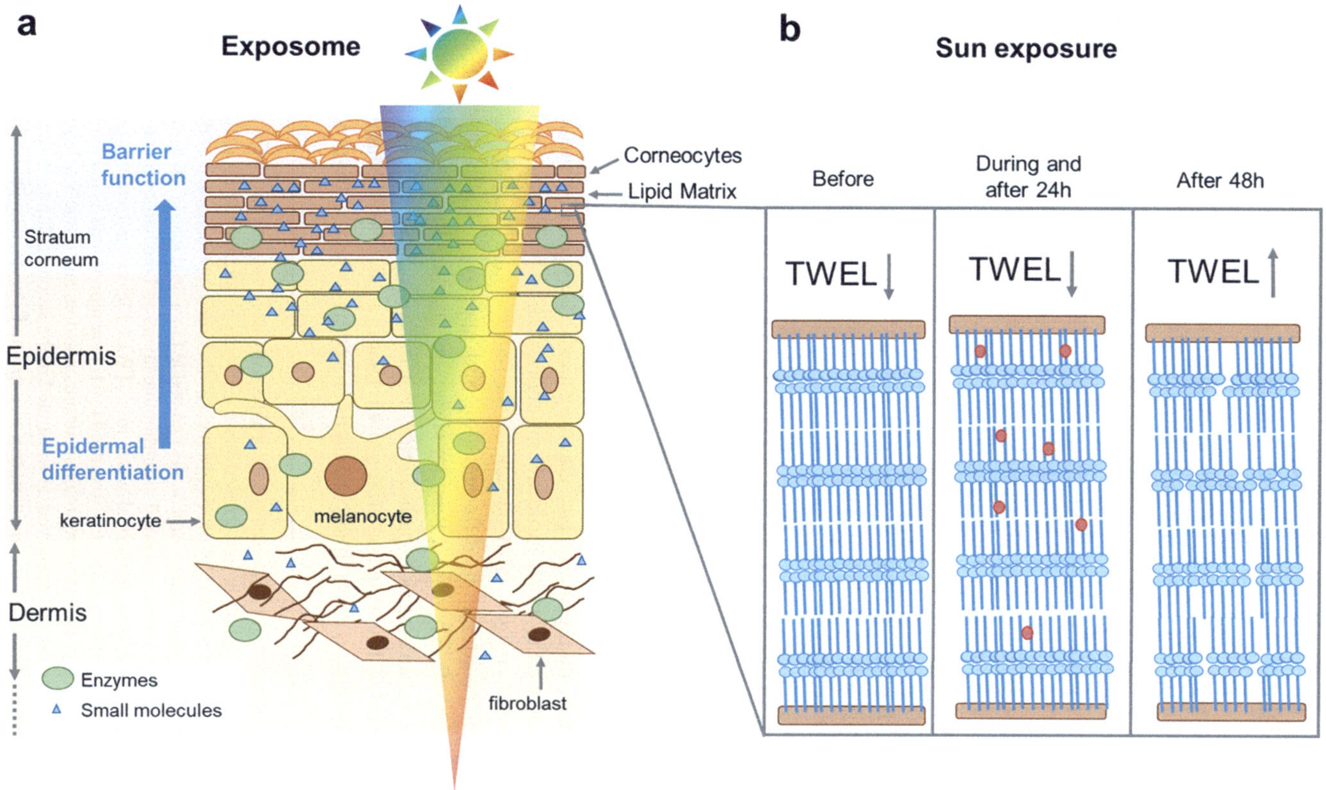

Fig. 8 Skin structure and effect of the exposome in its barrier function. (**a**) Schematic representation of the skin, showing the general features of the dermis and the epidermis, displaying characteristic structures of these layers, emphasizing the main cell types (fibroblasts, keratinocytes/corneocytes and melanocytes) and portraying the concentration gradient (increasing concentration in the direction of the external media) of molecular and enzymatic anti-oxidants. Differentiation of epithelium cells (mainly keratinocytes, but also including melanocytes) leads to the formation of the stratum corneum (SC), which offers a chemical and

The lipids present in the skin, which contains a rich and unique mixture of unsaturated and saturated lipids, vary considerably across the thin epidermal layer. The basal layer of the epidermis (deepest layer) is mostly made of keratinocytes (KCs) and its lipid composition is very similar to that of a cultured monolayer of KCs, which is made of phospholipids (70%, by mass), cholesterol (13%) and triacylglycerides (TAG, 11%). KCs differentiate, transforming themselves and the surrounding mixture of lipids, which includes lipids that are excreted by the sebum glandule. This phenomenon drastically changes the lipid composition of the SC to a mixture of ceramides (about 50%, mostly unsaturated), free fatty acids (10–20%, mostly saturated), and cholesterol (25%), with small amounts of cholesterol esters and sulfates. These lipids, which are differ significantly in terms of chain size, unsaturation and format, perfectly fit and self-organize into very dense orthorhombic lamellar phases. Ceramides found in the SC are highly enriched in linoleic acid (20–30%), an essential fatty acid. The presence of linoleic acid in the skin is key to maintain barrier function, and a deficiency of linoleic acid will most certainly affect this function. Besides the barrier function, the SC also works as a very sensitive sensor that responds to different types of environmental challenges (Harding 2004; Elias and Menon 1991).

Lipid peroxidation occurs as a consequence of natural processes in the skin, but it is greatly amplified by the interaction of the skin with challenge-factors present in the environment (e.g., light, pollution, microorganisms), jointly called the exposome (Krutmann et al. 2017). The oxidation of lipids in the skin can impair SC functions and induce photoaging, as well as several diseases such as skin cancer (Yamawaki et al. 2019; Lippman 1985). In addition to direct damage to the structures responsible for maintaining the integrity of the skin, reactive lipid peroxidation products can also cause cell death, mutagenicity and carcinogenesis. Oxygenated products of lipid peroxidation, as well as small-molecule initiators such as hydrogen peroxide, superoxide and singlet oxygen, also participate in redox signal transduction cascades, control of cell proliferation, differentiation and apoptosis (Kalinich et al. 2000; Barrera et al. 2004). Therefore, it is not surprising that lipid peroxidation products are implicated in several skin pathologies. For example, acne patients accumulate squalene peroxides, favoring bacterial growth (Sarici et al. 2009). Malonaldehyde, a short aldehyde derived from lipid peroxidation, seems to be a widespread marker of

---

Fig. 8 (continued) physical barrier against external agents and prevents loss of water molecules. The effect of external agents is exemplified here solely by sunlight; however, the SC protects humans against all injuries posed by the exposome. (**b**) Enlarged representation of the main futures of the SC, consisting of dead cells (corneocytes) embedded in a lipid matrix. In the preserved skin, the lipids are arranged in densely ordered orthorhombic layers and the barrier function is preserved as quantified by the low trans-epidermal water loss (TWEL) level (first column). Excessive sun exposure oxidizes the lipids in the matrix (oxidized lipids here represented as red circles), increasing fluidity of the lipid matrix and affecting fracture strain, cohesion, and tissue's delamination energy. Yet, the barrier function is still preserved, as illustrated in the second column. A few days after sun exposure, the reconstitution of the lipid matrix is corrupted due to damaged keratinocytes and melanocytes (third column). These damaged cells can no longer reset the proper lipid composition of the SC, which becomes leaky to water and other small molecules, thus increasing TWES level

several skin disorders, such as vitiligo, psoriasis and other dermatitis (Speeckaert et al. 2018; Szél et al. 2019). In fact, the lack of barrier function in psoriasis patients has been connected to an expressive decrease in the proportion of long-chain oxidizable ceramides (Smirnov et al. 2019).

Excess of sun exposure is the most important factor that can overflow the network of redox homeostasis in the skin. Upon exposure to solar radiation and depending on the intrinsic electronic properties of endogenous chromophores, photosensitizer reactive excited states will be formed (de Assis et al. 2021). The resulting formation of free radicals and reactive oxidants may deplete molecular and enzymatic antioxidants, leading to oxidation of glycerophospholipids, sphingolipids, unsaturated fatty acids and cholesterol, and accumulation of damaged biomolecules (Thiele et al. 1998). While electron paramagnetic resonance (EPR) has been used to qualify and quantify radicals generated in the skin after light exposure (Lohan et al. 2015), photoinduced lipid oxidation products are detected in human skin and associated with various skin diseases (Schalka et al. 2021).

Reactive aldehydes and lipid peroxidation end products (ALEs) (Tonolli et al. 2017; Vistoli et al. 2013) can react with other molecules forming advanced glycation end products (AGEs) (Gkogkolou and Böhm 2012; Delgado-Andrade 2016). The formation of AGE, ALE, carbonylated proteins and other oxidative end-products severely affects the structure and function of the skin. In KCs and melanocytes, the accumulation of these products is connected to a decrease in metabolic efficiency and cell aging, potentially leading to genomic instability (Martins et al. 2017). It is important to mention that photoinduced lipid oxidation in the skin is not a peculiarity of the much-feared ultraviolet (UV) radiation, being instead mainly caused by visible light, which deeply penetrates the dermis (de Assis et al. 2021). Indeed, visible light is responsible for more than 50% of the free radicals generated in the skin during sun exposure (Lohan et al. 2016). It is also noteworthy that short and controlled exposure to sunlight has many benefits (de Assis et al. 2021), including the production of vitamin D, which prevents autoimmune, bone, cancer and cardiovascular diseases. In this case, 7-dehydrocholesterol absorbs ultraviolet B photons that reach the skin, leading to its transformation to previtamin D3, which is rapidly converted to vitamin D3 (Holick 2004). This photochemical event is driven by an electrocyclic conrotatory ring-opening reaction. This is not an oxidation event but is certainly one of the most important photochemical transformations of lipids in human skin (de Assis et al. 2021).

As the first line of skin protection, the SC experiences several changes after exposure to sunlight. Oxidation of SC lipids was studied by molecular dynamic simulations. The behavior of SC oxidized lipids within a bilayer was similar to that of oxidized lipids in less complex model systems such as giant unilamellar vesicles, displaying a significant increase in the permeability of reactive oxygen species and consequently creating a self-feeding oxidation cycle (Yadav et al. 2019). Skin biopsies of human subjects indicate a significant decrease in the level of free cholesterol 24 h after UV exposure (Rauschkolb et al. 1967). Experiments in model membranes indicate that cholesterol scavenges skin oxidants, protecting other SC lipids from photoinduced oxidation. Cholesterol is present at high concentrations in the lamellar lipids of the SC (up to 50 mol%), and in combination with a low bond dissociation energy (BDE) value of 277 kJ mol^{-1} for the weakest C-H bond, effectively competes with the peroxidation of other polyunsaturated lipids such as the ceramides. Moreover, endoperoxide derivatives of cholesterol are relatively stable compared with other lipid hydroperoxides and other oxidation products (Lasch et al. 1997). Furthermore, the endoperoxide derivatives of cholesterol do not destabilize the lipid packing and cause little disturbance in lipid-lipid interactions in membrane models (Haluska et al. 2012).

SC lipids are also rich in sebum-delivered squalene, which is a highly oxidizable lipid (Shimizu et al. 2019). The second-order rate constant of singlet oxygen quenching by squalene ($3 \times 10^6\ M^{-1}\ s^{-1}$) is two orders of magnitude larger than that of another easily oxidized polyunsaturated lipid, ethyl linoleate ($3 \times 10^4\ M^{-1}\ s^{-1}$) (Kohno et al. 1995). Mudiyanselage and co-workers have shown that squalene was depleted from the skin in a light-dose-dependent manner, yielding a mixture of squalene monohydroperoxide isomers, and suggesting that squalene monohydroperoxides could be used as markers of photooxidative SC stress (Ekanayake Mudiyanselage et al. 2003). Squalene monohydroperoxide is cytotoxic to KCs and can trigger cytokine release, indicating that it is one of the signaling molecules that intensify the inflammatory responses connected to sun exposure (Uchino et al. 2002).

α-Tocopherol (vitamin E; E' = 0.48 V *vs.* NHE (Jovanovic et al. 1996)), the most abundant lipophilic antioxidant in the SC, is present at high concentrations in the deepest layers of SC, with a decreasing concentration gradient towards the skin surface. This antioxidant protects the propagation of lipid peroxidation reactions by forming a relatively stable α- tocopheroxyl radical. A single sub-erythemogenic UV dose (i.e., a dose that barely gives a clear redness response) decreases α-tocopherol concentrations by *ca.* 50% (Thiele et al. 1998). Yet, the redox signaling networks in the skin can quickly replenish this antioxidant by activating several antioxidant enzymatic routes (Thiele et al. 1998; Stoyanovsky et al. 1995; Podda and Grundmann-Kollmann 2001). UV and blue-violet irradiation also reduce the concentration of skin carotenoids (20–40%) (Vandersee et al. 2015). This decrease does not occur during the period of light exposure itself, but happens instead with *ca.* 90 min delay, excluding the possibility of the depletion occurring due to direct reactions with excited-state species, such as singlet oxygen (Darvin et al. 2006). Not only that, the depletion of carotenoids parallels the build-up of free radicals in the skin (Meinke et al. 2015). These data further endorse that the depletion occurs due to the scavenging of radicals, rather than due to the reaction with singlet oxygen. Therefore, both α-tocopherol and carotenoids seem to work as sacrificial antioxidants in the skin exposed to sunlight.

In general, topical application of antioxidants as dermo-cosmetics or consumption through diet is recommended to protect skin from the effects of light exposure (Afaq et al. 2005). Current epidemiologic studies show that coffee consumption and increased caffeine intake are associated with reduced melanoma risk (Song et al. 2012). Although caffeine itself is not a good antioxidant, its metabolites, methyl urates, are good electron donors (E' < 0.9 V *vs.* NHE). Due to their partition into skin and eyes, these compounds may successfully deactivate excited photosensitizer states and scavenge reactive free radicals, protecting tissues from oxidative stress (Scurachio et al. 2016). Likewise, a diet rich in polyunsaturated fatty acids (olive oil and fish) is associated with a lower risk of severe photoaging (Latreille et al. 2012) and a protective effect against advanced age-related macular degeneration in the eyes (Seddon 2001). This observation suggests that ω-3 lipids with C-H BDE as low as 267 kJ mol^{-1} act as sacrificial antioxidants, protecting sensitive protein structures in the skin and eye from light-induced oxidation (Cardoso et al. 2012).

Not all SC properties are affected by light exposure. Stiffness, for example, which is dependent on the structure of the keratin fibers, is kept unchanged upon light irradiation. However, several other SC properties, such as fracture strain, cohesion, and tissue's delamination energy significantly decrease, while lipid fluidity of the SC significantly increases. These mechanical changes have been correlated with oxidative-induced molecular changes in the lipids found in the orthorhombic lamella and the corneo-desmosomes (Biniek et al. 2012). Lipid packing and order are very important for the function of the multilayer orthorhombic lamellar phases of the SC. The organization of these phases is a direct consequence of how well the lipids fit each other based on specific lipid-lipid intermolecular interactions, which are highly disturbed by oxidation (Tsubone et al. 2021).

One of the best and easiest ways to evaluate epidermal barrier function, which is a direct indication of how intact is the SC, relies on measuring trans-epidermal water loss (TEWL). Avoiding water leakage from the body to the exterior medium is such a challenging function of the SC that small disturbances in the organization of the lamellar lipid structure can be detected by this method. Several authors have shown that UV radiation causes a significant increase in TEWL, both in humans and in laboratory animals. However, barrier function was shown to remain stable after reasonable sunlight doses and only showed an expressive decrease 48–72 h after photodamage. Early results indicate that the alteration of lipid composition associated with the increase in the TEWL was not due to the photochemical reactions *per se*, but instead due to impairments in lipid repositioning (Black and Rauschkolb 1971). Interestingly, the intrinsically aged epidermis shows less pronounced irradiation-induced alterations in barrier function (i.e., displays smaller increases in TEWL) (Haratake et al. 1997a). SC lipids, including ceramide subfractions, were shown to have their concentration increased a few days after light exposure (Lehmann et al. 1991). SC changes, which are connected to the increase in TWEL seem to be correlated with KC hyperproliferation (Haratake et al. 1997b). By evaluating the rate of TEWL increase, lipid production and keratinocyte functionality/differentiation in the different layers of the epidermis, Lehmann and co-workers proposed that the appearance of the irradiation-induced barrier abnormality (TEWL increase) is mainly caused by the photoinduced damage in KCs at the basal layers. Extensively damaged KCs accumulate at the SC/stratum granulosum interface, resulting in abnormal lamellar structures in the lower SC (Lehmann et al. 1991) (Fig. 8).

Definitely, excess sunlight exposure affects the viability of both KCs, melanocytes and fibroblasts. These cells undergo an extensive change in their lipid composition upon chronic exposure to sunlight (Gruber et al. 2019), favoring the accumulation of hydroperoxides and truncated lipids. Gruber and coworkers identified 173 oxidized phospholipids in irradiated KCs, being mostly hydroperoxides, truncated lipid aldehydes and carboxylic acids (Gruber et al. 2012). As mentioned earlier in this chapter, these compounds are formed both by free radicals or by singlet oxygen (Niki 2015) and cause severe changes in the biophysical properties of the membranes, including area per lipid, phase separation dynamics and membrane permeability (Tsubone et al. 2019b). Changes in the lipid composition of irradiated KCs are critical because they affect the lipid composition of the SC and alter several of the fundamental properties of the human skin (Holleran et al. 1997).

4 Photosensitized Lipid Oxidation in Photodynamic Therapy

Photosensitization reactions play a fundamental role in our interaction with the Earth's environment, as exemplified in the previous section by the effects of sunlight on the skin. The basic concepts of these reactions have been long ago applied to the medical field, in a procedure known as photodynamic therapy (PDT), which can treat a variety of malignant tumors and many non-oncological diseases such as age-related macular degeneration, psoriasis and arthritis, additionally to photoinactivating viruses and bacteria. As mentioned above, the mechanisms of photosensitized oxidation start with photosensitizers absorbing light and forming excited states. These excited species trigger reactions that modify the structure and function of many biomolecules, including lipids, and consequently affect biological homeostasis and may lead to cell death. As a result, PDT uses light and photosensitizers to induce localized redox misbalances and to reduce the viability of diseased cells (Castano et al. 2005; Agostinis et al. 2011; Tardivo et al. 2005).

Because photosensitizers act as drugs only in tissues that are properly illuminated, PDT produces results with fewer side effects than conventional drugs. These reduced side effects, together with its extraordinary cosmetic outcome, are its major advantage compared with other treatment modalities because the structure of tissues is usually preserved during the PDT treatment (Tardivo et al. 2015a; Tardivo et al. 2006). However, commercially available drugs and specifically designed illumination devices can make the treatment very expensive. To make this technology available for lower-income populations, Tardivo and co-workers combined home-made light sources to widely available and inexpensive photosensitizer molecules, such as methylene blue (Tardivo et al. 2004). The results obtained are remarkable, with hundreds of diabetic patients having their peripheral parts preserved in the public medical services in Brazil, without any cost to the patients (Tardivo et al. 2015b) and demonstrating that PDT can indeed be adopted by healthcare systems worldwide.

Understanding the details of the mechanisms triggering photoinduced cell death is critical to the development of more efficient photosensitizers. Even though several parameters, such as lower tendency of photosensitizer aggregation, higher absorption in the red region of the electromagnetic spectrum and higher efficiency of generation of singlet oxygen are important for the optimization of PDT efficiency (Tardivo et al. 2005), structure-activity studies indicate that the yield of reactive photooxidants generated during irradiation is less important that two other factors, namely: intracellular location and affinity to lipid membranes (Bacellar et al. 2015). When the role of intracellular location and singlet oxygen generation quantum yield were compared in terms of the final PDT outcome (cell death), it became evident that photosensitizer mitochondrial localization was significantly more relevant than their singlet oxygen generation quantum yield (Oliveira et al. 2011). Further studies indicated that damage in lysosomes was even more detrimental than a similar level of mitochondrial damage in terms of phototoxicity because the former causes a delayed and more effective cell death by blocking autophagy (Tsubone et al. 2017). Recent evidence indicates that parallel damage in mitochondria and lysosomes further amplifies cell death efficiency, since autophagy is activated by mitochondrial damage and at the same time is blocked by lysosomal damage, causing cell death at very a gentle condition in terms of photosensitizer concentration and light dose (Martins et al. 2018).

In all the mentioned examples, the extent of membrane interaction and membrane photodamage are critical to the initiation and progression of cell death. The reasons behind this fact were unclear for a long time, since the species generated during PDT (e.g., singlet oxygen) diffuse and can react with targets further apart (as much as 100 nm away in water), while lipid membranes are only 5 nm thick. The explanation came with the understanding of the molecular details of membrane permeabilization, as described in the previous sections. Indeed, photosensitizers can only induce radical-mediated lipid oxidation and membrane permeabilization by being in physical proximity with the membrane, so that their excited states may directly collide and react with lipids. In this way, photosensitizers can transform lipids into peroxyl and alkoxyl radicals, which suffer alkoxyl β-scission reaction and form lipid truncated aldehydes, which in turn allow membranes to leak (Bacellar et al. 2018). This effect may be amplified by lipid hydroperoxides (e.g., cholesterol hydroperoxides) undergoing spontaneous or protein-mediated transfer from one membrane to another (Girotti 2008). Although several membrane properties are affected by the formation of other lipid oxidation products, membrane permeabilization is fatal, fully compromising the homeostasis of cells and organelles.

For decades, researchers have focused on a single action mechanism of PDT, namely singlet oxygen generation. We have recently proposed that direct reactions with excited photosensitizers must be considered for the development of more efficient photosensitizers, meaning that other factors such as sub-cellular location, membrane binding and excited-state reactivity must be taken into account to optimize photosensitizer efficiency (Bacellar et al. 2018). From a singlet oxygen generation perspective, photosensitizer photobleaching decreases photodynamic efficiency. This is the case of radical-mediated peroxidation of protoporphyrin IX (post-irradiation), which contributes to protoporphyrin IX bleaching and decreases singlet oxygen generation. Interestingly, this effect can be protected by the presence of another radical (nitric oxide, NO) acting as a chain-breaking antioxidant (Niziolek et al. 2007). Yet, our recent work breaks the paradigm that photobleaching is necessarily associated with a decrease in photodynamic efficiency (Bacellar et al. 2018; Tasso et al. 2019). When we compared the efficiency of different compounds having similar membrane affinity, photosensitizer redox properties were shown to play a decisive role: the compounds that are stronger oxidants are the ones inducing the stronger membrane permeabilization, yet in parallel undergoing the greatest photobleaching. Therefore, relevant photodamage that drastically changes cell fate is the precise result of photosensitizer location and chemical reactivity, justifying the search for molecular-specific oxidative photodamage. While photobleaching may indeed be undesirable when singlet oxygen generation is the only mechanism of action, in conditions where excited photosensitizers trigger photooxidation, it may simply be a consequence of desired photosensitizer reactivity. Therefore, we suggest that photosensitizer regeneration strategies should be exploited as an effective tool to maximize and extend the effects of photosensitized oxidations (Tasso et al. 2019). It is still unclear how regeneration can be achieved in such a hostile oxidant environment, but we believe that solving this challenge could lead to substantial improvements in photosensitizer efficiency.

5 Conclusions

Although being fundamental to life on Earth, excessive light inflicts numerous detrimental biological effects, e.g., skin aging, cancer, inhibition of photosynthesis and food waste. The main mechanisms of photoinduced lipid oxidation are enabled by the higher reactivity of excited states, which can oxidize double bonds and start lipid peroxidation reactions even in the absence of polyunsaturated lipids. By providing an in-depth analysis of the major reaction pathways (i.e., direct reactions by photosensitizers *vs.* singlet-oxygen mediated reaction), we connected molecular mechanisms to the impact of photoinduced lipid oxidation in membrane biophysics (e.g., membrane permeabilization, thinning, and domain reorganization). We progressed the text to evaluate the consequences of these reactions to human skin and for photomedicine, showing that photoinduced lipid oxidation is key to all these processes. We anticipate that understanding these molecular mechanisms in detail is essential to improve the treatment of several diseases by light technologies, as well as for the development of comprehensive photoprotection strategies.

Acknowledgments Fundação de Amparo à Pesquisa do Estado de São Paulo (FAPESP) is acknowledged for financial support (CEPID-REDOXOMA 2013/07937-8 and grant 2017/01189-0). Conselho Nacional de Desenvolvimento Científico e Tecnológico (CNPq) is acknowledged for financial support (grant 309212/2019-7) and for research fellowships (D.R.C., M.S.B., and R.I.).

References

Afaq F, Adhami VM, Mukhtar H (2005, Apr 1) Photochemoprevention of ultraviolet B signaling and photocarcinogenesis. Mutat Res Mol Mech Mutagen [Internet] 571(1–2):153–173. Available from: https://linkinghub.elsevier.com/retrieve/pii/S0027510704004920

Agostinis P, Berg K, Cengel KA, Foster TH, Girotti AW, Gollnick SO et al (2011) Photodynamic therapy of cancer: an update. Am Cancer Soc 61:250–281

Alberti MN, Orfanopoulos M (2010, Aug 16) Unraveling the mechanism of the singlet oxygen Ene reaction: recent computational and experimental approaches. Chem – A Eur J [Internet]. 16(31): 9414–9421. Available from: http://doi.wiley.com/10.1002/chem.201000752

Anthonymuthu TS, Kim-Campbell N, Bayır H (2017, Aug) Oxidative lipidomics. Curr Opin Crit Care [Internet] 23(4):251–256. Available from: http://journals.lww.com/00075198-201708000-00002

Bacellar IOL, Pavani C, Sales EM, Itri R, Wainwright M, Baptista MS (2014) Membrane damage efficiency of phenothiazinium photosensitizers. Photochem Photobiol 90(4):801–813

Bacellar IOL, Tsubone TM, Pavani C, Baptista MS (2015, Aug 31) Photodynamic efficiency: from molecular photochemistry to cell death. Int J Mol Sci [Internet]. 16(9):20523–20559. Available from: http://www.mdpi.com/1422-0067/16/9/20523/

Bacellar IOL, Oliveira MC, Dantas LS, Costa EB, Junqueira HC, Martins WK, et al. (2018, Aug 10) Photosensitized membrane Permeabilization requires contact-dependent reactions between photosensitizer and lipids. J Am Chem Soc [Internet].. 140(30):9606–9615. Available from: http://pubs.acs.org/doi/10.1021/jacs.8b05014

Bacellar IOL, Cordeiro RM, Mahling P, Baptista MS, Röder B, Hackbarth S (2019) Oxygen distribution in the fluid/gel phases of lipid membranes. Biochim Biophys Acta – Biomembr 1861(4):879–886

Baptista MS, Cadet J, Di Mascio P, Ghogare AA, Greer A, Hamblin MR, et al. (2017, Jul) Type I and type II photosensitized oxidation reactions: guidelines and mechanistic pathways. Photochem Photobiol [Internet]. 93(4):912–919. Available from: http://doi.wiley.com/10.1111/php.12716

Barrera G, Pizzimenti S, Dianzani MU (2004, Sep) 4-hydroxynonenal and regulation of cell cycle: effects on the pRb/E2F pathway. Free Radic Biol Med [Internet]. 37(5):597–606. Available from: https://linkinghub.elsevier.com/retrieve/pii/S0891584904004277

Baumgart T, Hammond AT, Sengupta P, Hess ST, Holowka DA, Baird BA, et al. (2007, Feb 27) Large-scale fluid/fluid phase separation of proteins and lipids in giant plasma membrane vesicles. Proc Natl Acad Sci [Internet] 104(9):3165–3170. Available from: http://www.pnas.org/cgi/doi/10.1073/pnas.0611357104

Biniek K, Levi K, Dauskardt RH (2012, Oct 16) Solar UV radiation reduces the barrier function of human skin. Proc Natl Acad Sci [Internet]. 109(42):17111–17116. Available from: http://www.pnas.org/cgi/doi/10.1073/pnas.1206851109

Black HS, Rauschkolb EW (1971, May)) Effects of light on skin lipid metabolism. J Invest Dermatol [Internet]. 56(5):387–391. Available from: https://linkinghub.elsevier.com/retrieve/pii/S0022202X15479977

Bochkov V, Gesslbauer B, Mauerhofer C, Philippova M, Erne P, Oskolkova OV (2016, Dec) Pleiotropic effects of oxidized phospholipids. Free Radic Biol Med [Internet]. Available from: http://linkinghub.elsevier.com/retrieve/pii/S0891584916311352

Boonnoy P, Jarerattanachat V, Karttunen M, Wong-ekkabut J (2015, Dec 17) Bilayer deformation, pores, and Micellation induced by oxidized lipids. J Phys Chem Lett [Internet]. 6(24): 4884–4888. Available from: http://pubs.acs.org/doi/10.1021/acs.jpclett.5b02405

Braslavsky SE (2007) Glossary of terms used in photochemistry. Pure Appl Chem 79:293–465

Cardoso DR, Olsen K, Møller JKS, Skibsted LH (2006, Jul) Phenol and terpene quenching of singlet- and triplet-excited states of riboflavin in relation to light-struck flavor formation in beer. J Agric Food Chem [Internet]. 54(15):5630–5636. Available from: https://pubs.acs.org/doi/10.1021/jf060750d

Cardoso DR, Libardi SH, Skibsted LH (2012) Riboflavin as a photosensitizer. Effects on human health and food quality. Food Funct [Internet]. 3(5):487. Available from: http://xlink.rsc.org/?DOI=c2fo10246c

Castano AP, Demidova TN, Hamblin MR (2005) Mechanisms in photodynamic therapy: part two – cellular signaling, cell metabolism and modes of cell death. Photodiagn Photodyn Ther 2(1): 1–23

Catalá A (2009, Jan) Lipid peroxidation of membrane phospholipids generates hydroxy-alkenals and oxidized phospholipids active in physiological and/or pathological conditions. Chem Phys Lipids [Internet]. 157(1):1–11. Available from: https://linkinghub.elsevier.com/retrieve/pii/S0009308408003708

Corvalán NA, Caviglia AF, Felsztyna I, Itri R, Lascano R (2020, Aug 4) Lipid Hydroperoxidation effect on the dynamical evolution of the conductance process in bilayer lipid membranes: a condition toward criticality. Langmuir [Internet]. 36(30):8883–8893. Available from: https://pubs.acs.org/doi/10.1021/acs.langmuir.0c01243

Darvin ME, Gersonde I, Albrecht H, Sterry W, Lademann J (2006, May) In vivo Raman spectroscopic analysis of the influence of UV radiation on carotenoid antioxidant substance degradation of the human skin. Laser Phys [Internet]. 16(5):833–837. Available from: http://link.springer.com/10.1134/S1054660X06050148

de Almeida NEC, de Aguiar I, de Zawadzki A, Cardoso DR (2014, Dec 11) Kinetics and thermodynamics of 1-hydroxyethyl radical reaction with unsaturated lipids and Prenylflavonoids. J Phys Chem B [Internet]. 118(49):14278–14287. Available from: https://pubs.acs.org/doi/10.1021/jp509125b

de Assis LVM, Tonolli PN, Moraes MN, Baptista MS, de Lauro Castrucci AM (2021, Jun) How does the skin sense sun light? An integrative view of light sensing molecules. J Photochem Photobiol C Photochem Rev [Internet]. 47:100403. Available from: https://linkinghub.elsevier.com/retrieve/pii/S1389556721000022

Rosa R De, Spinozzi F, Itri R (2018, Nov) Hydroperoxide and carboxyl groups preferential location in oxidized biomembranes experimentally determined by small angle X-ray scattering: implications in membrane structure. Biochim Biophys Acta – Biomembr [Internet]. 1860(11): 2299–2307. Available from: https://linkinghub.elsevier.com/retrieve/pii/S0005273618301536

Delgado-Andrade C (2016) Carboxymethyl-lysine: thirty years of investigation in the field of AGE formation. Food Funct [Internet]. 7(1):46–57. Available from: http://xlink.rsc.org/?DOI=C5FO00918A

Doleiden FH, Fahrenholtz SR, Lamola AA, Trozzolo AM (1974, Dec) Reactivity of cholesterol and some fatty acids TOWARD singlet oxygen. Photochem Photobiol [Internet]. 20(6):519–521. Available from: http://doi.wiley.com/10.1111/j.1751-1097.1974.tb06613.x

Ekanayake Mudiyanselage S, Hamburger M, Elsner P, Thiele JJ (2003, Jun) Ultraviolet a induces generation of squalene monohydroperoxide isomers in human sebum and skin surface lipids in vitro and in vivo. J Invest Dermatol [Internet]. 120(6):915–22. Available from: http://www.ncbi.nlm.nih.gov/pubmed/12787115

Elias PM (1991, Jun) Epidermal barrier function: intercellular lamellar lipid structures, origin, composition and metabolism. J Control Release [Internet]. 15(3):199–208. Available from: https://linkinghub.elsevier.com/retrieve/pii/016836599190111P

Elias PM, Menon GK (1991) Structural and lipid biochemical correlates of the epidermal permeability barrier. Adv Lipid Res 24:1–26. Available from: https://linkinghub.elsevier.com/retrieve/pii/B9780120249244500055

Foote CS (1968) Mechanisms of photosensitized oxidation. Science (80-) [Internet]. 162(3857): 963–70. Available from: http://www.sciencemag.org/cgi/doi/10.1126/science.162.3857.963

Frankel EN (1984) Chemistry of free radical and singlet oxidation of lipids. Prog Lipid Res 23(4): 197–221

Garcez AS, Núñez SC, Baptista MS, Daghastanli NA, Itri R, Hamblin MR, et al. (2011) Antimicrobial mechanisms behind photodynamic effect in the presence of hydrogen peroxide.

Photochem Photobiol Sci [Internet]. 10(4):483–490. Available from: http://xlink.rsc.org/?DOI=C0PP00082E

Gardner HW (1989, Jan) Oxygen radical chemistry of polyunsaturated fatty acids. Free Radic Biol Med [Internet]. 7(1):65–86. Available from: http://linkinghub.elsevier.com/retrieve/pii/0891584989901020

Garrec J, Monari A, Assfeld X, Mir LM, Tarek M (2014, May 15) Lipid peroxidation in membranes: the peroxyl radical does not "float." J Phys Chem Lett [Internet]. 5(10):1653–1658. Available from: http://pubs.acs.org/doi/abs/10.1021/jz500502q

Girotti AW (1992, Apr) New trends in photobiology. J Photochem Photobiol B Biol [Internet]. 13(2):105–118. Available from: https://linkinghub.elsevier.com/retrieve/pii/1011134492850505

Girotti AW (2001) Photosensitized oxidation of membrane lipids: reaction pathways, cytotoxic effects, and cytoprotective mechanisms. J Photochem Photobiol B Biol 63(1–3):103–113

Girotti AW (2008, Mar) Translocation as a means of disseminating lipid hydroperoxide-induced oxidative damage and effector action. Free Radic Biol Med [Internet]. 44(6):956–968. Available from: https://linkinghub.elsevier.com/retrieve/pii/S089158490700809X

Gkogkolou P, Böhm M (2012, Jul 27) Advanced glycation end products. Dermatoendocrinol [Internet]. 4(3):259–270. Available from: http://www.tandfonline.com/doi/abs/10.4161/derm.22028

Gollnick K, Franken T, Schade G, Dörhöfer G (1970, Oct) Photosensitized oxygenation as a function of the triplet energy of sensitizers. Ann N Y Acad Sci [Internet]. 171-(1 International):89–107. Available from: http://doi.wiley.com/10.1111/j.1749-6632.1970.tb39307.x

Gruber F, Bicker W, Oskolkova OV, Tschachler E, Bochkov VN (2012, Jun 1) A simplified procedure for semi-targeted lipidomic analysis of oxidized phosphatidylcholines induced by UVA irradiation. J Lipid Res [Internet]. 53(6):1232–42. Available from: http://www.jlr.org/cgi/doi/10.1194/jlr.D025270

Gruber F, Kremslehner C, Narzt M-S (2019, Nov) The impact of recent advances in lipidomics and redox lipidomics on dermatological research. Free Radic Biol Med [Internet]. 144:256–265. Available from: https://linkinghub.elsevier.com/retrieve/pii/S0891584919301595

Hackbarth S, Röder B (2015) Singlet oxygen luminescence kinetics in a heterogeneous environment – identification of the photosensitizer localization in small unilamellar vesicles. Photochem Photobiol Sci [Internet]. 14(2):329–334. Available from: http://xlink.rsc.org/?DOI=C4PP00229F

Hackbarth S, Bornhütter T, Röder B (2016) Chapter 26. Singlet oxygen in heterogeneous systems. In: Singlet oxygen: applications in biosciences and nanosciences [Internet]. RSC, London, pp 27–42. Available from: http://ebook.rsc.org/?DOI=10.1039/9781782626992-00027

Halliwell B, Gutteridge JMC (2015) Free radicals in biology and medicine, 5th edn. Oxford University Press, Oxford

Haluska CK, Baptista MS, Fernandes AU, Schroder AP, Marques CM, Itri R (2012) Photo-activated phase separation in giant vesicles made from different lipid mixtures. Biochim Biophys Acta Biomembr 1818(3):666–672

Haratake A, Uchida Y, Mimura K, Elias PM, Holleran WM (1997a, Mar) Intrinsically aged epidermis displays diminished UVB-induced alterations in barrier function associated with decreased proliferation. J Invest Dermatol [Internet]. 108(3):319–323. Available from: https://linkinghub.elsevier.com/retrieve/pii/S0022202X15428325

Haratake A, Uchida Y, Schmuth M, Tanno O, Yasuda R, Epstein JH, et al. (1997b, May) UVB-induced alterations in permeability barrier function: roles for epidermal Hyperproliferation and Thymocyte-mediated response. J Invest Dermatol [Internet]. 108(5):769–775. Available from: https://linkinghub.elsevier.com/retrieve/pii/S0022202X15530224

Harding CR (2004, Jan) The stratum corneum: structure and function in health and disease. Dermatol Ther [Internet] 17(s1):6–15. Available from: http://doi.wiley.com/10.1111/j.1396-0296.2004.04S1001.x
Hill S, Lamberson CR, Xu L, To R, Tsui HS, Shmanai VV, et al (2012, Aug) Small amounts of isotope-reinforced polyunsaturated fatty acids suppress lipid autoxidation. Free Radic Biol Med [Internet]. 53(4):893–906. Available from: https://linkinghub.elsevier.com/retrieve/pii/S0891584912003395
Holick MF (2004, Dec 1) Sunlight and vitamin D for bone health and prevention of autoimmune diseases, cancers, and cardiovascular disease. Am J Clin Nutr [Internet] 80(6):1678S–1688S. Available from: https://academic.oup.com/ajcn/article/80/6/1678S/4690512
Holleran WM, Uchida Y, Halkier-Sorensen L, Haratake A, Hara M, Epstein JH, et al. (1997, Aug) Structural and biochemical basis for the UVB-induced alterations in epidermal barrier function. Photodermatol Photoimmunol Photomed [Internet]. 13(4):117–128. Available from: http://doi.wiley.com/10.1111/j.1600-0781.1997.tb00214.x
Howard JA, Ingold KU (1968, Feb) Self-reaction of sec-butylperoxy radicals. Confirmation of the Russell mechanism. J Am Chem Soc [Internet]. 90(4):1056–8. Available from: http://pubs.acs.org/doi/abs/10.1021/ja01006a037
Hu C, Wang M, Han X (2017, Aug) Shotgun lipidomics in substantiating lipid peroxidation in redox biology: methods and applications. Redox Biol [Internet]. 12:946–955. Available from: https://linkinghub.elsevier.com/retrieve/pii/S2213231717302227
Huvaere K, Cardoso DR, Homem-De-Mello P, Westermann S, Skibsted LH (2010) Light-induced oxidation of unsaturated lipids as sensitized by flavins. J Phys Chem B 114(16):5583–5593
Jovanovic SV, Steenken S, Hara Y, Simic MG (1996) Reduction potentials of flavonoid and model phenoxyl radicals. Which ring in flavonoids is responsible for antioxidant activity? J Chem Soc Perkin Trans [Internet]. 2(11):2497–2504. Available from: http://xlink.rsc.org/?DOI=p29960002497
Junqueira HC, Severino D, Dias LG, Gugliotti MS, Baptista MS (2002) Modulation of methylene blue photochemical properties based on adsorption at aqueous micelle interfaces. Phys Chem Chem Phys 4(11):2320–2328
Kalinich JF, Ramakrishnan R, McClain DE, Ramakrishnan N (2000, Jan 7) 4-Hydroxynonenal, an end-product of lipid peroxidation, induces apoptosis in human leukemic T- and B-cell lines. Free Radic Res [Internet]. 33(4):349–358. Available from: http://www.tandfonline.com/doi/full/10.1080/10715760000300891
Khandelia H, Loubet B, Olżyńska A, Jurkiewicz P, Hof M (2014) Pairing of cholesterol with oxidized phospholipid species in lipid bilayers. Soft Matter [Internet] 10(4):639–647. Available from: http://xlink.rsc.org/?DOI=C3SM52310A
Kohno Y, Egawa Y, Itoh S, Nagaoka S, Takahashi M, Mukai K (1995, Apr) Kinetic study of quenching reaction of singlet oxygen and scavenging reaction of free radical by squalene in n-butanol. Biochim Biophys Acta – Lipids Lipid Metab [Internet]. 1256(1):52–56. Available from: https://linkinghub.elsevier.com/retrieve/pii/000527609500005W
Krasnovsky AA, Kagan VE, Minin AA (1983, May 8) Quenching of singlet oxygen luminescence by fatty acids and lipids. FEBS Lett [Internet]. 155(2):233–236. Available from: http://doi.wiley.com/10.1016/0014-5793%2882%2980610-8
Krutmann J, Bouloc A, Sore G, Bernard BA, Passeron T (2017, Mar) The skin aging exposome. J Dermatol Sci [Internet] 85(3):152–161. Available from: https://linkinghub.elsevier.com/retrieve/pii/S0923181116308167
Lasch J, Schönfelder U, Walke M, Zellmer S, Beckert D (1997, Nov) Oxidative damage of human skin lipids. Biochim Biophys Acta – Lipids Lipid Metab [Internet]. 1349(2):171–181. Available from: https://linkinghub.elsevier.com/retrieve/pii/S0005276097000933
Latreille J, Kesse-Guyot E, Malvy D, Andreeva V, Galan P, Tschachler E, et al (2012, Sep 6) Dietary monounsaturated fatty acids intake and risk of skin photoaging. Soyer HP, editor. PLoS One [Internet]. 7(9):e44490. Available from: https://dx.plos.org/10.1371/journal.pone.0044490

Lehmann P, Hölzle E, Melnik B, Plewig G (1991, Jun) Effects of ultraviolet A and B on the skin barrier: a functional, electron microscopic and lipid biochemical study. Photodermatol Photoimmunol Photomed [Internet]. 8(3):129–34. Available from: http://www.ncbi.nlm.nih.gov/pubmed/1804292

Leonarduzzi G, Arkan MC, Başağa H, Chiarpotto E, Sevanian A, Poli G (2000, May) Lipid oxidation products in cell signaling. Free Radic Biol Med [Internet]. 28(9):1370–1378. Available from: https://linkinghub.elsevier.com/retrieve/pii/S0891584900002161

Libardo MDJ, Wang T-Y, Pellois J-P, Angeles-Boza AM (2017, Aug) How does membrane oxidation affect cell delivery and cell killing? Trends Biotechnol [Internet]. 35(8):686–690. Available from: https://linkinghub.elsevier.com/retrieve/pii/S0167779917300641

Lingwood D, Simons K (2010, Jan 1) Lipid rafts as a membrane-organizing principle. Science (80-) [Internet]. 327(5961):46–50. Available from: http://www.sciencemag.org/cgi/doi/10.1126/science.1174621

Lippman RD (1985, Jan) Rapid in vivo quantification and comparison of hydroperoxides and oxidized collagen in aging mice, rabbits and man. Exp Gerontol [Internet]. 20(1):1–5. Available from: https://linkinghub.elsevier.com/retrieve/pii/0531556585900038

Lohan S, Lauer A-C, Arndt S, Friedrich A, Tscherch K, Haag S, et al. (2015, Aug 19) Determination of the antioxidant status of the skin by in vivo-electron paramagnetic resonance (EPR) spectroscopy. Cosmetics [Internet]. 2(3):286–301. Available from: http://www.mdpi.com/2079-9284/2/3/286

Lohan SB, Müller R, Albrecht S, Mink K, Tscherch K, Ismaeel F, et al. (2016, May) Free radicals induced by sunlight in different spectral regions – in vivo versus ex vivo study. Exp Dermatol [Internet]. 25(5):380–385. Available from: http://doi.wiley.com/10.1111/exd.12987

Madison KC, Swartzendruber DC, Wertz PW, Downing DT (1987, Jun) Presence of intact intercellular lipid lamellae in the upper layers of the stratum Corneum. J Invest Dermatol [Internet]. 88(6):714–718. Available from: https://linkinghub.elsevier.com/retrieve/pii/S0022202X87902648

Martins WK, Gomide AB, Costa ÉT, Junqueira HC, Stolf BS, Itri R, et al. (2017, Jan) Membrane damage by betulinic acid provides insights into cellular aging. Biochim Biophys Acta – Gen Subj [Internet]. 1861(1):3129–3143. Available from: https://linkinghub.elsevier.com/retrieve/pii/S0304416516303932

Martins WK, Santos NF, Rocha CDS, IOL B, Tsubone TM, Viotto AC et al (2018) Parallel damage in mitochondria and lysosomes is an efficient way to photoinduce cell death. Autophagy 15: 259–279. https://doi.org/10.1080/15548627.2018

McQuaid R, Mrochen M, Vohnsen B (2016, Mar) Rate of riboflavin diffusion from intrastromal channels before corneal crosslinking. J Cataract Refract Surg [Internet]. 42(3):462–468. Available from: https://journals.lww.com/02158034-201603000-00016

Meinke MC, Müller R, Bechtel A, Haag SF, Darvin ME, Lohan SB, et al. (2015, Mar) Evaluation of carotenoids and reactive oxygen species in human skin after UV irradiation: a critical comparison between in vivo and ex vivo investigations. Exp Dermatol [Internet]. 24(3):194–197. Available from: http://doi.wiley.com/10.1111/exd.12605

Neff WE, Frankel EN, Weisleder D (1982, Nov) Photosensitized oxidation of methyl linolenate. Secondary products. Lipids [Internet]. 17(11):780–90. Available from: http://link.springer.com/10.1007/BF02535354

Neff WE, Frankel EN (1984, Dec) Photosensitized oxidation of methyl linolenate monohydroperoxides: Hydroperoxy cyclic peroxides, dihydroperoxides and hydroperoxy bis-cyclic peroxides. Lipids [Internet] 19(12):952–957. Available from: http://link.springer.com/10.1007/BF02534731

Nicolson GL, Ash ME (2014, Jun) Lipid replacement therapy: a natural medicine approach to replacing damaged lipids in cellular membranes and organelles and restoring function. Biochim Biophys Acta – Biomembr [Internet] 1838(6):1657–1679. Available from: https://linkinghub.elsevier.com/retrieve/pii/S0005273613004070

Niki E (2015, Jul 3) Lipid oxidation in the skin. Free Radic Res [Internet]. 49(7):827–834. Available from: http://www.tandfonline.com/doi/full/10.3109/10715762.2014.976213

Niziolek M, Korytowski W, Girotti AW (2007, Apr 30) Self-sensitized Photodegradation of membrane-bound Protoporphyrin mediated by chain lipid peroxidation: inhibition by nitric oxide with sustained singlet oxygen damage. Photochem Photobiol [Internet]. 81(2):299–305. Available from: http://doi.wiley.com/10.1111/j.1751-1097.2005.tb00187.x

Oliveira CS, Turchiello R, Kowaltowski AJ, Indig GL, Baptista MS (2011) Major determinants of photoinduced cell death: subcellular localization versus photosensitization efficiency. Free Radic Biol Med [Internet]. 51(4):824–833. Available from: https://doi.org/10.1016/j.freeradbiomed.2011.05.023

Pavani C, Uchoa AF, Oliveira CS, Iamamoto Y, Baptista MS (2009) Effect of zinc insertion and hydrophobicity on the membrane interactions and PDT activity of porphyrin photosensitizers. Photochem Photobiol Sci [Internet]. 8(2):233–40. Available from: http://www.ncbi.nlm.nih.gov/pubmed/19247516

Podda M, Grundmann-Kollmann M (2001, Oct) Low molecular weight antioxidants and their role in skin ageing. Clin Exp Dermatol [Internet]. 26(7):578–582. Available from: http://doi.wiley.com/10.1046/j.1365-2230.2001.00902.x

Quina FH, Silva GTM (2021, Sep) The photophysics of photosensitization: A brief overview. J Photochem Photobiol [Internet]. 7:100042. Available from: https://linkinghub.elsevier.com/retrieve/pii/S2666469021000270

Rauschkolb EW, Farrell G, Knox JM (1967, Dec) Effects of ultraviolet light on skin cholesterol. J Invest Dermatol [Internet]. 49(6):632–636. Available from: https://linkinghub.elsevier.com/retrieve/pii/S0022202X15474299

Reis A (2017, Jan) Oxidative Phospholipidomics in health and disease: Achievements, challenges and hopes. Free Radic Biol Med [Internet]. Available from: http://linkinghub.elsevier.com/retrieve/pii/S089158491730014X

Rems L, Viano M, Kasimova MA, Miklavčič D, Tarek M (2019, Feb) The contribution of lipid peroxidation to membrane permeability in electropermeabilization: a molecular dynamics study. Bioelectrochemistry [Internet]. 125:46–57. Available from: https://linkinghub.elsevier.com/retrieve/pii/S1567539418300938

Riske KA, Sudbrack TP, Archilha NL, Uchoa AF, Schroder AP, Marques CM et al (2009) Giant vesicles under oxidative stress induced by a membrane-anchored photosensitizer. Biophys J 97(5):1362–1370

Runas KA, Malmstadt N (2015) Low levels of lipid oxidation radically increase the passive permeability of lipid bilayers. Soft Matter [Internet]. 11(3):499–505. Available from: http://xlink.rsc.org/?DOI=C4SM01478B

Runas KA, Acharya SJ, Schmidt JJ, Malmstadt N (2016, Jan 26) Addition of cleaved tail fragments during lipid oxidation stabilizes membrane permeability behavior. Langmuir [Internet] 32(3): 779–786. Available from: http://pubs.acs.org/doi/abs/10.1021/acs.langmuir.5b02980

Russell GA (1957, Jul) Deuterium-isotope Effects in the Autoxidation of Aralkyl Hydrocarbons. Mechanism of the Interaction of Peroxy Radicals. J Am Chem Soc [Internet]. 79(14):3871–7. Available from: http://pubs.acs.org/doi/abs/10.1021/ja01571a068

Sankhagowit S, Wu SH, Biswas R, Riche CT, Povinelli ML, Malmstadt N (2014) The dynamics of giant unilamellar vesicle oxidation probed by morphological transitions. Biochim Biophys Acta Biomembr 1838(10):2615–2624

Sarici G, Cinar S, Armutcu F, Altınyazar C, Koca R, Tekin N (2009, Nov 23) Oxidative stress in acne vulgaris. J Eur Acad Dermatology Venereol [Internet] 24(7):763–767. Available from: http://doi.wiley.com/10.1111/j.1468-3083.2009.03505.x

Schalka S, Silva MS, Lopes LF, Freitas LM, Baptista MS (2021) The skin redoxome. J Eur Acad Dermatology Venereol. https://doi.org/10.1111/jdv.17780

Schweitzer C, Schmidt R (2003) Physical mechanisms of generation and deactivation of singlet oxygen. Chem Rev 103(5):1685–1757

Scurachio RS, Mattiucci F, Santos WG, Skibsted LH, Cardoso DR (2016, Oct) Caffeine metabolites not caffeine protect against riboflavin photosensitized oxidative damage related to skin and eye health. J Photochem Photobiol B Biol [Internet]. 163:277–83. Available from: https://linkinghub.elsevier.com/retrieve/pii/S1011134416301609

Seddon JM (2001, Aug 1) Dietary fat and risk for advanced age-related macular degeneration. Arch Ophthalmol [Internet]. 119(8):1191. Available from: http://archopht.jamanetwork.com/article.aspx?doi=10.1001/archopht.119.8.1191

Severino D, Junqueira HC, Gugliotti M, Gabrielli DS, Baptista MS (2003) Influence of negatively charged interfaces on the ground and excited state properties of methylene blue. Photochem Photobiol [Internet]. 77(5):459–468. Available from: http://www.ncbi.nlm.nih.gov/pubmed/12812286%5Cnhttp://doi.wiley.com/10.1562/0031-8655(2003)0770459IONCIO2.0.CO2

Sezgin E, Levental I, Mayor S, Eggeling C (2017, Mar 30) The mystery of membrane organization: composition, regulation and roles of lipid rafts. Nat Rev Mol Cell Biol [Internet]. 18(6): 361–374. Available from: http://www.nature.com/doifinder/10.1038/nrm.2017.16

Shimizu N, Ito J, Kato S, Eitsuka T, Saito T, Nishida H, et al. (2019, Dec 26) Evaluation of squalene oxidation mechanisms in human skin surface lipids and shark liver oil supplements. Ann N Y Acad Sci [Internet]. 1457(1):158–165. Available from: https://onlinelibrary.wiley.com/doi/abs/10.1111/nyas.14219

Siani P, de Souza RM, Dias LG, Itri R, Khandelia H (2016, Oct) An overview of molecular dynamics simulations of oxidized lipid systems, with a comparison of ELBA and MARTINI force fields for coarse grained lipid simulations. Biochim Biophys Acta – Biomembr [Internet]. 1858(10):2498–2511. Available from: http://linkinghub.elsevier.com/retrieve/pii/S0005273616301171

Silva AV, López-Sánchez A, Junqueira HC, Rivas L, Baptista MS, Orellana G (2015, Jan) Riboflavin derivatives for enhanced photodynamic activity against Leishmania parasites. Tetrahedron [Internet]. 71(3):457–462. Available from: https://linkinghub.elsevier.com/retrieve/pii/S0040402014016767

Simons K, Ikonen E (1997, Jun 5) Functional rafts in cell membranes. Nature [Internet] 387(6633): 569–572. Available from: http://www.nature.com/articles/42408

Smirnov VV, Egorenkov EA, Myasnikova TN, Petukhov AE, Gegechkori VI, Sukhanova AM, et al (2019) Lipidomic analysis as a tool for identifying susceptibility to various skin diseases. Medchemcomm [Internet]. 10(11):1871–4. Available from: http://xlink.rsc.org/?DOI=C9MD00364A

Song F, Qureshi AA, Han J (2012, Jul 1) Increased caffeine Intake is associated with reduced risk of basal cell carcinoma of the skin. Cancer Res [Internet]. 72(13):3282–3289. Available from: http://cancerres.aacrjournals.org/lookup/doi/10.1158/0008-5472.CAN-11-3511

Speeckaert R, Dugardin J, Lambert J, Lapeere H, Verhaeghe E, Speeckaert MM, et al. (2018, Jul) Critical appraisal of the oxidative stress pathway in vitiligo: a systematic review and meta-analysis. J Eur Acad Dermatology Venereol [Internet]. 32(7):1089–1098. Available from: http://doi.wiley.com/10.1111/jdv.14792

Stoyanovsky DA, Osipov AN, Quinn PJ, Kagan VE (1995, Nov) Ubiquinone-dependent recycling of vitamin E radicals by superoxide. Arch Biochem Biophys [Internet]. 323(2):343–351. Available from: https://linkinghub.elsevier.com/retrieve/pii/S0003986185799559

Szél E, Bozó R, Hunyadi-Gulyás É, Manczinger M, Szabó K, Kemény L, et al (2019, Dec 6) Comprehensive proteomic analysis reveals intermediate stage of non-lesional psoriatic skin and points out the importance of proteins outside this trend. Sci Rep [Internet]. 9(1):11382. Available from: http://www.nature.com/articles/s41598-019-47774-5

Tardivo JPP, Del Giglio A, Paschoal LHCHC, Ito ASS, Baptista MSS (2004, Dec) Treatment of melanoma lesions using methylene blue and RL50 light source. Photodiagnosis Photodyn Ther [Internet]. 1(4):345–6. Available from: https://linkinghub.elsevier.com/retrieve/pii/S1572100005000050

Tardivo JP, Del Giglio A, de Oliveira CS, Gabrielli DS, Junqueira HC, Tada DB, et al (2005, Sep) Methylene blue in photodynamic therapy: from basic mechanisms to clinical applications.

Photodiagnosis Photodyn Ther [Internet]. 2(3):175–91. Available from: https://linkinghub.elsevier.com/retrieve/pii/S1572100005000979

Tardivo JP, Giglio A Del, Paschoal LH, Baptista MS (2006, Aug) New photodynamic therapy protocol to treat AIDS-related Kaposi's sarcoma. Photomed Laser Surg [Internet]. 24(4): 528–531. Available from: https://www.liebertpub.com/doi/10.1089/pho.2006.24.528

Tardivo JP, Wainwright M, Baptista M (2015a, Sep) Small scale trial of photodynamic treatment of onychomycosis in São Paulo. J Photochem Photobiol B Biol [Internet]. 150:66–68. Available from: https://linkinghub.elsevier.com/retrieve/pii/S1011134415000925

Tardivo JP, Baptista MS, Correa JA, Adami F, Pinhal MAS (2015b, Aug 17) Development of the tardivo algorithm to predict amputation risk of diabetic foot. Santanelli, di Pompeo d'Illasi F, editor. PLoS One [Internet]. 10(8):e0135707. Available from: https://dx.plos.org/10.1371/journal.pone.0135707

Tasso TT, Schlothauer JC, Junqueira HC, Matias TA, Araki K, Liandra-Salvador É, et al. (2019, Oct 2) Photobleaching efficiency parallels the enhancement of membrane damage for Porphyrazine photosensitizers. J Am Chem Soc [Internet]. 141(39):15547–15556. Available from: http://pubs.acs.org/doi/10.1021/jacs.9b05991

Thiele JJ, Traber MG, Packer L (1998, May) Depletion of human stratum Corneum vitamin E: an early and sensitive in vivo marker of UV induced photo-oxidation. J Invest Dermatol [Internet]. 110(5):756–761. Available from: https://linkinghub.elsevier.com/retrieve/pii/S0022202X15400764

Tonolli PN, Chiarelli-Neto O, Santacruz-Perez C, Junqueira HC, Watanabe I-S, Ravagnani FG, et al. (2017, Nov) Lipofuscin generated by UVA turns keratinocytes photosensitive to visible light. J Invest Dermatol [Internet]. 137(11):2447–2450. Available from: https://linkinghub.elsevier.com/retrieve/pii/S0022202X17318481

Tsubone TM, Baptista MS, Itri R (2019a, Nov) Understanding membrane remodelling initiated by photosensitized lipid oxidation. Biophys Chem [Internet]. 254:106263. Available from: https://linkinghub.elsevier.com/retrieve/pii/S0301462219302753

Tsubone TM, Martins WK, Pavani C, Junqueira HC, Itri R, Baptista MS (2017, Dec 27) Enhanced efficiency of cell death by lysosome-specific photodamage. Sci Rep [Internet]. 7(1):6734. Available from: http://www.nature.com/articles/s41598-017-06788-7

Tsubone TM, Junqueira HC, Baptista MS, Itri R (2019b, Mar) Contrasting roles of oxidized lipids in modulating membrane microdomains. Biochim Biophys Acta – Biomembr [Internet]. 1861(3):660–669. Available from: https://linkinghub.elsevier.com/retrieve/pii/S0005273618303730

Tsubone TM, Martins WK, Franco MSF, Silva MN, Itri R, Baptista MS (2021, Jan) Cellular compartments challenged by membrane photo-oxidation. Arch Biochem Biophys [Internet]. 697:108665. Available from: https://linkinghub.elsevier.com/retrieve/pii/S0003986120306743

Turro NJ, Ramamurthy V, Scaiano JC (2009) Principles of molecular photochemistry: an introduction, 1st edn. University Science Books, Sausalito

Uchino T, Tokunaga H, Onodera H, Ando M (2002) Effect of squalene Monohydroperoxide on cytotoxicity and cytokine release in a three-dimensional human skin model and human epidermal keratinocytes. Biol Pharm Bull [Internet] 25(5):605–610. Available from: http://www.jstage.jst.go.jp/article/bpb/25/5/25_5_605/_article

Uchoa AF, Konopko AM, Baptista MS (2015) Chlorophyllin Derivatives as Photosensitizers: Synthesis and Photodynamic Properties. J Braz Chem Soc [Internet]. Available from: http://www.gnresearch.org/doi/10.5935/0103-5053.20150290

Vandersee S, Beyer M, Lademann J, Darvin ME (2015) Blue-violet light irradiation dose dependently decreases carotenoids in human skin, which indicates the generation of free radicals. Oxid Med Cell Longev [Internet]. 2015:1–7. Available from: http://www.hindawi.com/journals/omcl/2015/579675/

Vistoli G, De Maddis D, Cipak A, Zarkovic N, Carini M, Aldini G (2013, Aug 17) Advanced glycoxidation and lipoxidation end products (AGEs and ALEs): an overview of their mechanisms of formation. Free Radic Res [Internet]. 47(Sup1):3–27. Available from: http://www.tandfonline.com/doi/full/10.3109/10715762.2013.815348

Volinsky R, Kinnunen PKJ (2013, Jun) Oxidized phosphatidylcholines in membrane-level cellular signaling: from biophysics to physiology and molecular pathology. FEBS J [Internet]. 280(12): 2806–2816. Available from: http://doi.wiley.com/10.1111/febs.12247

Weber G, Charitat T, Baptista MS, Uchoa AF, Pavani C, Junqueira HC, et al. (2014) Lipid oxidation induces structural changes in biomimetic membranes. Soft Matter [Internet]. 10(24): 4241–7. Available from.: http://www.ncbi.nlm.nih.gov/pubmed/24871383

Wei Y, Zhou Y-M, Li Y-Q, Gao R-Y, Fu L-M, Wang P, et al (2021, Aug) Spatial effects of photosensitization on morphology of giant unilamellar vesicles. Biophys Chem [Internet]. 275: 106624. Available from: https://linkinghub.elsevier.com/retrieve/pii/S030146222100106X

Wilkinson F, Helman WP, Ross AB (1995) Rate constants for the decay and reactions of the lowest electronically excited singlet state of molecular oxygen in solution. An expand and revised compilation. J Phys Chem Ref Data 24(2):663–677

Wong-Ekkabut J, Xu Z, Triampo W, Tang I-M, Tieleman DP, Monticelli L (2007) Effect of lipid peroxidation on the properties of lipid bilayers: a molecular dynamics study. Biophys J 93(12): 4225–4236

Yadav DK, Kumar S, Choi E-H, Chaudhary S, Kim M-H (2019 Dec 14) Molecular dynamic simulations of oxidized skin lipid bilayer and permeability of reactive oxygen species. Sci Rep [Internet]. 9(1):4496. Available from: http://www.nature.com/articles/s41598-019-40913-y

Yamawaki Y, Mizutani T, Okano Y, Masaki H (2019 Feb) The impact of carbonylated proteins on the skin and potential agents to block their effects. Exp Dermatol [Internet]. 28:32–7. Available from: http://doi.wiley.com/10.1111/exd.13821

Yin H, Xu L, Porter NA (2011, Oct 12) Free radical lipid peroxidation: mechanisms and analysis. Chem Rev [Internet]. 111(10):5944–5972. Available from: http://pubs.acs.org/doi/abs/10.1021/cr200084z

Ytzhak S, Ehrenberg B (2014) The effect of photodynamic action on leakage of ions through liposomal membranes that contain oxidatively modified lipids. Photochem Photobiol 90(4): 796–800

Yusupov M, Van der Paal J, Neyts EC, Bogaerts A (2017, Apr) Synergistic effect of electric field and lipid oxidation on the permeability of cell membranes. Biochim Biophys Acta – Gen Subj [Internet]. 1861(4):839–847. Available from: http://linkinghub.elsevier.com/retrieve/pii/S0304416517300387

Zhou Y-M, Zhang Y, Gao R-Y, Liu W, Wei Y, Han R-M, et al (2021, Sep) Primary reaction intermediates of Type-I photosensitized lipid oxidation as revealed by time-resolved optical spectroscopies. J Photochem Photobiol A Chem [Internet]. 418:113376. Available from: https://linkinghub.elsevier.com/retrieve/pii/S1010603021002483

Part III
Health, Biological and Nutritional Effects

Antioxidants in Dentistry: Oxidative Stress and Periodontal Diseases

Sofía Elvira Fernández-Bravo

1 Introduction

1.1 Periodontal Diseases

Periodontal disorders are a group of inflammatory and destructive diseases of periodontium affecting a large portion (20–50%) of the adult population, produced through the complex interaction between periodontal pathogens and components of the host defense mechanism (Kinane et al. 2017; Zheng et al. 2021; Pihlstrom et al. 2005; Albandar and Rams 2002). It is considered a global burden and, together with caries, one of the two biggest threats to oral health and a prime cause of tooth loss (Zheng et al. 2021; Albandar and Rams 2002; Sheiham and Netuveli 2002).

The clinical diagnosis of periodontal disease is based on visual and radiographic assessment of the periodontal tissues and measurements of the supporting tissues around the tooth (Armitage 2004). Two main forms of non -reversible periodontitis are currently recognized, chronic (CP) and aggressive periodontitis (AgP) (Acquier et al. 2017). The most common is the chronic periodontal disease (CP), and can develop different degrees of severity, Fig. 1, meanwhile the aggressive periodontitis (AgP) is less frequent but causes more severe symptoms. The disease starts with reversible gingival inflammation and progressively destroys the connective tissue known as periodontal ligament and the alveolar bone that surrounds the tooth. Figure 1 illustrates the onset of the disease showing a healthy gum (A) and various degrees of periodontitis as observed with the naked eye. The initial lesion begins 2–4 days after the accumulation of the microbial plaque. The early injury develops within 4–10 days and is characterized by a pathological alteration of fibroblasts. Subsequently, the lesion develops within 2–3 weeks and is dominated by further loss

S. E. Fernández-Bravo (✉)
Health Department, IES Leixa, A Coruña, Spain
e-mail: sofia.fernandez2@udc.es

C. Bravo-Diaz (ed.), *Lipid Oxidation in Food and Biological Systems*,
https://doi.org/10.1007/978-3-030-87222-9_15

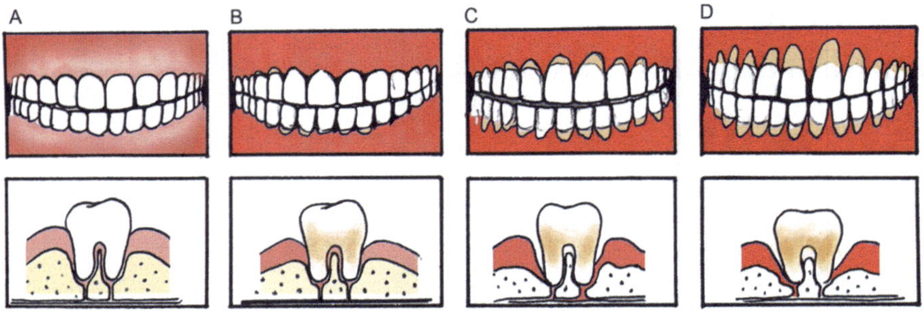

Fig. 1 Histopathological features of chronical periodontal disease. (**a**) Tissues that support the tooth including healthy supporting tissues (**b**) Mild periodontitis (**c**) Moderate (**d**) Severe periodontitis

Table 1 Risk factors for periodontal disease

Modifiable	Non-modifiable
Poor oral hygiene	Ageing
Smoking	Gender
Stress	Hormonal changes
Nutritional status	Genetic predisposition
Obesity/overweight	Association with other Medical conditions
Inmune status	Host response
Uncontrolled diabetes mellitus	Osteoporosis

of the marginal gingival connective tissue matrix to finally destroy the alveolar bone and periodontal ligament, loss of collagen attachment to the root surface and resorption of the alveolar bone (Muñoz-Carrillo et al. 2019).

AgP and CP cannot be distinguished on the basis of histopathologic or pathophysiological features (Smith et al. 2010) in spite of the differences in their clinical phenotypes (Armitage and Cullinan 2010) or on the basis of microbial colonization profiles, (Armitage 2010) but it has been observed that AgP mostly affect younger individuals and usually has a higher rate of progression and patterns of tissue destruction than CP. There is, however, some evidence of immunological differences, including the presence of neutrophil abnormalities in AgP (Acquier et al. 2017; Nussbaum and Shapira 2011).

The inflammatory and immune response induced by subgingival plaque is the most important factor in the progress of this disease, but the development of the disease in a particular individual depends on a variety of factors including genetic predisposition, bad oral hygiene, smoking, malnutrition and an exaggerated inflammatory response (Woelber et al. 2016; Woelber and Tennert 2020; Hujoel and Lingström 2017; Reynolds 2014). Table 1 shows some of the most common modifiable and non-modifiable risk factors that are directly related to periodontal diseases, (Reynolds 2014; Borgnakke 2016) which have been recognized as a risk factor for a number of important systemic diseases, including cardiovascular

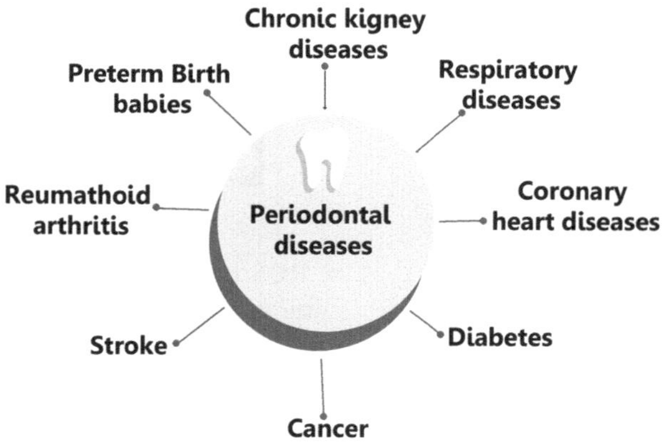

Scheme 1 Some common health problems related to periodontal diseases

diseases, diabetes and rheumatoid arthritis, Scheme 1 (Pihlstrom et al. 2005; Muñoz-Carrillo et al. 2019; Aksakalli 2013).

Oxidative stress was found to be involved in many inflammatory diseases, which in turn can trigger more damage to gingival and other tissues, worsening periodontitis (Tóthová and Celec 2017). An increasing body of evidence implicates reactive oxygen species (ROS) in the pathogenesis of various diseases, including AIDS, cancer, neurodegenerative diseases and aging processes (Schieber and Chandel 2014; Losada-Barreiro and Bravo-Díaz 2017). Periodontal diseases are not an exception to the potential harmful effects of ROS, even though it is believed that bacteria and hosts enzymes (proteases, metalloproteases and glycosidades) are involved in the destruction of periodontal tissues (Waddington et al. 2000; Halliwell and Gutteridge 2007; Halliwell et al. 1995; Ding et al. 1995; Lamont and Jenkinson 1998). ROS, and particularly the extremely reactive $^{\bullet}OH$ radical, can degrade structurally and metabolically important macromolecules (proteins, lipids, carbohydrates) resulting in severe cellular damage (Schieber and Chandel 2014; Losada-Barreiro and Bravo-Díaz 2017; Zhang et al. 2016; Lee et al. 2004).

There is, therefore, a great interest in describing the relationships between oxidative stress and periodontitis. This will permit to gain insights into the pathogenesis of periodontitis, into the relationships between periodontitis and systemic inflammation, and to propose potential therapeutic strategies to minimize the disease and its harmful effects.

1.2 Oxidative Stress

Oxidative stress is a term commonly employed to describe the balance (or imbalance) between the production and accumulation of oxygen reactive species (ROS) in cells and tissues and the ability of a biological system to detoxify these reactive products (Tóthová and Celec 2017; Losada-Barreiro and Bravo-Díaz 2017; Pizzino et al. 2017; Sezer et al. 2016). It is considered both as a pathomechanism

Scheme 2 Host response to periodontal patogens

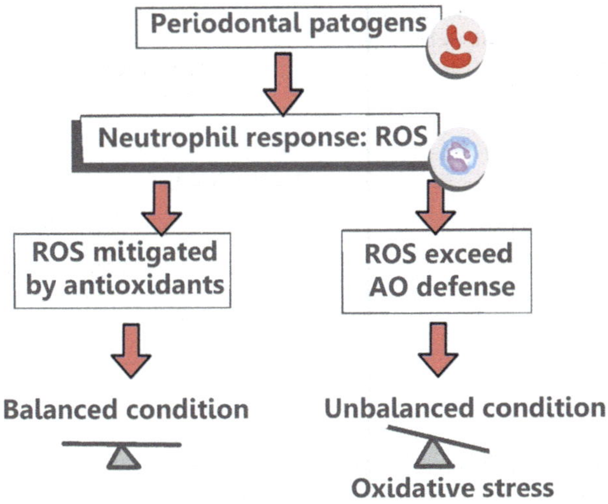

involved in diseases causing damage to lipids, nucleic acids and proteins, and as an important physiological process that enables the immune system to cope with microorganisms and intracellular cell signaling—oxidative stress (Sies et al. 2017; Alfadda and Sallam 2012).

The dual beneficial and destructive role attributed to ROS depends, among others, on their concentrations at the reaction site (Losada-Barreiro and Bravo-Díaz 2017). When ROS are present at low or moderate concentrations, their effects are beneficial because they are associated to cellular signaling and cellular responses to noxia (e.g., defense against infectious agents) (Zhang et al. 2016; Schieber and Navdeep 2014; Brown and Griendling 2015). On the contrary, when ROS are present at elevated concentrations because of an overproduction or because of a decrease in the antioxidant defense, ROS may cause biological damage (Di Domenico et al. 2015; Rosini et al. 2014). The subtle balance between harmful and beneficial effects of ROS results from the metabolic reactions that use oxygen and maintain the "redox homeostasis" by controlling the redox regulation in vivo, Scheme 2 (Losada-Barreiro and Bravo-Díaz 2017; Wang et al. 2013; Calabrese et al. 2009; Ray et al. 2012).

When an inflammatory process occurs, cells of the endogenous defense system (neutrophils and macrophages) become hyperactive and overproduce ROS, unbalancing the scale and contributing to the tissue damage and loss of teeth commonly associated with periodontitis (Muñoz-Carrillo et al. 2019; Battino et al. 1999; Bobyrev et al. 1994; Silva et al. 2015).

1.3 *Reactive Oxygen Species (ROS)*

Free radicals can be defined as reactive chemical species having unpaired electrons in the outermost shell (Losada-Barreiro and Bravo-Díaz 2017). These molecules are continuously produced during the normal function of living organisms, particularly during the respiratory chain (Tóthová and Celec 2017; Krumova and Cosa 2016; Kaludercic et al. 2014; Banerjee 2012). Ground state molecular oxygen is a stable radical necessary for normal life and cell function. Different ROS are produced during metabolization of $^3O_2$, Scheme 3. Among them, superoxide, peroxide anion, $O_2^{\bullet-}$, hydrogen peroxide, H_2O_2, and the hydroxyl radical, HO are considered the primary ROS and have sparked major research on the role of free radicals in biology and medicine. The toxic effects of ROS in biological systems, such as the oxidation of lipids, inactivation of enzymes, alterations in the DNA, and destruction of cell membranes and, ultimately, cells—are attributable to the reduction of the O_2 in the respiratory chain.

Table 2 shows some of the most important ROS. $^3O_2$ has two unpaired electrons and cannot react directly with most biomolecules (BM) because they have paired electrons. Such reactions are extremely slow because they require spin inversion.

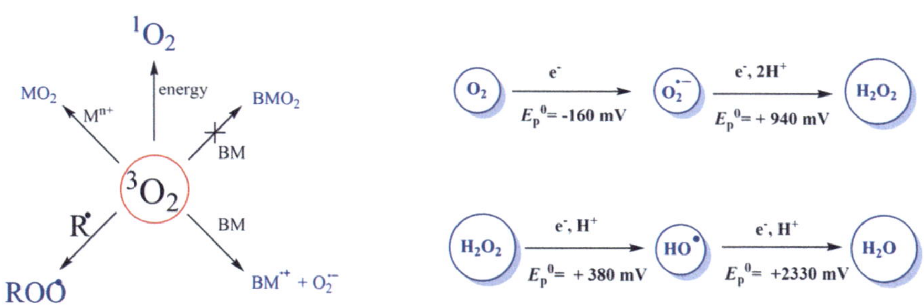

Scheme 3 Some of the main reactions of molecular oxygen (left) and redox states of ROS with their standard redox potentials (oxygen concentration = 1 M) for their formation through electron-transfer reactions

Table 2 Main reactive oxygen species (ROS) (Di Domenico et al. 2015; Lushchak 2014)

Non-radical	Hydrogen peroxide	H_2O_2
	Singlet oxygen	$^1O_2$
	Ozone	O_3
	Hypoclorous acid	HOCl
Radicals	Superoxide anion	$O_2^{\bullet-}$
	Hydroxyl	HO
	Hydroperoxyl	HO_2
	Perhydroxyl	$HO_2^{\bullet-}$
	Lipid alkoxyl	$Ro^{\bullet}$
	Lipid peroxyl	$ROO^{\bullet}$
	Alkoxyl	$ArO^{\bullet}$

However, it can produce stable radicals by one-electron transfer reactions ($BM^{+\bullet}$) or by reacting with molecules with unpaired electrons, such as lipid radicals ($R^{\bullet}$), transition metal ions (M^{n+}). UV radiation can supply energy to $^3O_2$ and excite it to singlet oxygen ($^1O_2$), that is much more reactive towards biomolecules because there is no spin restriction.

1.4 Sources of ROS in the Oral Cavity

ROS are generated by polymorphonucelar leukocytes (PMN), predominantly neutrophils, during an inflammatory response, and are regarded as very destructive in nature. PMN are believed to be the initial and predominant defense against bacterial pathogens in a variety of pathological diseases including the periodontal one (Waddington et al. 2000; Miller et al. 1984). The main source of free radicals in the body is physiological metabolism, (Tóthová and Celec 2017; Pizzino et al. 2017; Birch-Machin and Bowman 2016) but ROS can also be generated through exposure to external factors as shown in Table 3 (Ghosh et al. 2018). It is, thus, important to minimize the exposure to factors that may eventually generate ROS, to prevent the oxidative damage to cellular DNA, lipids and proteins and, in particular, to those that affect periodontal health (Muñoz-Carrillo et al. 2019; Tóthová and Celec 2017; Almerich-Silla et al. 2015).

1.5 Molecular Targets of ROS: Typical Lifetimes

ROS may display a broad range of reactivities because reactions of radical ROS have low activation energies, (Wang et al. 2013; Sima and Glogauer 2014) meanwhile those involving non-radical ROS species are much slower because of their higher activation energy (Losada-Barreiro and Bravo-Díaz 2017; Mikkelsen and Wardman 2003). Because ROS are highly reactive species, they may undergo self-quenching (e.g., $ROO^{\bullet} + ROO^{\bullet} \rightarrow$ products), but their concentrations in most biological reactions are much lower than those of radical scavengers, and therefore second

Table 3 Potential sources of ROS in dentistry

Xenobiotics	Ethanol, cigarrette smoke, drugs
Stress conditions	Physiological and intense physical stress
Food	High-fat and high protein diets, acrolein
Dental treatments	Ozone, ultrasounds, non-thermal plasma, laser and UV lights
Dental materials	Bleaching agents (H_2O_2), composite filling resins, dental cements, ceramic and metals, orthodontic, prothetic, TiO_2 implant coatings

Table 4 Half-lives and experimental rate constants k_{exp} of various radicals. L: linolenic acid

Radical	$t_{1/2}$ (s)[a]	k_{exp} (M^{-1} s^{-1})
$Ho^{\bullet}$	10^{-9}	10^{9}–10^{10}
$Ro^{\bullet}$	10^{-6}	10^{6}–10^{8}
$ROO^{\bullet}$	10	10^{1}–10^{3}
$L^{\bullet}$	10^{-8}	10^{4}–10^{8}
$O_2^{\bullet -}$	–	~1
$HOO^{\bullet}$	–	10^{0}–10^{3}
$^{1}O_2$	10^{-6}	

[a] Half-life values estimated by assuming a typical substrate concentration of 10^{-3} M at T = 37 °C. Values extracted from ref. (Losada-Barreiro and Bravo-Díaz 2017)

order reactions involving the encounter of two identical ROS are frequently rare. Therefore, the main reactions of ROS are with target molecules in the surrounding tissues. Estimates for the typical diffusion distance (distance that the radical can move prior to reacting with another radical or quencher) can be calculated from the lifetimes of the various species by employing the Einstein equation. Diffusion coefficients depend on the viscosity of the medium, and will generally be smaller in biological matrices than in pure water, but for small radicals as ROS, average values of the lifetimes can be estimated by employing average diffusion coefficients (Losada-Barreiro and Bravo-Díaz 2017; Mikkelsen and Wardman 2003). Table 4 shows the half-lives and rate constants of some ROS. Kinetic data for reactions of oxygen (and other) radicals have been published in the NDRL/NIST Solution Kinetics Database (http://kinetics.nist.gov/solution/) and the interested reader is referred to it for further information (Anbar et al. 1973; Ross and Ross 1977).

ROS with short half-lives react immediately with any nearby molecule located at short distances (a few nanometers), making them to be little selective and having a broad range of nonspecific targets (Alfadda and Sallam 2012; Burton and Ingold 1986). Low reactive ROS such as H_2O_2 may exert a long-range effect diffusing a few tens of micrometers (Anbar et al. 1973; Choe and Min 2006). In summary, highly reactive ROS such as $^{\bullet}OH$ have low target selectivity and diffusion distance, in contrast to those less reactive such as $ROO^{\bullet}$, whose harmful effects may be noticeable at longer distances. Therefore, preventing the formation of ROS is one of various commonly employed strategies to minimize their effects on periodontitis.

2 Mechanisms of Development and Control of Periodontitis

A broad range of therapies exist in periodontics (Woelber et al. 2016; Vo et al. 2020; San Miguel et al. 2011; Kassebaum et al. 2014). However, no single treatment approach can provide the only means of treating any one or all periodontal diseases. The conventional treatment for periodontitis focuses in patient education in conjunction with removal of supragingival and accessible subgingival bacterial plaque and calculus by periodontal scaling and antimicrobial therapy (Pihlstrom et al. 2005;

Muñoz-Carrillo et al. 2019; Hujoel and Lingström 2017; Basu et al. 1999). In more complex cases, surgical treatment is needed (resective procedures, periodontal regenerative procedures and replacement of teeth by dental implants) (Kinane et al. 2017; Smith et al. 2010; Armitage and Cullinan 2010). Control of risk factors is, therefore, a cornerstone of the periodontal treatment and, particularly, control of ROS balance and cell homeostasis is crucial to prevent the oxidative damage to cellular DNA, lipids and proteins.

2.1 Clinical Implications for ROS Involvement in Periodontal Diseases

Neutrophils can phagocyte bacteria in the gingival clavicle fluid, but this is not the predominant mechanism of protection (Silva et al. 2015; Sima and Glogauer 2014). It is believed that neutrophils play a central role in the initial host inflammatory response to the periodontal pathogens, protecting the host tissues by eradicating various pathogenic bacteria by either oxidative and/or non-oxidative mechanisms (Kinane et al. 2017; Muñoz-Carrillo et al. 2019; Hujoel and Lingström 2017; Tóthová and Celec 2017). The non-oxidative pathway consists in recognition and phagocytosis of opsonized pathogens, mediated by various lysosomal enzymes, peptides and proteins (Kinane et al. 2017). The oxidative mechanism is mainly mediated by neutrophiles that produce ROS in the presence of pathogens (Sima and Glogauer 2014). The oxidative killing of bacteria associated to the generation of ROS is a potential defense mechanism, but needs to be considered with caution because an overproduction of ROS may create an imbalance in the homeostasis leading to tissue damage, Scheme 4 (Losada-Barreiro and Bravo-Díaz 2017; Pavlin et al. 2016). Because, the released ROS are not target-specific and frequently react with nearby molecules (see Sect. 1.5), damage to host tissue occurs in different extents to lipids (peroxidation), proteins, enzymes and DNA (Tóthová and Celec 2017; Krumova and Cosa 2016; Birch-Machin and Bowman 2016; Ghosh et al. 2018; Pisoschi and Pop 2015; Ozcan and Ogun 2015).

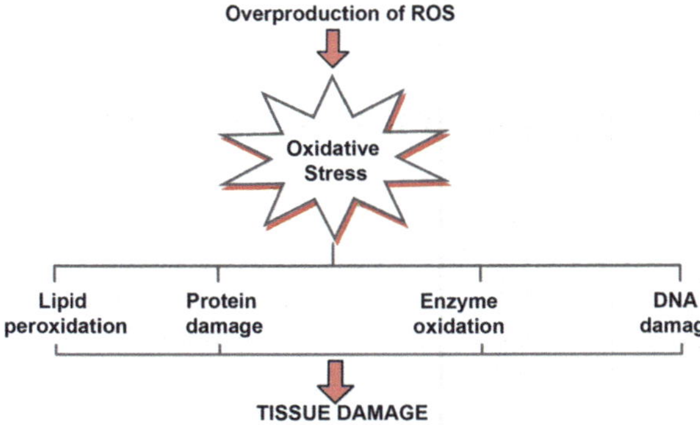

Scheme 4 Potential effects of the imbalance between the production and destruction of ROS

Overproduction of ROS plays an important role in during the osteoclast (type of bone cell that breaks down bone tissue) stimulation in active periodontal disease, leading to bone destruction because of the overproduction of ROS rather than direct degradation of the bone matrix (Kinane et al. 2017; Muñoz-Carrillo et al. 2019; Waddington et al. 2000; Vo et al. 2020; Hall et al. 1995; Bax et al. 1992).

Studies by Guarnieri et al. (Guarnieri et al. 1991) and Kimura et al. (Kimura et al. 1993) demonstrated that neutrophils present in the gingival crevicular fluid (GCF) and in blood of patients with adult periodontitis show an enhancement in the production of $O_2^{\bullet -}$ after stimulation compared with PMN isolated from the GCF and blood of a 'healthy' group (control). In addition, they also reported that the circulating PMN of patients with periodontitis produced spontaneously low levels of $O_2^{\bullet -}$, meanwhile no production was observed in the control group.

Other studies have also investigated the levels of metal ions in periodontal pockets as the levels of iron and copper within the gingival sulcus are also likely to play a role in the production of ROS via the Fenton and Haber-Weiss reactions (Waddington et al. 2000; Bhattacharya 2015).

2.2 *Pathogenesis of Periodontitis: Biomarkers of Oxidative Stress*

Virtually, all inflammatory diseases lead to an increase in the oxidative stress, which, in turn, can induce damage to surrounding tissues including the gingival one (Woelber et al. 2016; Woelber and Tennert 2020; Battino et al. 1999; Ahmadi-Motamayel et al. 2013). The oxidative stress levels in patients with periodontitis can be monitored by measuring the values of oxidative stress markers in either saliva, gingival crevicular fluid or blood and compare them with the values under the normal antioxidant status of the tissue. Overall results support the idea that the use of body fluids, especially the non-invasive diagnostic fluid saliva, as suitable sample types for diagnostics or monitoring the course of periodontitis (Muñoz-Carrillo et al. 2019; Ahmadi-Motamayel et al. 2013, 2018; Rahmani et al. 2015; Marrocco et al. 2017; da Silva and Muniz 2018; Gharbi et al. 2019).

Current data support the use of a set of markers, covering both, oxidative damage and antioxidative levels to monitor the course of periodontitis (Marrocco et al. 2017; Wang et al. 2017; Nguyen et al. 2017). However, cautions need to be taken when interpreting the results because of the low specificity of oxidative stress markers (even if more than one marker is used) and because the inter-individual and intra-individual variability of the analyzed markers is very high, preventing their use at the level of individual diagnostics (Acquier et al. 2017; Almerich-Silla et al. 2015; Sawamoto et al. 2005; Takane et al. 2002). Table 5 shows some of the most common oxidative markers employed. Nitric oxide, lipid peroxidation, protein, DNA oxidative markers, malondialdehyde, 8-hydroxydeoxyguanosine, 8-isprostanes, and substances reacting with thiobarbituric acid (TBARS) are commonly employed as

Table 5 Some common oxidative markers employed to monitor periodontitis

Reaction	Marker
Lipid peroxidation	Malondialdhyde, 4-hydroxyl-2-115 nonenal, Isoprostane, TBARS
Protein damage	Protein carbonyl, advanced oxidation products
DNA damage	Hydroxi-deoxyguanosine

oxidative markers, and in most cases, their values were higher in patients with periodontitis compared to those in the control group (Acquier et al. 2017; Almerich-Silla et al. 2015; Takane et al. 2002; Novakovic et al. 2014).

2.3 *Antioxidants in the Oral Cavity: Antioxidant Treatment of Periodontal Diseases*

The human body has developed an anti-oxidant defense system that functions to detoxify ROS and modify them to form less reactive species (Losada-Barreiro and Bravo-Díaz 2017; Birch-Machin and Bowman 2016; Jeeva et al. 2015; Rani and Yadav 2014). They halt oxidation reactions by donating their own electrons to neutralize the adverse effects of ROS. Antioxidants present in the human body can be supplied by the endogenous defense system or supplied from external sources such as foods and/or supplements.

Endogenous antioxidants include enzymatic (e.g., superoxide dismutase, glutathione and catalase peroxidases) and chain breaking molecules such as ascorbic acid (Vitamin C) and α-tocopherol (Vitamin E) (Halliwell and Gutteridge 2007; Jeeva et al. 2015; Halliwell and Gutteridge 1999; Krishnaiah et al. 2011). Foods and other natural products such as herbs and spices are the typical sources of exogenous antioxidants such as polyphenols, phenolic acids, coumarins, anthocyanins, flavonoids, etc. (Losada-Barreiro and Bravo-Díaz 2017; Watson et al. 2013; Salonen 1998)

Figure 2 shows some of the most important roles of antioxidants in the human body. Antioxidants can be (broadly) classified on the basis of their mechanism of action against ROS as primary and secondary antioxidants. Primary antioxidants are those that break the chain reaction of oxidation by hydrogen donation and generation of more stable radicals. For this reason, they are also usually called chain-breaking or primary antioxidants. Secondary antioxidants slow the oxidation rate by several mechanisms, including chelation of metals, regeneration of primary antioxidants, decomposition of hydroperoxides, and scavenging of oxygen, among others (Losada-Barreiro and Bravo-Díaz 2017; Wanasundara and Shahidi 2005). They work at three three different levels: (1) prevention, maintaining formation of ROS to a minimum level, (2) **interception**, scavenging ROS either by using catalytic and non-catalytic molecules, e.g., ascorbic acid and alpha-tocopherol; and (3) **repair**, restoring damaged molecules, for instance, regenerating lipid radicals (Losada-Barreiro and Bravo-Díaz 2017; Birben et al. 2012).

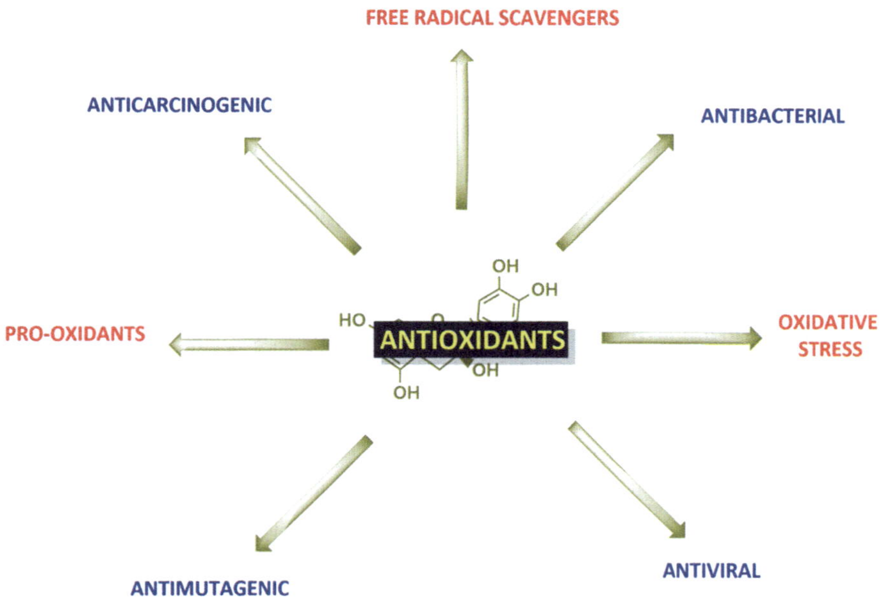

Fig. 2 Illustrative representation of the various roles of antioxidants beyond the modulation of oxidative stress

2.4 Effects of Diet on Periodontal Diseases and Oxidative Stress

As shown in previous sections, etiology of periodontitis and periodontal health is affected by a number of factors such as a bad oral hygiene, and a number of systemic factors including epigenetic, diabetes, pregnancy and cardiovascular diseases. The role of nutrition and diet has also been investigated in relation to the development of periodontal diseases, and a number of studies concluded that a balanced diet has an essential role in maintaining periodontal health. In addition, nutritional supplements and dietary components also promote healing after periodontal surgery (Woelber et al. 2016; Woelber and Tennert 2020; da Silva and Muniz 2018; Najeeb et al. 2016; Sidi and Ashley 1984). For instance, an inappropriate nutrition augments periodontitis through increased blood oxidative stress, (Tomofuji et al. 2009) and high cholesterol diets decrease alveolar bone density, increasing the serum level for oxidative DNA damage (Tomofuji et al. 2009).

2.5 Role of Nutrition in Periodontal Health: An Update

Micro and macro nutrients have a major impact on periodontal health (Hujoel and Lingström 2017; Tóthová and Celec 2017; San Miguel et al. 2011; Wang et al. 2017; Najeeb et al. 2016). Typical healthy diets also include antioxidants such as Vitamins C (ascorbic acid) and E (α-tocopherol), carotenoids and β-carotene (Vitamin A), glutathione and melatonin. All these antioxidants may help to control the oxidative stress caused by the inflammation of periodontal tissues.

- **Carbohydrates** (Keukenmeester et al. 2014; Palmer et al. 2010). Sugars contribute to periodontal disease because bacteria ferment them and produce acid, leading to the demineralization of the tooth structure. Hence, a reduction of sugar intake, coupled with scaling, root planing, and the use of xylitol- and maltitol-containing gums have the potential to improve the periodontal health of the general population.
- **Vitamins** (Dodington et al. 2015; Gunther 1976; Neiva et al. 2005). Apart from playing a vital role in cell metabolism, vitamins have potent antioxidant properties. Vitamins A, C, and E have all been observed to modulate the anti-oxidant defense system. Vitamins are necessary for the normal functioning of the human body and a nutritional deficiency of vitamins results scurvy and rickets. Scurvy is a disease caused by deficiency of Vitamin C, which also prevents oxidative damage by acting as a ROS scavenger and as a synergistic antioxidant when in presence of Vitamin E, regenerating oxidized α-tocopherol. Vitamin C can also be used in coatings and/or gel forms to enhance the osseointegration of dental implants and to improve post-surgical periodontal healing.

  Vitamin E (tocopherol) is a fat-soluble vitamin that is considered one of the key extracellular antioxidants that stabilizes the structure of cell membranes by terminating the propagation of free radicals during oxidative processes (i.e., lipid peroxidation). A reduction of vitamin E was observed in patients with periodontal diseases compared to healthy individuals, and thus their supplement may be advisable, even though the mechanisms of action for periodontal health remain a matter of debate because they are not well understood.
- **Dietary Minerals and trace metals** (Garcia et al. 2011; Chakraborty and Tewari 2014; Maguire 2014). A reduced level of micronutrients compromises the periodontal health. Elements such as Ca, P, K, S, Na, Cl, and Mg are normally required in abundance to perform normal physiological functions. Alternatively, oligoelements such as Fe, Co, Cu, Zn, Mn, Mb, Mo and Se are required in small concentrations, but their deficit compromises the periodontal health. A number of factors are involved in reducing the serum level of micronutrients such as genetic or gastrointestinal disorders (affects absorption and bioavailability), poor diet, or lifestyle (Najeeb et al. 2016). In addition, certain physiological changes such as pregnancy and aging may lower the daily requirement of various nutrients.

Table 6 shows the main sources and effects of various nutrients in periodontal diseases.

Table 6 Major vitamins and minerals, their dietary sources, daily requirements, deficiency diseases, and reported importance in periodontal health. Adapted from ref. (Najeeb et al. 2016)

Nutrient	Dietary source	Importance in periodontal health
Vitamin A (Dodington et al. 2015)	Cod liver oil, carrots, liver, sweet potato, broccoli,	Not clear. Research indicates insignificant improvement in periodontal health upon supplementation
B-vitamins (Neiva et al. 2005)	B1—Liver, oats, pork, potatoes, eggs	Supplementation may accelerate post-surgical healing.
	B2—Bananas, dairy, green beans	
	B3—Eggs, fish, meat, mushrooms, nuts	
	B5—Avocados, meat, broccoli	
	B7—Raw egg, liver, leafy vegetables,	
	B9—Cereals, leafy vegetables	
Vitamin C (Gunther 1976; Tada and Miura 2019)	Citrus fruits, vegetables, liver	Gingival bleeding and inflammation are hallmarks of scurvy. Supplementation may improve outcomes of periodontal therapy
Vitamin D (Jagelavičienė et al. 2018; Bashutski et al. 2011)	Fish eggs, mushrooms, liver, milk	Deficiency may lead to delayed post-surgical healing. Local application may accelerate post-surgical healing and osseointegration
Vitamin E (Kaur et al. 2016; Kim and Shklar 1983)	Poultry, meat, fish, nuts, seeds and cereals	Impaired gingival wound healing inactivation of ROS
Vitamin K (Aral et al. 2015)	Green vegetables, egg yolk	Deficiency may lead to gingival bleeding. No known effects on periodontal therapy if supplementation used as an adjunct.
Minerals		
Calcium (Garcia et al. 2011; Tanaka et al. 2014)	Milk products, eggs, canned bony fish, leafy vegetables, nuts, seeds	Required for formation of teeth and bones. Supplementation improves outcomes of non-surgical periodontal therapy. Local application enhances osseointegration.
Magnesium (Meisel et al. 2005)	Cocoa, soybeans, nuts, spinach, marine vegetables, tomatoes	Required for cell metabolism and bone formation. Supplementation may improve outcomes on on-surgical periodontal therapy
Iron (Enhos et al. 2009; Wu et al. 2020)	Red meat, tuna, dry beans, spinach	Possible antioxidant effects on periodontium
Zinc (Orbak et al. 2007; Apon and Kamble 2019)	Protein-rich foods, spinach, grains	Possible antioxidant effects on periodontium. Reduces severity of diabetes-induced periodontitis

3 Conclusions and Future Perspectives

Periodontal disease is one of the most prevalent oral diseases, affecting the surrounding and supporting structures of the teeth. Periodontal health is affected by a number of factors such as oral hygiene, genetic and epigenetic factors, systemic health, and nutrition. Periodontal disease is related with serious systemic diseases and that is why so much effort is made to control it. A balanced diet is crucial in maintaining periodontal health. Bone formation and periodontal regeneration are also affected by numerous vitamins, minerals, and trace elements, and attempts to find a correlation between tooth loss, periodontal health, and nutrition have been carried out.

Over the past few decades, the role of free radicals, particularly ROS, and antioxidants have attracted tremendous importance in the field of dentistry. ROS are highly reactive molecules derived from oxygen metabolism and are well recognized for playing a dual role as both deleterious and beneficial species, depending on its level in the body. In vivo, ROS may play constructive roles in cell physiology; however, they may also cause harmful effects on cell membranes and DNA by causing membrane lipid peroxidation, decreased membrane fluidity, and DNA mutations leading to cancer, degenerative, and other diseases. The prognosis of the disease depends, therefore, on the capacity of the host to neutralize the ROS produced by neutrophils at the site of infection.

The oral cavity is the first – and main - port of entry for foods and beverages, and is highly susceptible to negative effects that contribute to periodontitis. Current data from human clinical and biological studies have demonstrated that selected dietary polyphenols have important antimicrobial, antioxidant, and anti-inflammatory properties resulting in improved clinical markers in periodontitis (Basu et al. 2018; Koch 2019; Fresco et al. 2010). Therefore, a proper ingestion of polyphenols naturally present in foods in combination with an adequate oral hygiene care may certainly help in the prevention of periodontitis as well as other chronic inflammatory diseases that encompass this collection of co-morbid conditions (Halliwell 2007; Scalbert et al. 2005). Dietary polyphenols have been shown to be effective in improving gingival bleeding and in decreasing alveolar bone loss in animals and human clinical studies by suppressing osteoclastogenesis and inhibiting inflammatory cytokines (Basu et al. 2018; Koch 2019). Polyphenol-containing beverages, foods and some fruit and vegetable extracts have bacteriostatic/bactericidal activity against microbial species such as *P. gingivalis* and shown total bacterial burden in clinical studies (Basu et al. 2018). Polyphenols also exhibit anti-inflammatory and antioxidant effects, (Losada-Barreiro and Bravo-Díaz 2017) which may modify the various biological mechanisms for reducing the initiation and progression of periodontitis. While data from molecular studies are indeed promising, further research is needed to improve our knowledge on the effects of polyphenols for prevention and treatment of periodontal diseases.

Conflict of Interest The author declares no conflict of interest.

References

Acquier AB et al (2017) Parameters of oxidative stress in saliva from patients with aggressive and chronic periodontitis. Redox Rep 22(3):119–126

Ahmadi-Motamayel F et al (2013) Total antioxidant capacity of saliva and dental caries. Med Oral Patol Oral Cir Bucal 18(4):e553–e556

Ahmadi-Motamayel F et al (2018) Salivary and serum antioxidant and oxidative stress markers in dental caries. Caries Res 52(6):565–569

Aksakalli S (2013) Antioxidants in dentistry: review of literature. Dentistry 4:181

Albandar JM, Rams TE (2002) Global epidemiology of periodontal diseases: an overview. Periodontol 2000 29:7–10

Alfadda AA, Sallam RM (2012) Reactive oxygen species in Health and Disease. J Biomed Biotechnol 2012:1–14. https://doi.org/10.1155/2012/936486

Almerich-Silla JM et al (2015) Oxidative stress parameters in saliva and its association with periodontal Disease and types of bacteria. Dis Markers 2015:653537

Anbar M, Bambenek M, Ross AB (1973) Selected specific rates of reactions of transients from water in aqueous solution. 1. Hydrated electron. Stanford Research Inst., Menlo Park, Calif. (USA); Notre Dame Univ., Ind. (USA). Radiation Lab. p 67

Apon A, Kamble P (2019) Role of trace mineral in periodontal health: a review. Clin Trials Degener Dis 4(2):30–36

Aral K et al (2015) Therapeutic effects of systemic vitamin k2 and vitamin d3 on gingival inflammation and alveolar bone in rats with experimentally induced periodontitis. J Periodontol 86(5):666–673

Armitage GC (2004) The complete periodontal examination. Periodontol 2000 34:22–33

Armitage GC (2010) Comparison of the microbiological features of chronic and aggressive periodontitis. Periodontology 2000 53(1):70–88

Armitage GC, Cullinan MP (2010) Comparison of the clinical features of chronic and aggressive periodontitis. Periodontology 2000 53(1):12–27

Banerjee R (2012) Redox outside the box: linking extracellular redox remodeling with intracellular redox metabolism. J Biol Chem 287(7):4397–4402

Bashutski JD et al (2011) The impact of vitamin D status on periodontal surgery outcomes. J Dent Res 90(8):1007–1012

Basu T, Temple NJ, Garg ML (1999) In: Basu TK et al (eds) Antioxidants in human health and disease. CABI Publisher, Wallingford

Basu A, Masek E, Ebersole JL (2018) Dietary polyphenols and periodontitis—a mini-review of literature. Molecules 23(7):1786

Battino M et al (1999) Oxidative injury and inflammatory periodontal diseases: the challenge of anti-oxidants to free radicals and reactive oxygen species. Crit Rev Oral Biol Med 10 (4):458–476

Bax BE et al (1992) Stimulation of osteoclastic bone resorption by hydrogen peroxide. Biochem Biophys Res Commun 183(3):1153–1158

Bhattacharya S (2015) Reactive oxygen species and cellular defense system. In: Rani V, Yadav UCS (eds) Free radicals in human health and disease. Springer, India. https://doi.org/10.1007/978-81-322-2035-0_2

Birben E et al (2012) Oxidative stress and antioxidant defense. World Allergy Organ J 5(1):9

Birch-Machin MA, Bowman A (2016) Oxidative stress and ageing. Br J Dermatol 175(S2):26–29

Bobyrev VN et al (1994) Biochemical and ultrastructural changes in the periodontium during the chronic administration of pro-oxidant xenobiotics. Stomatologiia (Mosk) 73(4):57–61

Borgnakke WS (2016) "Non-modifiable" risk factors for periodontitis and diabetes. Current Oral Health Rep 3(3):270–281

Brown DI, Griendling KK (2015) Regulation of signal transduction by reactive oxygen species in the cardiovascular system. Circ Res 116(3):531–549

Burton GW, Ingold KU (1986) Vitamin E: application of the principles of physical organic chemistry to the exploration of its structure and function. Acc Chem Res 19:194–201

Calabrese V et al (2009) Phenolics in aging and neurodegenerative disorders. In: Plant phenolics and human health. John Wiley & Sons, New York, pp 427–451

Chakraborty S, Tewari S (2014) Impact of iron deficiency anemia on chronic periodontitis and superoxide dismutase activity: a cross-sectional study. J Periodontal Implant Sci 44(2):57–64

Choe E, Min DB (2006) Chemistry and reactions of reactive oxygen species in foods. Critic Rev Food Sci Nutrition 46(1):1–22

da Silva JC, Muniz F (2018) The effect of periodontal therapy on oxidative stress biomarkers: a systematic review. J Clin Periodontol 45(10):1222–1237

Di Domenico F et al (2015) Strategy to reduce free radical species in Alzheimer's disease: an update of selected antioxidants. Expert Rev Neurother 15:19–40

Ding Y et al (1995) Modulation of host matrix metalloproteinases by bacterial virulence factors relevant in human periodontal diseases. Oral Dis 1(4):279–286

Dodington DW et al (2015) Higher intakes of fruits and vegetables, β-carotene, Vitamin C, α-Tocopherol, EPA, and DHA are positively associated with periodontal healing after nonsurgical periodontal therapy in nonsmokers but not in smokers. J Nutrition 145 (11):2512–2519

Enhos S et al (2009) Relationship between iron-deficiency anemia and periodontal status in female patients. J Periodontol 80(11):1750–1755

Fresco P et al (2010) The anticancer properties of dietary polyphenols and its relation with apoptosis. Curr Pharm Des 16:114

Garcia MN et al (2011) One-year effects of vitamin D and calcium supplementation on chronic periodontitis. J Periodontol 82(1):25–32

Gharbi A et al (2019) Biochemical parameters and oxidative stress markers in Tunisian patients with periodontal disease. BMC Oral Health 19(1):225

Ghosh N et al (2018) Chapter 4 – reactive oxygen species, oxidative damage and cell death. In: Chatterjee S, Jungraithmayr W, Bagchi D (eds) Immunity and inflammation in health and disease. Academic Press, Amsterdam, pp 45–55

Guarnieri C et al (1991) Enhanced superoxide production with no change of the antioxidant activity in gingival fluid of patients with chronic adult periodontitis. Free Radic Res Commun 15 (1):11–16

Gunther S (1976) Vitamin a acid: clinical investigations with 405 patients. Cutis 17(2):287–290

Hall TJ et al (1995) The role of reactive oxygen intermediates in osteoclastic bone resorption. Biochem Biophys Res Commun 207(1):280–287

Halliwell B (2007) Dietary polyphenols: good, bad, or indifferent for your health? Cardiovasc Res 73(2):341–347

Halliwell B, Gutteridge JMC (1999) Free radicals in biology and medicine. Oxford University Press, New York

Halliwell B, Gutteridge J (2007) Free radicals in biology and medicine. Oxford University Press, New York

Halliwell B et al (1995) Free radicals and antioxidants in food and in vivo: what they do and how they work. Crit Rev Food Sci Nutr 35(1–2):7–20

Hujoel PP, Lingström P (2017) Nutrition, dental caries and periodontal disease: a narrative review. J Clin Periodontol 44(Suppl 18):S79–s84

Jagelavičienė E et al (2018) The Relationship between Vitamin D and periodontal pathology. Medicina (Kaunas, Lithuania) 54(3):45

Jeeva JS et al (2015) Enzymatic antioxidants and its role in oral diseases. J Pharm Bioallied Sci 7 (Suppl 2):S331–S333

Kaludercic N, Deshwal S, Di Lisa F (2014) Reactive oxygen species and redox compartmentalization. Front Physiol 5(285):1–15

Kassebaum NJ et al (2014) Global burden of severe periodontitis in 1990-2010: a systematic review and meta-regression. J Dent Res 93(11):1045–1053
Kaur G et al (2016) Dietary antioxidants and their indispensable role in periodontal health. J Food Drug Anal 24(2):239–246
Keukenmeester RS et al (2014) Effects of sugar-free chewing gum sweetened with xylitol or maltitol on the development of gingivitis and plaque: a randomized clinical trial. Int J Dent Hyg 12(4):238–244
Kim JE, Shklar G (1983) The effect of vitamin E on the healing of gingival wounds in rats. J Periodontol 54(5):305–308
Kimura S, Yonemura T, Kaya H (1993) Increased oxidative product formation by peripheral blood polymorphonuclear leukocytes in human periodontal diseases. J Periodontal Res 28(3):197–203
Kinane DF, Stathopoulou PG, Papapanou PN (2017) Periodontal diseases. Nat Rev Dis Primers 3 (1):17038
Koch W (2019) Dietary polyphenols—important non-nutrients in the prevention of chronic noncommunicable diseases. A systematic review. Nutrients 11:1039–1074
Krishnaiah D, Sarbatly R, Nithyanandam R (2011) A review of the antioxidant potential of medicinal plant species. Food Bioprod Process 89(3):217–233
Krumova K, Cosa G (2016) Overview of reactive oxygen species. In: Nonell S, Flors C (eds) Singlet oxygen: applications in biosciences and Nanosciences. Royal Society of Chemistry, London
Lamont RJ, Jenkinson HF (1998) Life below the gum line: pathogenic mechanisms of Porphyromonas gingivalis. Microbiol Mol Biol Rev: MMBR 62(4):1244–1263
Lee J, Koo N, Min DB (2004) Reactive oxygen species, aging, and Antioxidative nutraceuticals. Compr Rev Food Sci Food Saf 3:21–33
Losada-Barreiro S, Bravo-Díaz C (2017) Free radicals and polyphenols: the redox chemistry of neurodegenerative diseases. Eur J Med Chem 133:379–402
Lushchak VI (2014) Free radicals, reactive oxygen species, oxidative stress and its classification. Chem Biol Interact 224:164–175
Maguire A (2014) ADA clinical recommendations on topical fluoride for caries prevention. Evid Based Dent 15(2):38–39
Marrocco I, Altieri F, Peluso I (2017) Measurement and clinical significance of biomarkers of oxidative stress in humans. Oxidative Med Cell Longev 2017:6501046
Meisel P et al (2005) Magnesium deficiency is associated with periodontal disease. J Dent Res 84 (10):937–941
Mikkelsen RB, Wardman P (2003) Biological chemistry of reactive oxygen and nitrogen and radiation-induced signal transduction mechanisms. Oncogene 22(37):5734–5754
Miller DR, Lamster IB, Chasens AI (1984) Role of the polymorphonuclear leukocyte in periodontal health and disease. J Clin Periodontol 11(1):1–15
Muñoz-Carrillo JL et al (2019) Pathogenesis of periodontal disease. In: Mohammed N, Yussif A (eds) Pathogenesis of periodontal Disease, periodontal Disease – diagnostic and adjunctive non-surgical considerations. Intech Open, London
Najeeb S et al (2016) The role of nutrition in periodontal health: an update. Nutrients 8(9):530
Neiva RF et al (2005) Effects of vitamin-B complex supplementation on periodontal wound healing. J Periodontol 76(7):1084–1091
Nguyen TT et al (2017) Salivary oxidative stress biomarkers in chronic periodontitis and acute coronary syndrome. Clin Oral Investig 21(7):2345–2353
Novakovic N et al (2014) Salivary antioxidants as periodontal biomarkers in evaluation of tissue status and treatment outcome. J Periodontal Res 49(1):129–136
Nussbaum G, Shapira L (2011) How has neutrophil research improved our understanding of periodontal pathogenesis? J Clin Periodontol 38(s11):49–59
Orbak R et al (2007) Effects of zinc deficiency on oral and periodontal diseases in rats. J Periodontal Res 42(2):138–143

Ozcan A, Ogun M (2015) Biochemistry of reactive oxygen and nitrogen species. In: Joghi S, Gowder T (eds) Basic principles and clinical significance of oxidative stress. INTECH. isbn:978-953-51-2200-5

Palmer CA, Burnett DJ, Dean B (2010) It's more than just candy: important relationships between nutrition and oral health. Nutr Today 45(4):154–164

Pavlin M et al (2016) The chemistry of neurodegeneration: kinetic data and their implications. Mol Neurobiol 53(5):3400–3415

Pihlstrom BL, Michalowicz BS, Johnson NW (2005) Periodontal diseases. Lancet 366 (9499):1809–1820

Pisoschi AM, Pop A (2015) The role of antioxidants in the chemistry of oxidative stress: a review. Eur J Med Chem 97:55–74

Pizzino G et al (2017) Oxidative stress: harms and benefits for Human Health. Oxidative Med Cell Longev 2017:8416763–8416763

Rahmani M et al (2015) Evaluation of Total antioxidant capacity of saliva in high school students. Glob J Health Sci 8(4):89–94

Rani V, Yadav U (2014) Free radicals in human health and disease. Springer, New York

Ray PD, Huang B-W, Tsuji Y (2012) Reactive oxygen species (ROS) homeostasis and redox regulation in cellular signaling. Cell Signal 24(5):981–990

Reynolds MA (2014) Modifiable risk factors in periodontitis: at the intersection of aging and disease. Periodontology 2000 64(1):7–19

Rosini M et al (2014) Oxidative stress in Alzheimer's Disease: are we connecting the dots? J Med Chem 57:2821–2831

Ross F, Ross AB (1977) Selected specific rates of reactions of transients from water in aqueous solution III Hydroxyl radical and perhydroxyl radical and their radical ions. United States, p 122

Salonen JT (1998) Epidemiological studies on antioxidants, lipid peroxidation and atherosclerosis. In: Seiler JP, Autrup JL, Autrup H (eds) Diversification in toxicology — man and environment: proceedings of the 1997 EUROTOX congress meeting held in århus, Denmark, June 25–28, 1997. Springer, Berlin, Heidelberg, pp 249–267

San Miguel SM et al (2011) Use of antioxidants in oral healthcare. Compend Contin Educ Dent 32 (9):E156–E159

Sawamoto Y et al (2005) Detection of periodontopathic bacteria and an oxidative stress marker in saliva from periodontitis patients. Oral Microbiol Immunol 20(4):216–220

Scalbert A et al (2005) Dietary polyphenols and the prevention of diseases. Crit Rev Food Sci Nutr 45:287–306

Schieber M, Chandel NS (2014) ROS function in redox signaling and oxidative stress. Curr Biol 24:453–462

Schieber M, Navdeep S (2014) Chandel, ROS function in redox signaling and oxidative stress. Curr Biol 24(10):R453–R462

Sezer U et al (2016) Effect of chronic periodontitis on oxidative status in patients with psoriasis and psoriatic arthritis. J Periodontol 87(5):557–565

Sheiham A, Netuveli GS (2002) Periodontal diseases in Europe. Periodontol 2000 29:104–121

Sidi AD, Ashley FP (1984) Influence of frequent sugar intakes on experimental gingivitis. J Periodontol 55(7):419–423

Sies H, Berndt C, Jones DP (2017) Oxidative stress. Annu Rev Biochem 86(1):715–748

Silva N et al (2015) Host response mechanisms in periodontal diseases. J Appl Oral Sci 23 (3):329–355

Sima C, Glogauer M (2014) Neutrophil dysfunction and host susceptibility to periodontal inflammation: current state of knowledge. Current Oral Health Rep 1(2):95–103

Smith M, Seymour GJ, Cullinan MP (2010) Histopathological features of chronic and aggressive periodontitis. Periodontology 2000 53(1):45–54

Tada A, Miura H (2019) The relationship between vitamin C and periodontal diseases: a systematic review. Int J Environ Res Public Health 16(14):2472

Takane M et al (2002) New biomarker evidence of oxidative DNA damage in whole saliva from clinically healthy and periodontally diseased individuals. J Periodontol 73(5):551–554
Tanaka K et al (2014) Calcium intake is associated with decreased prevalence of periodontal disease in young Japanese women. Nutr J 13(1):109
Tomofuji T et al (2009) Periodontitis and increase in circulating oxidative stress. Jap Dent Sci Rev 45(1):46–51
Tóthová L, Celec P (2017) Oxidative stress and Antioxidants in the diagnosis and therapy of periodontitis. Front Physiol 8:1055
Vo TTT et al (2020) The promising role of antioxidant phytochemicals in the prevention and treatment of periodontal Disease via the inhibition of oxidative stress pathways: updated insights. Antioxidants 9(12):1211
Waddington RJ, Moseley R, Embery G (2000) Reactive oxygen species: a potential role in the pathogenesis of periodontal diseases. Oral Dis 6(3):138–151
Wanasundara PKJPD, Shahidi F (2005) Antioxidants: science, technology, and applications. In: Shahidi F (ed) Bailey's industrial oil and fat products. John Wiley & Sons, New York, pp 431–489
Wang K et al (2013) Redox homeostasis: the linchpin in stem cell self-renewal and differentiation. Cell Death Dis 4:e537. https://doi.org/10.1038/cddis.2013.5
Wang Y, Andrukhov O, Rausch-Fan X (2017) Oxidative stress and antioxidant system in periodontitis. Front Physiol 8:910
Watson RR, Preedy VR, Zibadi S (2013) Polyphenols in human health and disease. Elsevier Science, Amsterdam
Woelber JP, Tennert C (2020) Chapter 13: diet and periodontal diseases. Monogr Oral Sci 28:125–133
Woelber JP et al (2016) An oral health optimized diet can reduce gingival and periodontal inflammation in humans – a randomized controlled pilot study. BMC Oral Health 17(1):28
Wu D et al (2020) Decreased hemoglobin concentration and iron metabolism disorder in periodontitis: systematic review and meta-analysis. Front Physiol 10:1620
Zhang J et al (2016) ROS and ROS-mediated cellular signaling. Oxid Med Cell Longev 2016:4350965. https://doi.org/10.1155/2016/4350965
Zheng DX et al (2021) Periodontal disease and emotional disorders: a meta-analysis. J Clin Periodontol 48(2):180–204

Phytosterols as Functional Compounds and Their Oxidized Derivatives

Magdalena Rudzińska

1 Introduction

Sterols are alicyclic alcohols of the steroid group. These are crystalline compounds with high melting temperatures. In fats, they occur as free compounds, fatty acid esters, glycosides, and acyl derivatives. Sterols are insoluble in water but soluble in hydrophobic solvents, so they can be classed as lipids. They are the main component (making up over 50%) of the unsaponifiable fraction of many fats and oils. Sterols have a cyclopentanoperhydrophenanthrene ring with a hydroxyl group at C3 in the β-position to the methyl group at C10, and a side chain at C17 (Figs. 1, 2, and 3). They differ in the presence of a double bond in the B-ring and in the side chain. With their system of two conjugated double bonds in the B-ring, they are provitamins of D vitamins.

Epidemiological studies have established that a diet rich in plant sterols or its esters has protective properties against cardiovascular diseases (CVD), and may alleviate breast, colon, and lung carcinogenesis (Awad et al. 2001; Plat et al. 2015; Ramprasath and Awad 2015). Plant sterols act through multiple modes of action, and can promote cancer cell apoptosis and inhibit their growth, invasiveness, through reducing the angiogenesis (Ramprasath and Awad 2015; Woyengo et al. 2009).

Phytosterols undergo autoxidative degradation which may occur inside or outside of the human body, leading to the formation of phytosterol oxidation products (POPs), known also as oxyphytosterols (Hovenkamp et al. 2008). The most common pathway of autoxidative degradation occurs via free radical mechanisms, which includes initiation by reactive oxygen species (Lengyel et al. 2012). These compounds are generated more rapidly during thermal processes and storage of food

M. Rudzińska (✉)
Department of Food Technology of Plant Origin, Faculty of Food Science and Nutrition, Poznań University of Life Sciences, Poznań, Poland
e-mail: magdalena.rudzinska@up.poznan.pl

C. Bravo-Diaz (ed.), *Lipid Oxidation in Food and Biological Systems*,
https://doi.org/10.1007/978-3-030-87222-9_16

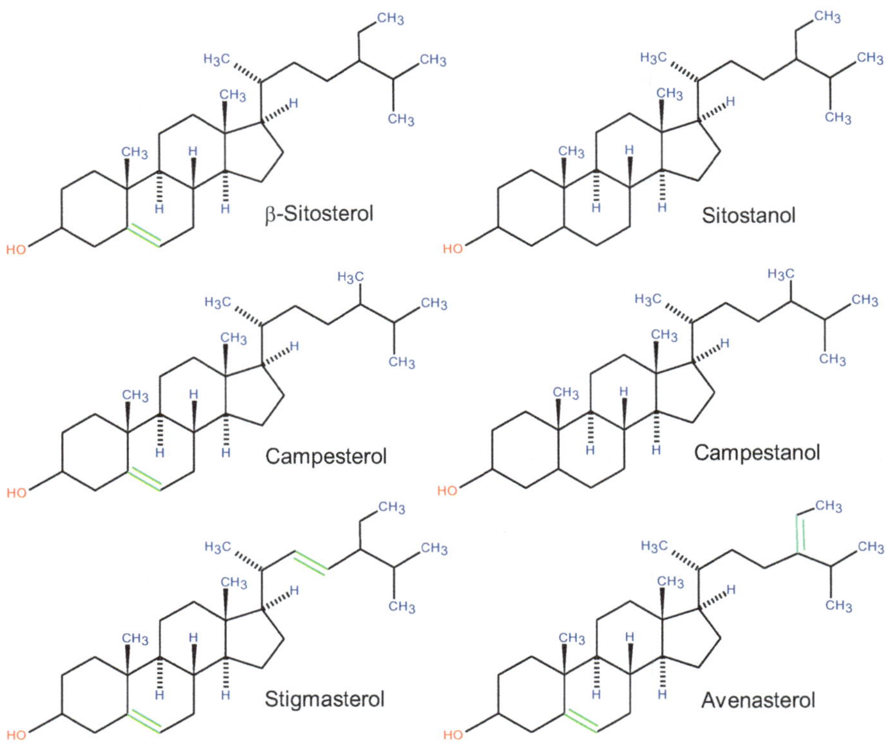

Fig. 1 The main 4-desmethylsterols/stanols

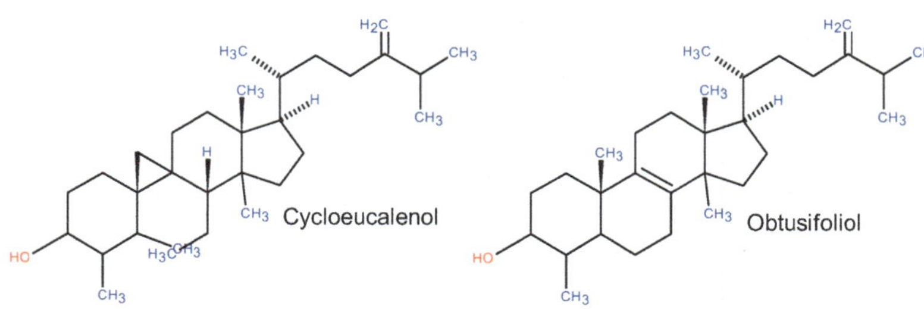

Fig. 2 The main 4-monomethylsterols

products (Barriuso et al. 2015; Rudzińska et al. 2005, 2014). Structurally, phytosterol oxidation products are similar to cholesterol oxides, which have been shown to have an adverse effect on the human body (Poli et al. 2009; Lin et al. 2016). The biological properties of phytosterol oxidation products, such as proatherogenicity and cytotoxicity, have been described, but the data are controversial and more research is needed to firmly establish their detrimental properties (Wang and Lu 2018).

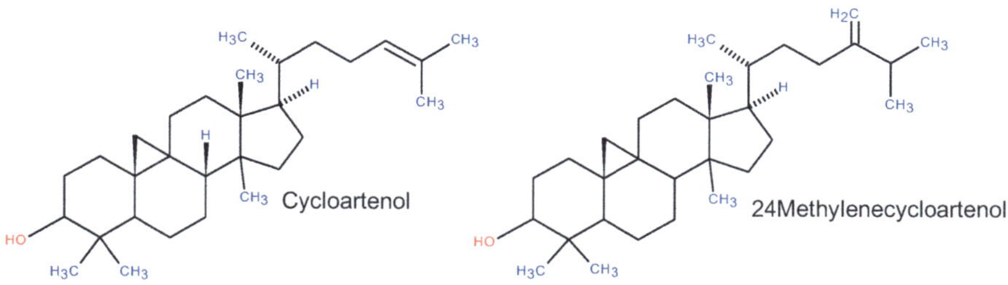

Fig. 3 The main 4.4′-dimethylsterols

This review aims to provide an overview of studies investigating plant sterols in food products, their degradation during thermo-oxidation, and the formation of oxyphytosterols.

2 Phytosterols in Food Products

The main sources of plant sterols are vegetable oils and nuts, as well as fruits, vegetables, legumes and cereals (da Costa et al. 2010; Oliveira et al. 2008). Wild rice lipids contain very large amounts of phytosterols, from 70 mg to 145 mg/g (Przybylski et al. 2009), followed by wheat germ oil at 9 mg/g (Ying et al. 2018), rapeseed oil at 8 mg/g (Kasprzak et al. 2020), and safflower and soybean oils at 4 mg/g (Xu et al. 2018). The main factor in vegetable oil quality, especially in terms of bioactive compound content, is the maturity and quality of the oilseeds. Apart from the cultivar, the quantity of phytosterols in vegetable oils is influenced by seed moisture, maturity and storage conditions (Vlahakis and Hazebroek 2000). The rate of phytosterol degradation in rapeseed is affected by moisture content and temperature during storage. Seeds with moisture contents of 10 and 12.5% stored at 30 °C were shown to undergo a 13–16% reduction in phytosterol after 18 days of storage. Much greater losses of 25 and 58% were seen when the seeds were stored at 15.5% moisture and at 25 and 30 °C (Gawrysiak-Witulska and Rudzińska 2012).

Various seeds, kernels, and nuts are used as sources in the production of edible cold-pressed oils. The content and composition of phytosterols in these oils depend on the variety of the source. Alongside traditional oils from olives, rapeseed, linseed, and pumpkin, unconventional oils from melon, apricot, and plum kernels, and from apple, pear, and cherry seeds, and others are produced and are consumed by customers as “healthy” food (Zevenbergen et al. 2009). In recent years, the range of cold-pressed vegetable oils available on the market has increased significantly. The content of phytosterols in these oils varies; for example, *Camelina sativa* L. seed oils contained 3.3–4.4 mg/g phytosterols (Ratusz et al. 2018), blackcurrant seed oil contained 2.5 mg/g phytosterols, and argan oil 1.6 mg/g (Ying et al. 2018). It is understood that the content of phytosterols in cold-pressed vegetable oils is higher than in refined oils because these compounds are lost during the refining process.

Van Hoed et al. (Van Hoed et al. 2010) demonstrated that the content of minor compounds was slightly higher in extracted oils than in pressed oils. This can be explained by the more efficient extraction of those compounds into the solvent. The roasting of oilseeds prior to cold-pressing leads to the formation of a pleasant flavor and is also reported to improve the yield of oil and its oxidative stability (Mildner-Szkudlarz et al. 2019). The phytosterol content in oils pressed from berry seeds decreased by 26% in blackcurrant and 19% in chokeberry oils, but significant losses were not detected for rapeseed oil (Mildner-Szkudlarz et al. 2019; Siger et al. 2015).

Beside vegetable oils, other good sources of plant sterols include nut, seeds, cereals, vegetables, and fruit. Sesame contains 3.6 mg of phytosterols per gram of seed, wheat germ contains 3.4 mg/g, beans contain 76 mg/g, and passion fruit contains 0.4 mg/g (Gupta et al. 2011). The content of total phytosterols in artichoke, cauliflower, and cabbage is 0.5, 0.4 and 0.3 mg/g of fresh weight, respectively. Legumes, chickpeas, and lentil contained about 1.2 mg/g fresh materials, while pistachios had about 2.4 mg of phytosterols per gram of fresh material (Jiménez-Escrig et al. 2006).

In addition to products that are natural sources of phytosterols, foods enriched with phytosterols or their esters are found on the market. Commercial spreadable fats, margarines, milk, and yoghurt formulated with phytosterols are available. The concentration of phytosterols in these items ranges from 80 mg of free phytosterol/stanols to 90 mg of phytosteryl/stanyl esters/g of product, which is equivalent to 54 mg free phytosterols/stanols per gram of product (Rudzińska et al. 2014; Raczyk et al. 2018a). Phytosterols are used as functional additives in different food products, and are obtained from deodorized distillates of vegetable oils or tall oil soap, a waste product of the sulphate cellulose production process (Kasim et al. 2010; Rousková et al. 2011).

The chemical structure of phytosterols defines their chemical activity. In order to fully utilize their healthful properties, it is necessary to minimize losses during processing. The first stage where phytosterols undergo degradation is the post-harvest processing. The rate of phytosterol losses is here dependent on the moisture content of seeds and the temperature during storage. Seeds with moisture contents of 10% and 12.5% stored at 25 °C showed a reduction of 11–12% in total sterol contents after 18 days of storage (Gawrysiak-Witulska et al. 2012). After air drying of rapeseeds at 40–100 °C, the degradation of phytosterols did not exceed 5%, while after drying at 120 and 140 °C these losses amounted to 17–50% (Gawrysiak-Witulska et al. 2020). During storage of rapeseeds in a silo where self-heating phenomenon was detected, the decrease in phytosterols was 10% after 6 days and 22% after 13 days of storage (Gawrysiak-Witulska et al. 2018). The content of total phytosterols in cold-pressed rapeseed oils ranged from 5.4 to 11.9 mg/g (Mildner-Szkudlarz et al. 2019; Yang et al. 2013a). When rapeseed oil was cold pressed from roasted seeds, the content of phytosterols was unchanged or increased (Mildner-Szkudlarz et al. 2019; Azadmard-Damirchi et al. 2010). During refining of vegetable oils, losses of phytosterols are observed. Major degradation of phytosterols (by 15%) was observed during neutralization of crude soya bean oil, and during the refining process the decrease in total phytosterols was 20% (Costa et al. 2011). Although crude rice bran oil had a high level of phytosterols, at 136.2–137.6 mg/g, the physical refining process reduced this to 82.0–89.5 mg/g (Sawadikiat and Hongsprabhas 2014). These compounds evaporated and accumulated in the deodorizer distillate, leaving 15.9–33.9 mg/g in the product (Sawadikiat and Hongsprabhas 2014). The rate of phytosterol degradation in refined oil was slower than in crude oil during accelerated storage, amounting to 12% and 31%, respectively (Chew et al. 2017). During interesterification at 90 and 120 °C, the total sterol level in an olive oil and palm stearin blend (1:1 w/w) declined by 3 and 5% (Azadmard-Damirchi and Dutta 2008). The heating of pressed, refined, partially hydrogenated rapeseed oil at 170 °C for 48 h resulted in a loss of phytosterols (Kmiecik et al. 2020). The greatest degradation was determined in partially hydrogenated and pressed oil, at 21.2% and 24.5%, respectively.

The degradation of phytosterols during thermo-oxidation of model systems and food products has been investigated (Rudzińska et al. 2009, 2010).

The experiment carried out in the model system allowed for the elimination of many external factors influencing the course of the sterol degradation process (e.g. antioxidants, pro-oxidants, light, unsaturated matrix) and made it possible to monitor the transformations of the studied sterols under strictly controlled conditions. The obtained results made it possible to determine the rate constants of the sterol degradation reaction and to formulate kinetic equations that provide a lot of information on the behavior of these compounds at elevated temperatures (Table 1). Based on the literature data, it was assumed that the degradation process of the studied sterols follows the 1st order reactions (Hu and Chen 2002). There are no data in the literature that would allow a comparison of the degradation rate constants of various phytosterols. Rudzińska et al. (2009) showed that, after heating at 180 °C for 24 h, 66% of sitosterol was transformed into degradation products. The degradation of sterols during heating at 180 °C for 5 min and 30 min was 60% and 88%, respectively, and when time was extended to 360 min, 95% of the initial sterols disappeared (Barriuso et al. 2012). Raczyk et al. (2017a) showed that thermo-oxidative degradation of phytosterol esters is strongly affected by the unsaturation of fatty acids and by parameters affecting the chemical reaction rate, such as temperature and time. Julien-David et al. (2014) demonstrated that losses of sitosteryl oleate in Pro-Activ margarine during heating at 170 °C and 200 °C for 120 min were at 72% and 77%, respectively. The total levels of phytosterols and phytostanols in margarines enriched with plant sterols esters decreased by 20% and 30% when they were stored for 18 weeks at 4 °C and 20 °C, respectively (Rudzińska et al. 2014). The rate of phytosterol degradation depends on many factors, including chemical structure, temperature, heating duration, and oxygen access. A broad group of derivatives was formed. The first stage of sterol degradation was autoxidation, where phytosterol oxidation products (oxides, POPs) were created. Volatile compounds and fragmented sterols were identified, as were dimers, trimers, and oligomers.

Table 1 Modeled first-order equations and rate constants for degradation of sterols heated at 60, 120 and 180 °C (Rudzińska 2011)

Sterols	First-order degradation	Degradation rate constant (min^{-1})	Determination coefficient R^2
60 °C			
Brassicasterol	$C_t = 31.34[exp(-0.1765 \times 10^{-3}t)]$	0.1765×10^{-3}	0.8793
Campesterol	$C_t = 126.73[exp(-0.1384 \times 10^{-3}t)]$	0.1384×10^{-3}	0.8752
Stigmasterol	$C_t = 50.43[exp(-0.1770 \times 10^{-3}t)]$	0.1770×10^{-3}	0.8455
β-Sitosterol	$C_t = 532.82[exp(-0.1451 \times 10^{-3}t)]$	0.1451×10^{-3}	0.8852
Δ5-Avenasterol	$C_t = 47.39[exp(-0.6467 \times 10^{-3}t)]$	0.6467×10^{-3}	0.9636
120 °C			
Brassicasterol	$C_t = 31.34[exp(-0.6593 \times 10^{-3}t)]$	0.6593×10^{-3}	0.8764
Campesterol	$C_t = 126.73[exp(-0.6326 \times 10^{-3}t)]$	0.6326×10^{-3}	0.8942
Stigmasterol	$C_t = 50.43[exp(-0.6652 \times 10^{-3}t)]$	0.6652×10^{-3}	0.8704
β-Sitosterol	$C_t = 532.82[exp(-0.6012 \times 10^{-3}t)]$	0.6012×10^{-3}	0.9374
Δ5-Avenasterol	$C_t = 47.39[exp(-1.6147 \times 10^{-3}t)]$	1.6147×10^{-3}	0.9115
180 °C			
Brassicasterol	$C_t = 31.34[exp(-1.2249 \times 10^{-3}t)]$	1.2249×10^{-3}	0.9308
Campesterol	$C_t = 126.73[exp(-1.0517 \times 10^{-3}t)]$	1.0517×10^{-3}	0.9495
Stigmasterol	$C_t = 50.43[exp(-1.3361 \times 10^{-3}t)]$	1.3361×10^{-3}	0.9370
β-Sitosterol	$C_t = 532.82[exp(-1.1643 \times 10^{-3}t)]$	1.1643×10^{-3}	0.9627
Δ5-Avenasterol	$C_t = 47.39[exp(-2.6795 \times 10^{-3}t)]$	2.6795×10^{-3}	0.9821

3 Biological Properties of Phytosterols

3.1 Cholesterol-Lowering Properties of Phytosterols

Phytosterols are not synthesized in the human body and must be obtained from the diet. For decades it has been known that they can decrease the level of cholesterol in the blood.

The total estimated intake of plant sterols in the typical western diet is about 300 mg/day (Klingberg et al. 2012; Sioen et al. 2011), while vegetarian and plant-based diets deliver 300–600 mg of phytosterols/day (Jaceldo-Siegl et al. 2017). In 2010, Health Canada and the US FDA approved that plant sterol health claims that consumption of 2 g of phytosterols per day could significantly reduce serum LDL-cholesterol concentration, potentially reducing the risk of cardiovascular disease. The National Heart Foundation in Australia recommends that adults with a high risk of cardiovascular diseases consume 2–3 g of phytosterols per day in two or three serving of fortified foods. Phytosterols help decrease serum LDL cholesterol in people with hypercholesterolemia, dyslipidemia, and other cardiovascular risks, but it still has not been established whether dietary phytosterols from natural foods have beneficial effects on cardiovascular events (Kaur and Myrie 2020).

3.2 Neuroprotective Properties of Phytosterols

The central nervous system is the most cholesterol-rich part of the body in mammals. Cholesterol homeostasis is essential for proper brain functioning, and dysregulation of cholesterol metabolism can lead to neurological problems. Multiple sclerosis, Alzheimer's disease, and dementia are neurological diseases characterized by a disturbed cholesterol metabolism (Dierckx et al. 2018). The concentration of plant sterols in the brains of mice being fed a plant sterol ester-enriched diet for 6 weeks was found to double or triple (Vanmierlo et al. 2012). After blocking intestinal plant sterol uptake, the level of phytosterols did not changed for the next 6 months. Vanmierlo et al. (Vanmierlo et al. 2012) concluded that, over a 6-month period, plant sterol accumulation in the murine brain is virtually irreversible. Shuang et al. (2016) studied the molecular mechanism of phytosterols, which may activate peroxisome proliferator receptor Sirtuin1 (PPARs-SIRT-1), preventing dementia in the elderly patients.

3.3 Anticancer and Anti-inflammatory Properties of Phytosterols

Phytosterols are recommended for the reduction of cardiovascular disease risk, and are also promising anticancer compounds, which could reach a large population at low cost, with a well-understood safety profile and an understanding of the mechanism of action (Cioccoloni et al. 2020). Alvarez-Sala et al. (2018) demonstrated that β-sitosterol, the main plant sterol in food products, can be regarded as an anticancer functional compound against breast cancer, colon cancer, and cervical cancer. They observed anticancer activity through the induction of DNA fragmentation and apoptosis of cancer cells. Other reviews of the anticancer properties of phytosterols have been published (Woyengo et al. 2009; Bradford and Awad 2010; Shahzad et al. 2017).

Stigmasterol shows anti-inflammatory potential and anticatabolic properties (Gabay et al. 2010). A regular intake of food products enriched with phytosterols was not noted to significantly change C-reactive protein (CRP) plasma levels, and overall did not change other representative inflammatory biomarkers (Rocha et al. 2016).

3.4 Mechanisms of Phytosterols Absorption

Several possible mechanisms have been proposed for the absorption of phytosterols and for their properties as lowering agents of the levels of total cholesterol (TC) and its low-density lipoprotein fraction (LDL). Competition between cholesterol and plant sterols in micellar space in the small bowel has been proposed, as has an effect on the epithelial cell level (Plat and Mensink 2002). The incorporation of plant sterols into mixed micelles in the intestinal tract has been shown to increase cholesterol excretion with the feces (Marangoni and Poli 2010). It has been hypothesized that plant sterols stimulate the ABCG-5/ABCG-8 complex and increase cholesterol excretion (Schmitz et al. 2001). The polymorphism of these complexes may be responsible for the ability of phytosterols to displace cholesterol from intestinal micelles. Brauner et al. (2012) suggested that phytosterols reduce 27-hydroxycholesterol formation through competitive inhibition of sterol 27-hydroxylase (CYP27), a member of the cytochrome P450, thus preventing cholesterol absorption. This favors sterol reexcretion into the gut lumen by ABCG5/G8. Thus, antagonism of phytosterols in enterocytes reduces the fractional cholesterol absorption by the ABCA1 pathway. This could explain the high degree of variability observed between individual levels of consumed plant sterols.

The absorption of phytosterols depends on the side chain length, ranging from about 0.5% for sitosterol to 1.9% for campesterol (Fernandez and Vega-López 2005), while cholesterol absorption averages $56.2 \pm 12\%$ in normal subjects (Bosner et al. 1999). Genetic factors, individual variability, and dietary conditions affect the absorption of these compounds in humans (Sanclemente et al. 2009).

4 Oxidation of Plant Sterols

4.1 Autoxidation

Like other unsaturated lipids, sterols are prone to free radical and non-radical mediated autoxidation (Figs. 4 and 5). Autoxidation of plant sterols can occur in the ring and side chain parts of the sterol molecule, producing a variety of oxides. Autoxidation is affected by external factors, such as access to oxygen, light, type of matrix, and the presence of metals. Most modeling studies that have been conducted show that these factors have a significant effect on the oxidative changes of sterols

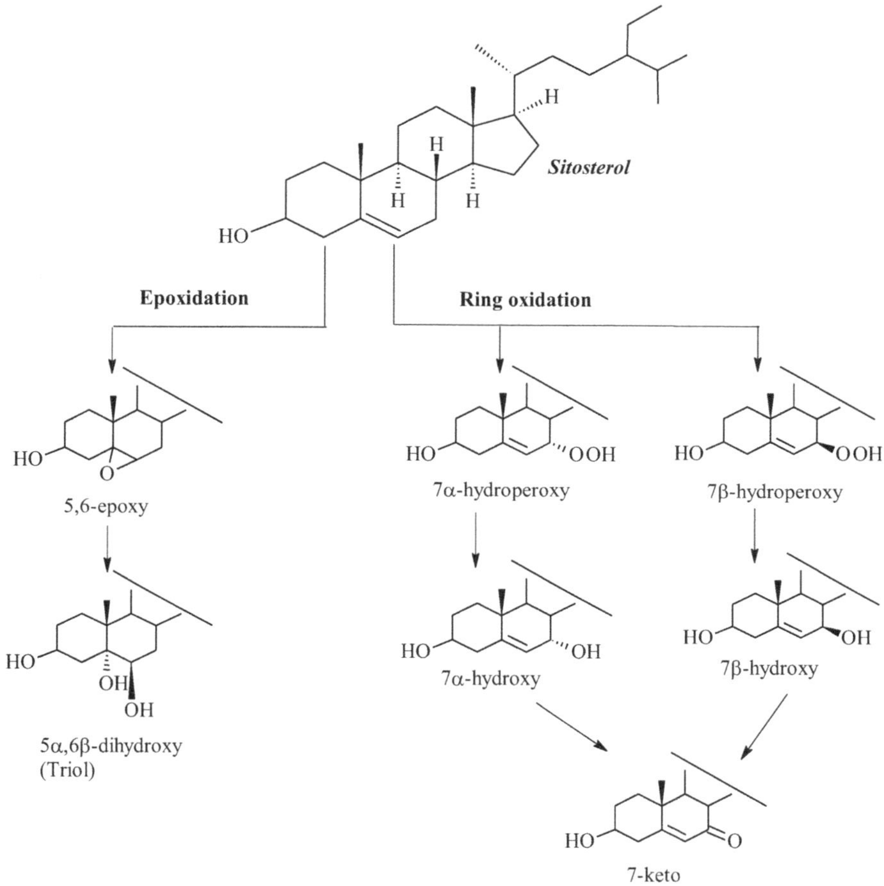

Fig. 4 Sterol oxidation products formed during autoxidation mediated by free radical reactions

and on the formation of their oxidized derivatives (Kemmo et al. 2005). The primary oxidation products are 7α- and 7β-hydroperoxides, which are unstable and decompose into the secondary oxidation products 7α- and 7β-hydroxysterols, which in turn generate 7-ketosterols (Lengyel et al. 2012). At the same time, the interaction between the sterol and its free radicals produces 5α,6α- and 5β,6β-epoxysterols, which are transformed into 3β,5,6β-triol (Iuliano 2011). Hydrogen abstraction, either by a primary initiating species or by secondary radicals formed during the chain-propagation steps, may also occur at carbon C25, producing unstable 25-hydroperoxide, which then form more stable 25-hydroxysterol (Iuliano 2011). However, the oxidation of phytosterol side chains is less well understood, and these derivatives are probably relatively rare compared with the 7-hydroxysterols and epoxysterols.

So far, studies of sterol oxidation have investigated the secondary oxidation products of the ring part of the molecule (Conchillo et al. 2005; García-Llatas and

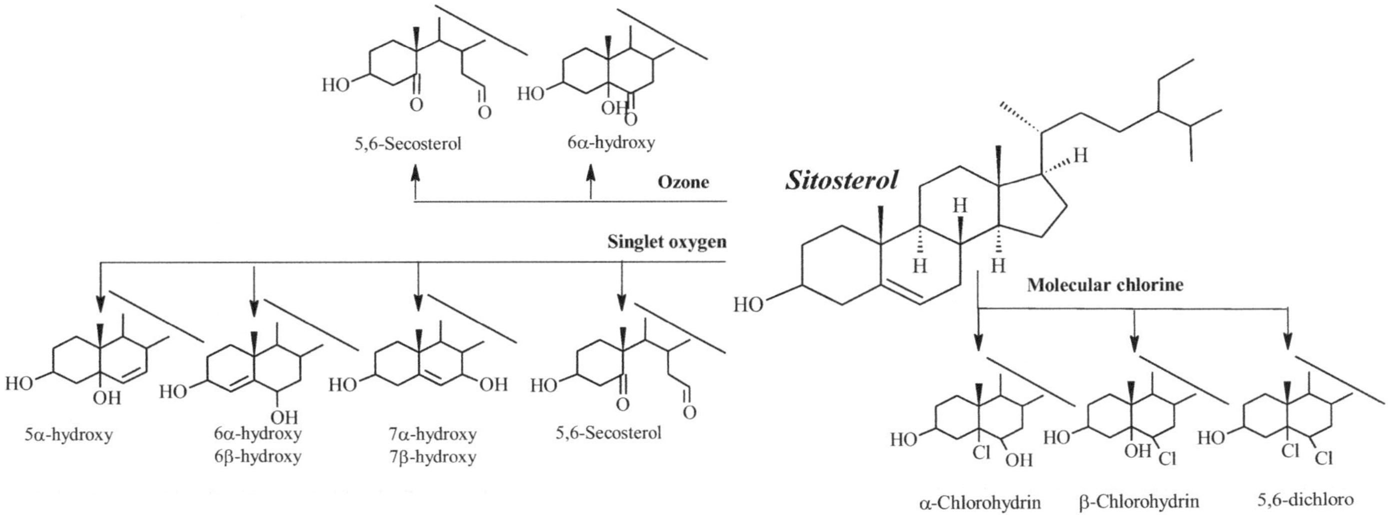

Fig. 5 Sterol oxidation products formed during autoxidation mediated by non-free radical reactions

Rodríguez-Estrada 2011). 25-hydroxycampesterol and 25-hyroxysitosterol were identified in the pan-frying of canola oil with the addition of free phytosterols (Soupas et al. 2007). The formation of side-chain phytosterol oxides probably depends on the chemical structure of the C-24 group (Lengyel et al. 2012).

During the thermo-oxidation of phytosterols, six oxyphytosterols are usually detected: 7α-hydroxysterol (7αOH), 7β-hydroxysterol (7βOH), β-epoxysterol (β-epoxy), α-epoxysterol (α-epoxy), triol, and 7-ketosterol (7-keto).

4.2 Photooxidation

Photooxidation of sterols is caused by a slow but self-sustained process of lipid photooxidation. However, information on phytosterol photooxidation is limited. Yang et al. (2020) identified phytosterol photooxidation products and suggested a mechanism of chemical reactions for light-induced oxidation. The photooxidation reaction of sterols can be divided into two types: (I) a process without singlet oxygen and (II) a process involving singlet oxygen (Fig. 5). In food products type (II) is the main mechanism.

The pathway of sterol oxidation via singlet oxygen was described by Iuliano (2011). Yang et al. (2020) proposed three possible reaction mechanisms for phytosterol photooxidation: concerted, diradical, and free radical reaction.

The photooxidation of phytosterols can be accelerated by an increase in light intensity and by the presence of photosensitizers, whereas the unsaturated matrix inhibited the formation of oxides (Zhao et al. 2019). 7β-hydroxyphytosterols were the main derivatives formed during photooxidation of phytosterols in model samples, which represented about one third of the total POPs, followed by 7α-hydroxy, β-epoxy, 7keto, α-epoxy, and 6β-hydroxy derivatives. Lengyel et al. (2012) suggested that the photooxidation of phytosterols is located at the C5 and C6, rather than at the C7, and that mainly 5α-, 6α and 6β-hydroperoxides are formed. The formation of these compounds is typical of the singlet oxygen reaction, which occurs when molecular oxygen receives energy from photoactivation (Iuliano 2011).

5 Phytosterol Oxidation Products in Foods

When free phytosterols were heated at 120 °C, the POP content was higher than after heating at 60 or 180 °C. When the duration of 60 °C heating was increased, the total POPs also increased. At a temperature of 180 °C, the quantity of oxidized sterols decreased when the heating time was increased. This indicates that oxidized sterols were the main precursors involved in the formation of other components during thermo-oxidative degradation (Rudzińska et al. 2009). The degradation of stigmasterol esters was examined after heating at 60 and 180 °C (Raczyk et al. 2017a). These data showed that heating at a lower temperature induced the formation of oxidative derivatives of stigmasterol, whereas heating at higher temperature caused faster degradation and stimulated interaction. This finding points to the instability and potential chemical reactivity of phytosterol oxidation products, which may lead to the formation of other products, such as low-molecular-weight compounds that could affect sensory properties, or oligomers formed as a result of condensation (Sosińska et al. 2013). Phytosterols, as surface active compounds, may be particularly prone to oxidation when incorporated in emulsions, such as margarine. Cercaci et al. (2007) suggested that phytosterols in an emulsion oxidize at the droplet interface. Phytosterol oxidation products have been detected in enriched margarines stored at 4 and 20 °C for 18 weeks (Rudzińska et al. 2014). 7-Hydroxy derivatives dominated among all the oxidized phytosterols, and their concentration had increased threefold by the end of storage. Epoxy derivatives reached their peak after 6 weeks of storage at 20 °C, and thereafter continually decreased. Menéndez-Carreño et al. (2016) found POPs in food products that had been cooked or baked using margarine with added phytosterols. The highest levels were detected in vegetables, potatoes, and meat, and ranged from 0.5 to 9.8 mg/100 g in fish (0.1–3.2 mg/100 g), eggs (0.6 mg/100 g), and bakery products (0.4–1.1 mg/100 g). Phytosteryl/stanyl ester-enriched margarines were heated in a microwave, pan, and oven and approximately 20% of the ester losses were explained by the formation of POPs (Scholz et al. 2016). The smallest increase in POPs was noted for microwave heating, and the highest was for the oven process.

The level of phytosterols in food products is much lower than in functional foods enriched with these compounds; as a result, the food products have lower levels of phytosterol oxidation products. POPs have been found in crude, refined, and hydrogenated vegetable oils (Kmiecik et al. 2020; Bortolomeazzi et al. 2003; Johnsson and Dutta 2006; Zhang et al. 2005). The quantity of POPs in hot-pressed rapeseed oil was 128.2 μg/g, and increased after refining to 166.6 μg/g; when partially hydrogenated with iodine values of 90 and 79, this oil contained 112.3 and 95.7 μg POP/g, respectively (Kmiecik et al. 2020). Fried potato chips contained 0.05–0.68 mg POP per 100 g (Tabee et al. 2008). French fries cooked at 225 °C for 15 min contained 9.0–68.8 μg POP/g fat, and 7-keto- and epoxy-derivatives of phytosterols were also detected (Derewiaka and Obiedziński 2012). Readymade fish products contained 2.5–13.9 μg of POPs/g of fat when pan-fried, but fish fillets contained 32.3–38.7 μg POP/g fat. Only 7-keto-phytosterols were identified in these products. Boiled noodles and pan-fried meat chops were found to contain only 7-keto-derivatives, with their total content at respectively 1.8–4.0 and 0.2–1.7 μg/g fat (Derewiaka and Obiedziński 2012).

This study has revealed the importance of different processes on the stability of sterols in food products, showing the need to monitor sterol oxidation products, given their potential negative health effects.

6 Biological Properties of Oxyphytosterols

Phytosterol oxidation products are present in food products, although at very low concentrations. In the recent years, specific and sensitive analytical techniques have allowed precise measurement of phytosterol oxidation products in human plasma and tissues, and opened new means to advance research in this field (Barriuso et al. 2014). POPs, like cholesterol oxidation products, show atherogenic properties (Tomoyori et al. 2004). Plat et al. (2014) have suggested that dietary POPs may increase the risk of severe atherosclerotic lesions. Luister et al. (2015) and Baumgartner et al. (2017) observed that patients with coronary problems are usually labelled with elevated levels of POPs in serum. The cytotoxicity of POPs was first demonstrated by Adcox et al. (2001). Ryan et al. (2005) analyzed the cytotoxicity of sitosterol oxidation products, while Kenny et al. (2012) and O'Callaghan et al. (2013) investigated the cytotoxicity of dihydrobrassicasterol and campesterol oxides. The inflammatory properties of POPs were assessed by Plat et al. (2014), Vejux et al. (2012), and Alemany et al. (2013). These compounds also play a role in the formation of reactive oxygen species during cell metabolism and affect this metabolism (Yang et al. 2013b). Oxidative degradation products formed from phytosterols increase oxidative stress, deplete glutathione, lead to mitochondrial dysfunction, and elevate caspase activity (O'Callaghan et al. 2013).

However, it has been shown that oxidized derivatives of β-sitosterol (the most abundant plant sterol, as well as cholesterol (the most abundant sterol in foods of animal origin), cause apoptosis, inducing caspase-3 activity in cells (Rubiś et al. 2008). Further tests have showed that degradation products formed from stigmasteryl linoleate show the greatest cytotoxicity (Raczyk et al. 2018b): heating the ester for 1 h caused it to degrade, and the mixture of degradation products had the strongest cytotoxicity towards lymphoblastic leukemia cell lines CCRF-CEM. Interestingly, extending the heating time lowered the cytotoxicity of the degradation products, probably due to the formation of nonpolar dimers and oligomers. From the two esters of stigmasterol, stigmasteryl linoleate was more potent, which is confirmed by lower IC50 values calculated for CCRFCEM and CCRF-VCR1000 cells (79.0 ± 2.1 and 41.3 ± 1.2 μg/mL, respectively) compared to IC50 values calculated for stigmasteryl linolenate.

The concentration of the compounds which reduce the cell viability to 50% was for stigmasteryl linoleate heated for 1 and 2 h 74 and 127 μg/mL, respectively. Our last study disclosed that stigmasterol degradation products significantly reduce the viability of: (1) normal human liver cells (THLE-2); (2) colon cells (CCD 841 CoN), and (3) intestine epithelial cells (FHs 74 Int). These effects are time-dependent and dose-dependent, but their esters produced degradation products with lower negative effectiveness. Western blot tests were used to observe changes in the level and activity of the proteins that regulate apoptosis in cells treated with stigmasterol degradation products. A caspase cascade was demonstrated by caspase-3 and PARP cleavage detection.

7 Other Products Formed During the Oxidation of Phytosterols

7.1 *Volatile Derivatives*

Oxidized sterols are the main precursors involved in the formation of volatile components and oligomers during thermo-oxidative degradation. The amount and type of these products formed is directly affected by both temperature and time. Diversified volatile compounds were observed during thermo-oxidative degradation of phytosterols, with most being typical compounds involved in off-flavor and rancidity formation (Rudzińska et al. 2009).

Aldehydes, ketones, alcohols, and hydrocarbons were found among the volatiles formed during the heating of stigmasteryl esters. The mechanism of formation of volatile compounds from sterol esters was associated with the oxidation of steryl and fatty acid moieties. In particular, 2-methyl-3-pentanone and 5-ethyl-6-methyl-3-hepten-2-one were identified as unique degradation products formed specifically from the degradation of the steryl moiety, and a formation mechanism was suggested. Both volatiles could serve as good indicators of the thermo-oxidative degradation of functional food products enriched with phytosterols and their esters (Raczyk et al. 2017b).

7.2 *Dimers, Trimers, Oligomers*

Oligomers were detected when the β-sitosterol standard was heated at 60, 120, and 180 °C for different durations.

These were products of POP condensation and polymerization; of these, we observed dimers (MW 780–860 Da), trimers (MW 1100–1800 Da), and tetramers (MW 1500–1800 Da). The dimers were the dominant oligomer, while the tetramers were detected after heating at 180 °C for 12 and 24 h (Rudzińska et al. 2009).

Sterol dimers are the main oxidation products formed during sterol degradation at elevated temperatures (Sosińska et al. 2013, 2014). The structure of the dimers with various polarities formed during thermo-oxidation of β-sitosterol was described. The 3,3′-disteryl ether was the most abundant compound in the nonpolar fraction, while 7-keto disteryl ether was the major dimer in the midpolar fraction. A 7-hydroxy disteryl ether was identified as a prominent dimeric component in the complex of polar phytosterol degradation products. The formation of the C7-linked disteryl ether is based on the free radical mechanism, while the C3-linked (A-ring) disteryl ether suggests the thermal dimerization of sterols. Steric hindrance around C7, compared to C3, coupled with the high concentration of unoxidized sterol substrate, suggests a preference for 3,3′-disteryl ether formation (Sosińska et al. 2014).

8 Conclusions

Phytosterols and their esters are commonly used to fortify foods, and mainly consist of β-sitosterol, campesterol, stigmasterol and brassicasterol. According to a new market report published by Transparency Market Research (http://www.transparencymarketresearch.com), entitled "Phytosterols Market (β-Sitosterol, Campesterol, Stigmasterol, Ergosterol)—Global Industry Analysis, Market Size, Share, Growth and Forecast, 2010–2018", the global market for phytosterols was estimated to be worth over US$391.5 million in 2010, and was expected to exceed US$887.8 million by 2018. Europe is the biggest market for phytosterols, accounting for approximately 51% of the global market in 2011, followed by North America.

Because food products containing phytosterols and its esters are recommended for direct consumption—as well as for cooking, baking, and frying—they could be a source of thermo-oxidative degradation products, including phytosterol oxidation products and low-molecular weight compounds, among them volatiles and oligomers. While the health-promoting properties of phytosterols are well known and widely studied, we know very little about the effects on the human body of their oxidized derivatives and other products formed during thermal degradation. Further study is thus required to better understand the prevailing mechanism of phytosterol degradation product formation, as well as the biological properties of these compounds.

Acknowledgements This research was supported by the National Science Centre, Poland, grant number 2018/31/B/NZ9/00602.

References

Adcox C, Boyd L, Oehrl L, Allen J, Fenner G (2001) Comparative effects of phytosterol oxides and cholesterol oxides in cultured macrophage-derived cell lines. J Agric Food Chem 49 (4):2090–2095

Alemany L, Laparra JM, Barberá R, Alegría A (2013) Relative expression of cholesterol transport-related proteins and inflammation markers through the induction of 7-ketosterol-mediated stress in Caco-2 cells. Food Chem Toxicol [Internet] 56:247–253. Available from https://doi.org/10.1016/j.fct.2013.02.040

Alvarez-Sala A, Blanco-Morales V, Cilla A, Garcia-Llatas G, Sánchez-Siles LM, Barberá R et al (2018) Safe intake of a plant sterol-enriched beverage with milk fat globule membrane: bioaccessibility of sterol oxides during storage. J Food Compos Anal [Internet] 68:111–117. Available from https://doi.org/10.1016/j.jfca.2017.03.011

Awad AB, Smith AJ, Fink CS (2001) Plant sterols regulate rat vascular smooth muscle cell growth and prostacyclin release in culture. Prostaglandins Leukot Essent Fatty Acids 64(6):323–330

Azadmard-Damirchi S, Dutta PC (2008) Stability of minor lipid components with emphasis on phytosterols during chemical interesterification of a blend of refined olive oil and palm stearin. JAOCS J Am Oil Chem Soc 85(1):13–21

Azadmard-Damirchi S, Habibi-Nodeh F, Hesari J, Nemati M, Achachlouei BF (2010) Effect of pretreatment with microwaves on oxidative stability and nutraceuticals content of oil from rapeseed. Food Chem [Internet] 121(4):1211–1215. Available from https://doi.org/10.1016/j.foodchem.2010.02.006

Barriuso B, Otaegui-Arrazola A, Menéndez-Carreño M, Astiasarán I, Ansorena D (2012) Sterols heating: degradation and formation of their ring-structure polar oxidation products. Food Chem [Internet] 135(2):706–712. Available from https://doi.org/10.1016/j.foodchem.2012.05.027

Barriuso B, Ansorena D, Poyato C, Astiasarán I, Ansorena D, Berger A et al (2014) Identification of fatty acid steryl esters in margarine and corn using direct flow injection ESI-MSn ion trap-mass spectrometry. J Agric Food Chem [Internet] 99(1):24–31. Available from https://doi.org/10.1016/j.biochi.2012.09.021

Barriuso B, Ansorena D, Poyato C, Astiasarán I (2015) Cholesterol and stigmasterol within a sunflower oil matrix: thermal degradation and oxysterols formation. Steroids [Internet] 99 (PB):155–160. Available from https://doi.org/10.1016/j.steroids.2015.02.009

Baumgartner S, Mensink RP, De Smet E, Konings M, Fuentes S, de Vos WM et al (2017) Effects of plant stanol ester consumption on fasting plasma oxy(phyto)sterol concentrations as related to fecal microbiota characteristics. J Steroid Biochem Mol Biol 169:46–53

Bortolomeazzi R, Cordaro F, Pizzale L, Conte LS (2003) Presence of phytosterol oxides in crude vegetable oils and their fate during refining. J Agric Food Chem 51(8):2394–2401

Bosner MS, Lange LG, Stenson WF, Ostlund RE (1999) Percent cholesterol absorption in normal women and men quantified with dual stable isotopic tracers and negative ion mass spectrometry. J Lipid Res [Internet] 40(2):302–308. Available from https://doi.org/10.1016/S0022-2275(20)33370-8

Bradford PG, Awad AB (2010) Modulation of signal transduction in cancer cells by phytosterols. Biofactors 36(4):241–247

Brauner R, Johannes C, Ploessl F, Bracher F, Lorenz RL (2012) Phytosterols reduce cholesterol absorption by inhibition of 27-hydroxycholesterol generation, liver X receptor α activation, and expression of the basolateral sterol exporter atp-binding cassette A1 in caco-2 enterocytes. J Nutr 142(6):981–989

Cercaci L, Rodriguez-Estrada MT, Lercker G, Decker EA (2007) Phytosterol oxidation in oil-in-water emulsions and bulk oil. Food Chem 102(1):161–167

Chew SC, Tan CP, Nyam KL (2017) Comparative study of crude and refined kenaf (Hibiscus cannabinus L.) seed oil during accelerated storage. Food Sci Biotechnol 26(1):63–69

Cioccoloni G, Soteriou C, Websdale A, Wallis L, Zulyniak MA, Thorne JL (2020) Phytosterols and phytostanols and the hallmarks of cancer in model organisms: a systematic review and meta-analysis. Crit Rev Food Sci Nutr [Internet] 1–21. Available from https://doi.org/10.1080/10408398.2020.1835820

Conchillo A, Cercaci L, Ansorena D, Rodriguez-Estrada MT, Lercker G, Astiasarán I (2005) Levels of phytosterol oxides in enriched and nonenriched spreads: application of a thin-layer chromatography-gas chromatography methodology. J Agric Food Chem 53(20):7844–7850

Costa J, Amaral JS, Mafra I, Oliveira MBPP (2011) Refining of Roundup Ready® soya bean oil: effect on the fatty acid, phytosterol and tocopherol profiles. Eur J Lipid Sci Technol 113 (4):528–535

da Costa PA, Ballus CA, Teixeira-Filho J, Godoy HT (2010) Phytosterols and tocopherols content of pulps and nuts of Brazilian fruits. Food Res Int [Internet] 43(6):1603–1606. Available from https://doi.org/10.1016/j.foodres.2010.04.025

Derewiaka D, Obiedziński M (2012) Phytosterol oxides content in selected thermally processed products. Eur Food Res Technol 234(4):703–712

Dierckx T, Bogie JFJ, Hendriks JJA (2018) The impact of phytosterols on the healthy and diseased brain. Curr Med Chem 26(37):6750–6765

Fernandez ML, Vega-López S (2005) Efficacy and safety of sitosterol in the management of blood cholesterol levels. Cardiovasc Drug Rev 23(1):57–70

Gabay O, Sanchez C, Salvat C, Chevy F, Breton M, Nourissat G et al (2010) Stigmasterol: a phytosterol with potential anti-osteoarthritic properties. Osteoarthr Cartil [Internet] 18 (1):106–16. Available from https://doi.org/10.1016/j.joca.2009.08.019

García-Llatas G, Rodríguez-Estrada MT (2011) Current and new insights on phytosterol oxides in plant sterol-enriched food. Chem Phys Lipids 164(6):607–624

Gawrysiak-Witulska M, Rudzińska M (2012) Degradation of phytosterols during near-ambient drying of rapeseeds in a thick immobile layer. JAOCS J Am Oil Chem Soc 89(9)

Gawrysiak-Witulska M, Rudzińska M, Wawrzyniak J, Siger A (2012) The effect of temperature and moisture content of stored rapeseed on the phytosterol degradation rate. JAOCS J Am Oil Chem Soc 89(9)

Gawrysiak-Witulska M, Siger A, Rudzińska M, Stuper-Szablewska K, Rusinek R (2018) Effect of self-heating on the processing quality of rapeseed. Int Agrophys 32(3):313–323

Gawrysiak-Witulska M, Siger A, Rudzińska M, Bartkowiak-Broda I (2020) The effect of drying on the native tocopherol and phytosterol content of Sinapis alba L. seeds. J Sci Food Agric 100 (1):354–361

Gupta AK, Savopoulos CG, Ahuja J, Hatzitolios AI (2011) Role of phytosterols in lipid-lowering: current perspectives. QJM 104(4):301–308

Hovenkamp E, Demonty I, Plat J, Lütjohann D, Mensink RP, Trautwein EA (2008) Biological effects of oxidized phytosterols: a review of the current knowledge. Prog Lipid Res 47(1):37–49

Hu PC, Chen BH (2002) Effects of riboflavin and fatty acid methyl esters on cholesterol oxidation during illumination. J Agric Food Chem 50(12):3572–3578

Iuliano L (2011) Pathways of cholesterol oxidation via non-enzymatic mechanisms. Chem Phys Lipids [Internet] 164(6):457–468. Available from https://doi.org/10.1016/j.chemphyslip.2011.06.006

Jaceldo-Siegl K, Lütjohann D, Sirirat R, Mashchak A, Fraser GE, Haddad E (2017) Variations in dietary intake and plasma concentrations of plant sterols across plant-based diets among North American adults. Mol Nutr Food Res 61(8):1–10

Jiménez-Escrig A, Santos-Hidalgo AB, Saura-Calixto F (2006) Common sources and estimated intake of plant sterols in the Spanish diet. J Agric Food Chem 54(9):3462–3471

Johnsson L, Dutta PC (2006) Determination of phytosterol oxides in some food products by using an optimized transesterification method. Food Chem 97(4):606–613

Julien-David D, Zhao M, Geoffroy P, Miesch M, Raul F, Aoude-Werner D et al (2014) Analysis of sitosteryl oleate esters in phytosterols esters enriched foods by HPLC-ESI-MS2. Steroids [Internet] 84:84–91. Available from https://doi.org/10.1016/j.steroids.2014.03.013

Kasim NS, Gunawan S, Yuliana M, Ju YH (2010) A simple two-step method for simultaneous isolation of tocopherols and free phytosterols from soybean oil deodorizer distillate with high purity and recovery. Sep Sci Technol 45(16):2437–2446

Kasprzak M, Rudzińska M, Przybylski R, Kmiecik D, Siger A, Olejnik A (2020) The degradation of bioactive compounds and formation of their oxidation derivatives in refined rapeseed oil during heating in model system. LWT 123

Kaur R, Myrie SB (2020) Association of dietary phytosterols with cardiovascular disease biomarkers in humans. Lipids 55(6):569–584

Kemmo S, Soupas L, Lampi AM, Piironen V (2005) Formation and decomposition of stigmasterol hydroperoxides and secondary oxidation products during thermo-oxidation. Eur J Lipid Sci Technol 107(11):805–814

Kenny O, O'Callaghan Y, O'Connell NM, McCarthy FO, Maguire AR, O'Brien NM (2012) Oxidized derivatives of dihydrobrassicasterol: cytotoxic and apoptotic potential in U937 and HepG2 cells. J Agric Food Chem 60(23):5952–5961

Klingberg S, Ellegård L, Johansson I, Hallmans G, Weinehall L, Andersson H et al (2012) Inverse relation between dietary intake of naturally occurring plant sterols and serum cholesterol in northern Sweden (American Journal of Clinical Nutrition (2008) 87, (993-1001)). Am J Clin Nutr 96(3):680

Kmiecik D, Fedko M, Rudzińska M, Siger A, Gramza-Michałowska A, Kobus-Cisowska J (2020) Thermo-oxidation of phytosterol molecules in rapeseed oil during heating: the impact of unsaturation level of the oil. Foods 10(1):50

Lengyel J, Rimarčík J, Vagánek A, Fedor J, Lukeš V, Klein E (2012) Oxidation of sterols: energetics of C-H and O-H bond cleavage. Food Chem 133(4):1435–1440

Lin Y, Knol D, Menéndez-Carreño M, Blom WAM, Matthee J, Janssen HG et al (2016) Formation of plant sterol oxidation products in foods during baking and cooking using margarine without and with added plant sterol esters. J Agric Food Chem 64(3):653–662

Luister A, Schött HF, Husche C, Schäfers HJ, Böhm M, Plat J et al (2015) Increased plant sterol deposition in vascular tissue characterizes patients with severe aortic stenosis and concomitant coronary artery disease. Steroids [Internet] 99(PB):272–280. Available from https://doi.org/10.1016/j.steroids.2015.03.011

Marangoni F, Poli A (2010) Phytosterols and cardiovascular health. Pharmacol Res [Internet] 61 (3):193–199. Available from https://doi.org/10.1016/j.phrs.2010.01.001

Menéndez-Carreño M, Knol D, Janssen HG (2016) Development and validation of methodologies for the quantification of phytosterols and phytosterol oxidation products in cooked and baked food products. J Chromatogr A 1428:316–325

Mildner-Szkudlarz S, Różańska M, Siger A, Kowalczewski PŁ, Rudzińska M (2019) Changes in chemical composition and oxidative stability of cold-pressed oils obtained from by-product roasted berry seeds. LWT 111:541–547

O'Callaghan Y, Kenny O, O'Connell NM, Maguire AR, McCarthy FO, O'Brien NM (2013) Synthesis and assessment of the relative toxicity of the oxidised derivatives of campesterol and dihydrobrassicasterol in U937 and HepG2 cells. Biochimie [Internet] 95(3):496–503. Available from https://doi.org/10.1016/j.biochi.2012.04.019

Oliveira L, Freire CSR, Silvestre AJD, Cordeiro N (2008) Lipophilic extracts from banana fruit residues: a source of valuable phytosterols. J Agric Food Chem 56(20):9520–9524

Plat J, Mensink RP (2002) Increased intestinal ABCA1 expression contributes to the decrease in cholesterol absorption after plant stanol consumption. FASEB J 16(10):1248–1253

Plat J, Theuwissen E, Husche C, Lütjohann D, Gijbels MJJ, Jeurissen M et al (2014) Oxidised plant sterols as well as oxycholesterol increase the proportion of severe atherosclerotic lesions in female LDL receptor+/- mice. Br J Nutr 111(1):64–70

Plat J, Baumgartner S, Mensink RP (2015) Mechanisms underlying the health benefits of plant sterol and stanol ester consumption. J AOAC Int 98(3):697–700

Poli G, Sottero B, Gargiulo S, Leonarduzzi G (2009) Cholesterol oxidation products in the vascular remodeling due to atherosclerosis. Mol Aspects Med [Internet] 30(3):180–189. Available from https://doi.org/10.1016/j.mam.2009.02.003

Przybylski R, Klensporf-Pawlik D, Anwar F, Rudzinska M (2009) Lipid components of North American wild rice (Zizania palustris). JAOCS J Am Oil Chem Soc 86(6)

Raczyk M, Kmiecik D, Przybylski R, Rudzińska M (2017a) Effect of fatty acid unsaturation on phytosteryl ester degradation. JAOCS J Am Oil Chem Soc 94(5)

Raczyk M, Kmiecik D, Schieberle P, Przybylski R, Jeleń H, Rudzińska M (2017b) Model studies on the formation of volatile compounds generated by a thermal treatment of steryl esters with different fatty acid moieties. Food Res Int 97

Raczyk M, Bonte A, Matthäus B, Rudzińska M (2018a) Impact of added phytosteryl/phytostanyl fatty acid esters on chemical parameters of margarines upon heating and pan-frying. Eur J Lipid Sci Technol 120(2):1–11

Raczyk M, Paszel-Jaworska A, Rudzińska M (2018b) Cytotoxic activity of stigmasteryl esters and products of their thermo-oxidative degradation against drug sensitive and drug resistant human acute lymphoblastic leukemia cells. Acta Sci Pol Technol Aliment 17(1):11–18

Ramprasath VR, Awad AB (2015) Role of phytosterols in cancer prevention and treatment. J AOAC Int 98(3):735–738

Ratusz K, Symoniuk E, Wroniak M, Rudzińska M (2018) Bioactive compounds, nutritional quality and oxidative stability of cold-pressed camelina (Camelina sativa L.) oils. Appl Sci 8(12):1–17

Rocha VZ, Ras RT, Gagliardi AC, Mangili LC, Trautwein EA, Santos RD (2016) Effects of phytosterols on markers of inflammation: a systematic review and meta-analysis. Atherosclerosis [Internet] 248:76–83. Available from https://doi.org/10.1016/j.atherosclerosis.2016.01.035

Rousková M, Heyberger A, Tøíska J, Krtièka M (2011) Extraction of phytosterols from tall oil soap using selected organic solvents. Chem Pap 65(6):805–812

Rubiś B, Paszel A, Kaczmarek M, Rudzińska M, Jeleń H, Rybczynska M (2008) Beneficial or harmful influence of phytosterols on human cells? Br J Nutr 100(6)

Rudzińska M (2011) Products formed during thermo-oxidative degradation of sterols in model system and edible fats. Poznań University of Life Sciences, Poznań, 102 p

Rudzińska M, Korczak J, Wąsowicz E (2005) Changes in phytosterols and their oxidation products during frying of french fries in rapeseed oil. Pol J Food Nutr Sci 14(4):381–387

Rudzińska M, Przybylski R, Wąsowicz E (2009) Products formed during thermo-oxidative degradation of phytosterols. JAOCS J Am Oil Chem Soc 86(7)

Rudzińska M, Przybylski R, Zhao YY, Curtis JM (2010) Sitosterol thermo-oxidative degradation leads to the formation of dimers, trimers and oligomers: a study using combined size exclusion chromatography/mass spectrometry. Lipids 45(6)

Rudzińska M, Przybylski R, Wąsowicz E (2014) Degradation of phytosterols during storage of enriched margarines. Food Chem 142:294–298

Ryan E, Chopra J, McCarthy F, Maguire AR, O'Brien NM (2005) Qualitative and quantitative comparison of the cytotoxic and apoptotic potential of phytosterol oxidation products with their corresponding cholesterol oxidation products. Br J Nutr 94(3):443–451

Sanclemente T, Marques-Lopes I, Puzo J, García-Otín AL (2009) Role of naturally-occurring plant sterols on intestinal cholesterol absorption and plasmatic levels. J Physiol Biochem 65(1):87–98

Sawadikiat P, Hongsprabhas P (2014) Phytosterols and γ-oryzanol in rice bran oils and distillates from physical refining process. Int J Food Sci Technol 49(9):2030–2036

Schmitz G, Langmann T, Heimerl S (2001) Role of ABCG1 and other ABCG family members in lipid metabolism. J Lipid Res [Internet] 42(10):1513–1520. Available from https://doi.org/10.1016/S0022-2275(20)32205-7

Scholz B, Menzel N, Lander V, Engel KH (2016) Heating two types of enriched margarine: complementary analysis of phytosteryl/phytostanyl fatty acid esters and phytosterol/phytostanol oxidation products. J Agric Food Chem 64(13):2699–2708

Shahzad N, Khan W, Md S, Ali A, Saluja SS, Sharma S et al (2017) Phytosterols as a natural anticancer agent: current status and future perspective. Biomed Pharmacother 88:786–794

Shuang R, Rui X, Wenfang L (2016) Phytosterols and dementia. Plant Foods Hum Nutr [Internet] 71(4):347–354. Available from https://doi.org/10.1007/s11130-016-0574-1

Siger A, Kaczmarek A, Rudzińska M (2015) Antioxidant activity and phytochemical content of cold-pressed rapeseed oil obtained from roasted seeds. Eur J Lipid Sci Technol 117(8)

Sioen I, Matthys C, Huybrechts I, Van Camp J, De Henauw S (2011) Consumption of plant sterols in Belgium: consumption patterns of plant sterol-enriched foods in Flanders, Belgium. Br J Nutr 105(6):911–918

Sosińska E, Przybylski R, Hazendonk P, Zhao YY, Curtis JM (2013) Characterisation of non-polar dimers formed during thermo-oxidative degradation of β-sitosterol. Food Chem 139 (1–4):464–474

Sosińska E, Przybylski R, Aladedunye F, Hazendonk P (2014) Spectroscopic characterisation of dimeric oxidation products of phytosterols. Food Chem 151:404–414

Soupas L, Huikko L, Lampi AM, Piironen V (2007) Pan-frying may induce phytosterol oxidation. Food Chem 101(1):286–297

Tabee E, Jägerstad M, Dutta PC (2008) Lipids and phytosterol oxidation products in commercial potato crisps commonly consumed in Sweden. Eur Food Res Technol 227(3):745–755

Tomoyori H, Kawata Y, Higuchi T, Ichi I, Sato H, Sato M et al (2004) Erratum: Phytosterol oxidation products are absorbed in the intestinal lymphatics in rats but do not accelerate

atherosclerosis in apolipoprotein E-deficient mice (Journal of Nutrition (2004) 134 (1690-1696)). J Nutr 134(10):2738
Van Hoed V, Ali CB, Slah M, Verhé R (2010) Quality differences between pre-pressed and solvent extracted rapeseed oil. Eur J Lipid Sci Technol 112(11):1241–1247
Vanmierlo T, Weingärtner O, Van Der Pol S, Husche C, Kerksiek A, Friedrichs S et al (2012) Dietary intake of plant sterols stably increases plant sterol levels in the murine brain. J Lipid Res 53(4):726–735
Vejux A, Montange T, Martine L, Zarrouk A, Riedinger JM, Lizard G (2012) Absence of oxysterol-like side effects in human monocytic cells treated with phytosterols and oxyphytosterols. J Agric Food Chem 60(16):4060–4066
Vlahakis C, Hazebroek J (2000) Phytosterol accumulation in canola, sunflower, and soybean oils: effects of genetics, planting location, and temperature. JAOCS J Am Oil Chem Soc 77(1):49–53
Wang M, Lu B (2018) How do oxyphytosterols affect human health? Trends Food Sci Technol 79 (April):148–159
Woyengo TA, Ramprasath VR, Jones PJH (2009) Anticancer effects of phytosterols. Eur J Clin Nutr 63:813–820
Xu B, Zhang L, Ma F, Zhang W, Wang X, Zhang Q et al (2018) Determination of free steroidal compounds in vegetable oils by comprehensive two-dimensional gas chromatography coupled to time-of-flight mass spectrometry. Food Chem [Internet] 245:415–425. Available from https://doi.org/10.1016/j.foodchem.2017.10.114
Yang M, Zheng C, Zhou Q, Huang F, Liu C, Wang H (2013a) Minor components and oxidative stability of cold-pressed oil from rapeseed cultivars in China. J Food Compos Anal [Internet] 29 (1):1–9. Available from https://doi.org/10.1016/j.jfca.2012.08.009
Yang C, Chen ZY, Wong SL, Liu J, Liang YT, Lau CW et al (2013b) β-Sitosterol oxidation products attenuate vasorelaxation by increasing reactive oxygen species and cyclooxygenase-2. Cardiovasc Res 97(3):520–532
Yang B-W, Lu B-Y, Zhao Y-J, Luo J-Y, Hong X (2020) Formation of phytosterol photooxidation products: a chemical reaction mechanism for light-induced oxidation. Food Chem [Internet] 333:127430. Available from https://doi.org/10.1016/j.foodchem.2020.127430
Ying Q, Wojciechowska P, Siger A, Kaczmarek A, Rudzińska M (2018) Phytochemical content, oxidative stability, and nutritional properties of unconventional cold-pressed edible oils. J Food Nutr Res 6(7):476–485
Zevenbergen H, De Bree A, Zeelenberg M, Laitinen K, Van Duijn G, Flöter E (2009) Foods with a high fat quality are essential for healthy diets. Ann Nutr Metab 54(suppl 1):15–24
Zhang X, Julien-David D, Miesch M, Geoffroy P, Raul F, Roussi S et al (2005) Identification and quantitative analysis of β-sitosterol oxides in vegetable oils by capillary gas chromatography-mass spectrometry. Steroids 70(13):896–906
Zhao Y, Yang B, Xu T, Wang M, Lu B (2019) Photooxidation of phytosterols in oil matrix: effects of the light, photosensitizers and unsaturation degree of the lipids. Food Chem 288:162–169

Amino Acids, Amino Acid Derivatives and Peptides as Antioxidants

Luís S. Monteiro and Fátima Paiva-Martins

1 Introduction

The biological concept of antioxidant refers to any compound that is able to delay or prevent the oxidation of easily oxidizable substrates, such as DNA, proteins and lipids. Antioxidants reduce oxidative stress and mutations in DNA as well as other parameters associated with cell damage. Epidemiological studies have proven the ability of antioxidants to contain the effects of reactive oxygen species (ROS) and of reactive nitrogen species (RNS) activities and decrease the incidence of cancer, as well as other degenerative diseases (Morales-Gonzalez 2013).

The demand for natural antioxidants has gained great importance in recent years, since some synthetic antioxidants have health risks, mainly liver damage (Ndhlala et al. 2010; Bast and Haenen 2002). Phenolic compounds are the most abundant class of natural antioxidants (Shahidi and Ambigaipalan 2015). The potential of these compounds as antioxidants has long been recognized due to their great ability to break chains and eliminate radicals, protecting cells from the harmful effects of ROS. The presence and disposition of multiple hydroxyl groups in the chemical structure of polyphenols is important for their antioxidant capacity (Bast and Haenen 2002).

Phenolic amino acids or amino acids coupled with phenolic or catecholic groups are bioactive substances involved in suppressing the harmful effects caused by oxidative stress (Wei et al. 2012a; Kwak et al. 2009, 2012; Seo et al. 2010; Son

L. S. Monteiro (✉)
Chemistry Centre, University of Minho, Braga, Portugal
e-mail: monteiro@quimica.uminho.pt

F. Paiva-Martins
REQUIMTE-LAQV, Department of Chemistry and Biochemistry, Faculty of Sciences, University of Porto, Porto, Portugal
e-mail: mpmartin@fc.up.pt

C. Bravo-Diaz (ed.), *Lipid Oxidation in Food and Biological Systems*,
https://doi.org/10.1007/978-3-030-87222-9_17

and Lewis 2002), having biological activities such as anti-cancer (De Baltas and Bedos-Belval 2011), antimicrobial (Wei et al. 2012b; Narasimhan et al. 2004; Fu et al. 2010; Chochkova et al. 2012; Georgiev et al. 2012; Heijnen et al. 2001), anti-atherogenic (Wei et al. 2012b), among others. Studies confirm that the conjugation of amino acids with phenolic acids is useful as a strategy to improve antioxidant efficiency and bioactivity (Monteiro et al. 2019).

The mitochondria play a vital role in regulating cellular energy metabolism. Their ability to regulate the redox/oxidative balance is critical in controlling cellular life and death. Thus, mitochondrial dysfunction caused by oxidative damage has been implicated in several human pathologies such as neurodegenerative diseases and metabolic syndromes (Reddy et al. 2017; Teixeira et al. 2018). The development of new therapeutic strategies involving the minimization of mitochondrial dysfunction is of major importance. In fact, great progress has been made in the development and functional testing of mitochondria-targeted molecules. Special attention has been given to small peptides capable of regulating mitochondrial reactive oxygen species production and facilitating mitochondrial respiration and ATP synthesis (Sorriento et al. 2014; Apostolova and Victor 2015). Thus, in the last two decades, many structurally modified peptides with antioxidant properties and improved ability to cross the cell membrane while maintaining low toxicity and immunogenicity have been synthesized (Cerrato et al. 2015).

2 Antioxidant Amino Acids

2.1 Phenolic Amino Acids

2.1.1 L-Tyrosine and Its Derivatives

Natural phenols constitute a diverse class of compounds with biological interest. Phenolic compounds have been studied as antioxidants, due to their stable structure after the elimination of free radicals (Shahidi and Ambigaipalan 2015).

L-tyrosine, a phenolic proteinogenic amino acid, has an effective antioxidant activity which has been assessed by several in vitro antioxidant activity assays, such as the oxygen radical absorbance capacity assay (ORAC) and the trolox equivalent antioxidant capacity assay (TEAC) (Torkova et al. 2015).

Tyrosine residues have been shown to accumulate in the transmembrane domains of integral membrane proteins, particularly in the high lipid density region. This region is formed by the inner portion of the polar head groups and the beginning of the hydrocarbon tails. It is believed that these tyrosine residues play a vital role as antioxidants inside lipid bilayers, protecting cells against oxidative destruction (Ndhlala et al. 2010). Tyrosines acylated with long-chains are capable of inhibiting lipid peroxidation and thus reduce oxidative cell death. Low-protein neuronal membranes are observed in neurodegenerative disorders. These membranes have a high vulnerability to oxidative stress, which may be due to lack of tyrosine residues

(Moosmann and Behl 2000). This is further corroborated by the high vulnerability to oxidants of low-protein membranes and artificial lipid-only membranes. The higher resistance to oxidative stress of the inner membranes of mitochondria, which have a high protein content, when compared to membranes with low protein content such as myelin sheaths, further supports the importance of tyrosine residues as antioxidants (Moosmann and Behl 2000).

The capacity of the phenolic group to act as hydrogen radical donor inside lipidic phases and thus interfere with peroxidising free radical chain reactions, may explain the cytoprotective antioxidant effects of membrane-anchored tyrosine. Tyrosine residues, in the same way as many nonpeptide low molecular mass antioxidants, such as oestrogen, serotonin, and tocopherol are converted to nonreactive and relatively stable phenoxyl radicals. These more stable radicals have longer lifetimes than simple peroxyl radicals and can reverse reactions or inhibited propagation of the radical-mediated peroxidising chain reaction. When tyrosine is radicalized, it becomes more polar, facilitating its diffusion into zone one of the lipid bilayer where it is exposed to hydrophilic reducing molecules, such as, ascorbate or glutathione (Moosmann and Behl 2000). Thus, tyrosinyl lipids may become a new class of cytoprotective antioxidants, since the accumulation of tyrosine in transmembrane proteins protects the surrounding lipid bilayer from peroxidation.

L-3,4-diHydroxyphenylalanine (L-DOPA) is obtained from tyrosine and is the immediate precursor of the natural neurotransmitter dopamine. With the aim of alleviating the symptoms in Parkinson's disease caused by decreased dopamine levels in the brain, L-DOPA has been widely used as medication. In addition, the reduction of chronic diseases, mutagenesis and carcinogenesis have been shown to be due to the antioxidant activity of L-DOPA, which prevents hydrogen peroxide induced oxidative damage to DNA (Shi et al. 2002).

Several in vitro assays have shown that L-tyrosine and L-DOPA are effective antioxidants with activities that compare with the reference antioxidants butylated hydroxyanisol (BHA), butylated hydroxytoluene (BHT) and α-tocopherol (Gülçin 2007). Of these assays, the antilipid peroxidation test, the determination of reducing capacity, the elimination of radicals such as 2,2′-azino-bis(3-ethylbenzothiazoline-6-sulfonic acid (ABTS), 2,2-diphenyl-1-picryl hydrazyl ($DPPH^{•}$), superperoxide and hydrogen peroxide can be highlighted. Of these two compounds L-Dopa has shown to be much more effective than L-tyrosine. This indicates that the number of hydroxyl groups present in the aromatic moiety is fundamental in controlling radical elimination and antioxidant activity.

2,6-Dimethyl-L-tyrosine (Dmt) is a non-natural amino acid widely used in the synthesis of opioid peptides and other small molecules. Typically, opioid ligands containing Dmt instead of tyrosine at the amine-terminal exhibit a greater affinity for μ-opioid receptors (Schiller 2010). In addition, small peptides containing a Dmt residue can be absorbed by cells and help to mitigate oxidative stress (Sinha and Eudes 2015). Other studies with peptides containing this amino acid also highlight the antioxidant properties of Dmt (Cerrato et al. 2015) and will be discussed further on.

2.1.2 Hydroxyphenylglycines

Non-proteinogenic amino acids can have a variety of applications such as antiviral, antitumor, anti-inflammatory, immunosuppressor and antioxidant (Gilon et al. 2003; Kotha 2003). Non-natural α-amino acids have been used to modify the conformation and thus the activity of peptides and proteins. Non-proteinogenic phenolic amino acids play an important role in several natural peptide products such as antibiotics (Hubbard et al. 2000) and cell-penetrating peptides (CPPs) (Apostolova and Victor 2015).

The phenolic non-proteinogenic amino acid 4-hydroxyphenylglycine (HPG) is found in several natural glycopeptides, namely in some antibiotics, as is the case of vancomycin and its derivatives, in antimicrobial compounds, such as ramoplanin, and in calcium-dependent antibiotics (CDA) (Hubbard et al. 2000). The rigid structure characteristic of vancomycin, results from oxidative cross-linking of HPG with the aromatic rings of L-hydroxytyrosines and this rigidity plays an important role in the structure and function of the final molecule (Hubbard et al. 2000).

Catechols are a class of compounds endowed with a wide range of properties and important biochemical functions, which are conferred by the *ortho*-dihydroxyaryl moiety (Sedó et al. 2013), namely:

- At moderate redox potentials and pHs they are able to establish reversible equilibria;
- They cross-link irreversibly through complex oxidation mechanisms;
- They possess excellent chelating capacities;
- They interact with surfaces of different natures due to their vicinal hydroxyl groups.

Regardless of these properties, amino acids with the *ortho*-dihydroxyaryl function, such as 3,4-dihydroxyphenylglycine, have attracted reduced attention. This non-natural amino acid has been studied as copper ligand (Gordon and Jameson 1972) and as a substrate for tyrosinase, being converted to 3,4-dihydroxybenzaldehyde via spontaneous decarboxylation of the enzymatically generated *ortho*-quinone (Sugumaran et al. 1996).

The multicomponent Ugi reaction, which involves a carboxylic acid, an amine, an aldehyde and an isocyanide gives amino acids in the form of their bis-amide derivatives. 4-Hydroxyphenylglycine and 4-hydroxy-3-methoxyphenylglycine have been prepared through this reaction by using a hydroxyl substituted benzaldehyde as reactant and their effect on radical quenching and DNA oxidation determined (Wang and Liu 2013).

Recently, in addition to the above mentioned hydroxyphenylglycines, this methodology allowed the synthesis of derivatives of 3,4-dihydroxyphenylglycine (Fig. 1) (Monteiro et al. 2019). Their radical-scavenging activity was determined by the DPPH$^{\bullet}$ assay and the oxidation peak potentials by cyclic voltammetry. The results show that the 4-hydroxyphenylglycine derivative has radical scavenging activity

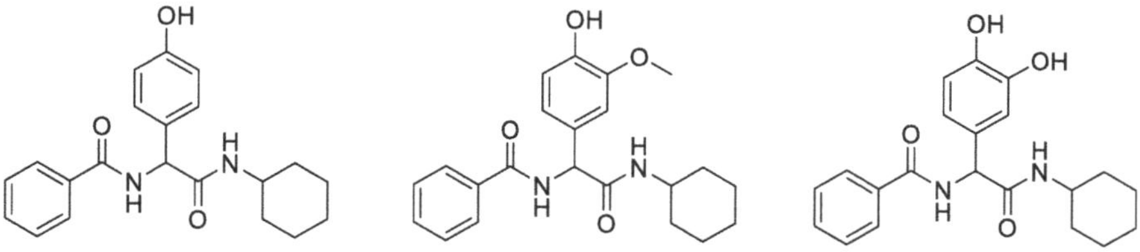

Fig. 1 Structure of hydroxyphenylglycines derivatives

(determined after 5 min) and first anodic peak potential comparable to that of tyrosine. In the case of 4-hydroxy-3-methoxyphenylglycine derivative, a significant rise in radical scavenging activity and decrease in first anodic peak potential was found. For the 3,4-dihydroxyphenylglycine derivative an approximately 100-fold rise in radical scavenging activity when compared to tyrosine was registered with a further decrease in first anodic peak potential.

2.2 *Other Natural Amino Acids with Antioxidant Properties*

In addition to tyrosine, other natural amino acids have antioxidant capacity (Davalos et al. 2004; Sarmadi and Ismail 2010). Amino acid antioxidant capacity has been investigated and compared with the chain-breaking antioxidant activity of known compounds such as, ascorbic acid and trolox (Meucci and Mele 1997). No radical scavenging activity has been observed for basic, acidic and most neutral amino acids. On the contrary, tryptophan, tyrosine, cysteine and homocysteine showed antiradical scavenging ability at concentrations which are within the usually reported physiological ranges. Davalos et al. established the following decreasing order in amino acid antioxidant activity: tryptophan (Trp), tyrosine (Tyr), methionine (Met), cysteine (Cys), histidine (His), phenylalanine (Phe) (Davalos et al. 2004).

The radical-scavenging properties of aromatic amino acids such as His, Trp and Phe is attributed to their proton donating capacity to electron deficient radicals (Rajapakse et al. 2005). Additionally, the imidazole ring of histidine has shown to have hydrogen donating, lipid peroxyl radical trapping and metal ion-chelating abilities (Wade and Tucker 1998; Chan et al. 1994). Cysteine contributes to antioxidant power since, due to its reducing power, the sulfydryl group can act as a radical scavenger (Patterson and Rhoades 1988).

The acidic amino acids, aspartic acid (Asp) and glutamic acid (Glu) and the basic amino acids, arginine (Arg) and lysine (Lys) can use their side chain carbonyl and amino groups as chelators of metal ions (Suetsuna et al. 2000). Although, having intrinsic antioxidant activities, these amino acids have not shown to be effective antioxidants in food and biological systems (Davalos et al. 2004). However, despite having little or no antioxidant effect as free amino acids, they can exhibit high antioxidant activity in peptides and this will be addressed later (Nagasawa et al. 2001; Kawashima et al. 1979).

2.3 *α,β-Dehydroamino Acids*

Non-proteinogenic amino acids containing a double bond between the α and the β carbons are designated as α,β-dehydroamino acids (Δaa) and have several important biological activities (Siodlak 2015). They are also intermediates in the synthesis of new amino acids and peptides (Siodlak 2015; Bierbaum et al. 1996; Dawid 2015; Gupta and Chauhan 2011). They appear in some bacteria, or yeasts and are constituents of many natural antibiotics (Jiang et al. 2015). They may also play a vital role in the active centre of some enzymes (Jain and Chauhan 1996). The most common α,β-dehydroamino acids are dehydroalanine (ΔAla), dehydroaminobutyric acid (ΔAbu), dehydrophenylalanine (ΔPhe), dehydrovaline (ΔVal) and dehydroleucine (ΔLeu) (Fig. 2) (Jiang et al. 2015).

Among the activities of dehydroamino acid are antioxidant properties. When reacting with oxygen or hydroxyl radicals, these compounds form stabilized free radical adducts and thus function as radical scavengers. In order to establish structure-antioxidant activity relationships that can lead to new analogues with increased activities, Suzen et al. prepared the amides of several *N*-acetyl dehydroalanine derivatives (Fig. 3) (Suzen et al. 2006). Their free radical scavenging activity against the DPPH radical and their antioxidant properties against rat liver lipid peroxidation were evaluated (Suzen et al. 2006). Little or no effect on DPPH$^{\bullet}$ was detected, however a strong inhibitory effect on rat liver lipid peroxidation was

Fig. 2 Structure of the most common α,β-dehydroamino acids

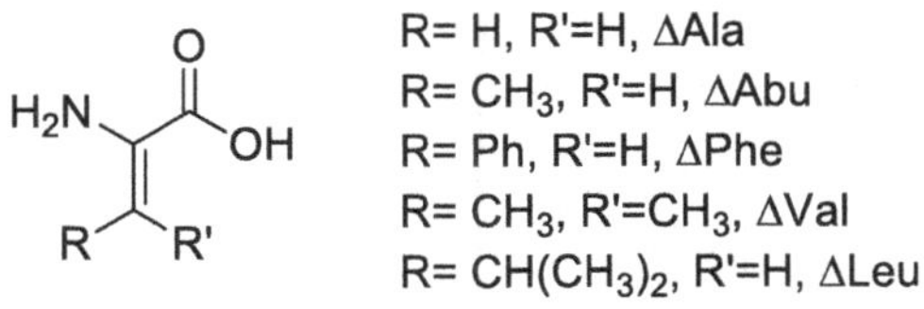

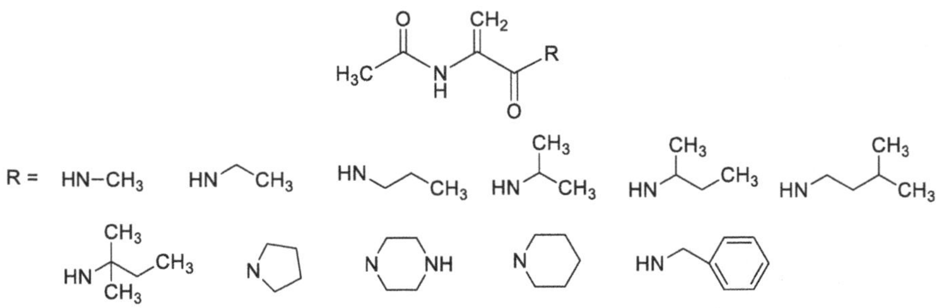

Fig. 3 Structure of *N*-acetyl dehydroalanine amides

observed. The highest effect was found for the amides of *N*-acetyl dehydroalanine with aliphatic chains of 3 or 4 carbons and cyclic 5 member rings.

Several methyl esters of dehydroalanine with various *N*-substituting groups were subject to a similar study using both DPPH$^{\bullet}$ and superoxide ($O_2^{\bullet-}$) radical scavenging activity assays (Ferreira et al. 2009). Again, no significant activity was observed. However, all the compounds were effective in lipid peroxidation experiments. These results led to the conclusion that these dehydroalanine derivatives are not able to scavenge the DPPH or superoxide radical but can scavenge the hydroxyl radical. In fact, for many compounds no radical scavenging activity for several radicals used to evaluate radical scavenging activity, such as DPPH$^{\bullet}$ or ABTS is observed. However, during oxidative injury, highly reactive radicals in vivo can be formed, and can be neutralized by these compounds and therefore protect cells (Senoner and Dichtl 2019). This is the case of tyrosol (Fernandes et al. 2020). Most ROS exert their pathological effects by giving rise to the hydroxyl radical or closely related species, the final mediators of most free radical induced tissue damage. This is due to the hydroxyl radical being capable of reacting, with extremely high rate constants, with almost every type of molecule found in living cells, such as lipids and nucleotides. Hydroxyl radical formation can occur in several ways, however, the transition metal catalysed decomposition of superoxide anion and hydrogen peroxide is likely to be most important mechanism in vivo (El Haouari 2019).

3 Amino Acids Coupled with Phenols

Several studies indicate that, due to the synergic effects between different types of molecules, a mixture of antioxidants, with different molecular structures and mechanisms of action, result more effective than a single antioxidant (Milde et al. 2004, 2007; Shi and Kakuda 2006; Gonzalez-Perez and Gonzalez-Castaneda 2006; Yogeeta et al. 2006; Trombino et al. 2004; Cirico and Omaye 2006). In order to better understand mechanistic aspects and possible synergic actions, the design of modified and/or dualistic molecules is an important approach. Coupling of compounds with different functionalities, such as, amino acids and phenolic acids is a strategy that can lead to improved antioxidant efficiency and bioactivity and also allow the establishment of structure-activity relationships (Silvia et al. 2012).

Fruits, vegetables and beverages are natural sources of phenolic acids coupled with amines or amino acids but they can also be obtained synthetically. The deleterious effects of oxidative stress (Kwak et al. 2009, 2012; Seo et al. 2010; Son and Lewis 2002) can be suppressed by these bioactive substances which also have a broad scope of other biological activities, such as, anticancer (De Baltas and Bedos-Belval 2011) and antimicrobial (Wei et al. 2012b; Narasimhan et al. 2004; Fu et al. 2010; Chochkova et al. 2012; Georgiev et al. 2012; Heijnen et al. 2001). For example, accumulation of hydroxycinnamic acid amides in plants occurs in a response to deleterious environmental stimuli such as wounding, fungal infection

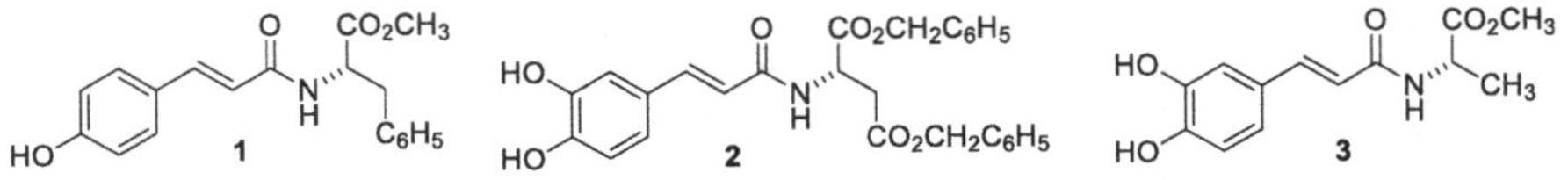

Fig. 4 Structure of *N*-(hydroxycinnamoyl) amino acid derivatives

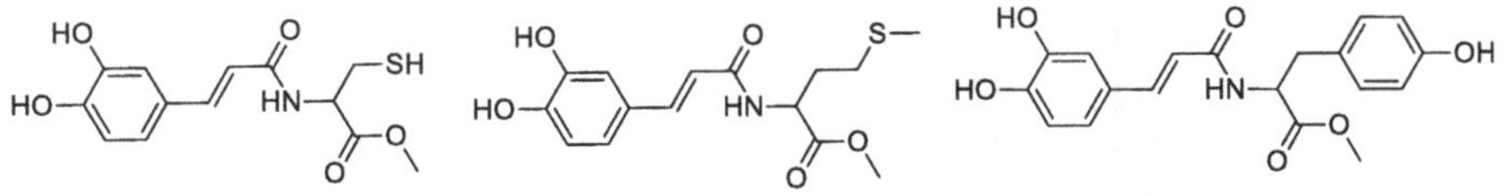

Fig. 5 Chemical structure of phenolic acid-amino acid conjugates with highest antioxidant activity

or heavy metal ions (Negrel et al. 1993; Peipp et al. 1997; Fink et al. 1990; Negrel et al. 1995).

Synthesis of a series of hydroxycinnamic amino acid derivatives and evaluation of their biological activities in lipoprotein metabolism was carried out by Lee et al. (2004) (Fig. 4).

These authors found that compounds **1** and **2** inhibited human acyl-CoA: cholesterol acyltransferase (ACAT) activities. These compounds also acted as antioxidants against copper induced low-density lipoprotein (LDL) oxidation. Compound **3** presented a potent in vivo activity. In hypercholesterolemic rabbits, this compound showed an important reduction in the formation of atherosclerotic type lesions, with an improvement in the serum lipid profile.

Wei et al. (2012a) also synthesized a series of *N*-hydroxycinnamoyl amino acid esters with the aim of finding more active antioxidants with these moieties. DPPH radical scavenging and human red blood cells haemolysis methods evaluated their antioxidative activities. It was found that:

- *N*-hydroxycinnamoyl amino acid derivatives exhibited stronger antioxidative activity than the free acids or their esters.
- Of the three hydroxycinnamamides studied (caffeoylamides, feruloylamides, and *p*-coumaroylamides), *N*-caffeoyl amino acid derivatives exhibited the highest DPPH radical scavenging activities, whereas *N*-feruloylamides had the highest antihaemolysis activities.

Silvia et al. proposed the synthesis of a series of natural amino acid derivatives with possibly enhanced antioxidant activities (Silvia et al. 2012). These authors prepared several combinations of amino acids with phenolic acids and some of these conjugates were additionally coupled with dopamine. With the conjugates it was possible to investigate the effect on the antioxidant activity of the different phenolic moieties (in particular caffeic acid and 3,4-dihydroxyphenylacetic acid). The results obtained indicate that, in order to observe significant antioxidant activity, the phenolic acid must have at least two hydroxyl groups and a conjugated spacer between the aromatic ring and the amide (Fig. 5).

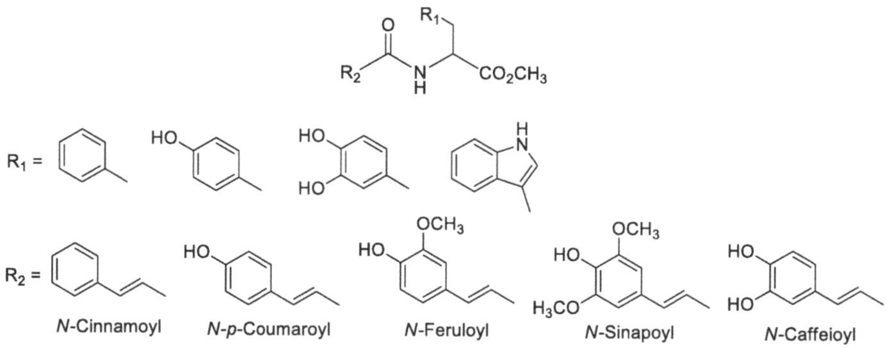

Fig. 6 Methyl esters of *N*-cinnamoylamino acids

Coupling of dopamine with the phenolic acid-amino acid conjugates did not improve significantly the antioxidant activity. This result suggests that, not only is the shape and complexity of the molecule important for its antioxidant activity, but that, also, this approach is limited by a "saturation effect" that limits the maximum potency achievable. However, the results obtained indicate that, combinations of molecules with different moieties increase the antioxidant efficiency of natural antioxidants, having the authors referred to this methodology as the "Centaurus tactic" (Silvia et al. 2012).

Georgiev et al. prepared a library of *N*-cinnamoylamino acids by combining cinnamic, *p*-coumaric, ferulic, sinapic and caffeic acids with the esters of phenylalanine, tyrosine, DOPA and tryptophan (Fig. 6) (Georgiev et al. 2013).

In vitro studies of the antioxidant activity of these compounds was carried out by two antioxidant assay systems, the DPPH$^\bullet$ assay and inhibition of lipid peroxidation (LPO) (Georgiev et al. 2013). From the results obtained, the authors could conclude that:

- The most active compounds contained the catechol moiety, whereas the presence of a methoxy groups decreased the activity;
- The conjugation of the catechol type amino acid DOPA to the less active cinnamic and *p*-coumaric acids leads to an increase in their radical scavenging activity;
- The radical scavenging activities of all compounds was dose-dependent and correlated positively with the concentrations, except for the *N*-caffeoyl-DOPA-OMe. This correlates with the saturation effect described above (Silvia et al. 2012).
- The radical scavenging activity of the aromatic amino acids (phenylalanine and tryptophan) showed that even for compounds with the weaker sinapoyl moiety, a positive influence of the phenyl and the indole moiety.

Using an innovative strategy, Monteiro et al. prepared *N*-phenolic and *N*-catecholic dehydroalanine and dehydrophenylalanine derivatives in order to study

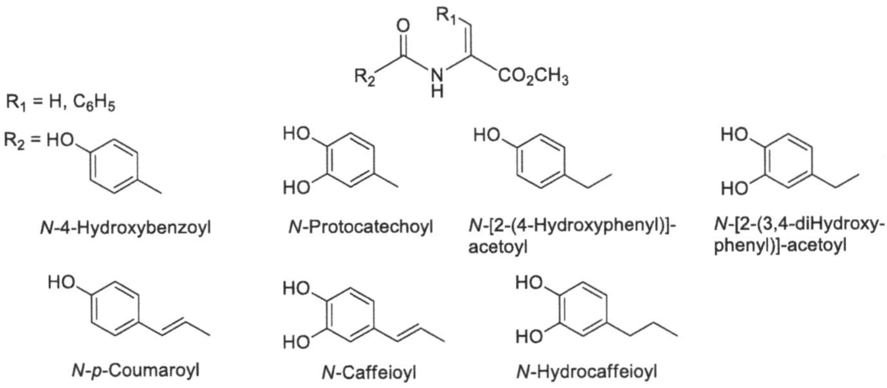

Fig. 7 Methyl esters of *N*-phenoyl and *N*-catechoyl dehydroalanines and dehydrophenylalanines

the effect of conjugation of dehydroamino acids with phenolic or catecholic acids (Fig. 7) (Monteiro et al. 2017).

The toxicity of the dehydroalanine and dehydrophenylalanine derivatives was evaluated using cancerous and non-cancerous cell cultures. These tests showed higher toxicity in relation to cancer cells than to non-cancer cells, being the most potent molecules the dehydrophenylalanine derivatives. This can be attributed to their greater liposolubility (Monteiro et al. 2017).

Phenolic conjugates of hydroxyphenylglycines could be obtained using the Ugi reaction in a similar fashion as described before to obtain hydroxyphenylglycines, but with the additional use of hydroxyl substituted benzoic acids. The effect of these hydroxylated bis-amides on the scavenging of radicals and in the protection of DNA oxidation was determined (Scheme 1) (Wang and Liu 2013). This allowed the study of the effects on antioxidant activity of hydroxyl groups attached to the different aromatic rings, as well as the influence of the isocyanide component, on the radical scavenging activity and protective effect of bis-amides.

The results showed a marked influenced of the structural feature derived from the isocyanide used in the Ugi reaction on the antioxidant effectiveness of the hydroxylated phenylglycines (Wang and Liu 2013). It was actually found that the isocyanide moiety at one end of the molecule strongly influences the antioxidant properties of the phenolic hydroxyl groups at the other end of the molecule. The most effective was the ferrocenylmethyl group, which enhances both radical scavenging activity and DNA oxidation inhibition of the bis-amides. Additionally, comparison of a series of bis-amide derivatives with this ferrocenylmethyl group indicate that the hydroxyl groups at phenyl ring C play the major role in inhibiting DNA oxidation, followed by the hydroxyl groups attached to aromatic rings B and A.

New therapeutic strategies based on multitarget-directed ligands have been proposed as a possible approach for the treatment of Alzheimer's disease. The goal is to bind simultaneously at diverse enzymatic systems or receptors involved in the progress and development of the disease. Lambruschini et al. (2017) using the Ugi

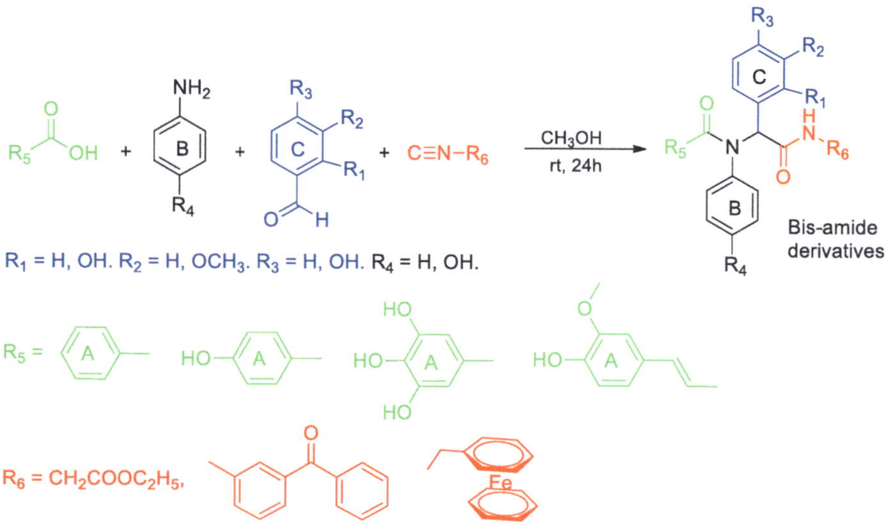

Scheme 1 Synthesis of hydroxylated bis-amides obtained by Ugi condensation

reaction, were able to prepare a series of complex polyphenols containing two to four hydroxy-substituted aryl groups linked to the main structure through linkers of different lengths. Some compounds showed highly promising capacity to inhibit aggregation of two β-amyloid peptides and potentially can be used as prevention or therapy for Alzheimer's disease.

Novel multifunctional tacrine derivatives (tacrine is the first FDA-approved drug for the treatment of Alzheimer's disease) have been obtained by the Ugi-reaction and proposed as a therapeutic strategy to modulate oxidative stress in Alzheimer's (Benchekroun et al. 2016). To inforce the antioxidant additive approach, reaction between ferulic or lipoic acid, a melatonin-like isocyanide, formaldehyde, and tacrine derivatives, gave glycine bis-amides. Thus, these glycine bis-amides contained two antioxidant motifs (the ferroloyl or lipoyl, and the melatonin fragments). Biological evaluation of these ferulic (or lipoic) acid-tacrine-melatonin glycine derivatives as antioxidants, and neuroprotective agents for the potential treatment of Alzheimer's disease were carried out. From these studies, a ferulic acid-tacrine-melatonin glycine derivative was identified as a potent cholinesterase inhibitor and a strong antioxidant, being likely to penetrate the blood brain barrier. Furthermore, this compound showed the best neuroprotective profile (Fig. 8) (Benchekroun et al. 2016).

Using the Ugi reaction to combine hydroxybenzaldehydes and phenolic or cathecolic acids with 4-methoxybenzylamine and cyclohexyl isocyanide, Monteiro et al. prepared a series of *N*-phenoyl and *N*-catechoyl hydroxyphenylglycine amides (Scheme 2) (Monteiro et al. 2019).

Radical-scavenging activities and anodic oxidation peak potentials of the compounds prepared were determined. The radical-scavenging activity studies showed a

Fig. 8 Structure of the most active ferulic acid-tacrine-melatonin glycine derivative

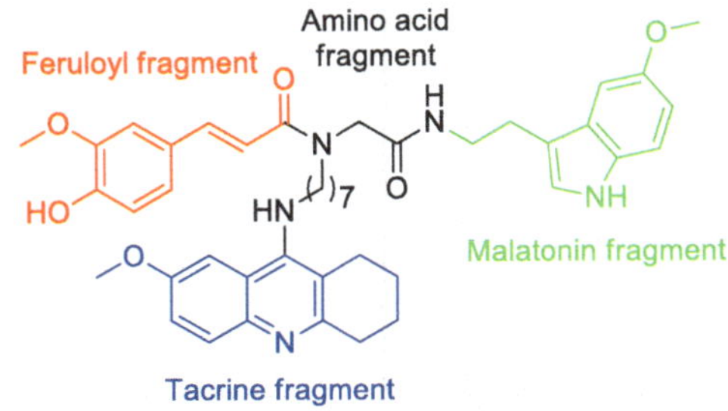

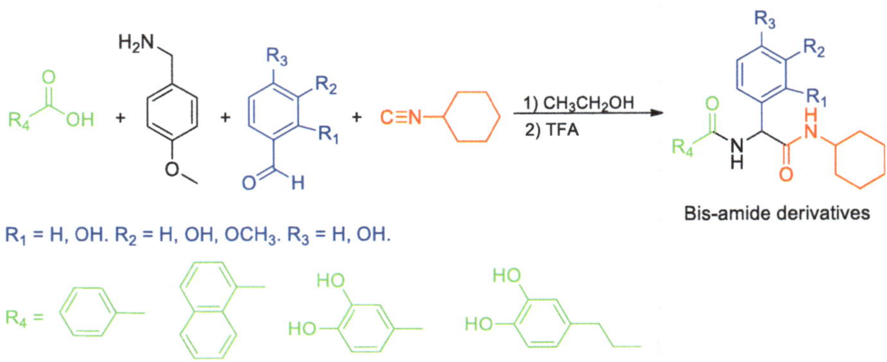

Scheme 2 Synthesis of hydroxylated bis-amides obtained by Ugi condensation

sharp increase in activity with the increase in number of hydroxyl or catechol groups present. On the other hand, a high correlation between the oxidation peak potentials determined cyclic voltammetry and radical-scavenging activity could be established (Monteiro et al. 2019).

4 Antioxidant Peptides

Antioxidant peptides are gradually being accepted as food ingredients that positively regulate oxidative stress in the human body against lipid and protein oxidation (Liu et al. 2016). The commercial use of antioxidant peptides covers several areas, such as, functional foods, nutraceuticals and cosmeceuticals (Liu et al. 2016).

In recent years, characterization of antioxidant peptides from several food sources, such as, meat (Liu et al. 2016), fish (Sila and Bougatef 2016), cereals (Esfandi et al. 2019), seeds (Ye et al. 2018; Yang et al. 2018), among others, has been conducted in a growing number of studies. Both in vitro and in vivo assay methods have been used for assessing these antioxidant properties of peptides (Liu et al. 2016). There is solid evidence that high antioxidant activities of purified peptides can be determined in in vitro assays, however, degradation and modification by the intestine, vascular system and liver, challenge whether these function in the human body. Thus, to confirm the bioavailability, in vivo assays, such as, animal studies and clinical trials must be carried out.

The determination of DPPH radical scavenging activity, hydroxyl radical scavenging, ferric-reducing antioxidant power, superoxide ion scavenging activity and linoleic acid peroxidation inhibition activity are among the in vitro methods used (Liu et al. 2016; Sila and Bougatef 2016). In vivo tests have included ethanol-induced cardiotoxicity in rats (Kamoun et al. 2012) and the study of the protective effects of peptides on mice skin subject to photoageing induced by UV irradiation (Sun et al. 2013).

Protein hydrolysates (peptides) are more potent antioxidants than free amino acids, which results from their chemical composition and physical properties (Liu et al. 2016; Elias et al. 2008). Even some free amino acids which have little or no effect individually can exhibit high antioxidant capacity in peptides (Nagasawa et al. 2001; Kawashima et al. 1979). Thus, the amino acid composition and sequence play important roles in determining the antioxidant activity of peptides. However, the relationship between structural characteristics of peptides, such as molecular size, amino acid composition and sequence, hydrophobicity and their activities as antioxidants is still unclear (Liu et al. 2016).

Several studies have shown that crude peptides of smaller molecular weights tend to have higher antioxidant activity. Peptide chains of 4–16 amino acids, corresponding to molecular weights of 0.5–2 kDa correspond to maximum activity (Meisel and FitzGerald 2003; Tang et al. 2009; Bougatef et al. 2012). Lower molecular weight peptides in the 1–3 kDa range were reported by Samaranayaka and Li-Chan (2011) to interact more effectively with radicals leading to termination of lipid peroxidation propagation cycles. The molecular weight ideal for antioxidant activities was further lowered by Chi et al. (2015) which studied smaller molecular size peptides with hydrophobic and/or aromatic amino acids in their sequences. The highest scavenging activities on $DPPH^{\bullet}$, $HO^{\bullet}$ and $O_2^{-\bullet}$. was found for a peptide with molecular weight 432.52 Da. Molecular weight also affects the antioxidant capacity in vivo since it influences the routes by which bioactive peptides are transferred into target sites (Rubas and Grass 1991). Peptides with 2–6 amino acids, when compared with proteins or single amino acids, can be more easily absorbed through the gastrointestinal barrier and enter peripheral blood, with an increase in bioavailability at tissue level (Roberts et al. 1999; Grimble 1972).

Histidine-containing peptides have also been reported to have antioxidative activity (Murase et al. 1993). This activity can be attributed to the, previously stated, hydrogen donating ability of the side chain imidazole ring, and also to metal ion-chelating and lipid peroxyl radical trapping properties (Chan et al. 1994). Peptides containing His in the N-terminus show higher scavenging ability on $DPPH^{\bullet}$, $OH^{\bullet}$ and superoxide suggesting that this feature contributes to their high antioxidant capacity (Liu et al. 2010; Lee et al. 2012). On the other hand, Yamaguchi et al. (1975) reported stronger antioxidant activity for dipeptides consisting of Tyr and Trp at the amino terminus, and His and Met at the carboxyl terminus.

Several studies of structure and activity of antioxidant peptides have reported that peptides with hydrophobic amino acids, Val, Leu or Ile and having Ala, Tyr, His, Pro or Met at the N-terminal have potent inhibitory activity (Liu et al. 2016; Chen et al. 1998).

In conclusion, the structure-antioxidant activity relationship of peptides has not been completely established and structural information of antioxidant peptides from various protein sources is still lacking. On the other hand, progress in testing the bioavailability and thus, the application in the consumer markets of antioxidant peptides has been limited due to limited animal model experiments and human clinical trials. Further research has to be conducted in vivo to establish the potential antioxidant effects of peptides.

5 Antioxidants Targeting the Mitochondria

The mitochondria is a cytoplasmic organelle that regulates cell metabolism (Nelson and Cox 2017). This organelle is involved in oxidative energy metabolism through the catabolism of nutrients, ATP synthesis by oxidative phosphorylation and heat production. Structurally, the mitochondria consists of two different components: the inner mitochondrial matrix and the intermembrane space, with an internal mitochondrial membrane that separates them. The intermembrane space is limited by the outer mitochondrial membrane, which plays the role of signal transport and transduction between the organelle and the cytosol.

Mitochondria are an important source of reactive oxygen species and reactive nitrogen species (Murphy 2009). Initial studies showed that the respiratory chain produces ROS. Subsequently it was determined that isolated mitochondria produce hydrogen peroxide and that this hydrogen peroxide comes from the dismutation of superperoxide generated in the mitochondria (Galley 2011).

Mitochondria have natural processes to eliminate ROS in order to keep the cell active. This network of antioxidant defence systems, consisting of a combination of enzyme and non-enzyme pathways is tightly regulated. Under normal conditions, the effect of ROS is cancelled out by antioxidants. When this balance is broken and the effect of ROS is more potent than antioxidant power, damage occurs in the mitochondria. This unbalance has been linked to several degenerative diseases, such as Alzheimer's, Parkinson's, neural death, cancer and cardiovascular diseases (Sorriento et al. 2014; Szeto 2006; Hoye et al. 2008; Anders et al. 2006; Victor and Rocha 2007).

5.1 Non-peptidic Mitochondrial Antioxidants

Several strategies to direct antioxidants to the mitochondria have been reported (Anders et al. 2006; Victor and Rocha 2007). One such strategy is covalently linking antioxidant molecules to lipid cations that accumulate in the mitochondria as a result of the potential of the mitochondrial membrane. MitoQ (Fig. 9) consists of binding

Fig. 9 Structure of MitoQ

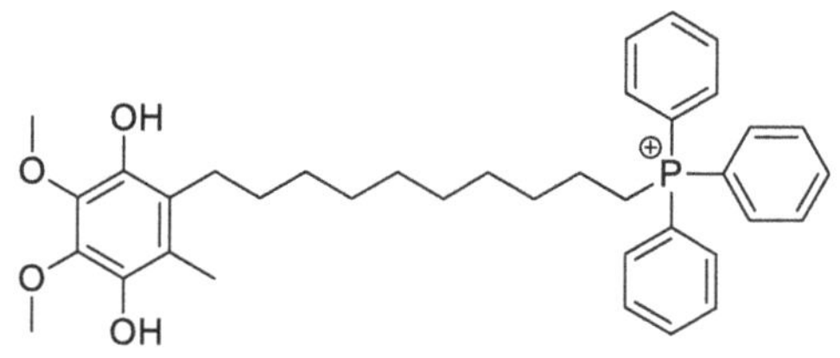

the ubiquinone antioxidant, also known as coenzyme 10, to the lipophilic triphenylphosphonium cation (TPP) (Galley 2011; Murphy 2008).

The negative charge present within the mitochondrial membrane results in the accumulation of MitoQ within the mitochondria in about 500 times the levels present in the cytoplasm (Galley 2011; Murphy 2008). The side chain of MitoQ allows it to penetrate deeply into the membrane. Thus, as soon as MitoQ enters the mitochondria, it is absorbed by the inner membrane of the mitochondria and is recycled as active uniquinol through the respiratory chain.

Tests in isolated mitochondria have verified the effectiveness of MitoQ against lipid peroxidation. Animal experiments indicate that MitoQ is efficient against sepsis-induced organ dysfunction, opening perspectives for the study of this substance in human diseases (Galley 2011).

Other antioxidants have been conjugated with TPP and tested against mitocondrial dysfunctions, such as, tocopherol (MitoVitE) (Minter et al. 2020) and lipoic acid (MitoLipoic acid) (Smith et al. 2008).

5.2 *Cell Penetrating Peptides*

The plasma membrane is a barrier that protects the cell from the unregulated influx of bioactive molecules and ions, thus controlling the stability of its internal environment (Nelson and Cox 2017). Most drugs need to cross one or more cell membranes to reach their targets and thus have a therapeutic effect. Small molecules are able to penetrate the membrane, however larger molecules, due to their physical-chemical properties, are unable to do so (Cerrato et al. 2015). The ability to transport macromolecules into cells is important both in drug administration and in biotechnological applications (therapy genetics) (Rodriguez-Plaza et al. 2014).

In the last decades, new biotherapeutic agents, such as peptides and proteins with cell-penetrating capability have contributed to the treatment of several diseases and have been designated as cell penetrating peptides (CPPs). These peptides have less than 10 amino acids and contain chemical features that permit these molecules to freely penetrate by passive diffusion into cells membranes (Zhao et al. 2004). CPPs have opened a new pathway for delivering a wide variety of bioactive compounds across the cell membrane—from proteins to therapeutic molecules—due to their high efficiency in the internalization of CPPs and low cytotoxicity (Derakhshankhah and Jafari 2018; Khafagy and Morishita 2012).

Although several studies on CPPs have been carried out, the mechanism by which they penetrate the cell is still not completely understood (Nasrollahi et al. 2012). It is suggested that their entry into the cell can be influenced by several factors, such as, length of the molecule, delocalization of charge, hydrophobicity and concentration (Jones et al. 2005). When linked to bioactive cargos the nature of this cargo (size and charge) also influences their internalization (Fonseca et al. 2009).

Two mechanisms have been proposed for the entry of CPPs into the cell: direct translocation across the cell membrane and through the endocytic route (Ram et al. 2008).

There are two types of CPPs that have been used to cross the cell membrane: (1) cationic CPPs, consisting of short chains with various amino acid residues such as arginine, lysine and histidine that provide positive charge to the peptides and allow their interaction with anionic structures, present in the plasma membrane; (2) amphipathic peptides, which contain lipophilic and hydrophilic tails responsible for a direct translocation mechanism of peptides across the cell membrane (Aroui and Kenani 2020).

CPPs have been used to deliver numerous classes of molecules, including DNA, proteins, drugs and nanoparticles, many of which are sometimes larger than the CPP itself (Langel 2019; Wagstaff and Jans 2006). However, it has not yet been determined whether the coupling of a CPP to a molecule of considerable size interferes with the translocation mechanism (Magzoub and Gräslund 2004).

The conjugation of bioactive charges with a CPP can occur in two ways: (1) covalent conjugation, where the bioactive molecules bind to the CPP through a covalent bond; and (2) through an electrostatic interaction, in which aggregates or nanoparticles are formed (Derakhshankhah and Jafari 2018; Foged and Nielsen 2008). These different conjugation processes can impact the route of administration, the mechanism of entry into the cell, the distribution within the cell and other different effects at the cellular level. Furthermore, based on the therapeutic use and the nature of the target in which the drug acts, the choice of the form of conjugation plays a very important role (Feni and Neundorf 2017).

5.3 Mitochondrial Penetrating Peptides

As stated before, mitochondria are an important target in the treatment of various diseases, due to their role in energy production and cell death (Sorriento et al. 2014; Hoye et al. 2008; Anders et al. 2006; Victor and Rocha 2007). Mitochondrial penetrating peptides (MPPs) represent a new direction for the development of vectors targeting the mitochondria. They are short peptides, with high uptake by mitochondria. MPPs are generally small synthetic, positively charged, basic, hydrophobic peptides of less than 10 amino acids that freely penetrate by passive diffusion into cells and are taken up into the mitochondria and accumulate in the matrix (Galley 2011). Additionally, MPPs exhibit low cytotoxicity (Zhao et al. 2019).

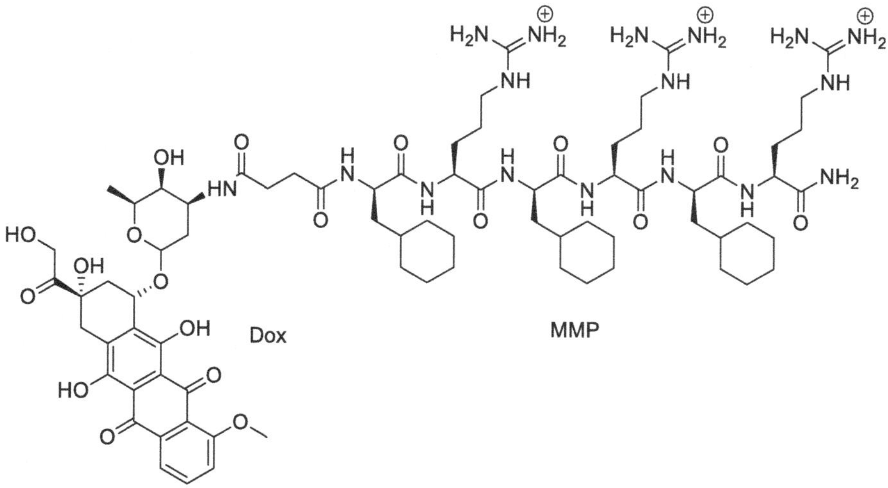

Fig. 10 Structure of the complex doxorubicin (Dox)-MPP

MPPs have been used as vehicles for the transport of bioactive molecules into the mitochondria.

With the aim of studying if the interference of nucleic acid synthesis in mitochondria would have significant cellular effects, Chamberlain et al. (2013) complexed doxorubicin (Dox) with a mitochondrial penetrating peptide (Fig. 10).

Doxorubicin is an anti-cancer drug, which inhibits the DNA topoisomerase II enzyme present in both the cell nucleus and the mitochondria. Although the potency of the Dox-MPP complex in the mitochondria decreased somewhat in sensitive cells, when compared to doxorubicin, the complex demonstrated an ability to overcome the mechanisms of resistance to multiple drugs (Chamberlain et al. 2013).

The existence of excess iron is a concern in a number of clinical conditions, as excess iron is involved in the production of pro-oxidant species by reaction with oxygen and nitrogen substrates giving rise to oxidative stress (Halliwell and Gutteridge 2015). One such disease is Friedreich's ataxia where excess iron is found in the mitochondria (Alta et al. 2017). Excess iron must be removed through the use of natural or synthetic iron chelators, with deferoxamine (DFO) being the most clinically used chelator (Kwiatkowski 2011).

Alta et al. (2017) synthesized four MPP-DFO conjugates using succinic acid as a binder, in order to have a strong iron chelator and be permeable to mitochondria. The four MPP-DFO conjugates were studied for their chelating and antioxidant properties. Results showed that the iron binding and affinity of the four conjugates are identical to free DFO and that they effectively suppress iron-catalysed oxidation (Alta et al. 2017). MPP-DFO conjugates were labelled to demonstrate mitochondrial localization in cells and all conjugates entered the mitochondria. In addition, MPP-DFO exhibited low levels of toxicity, cell cycle disruption, mitochondrial DNA damage and apoptosis (Alta et al. 2017).

The advance of research in the area of mitochondrial targeting and some positive results in clinical trials, make it necessary to continue studying the targeted delivery of different molecules to the mitochondria, in order to develop useful tools for therapy and research (Galley 2011).

5.4 Antioxidant Mitochondrial Penetrating Peptides

An alternative approach to targeting antioxidants to the mitochondria is the use of small positively charged peptides with antioxidant properties capable of protecting the mitochondria from oxidative stress. Peptides of this type were developed by Szeto and Schiller having acquired the designation of SS peptides (Szeto 2006, 2008; Rocha et al. 2010; Smith and Murphy 2011).

SS peptides are synthetic tetrapeptides originally developed as opioid analgesics with alternating aromatic residues and basic amino acids (Cerrato et al. 2015; Anders et al. 2006). Of these compounds the peptide SS-02 (Dmt-D-Arg-Phe-Lys-NH_2) and SS-31 (D-Arg-Dmt-Lys-Phe-NH_2) have been prepared and studied in cell cultures and isolated mitochondria (Szeto 2008).

Despite these peptides being water soluble, with a net charge of +3, they are readily taken up by all cells via passive diffusion (Szeto 2008). Both SS-02 and SS-31 have a 2,6-dimethyl-L-tyrosine residue (Dmt) but in a different position in the peptide chain. The presence of a D-amino acid renders them resistant to aminopeptidase activity.

These SS peptides may be viewed as "cloaked" or "stealth" as they can evade cellular membranes, even penetrating cell barriers with tight junctions including the blood-brain barrier (Sedó et al. 2013). Absorption studies with SS-02 show rapid absorption (<30 min) producing a beneficial effect, with a greater concentration in the mitochondria than in the cytosol. These results and the membrane's permeability to these peptides, suggest that they can freely pass the membrane in both directions (Cerrato et al. 2015; Rocha et al. 2010). The mechanism behind their cell permeability is unclear, but the aromatic rings may serve as electron cages to shield the cationic charges via cation-π interaction.

Studies with cell cultures and isolated mitochondria have proven that these peptides are capable of eliminating mitochondrial ROS production, reduce their production and inhibiting the mitochondrial permeability transition, and are therefore potent in preventing apoptosis induced by oxidative stress. These peptides have shown excellent efficacy in animal models of neurodegeneration and renal fibrosis, with no toxicity (Szeto 2006, 2008).

References

Alta RY, Vitorino HA, Goswami D, Liria CW, Wisnovsky SP, Kelley SO, Machini MT, Esposito BP (2017) Mitochondria-penetrating peptides conjugated to desferrioxamine as chelators for mitochondrial labile iron. PLoS One 12:e0171729

Anders M, Robotham JL, Sheu S-S (2006) Mitochondria: new drug targets for oxidative stress-induced diseases. Expert Opin Drug Metab Toxicol 2:71–79

Apostolova N, Victor VM (2015) Molecular strategies for targeting antioxidants to mitochondria: therapeutic implications. Antioxid Redox Signal 22:686–729

Aroui S, Kenani A (2020) Cell-penetrating peptides: a challenge for drug delivery. In: Cheminformatics and its applications. IntechOpen

Bast A, Haenen GR (2002) The toxicity of antioxidants and their metabolites. Environ Toxicol Pharmacol 11:251–258

Benchekroun M, Romero A, Egea J, Leon R, Michalska P, Buendia I, Jimeno ML, Jun D, Janockova J, Sepsova V, Soukup O, Bautista-Aguilera OM, Refouvelet B, Ouari O, Marco-Contelles J, Ismaili L (2016) The antioxidant additive approach for Alzheimer's disease therapy: new ferulic (lipoic) acid plus melatonin modified tacrines as cholinesterases inhibitors, direct antioxidants and nuclear factor (erythroid-derived 2)-like 2 activators. J Med Chem 59:9967–9973

Bierbaum G, Gotz F, Peschel A, Kupke T, VandeKamp M, Sahl HG (1996) Protein engineering of lantibiotics. Anton Leeuw Int J G 69:119–127

Bougatef A, Balti R, Haddar A, Jellouli K, Souissi N, Nasri M (2012) Protein hydrolysates from bluefin tuna (Thunnus thynnus) heads as influenced by the extent of enzymatic hydrolysis. Biotechnol Bioprocess Eng 17:841–852

Cerrato CP, Pirisinu M, Vlachos EN, Langel U (2015) Novel cell-penetrating peptide targeting mitochondria. FASEB J 29:4589–4599

Chamberlain GR, Tulumello DV, Kelley SO (2013) Targeted delivery of doxorubicin to mitochondria. ACS Chem Biol 8:1389–1395

Chan KM, Decker EA, Feustman DC (1994) Endogenous skeletal muscle antioxidants. Crit Rev Food Sci Nutr 34:403–426

Chen HM, Muramoto K, Yamauchi F, Fujimoto K, Nokihara K (1998) Antioxidative properties of histidine containing peptides designed from peptide fragments found in the digests of a soybean protein. J Agric Food Chem 46:49–53

Chi CF, Wang B, Hu FY, Wang YM, Zhang B, Deng SG, Wu CW (2015) Purification and identification of three novel antioxidant peptides from protein hydrolysate of bluefin leather jacket (Navodon septentrionalis) skin. Food Res Int 73:124–139

Chochkova MG, Chorbadzhiyska EY, Ivanova GI, Najdenski H, Ninova M, Milkova T (2012) Antimicrobial and radical scavenging activities of N-hydroxycinnamoyl-L-cysteine and-L-proline ethyl ester. Nat Prod J 2:50–54

Cirico TL, Omaye S (2006) Additive or synergetic effects of phenolic compounds on human low density lipoprotein oxidation. Food Chem Toxicol 44:510–516

Davalos A, Miguel M, Bartolome B, Lopez-Fandino R (2004) Antioxidant activity of peptides derived from egg white proteins by enzymatic hydrolysis. J Food Prot 67:1939–1944

Dawid S (2015) α,β-Dehydroamino acids in naturally occurring peptides. Amino Acids 47:1–17

De Baltas P, Bedos-Belval F (2011) Cinnamic acid derivatives as anticancer agents—a review. Curr Med Chem 18:1672–1703

Derakhshankhah H, Jafari S (2018) Cell penetrating peptides: a concise review with emphasis on biomedical applications. Biomed Pharmacother 108:1090–1096

El Haouari M (2019) Platelet oxidative stress and its relationship with cardiovascular diseases in type 2 diabetes mellitus patients. Curr Med Chem 26:4145–4165

Elias RJ, Kellerby SS, Decker EA (2008) Antioxidant activity of proteins and peptides. Crit Rev Food Sci Nutr 48:430–441

Esfandi R, Walters ME, Tsopmo A (2019) Antioxidant properties and potential mechanisms of hydrolyzed proteins and peptides from cereals. Heliyon 5:e01538

Feni L, Neundorf I (2017) The current role of cell-penetrating peptides in cancer therapy. Adv Exp Med Biol 1030:279–295

Fernandes S, Ribeiro C, Paiva-Martins F, Catarino C, Santos-Silva A (2020) Protective effect of olive oil polyphenol phase II sulfate conjugates on erythrocyte oxidative-induced hemolysis. Food Funct 11:8670–8679

Ferreira PMT, Monteiro LS, Coban T, Suzen S (2009) Comparative effect of N-substituted dehydroamino acids and α-tocopherol on rat liver lipid peroxidation activities. J Enzyme Inhib Med Chem 24:967–971

Fink W, Liefland M, Mendgen K (1990) Comparison of various stress responses in oat in compatible and nonhost resistant interactions with rust fungi. Physiol Mol Plant Pathol 37:309–321

Foged C, Nielsen HM (2008) Cell-penetrating peptides for drug delivery across membrane barriers. Expert Opin Drug Deliv 5:105–117

Fonseca SB, Pereira MP, Kelley SO (2009) Recent advances in the use of cell-penetrating peptides for medical and biological applications. Adv Drug Deliv Rev 61:953–964

Fu J, Cheng K, Zhang ZM, Fang RQ, Zhu HL (2010) Synthesis, structure and structure-activity relationship analysis of caffeic acid amides as potential antimicrobials. Eur J Med Chem 45:2638–2643

Galley HF (2011) Oxidative stress and mitochondrial dysfunction in sepsis. Br J Anaesth 107:57–64

Georgiev L, Chochkova M, Ivanova G, Najdenski H, Ninova M, Milkova T (2012) Radical scavenging and antimicrobial activities of cinnamoyl amides of biogenic monoamines. Riv Ital Sost Grasse 89:91–102

Georgiev L, Chochkova M, Totseva I, Seizova K, Marinova E, Ivanova G, Ninova M, Najdenski H, Milkova T (2013) Anti-tyrosinase, antioxidant and antimicrobial activities of hydroxycinnamoylamides. Med Chem Res 22:4173–4182

Gilon C, Dechantsreiter MA, Burkhart F, Friedler A, Kessler H (2003) Synthesis of N-alkylated peptides. In: Goodman M, Felix A, Moroder L, Toniolo C (eds) Houben-Weyl: methods of organic chemistry. Synthesis of peptides and peptidomimetics, vol E22c. Thieme, New York, pp 215–271

Gonzalez-Perez O, Gonzalez-Castaneda RE (2006) Therapeutic perspectives on the combination of alpha-lipoic acid and vitamin E. Nutr Res 26:1–5

Gordon JE, Jameson RF (1972) Complexes of doubly chelating ligands. Part III. Proton and copper (II) complexes of 3,4-dihydroxyphenylglycine (DOPG). J Chem Soc Dalton Trans 3:307–310

Grimble GK (1972) The significance of peptides in clinical nutrition. Rev Surg 29:222–223

Gülçin I (2007) Comparison of in vitro antioxidant and antiradical activities of L-tyrosine and L-dopa. Amino Acids 32:431–438

Gupta M, Chauhan VS (2011) De novo design of α,β-didehydrophenylalanine containing peptides: from models to applications. Biopolymers 95:161–173

Halliwell B, Gutteridge JM (2015) Free radicals in biology and medicine. Oxford University Press, New York

Heijnen CG, Haenen GR, Vekemans JA, Bast A (2001) Peroxynitrite scavenging of flavonoids: structure activity relationship. Environ Toxicol Pharmacol 10:199–206

Hoye AT, Davoren JE, Wipf P, Fink MP, Kagan VE (2008) Targeting mitochondria. Acc Chem Res 41:87–97

Hubbard BK, Thomas MG, Walsh C (2000) Biosynthesis of L-p-hydroxyphenylglycine, a non-proteinogenic amino acid constituent of peptide antibiotics. Chem Biol 7:931–942

Jain R, Chauhan VS (1996) Conformational characteristics of peptides containing α,-β-dehydroamino acid residues. Biopolymers 40:105–119

Jiang J, Ma Z, Castle SL (2015) Bulky α,β-dehydroamino acids: their occurrence in nature, synthesis, and applications. Tetrahedron 71:5431–5451

Jones SW, Christison R, Bundell K, Voyce CJ, Brockbank SM, Newham P, Lindsay MA (2005) Characterisation of cell-penetrating peptide-mediated peptide delivery. Br J Pharmacol 145:1093–1102
Kamoun Z, Kamoun AS, Bougatef A, Chtourou Y, Boudawara T, Nasri M, Zeghal N (2012) Efficacy of sardinelle protein hydrolysate to alleviate ethanol-induced oxidative stress in the heart of adult rats. J Food Sci 77:156–162
Kawashima K, Itoh H, Miyoshi M, Chibata I (1979) Antioxidant properties of branched-chain amino acid derivatives. Chem Pharm Bull 27:1912–1916
Khafagy ES, Morishita M (2012) Oral biodrug delivery using cell-penetrating peptide. Adv Drug Deliv Rev 64:531–539
Kotha S (2003) The building block approach to unusual α-amino acid derivatives and peptides. Acc Chem Res 36:342–351
Kwak S-Y, Seo H-S, Lee Y-S (2009) Synergistic antioxidative activities of hydroxycinnamoyl-peptides. J Pept Sci 15:634–641
Kwak S-Y, Lee S, Yang JK, Lee Y-S (2012) Antioxidative activities of caffeoyl-proline dipeptides. Food Chem 130:847–852
Kwiatkowski JL (2011) Real-world use of iron chelators. Hematol Am Soc Hematol Educ Prog 451–458
Lambruschini C, Galante D, Moni L, Ferraro F, Gancia G, Riva R, Traverso A, Banfi L, D'Arrigo C (2017) Multicomponent, fragment-based synthesis of polyphenol-containing peptidomimetics and their inhibiting activity on beta-amyloid oligomerization. Org Biomol Chem 15:9331–9351
Langel Ü (2019) Classes and applications of cell-penetrating peptides. In: CPP, cell-penetrating peptides. Springer, pp 29–82
Lee S, Han J-M, Kim H, Kim E, Jeong T-S, Lee WS, Cho K-H (2004) Synthesis of cinnamic acid derivatives and their inhibitory effects on LDL-oxidation, acyl-CoA:cholesterol acyltransferase-1 and -2 activity, and decrease of HDL-particle size. Bioorg Med Chem Lett 14:4677–4681
Lee S-J, Kim Y-S, Hwang J-W, Kim E-K, Moon S-H, Jeon B-T, Jeon Y-J, Kim JM, Park P-J (2012) Purification and characterization of a novel antioxidative peptide from duck skin by-products that protects liver against oxidative damage. Food Res Int 49:285–295
Liu Q, Kong B, Xiong YL, Xia X (2010) Antioxidant activity and functional properties of porcine plasma protein hydrolysate as influenced by the degree of hydrolysis. Food Chem 118:403–410
Liu R, Xing L, Fu Q, Zhou GH, Zhang WG (2016) A review of antioxidant peptides derived from meat muscle and by-products. Antioxidants 5:32–47
Magzoub M, Gräslund A (2004) Cell-penetrating peptides: small from inception to application. Q Rev Biophys 37:147–195
Meisel H, FitzGerald R (2003) Biofunctional peptides from milk proteins: mineral binding and cytomodulatory effects. Curr Pharm Des 9:1289–1296
Meucci E, Mele MC (1997) Amino acids and plasma antioxidant capacity. Amino Acids 12:373–377
Milde J, Elstner EF, Grassmann J (2004) Synergistic inhibition of low-density lipoprotein oxidation by rutin, gamma-terpinene, and ascorbic acid. Phytomedicine 11:105–113
Milde J, Elstner EF, Grassmann J (2007) Synergistic effects of phenolics and carotenoids on human low-density lipoprotein oxidation. Mol Nutr Food Res 51:956–961
Minter BE, Lowes DA, Webster NR, Galley HF (2020) Differential effects of MitoVitE, α-tocopherol and trolox on oxidative stress, mitochondrial function and inflammatory signalling pathways in endothelial cells cultured under conditions mimicking sepsis. Antioxidants 9:195–209
Monteiro LS, Oliveira S, Paiva-Martins F, Ferreira PMT, Pereira DM, Andrade PB, Valentão P (2017) Synthesis and preliminary biological evaluation of new phenolic and catecholic dehydroamino acid derivatives. Tetrahedron 73:6199–6209
Monteiro LS, Paiva-Martins F, Oliveira S, Machado I, Costa M (2019) An efficient one-pot synthesis of polyphenolic amino acids and evaluation of their radical-scavenging activity. Bioorg Chem 89:102983–102990

Moosmann B, Behl C (2000) Cytoprotective antioxidant function of tyrosine and tryptophan residues in transmembrane proteins. Eur J Biochem 267:5687–5692
Morales-Gonzalez JA (2013) Oxidative stress and chronic degenerative diseases, a role for antioxidants. IntechOpen
Murase H, Nagao A, Terao J (1993) Antioxidant and emulsifying activity of N-(long-chain-acyl) histidine and N-(long-chain-acyl)carnosine. J Agric Food Chem 41:1601–1604
Murphy MP (2008) Targeting lipophilic cations to mitochondria. Biochim Biophys Acta 1777:1028–1031
Murphy MP (2009) How mitochondria produce reactive oxygen species. Biochem J 417:1–13
Nagasawa T, Yonekura T, Nishizawa N, Kitts DD (2001) In vitro and in vivo inhibition of muscle lipid and protein oxidation by carnosine. Mol Cell Biochem 225:29–34
Narasimhan B, Belsare D, Pharande D, Mourya V, Dhake A (2004) Esters, amides and substituted derivatives of cinnamic acid: synthesis, antimicrobial activity and QSAR investigations. Eur J Med Chem 39:827–834
Nasrollahi SA, Taghibiglou C, Azizi E, Farboud ES (2012) Cell-penetrating peptides as a novel transdermal drug delivery system. Chem Biol Drug Des 80:639–646
Ndhlala AR, Moyo M, Van Staden J (2010) Natural antioxidants: fascinating or mythical biomolecules? Molecules 15:6905–6930
Negrel J, Javelle F, Paynot M (1993) Wound-induced tyramine hydroxycinnamoyl transferase in potato (Solanum tuberosum) tuber discs. J Plant Physiol 142:518–524
Negrel J, Lofty S, Javelle F (1995) Modulation of the activity of two hydroxycinnamoyl transferases in wound-healing potato tuber discs in response to pectinase or abscisic acid. J Plant Physiol 146:318–325
Nelson DL, Cox MM (2017) Lehninger principles of biochemistry, 7th edn. W.H. Freeman, New York
Patterson CE, Rhoades RA (1988) Protective role of sulfhydryl reagents in oxidant lung injury. Exp Lung Res 14:1005–1019
Peipp H, Maier W, Schmidt J, Wray V, Strac D (1997) Arbuscular mycorrhizal fungus-induced changes in the accumulation of secondary compounds in barley roots. Phytochemistry 44:581–587
Rajapakse N, Mendis E, Jung WK, Je JY, Kim SK (2005) Purification of a radical scavenging peptide from fermented mussel sauce and its antioxidant properties. Food Res Int 38:175–182
Ram N, Aroui S, Jaumain E, Bichraoui H, Mabrouk K, Ronjat M, Lortat-Jacob H, De Waard M (2008) Direct peptide interaction with surface glycosaminoglycans contributes to the cell penetration of Maurocalcine. J Biol Chem 283:24274–24284
Reddy PH, Manczak M, Kandimalla R (2017) Mitochondria-targeted small molecule SS31: a potential candidate for the treatment of Alzheimer's disease. Hum Mol Genet 26:1483–1496
Roberts PR, Burney JD, Black KW, Zaloga GP (1999) Effect of chain length on absorption of biologically active peptides from the gastrointestinal tract. Digestion 60:332–337
Rocha M, Hernandez-Mijares A, Garcia-Malpartida K, Banuls C, Bellod L, Victor VM (2010) Mitochondria-targeted antioxidant peptides. Curr Pharm Des 16:3124–3131
Rodriguez-Plaza JG, Morales-Nava R, Diener C, Schreiber G, Gonzalez ZD, Lara Ortiz MT, Ortega Blake I, Pantoja O, Volkmer R, Klipp E, Herrmann A, Del Rio G (2014) Cell penetrating peptides and cationic antibacterial peptides: two sides of the same coin. J Biol Chem 289:14448–14457
Rubas W, Grass GM (1991) Gastrointestinal lymphatic absorption of peptides and proteins. Adv Drug Deliv Rev 7:15–69
Samaranayaka AGP, Li-Chan ECY (2011) Food-derived peptidic antioxidants: a review of their production, assessment, and potential applications. J Funct Foods 3:229–254
Sarmadi BH, Ismail A (2010) Antioxidative peptides from food proteins: a review. Peptides 31:1949–1956
Schiller PW (2010) Bi- or multifunctional opioid peptide drugs. Life Sci 86:598–603

Sedó J, Saiz-Poseu J, Busqué F, Ruiz-Molina D (2013) Catechol-based biomimetic functional materials. Adv Mater 25:653–701
Senoner T, Dichtl W (2019) Oxidative stress in cardiovascular diseases: still a therapeutic target? Nutrients 11:2090
Seo H-S, Kwak S-Y, Lee Y-S (2010) Antioxidative activities of histidine containing caffeic acid-dipeptides. Bioorg Med Chem Lett 20:4266–4272
Shahidi F, Ambigaipalan P (2015) Phenolics and polyphenolics in foods, beverages and spices: antioxidant activity and health effects—a review. J Funct Foods 18:820–897
Shi J, Kakuda Y (2006) Bioavailability and synergistic effects of tea catechins as antioxidants in the human diet. Herbs Challenges Chem Biol 925:254–264
Shi Y-L, Benzie IFF, Buswell JA (2002) L-DOPA oxidation products prevent H_2O_2-induced oxidative damage to cellular DNA. Life Sci 71:3047–3057
Sila A, Bougatef A (2016) Antioxidant peptides from marine by-products: isolation, identification and application in food systems. A review. J Funct Foods 21:10–26
Silvia V, Baldisserotto A, Scalambra E, Malisardi G, Durini E, Manfredini S (2012) Novel molecular combination deriving from natural amino acids and polyphenols: design, synthesis and free-radical scavenging activities. Eur J Med Chem 50:383–392
Sinha RK, Eudes F (2015) Dimethyl tyrosine conjugated peptide prevents oxidative damage and death of triticale and wheat microspores. Plant Cell Tiss Org 122:227–237
Siodlak D (2015) alpha,beta-Dehydroamino acids in naturally occurring peptides. Amino Acids 47:1–17
Smith RA, Murphy MP (2011) Mitochondria-targeted antioxidants as therapies. Discov Med 11:106–114
Smith RAJ, Hartley RC, Murphy MP (2008) Mitochondria-targeted small molecule therapeutics and probes. Antioxid Redox Signal 13:3021–3038
Son S, Lewis BA (2002) Free radical scavenging and antioxidative activity of caffeic acid amide and ester analogues: structure–activity relationship. J Agric Food Chem 50:468–472
Sorriento D, Pascale AV, Finelli R, Carillo AL, Annunziata R, Trimarco B, Iaccarino G (2014) Targeting mitochondria as therapeutic strategy for metabolic disorders. Sci World J. https://doi.org/10.1155/2014/604685
Suetsuna K, Ukeda H, Ochi H (2000) Isolation and characterization of free radical scavenging activities peptides derived from casein. J Nutr Biochem 11:128–131
Sugumaran M, Tan S, Sun HL (1996) Tyrosinase-catalyzed oxidation of 3, 4-dihydroxyphenylglycine. Arch Biochem Biophys 329:175–180
Sun LP, Zhang Y, Zhuang YL (2013) Antiphotoaging effect and purification of an antioxidant peptide from tilapia (Oreochromis niloticus) gelatin peptides. J Funct Foods 5:154–162
Suzen S, Gurkok G, Coban T (2006) Novel N-acyl dehydroalanine derivatives as antioxidants: studies on rat liver lipid peroxidation levels and DPPH free radical scavenging activity. J Enzyme Inhib Med Chem 21:179–185
Szeto HH (2006) Cell-permeable, mitochondrial-targeted, peptide antioxidants. AAPS J 8:E277–E283
Szeto HH (2008) Mitochondria-targeted cytoprotective peptides for ischemia-reperfusion injury. Antioxid Redox Signal 10:601–619
Tang C-H, Peng J, Zhen D-W, Chen Z (2009) Physicochemical and antioxidant properties of buckwheat (Fagopyrum esculentum Moench) protein hydrolysates. Food Chem 115:672–678
Teixeira J, Deus CM, Borges F, Oliveira PJ (2018) Mitochondria: targeting mitochondrial reactive oxygen species with mitochondriotropic polyphenolic-based antioxidants. Int J Biochem Cell Biol 97:98–103
Torkova A, Koroleva O, Khrameeva E, Fedorova T, Tsentalovich M (2015) Structure-functional study of tyrosine and methionine dipeptides: an approach to antioxidant activity prediction. Int J Mol Sci 16:25353–25376

Trombino S, Serini S, Di Nicuolo F, Celleno L, Andò S, Picci N, Calviello G, Palozza P (2004) Antioxidant effect of ferulic acid in isolated membranes and intact cells: synergistic interactions with alpha-tocopherol, beta-carotene, and ascorbic acid. J Agric Food Chem 52:2411–2420
Victor V, Rocha M (2007) Targeting antioxidants to mitochondria: a potential new therapeutic strategy for cardiovascular diseases. Curr Pharm Des 13:845–863
Wade AM, Tucker HN (1998) Antioxidant characteristics of L-histidine. J Nutr Biochem 9:308–315
Wagstaff KM, Jans DA (2006) Protein transduction: cell penetrating peptides and their therapeutic applications. Curr Med Chem 13:1371–1387
Wang R, Liu Z-Q (2013) Ugi multicomponent reaction product: the inhibitive effect on DNA oxidation depends upon the isocyanide moiety. J Org Chem 78:8696–8704
Wei Q-Y, Jiang H, Zhang J-X, Guo P-F, Wang H (2012a) Synthesis of N-hydroxycinnamoyl amino acid ester analogues and their free radical scavenging and antioxidative activities. Med Chem Res 21:1905–1911
Wei Q-Y, Jiang H, Zhang J-X, Zhang C, Guo P-F (2012b) The antimicrobial activities of the cinnamoyl amide of amino acid derivatives. Asian J Chem 24:2383–2388
Yamaguchi N, Yokoo Y, Fujimaki M (1975) Studies on antioxidative activities of amino compounds on fats and oils, II: antioxidative activities of dipeptides and their synergistic effects on tocopherol. Nippon Shokuhin Kogyo Gakkaishi 22:425–430
Yang J, Hu L, Cai T, Chen Q, Ma Q, Yang J, Meng C, Hong J (2018) Purification and identification of two novel antioxidant peptides from perilla (Perilla frutescens L. Britton) seed protein hydrolysates. PLoS One 13:e0200021
Ye N, Hu P, Xu S, Chen M, Wang S, Hong J, Chen T, Cai T (2018) Preparation and characterization of antioxidant peptides from carrot seed protein. J Food Qual. Article ID 8579094
Yogeeta SK, Gnanapragasam A, Kumar SS, Subhashini R, Sathivel A, Devaki T (2006) Synergistic interactions of ferulic acid with ascorbic acid: its cardioprotective role during isoproterenol induced myocardial infarction in rats. Mol Cell Biochem 283:139–146
Zhao K, Zhao GM, Wu D, Soong Y, Birk AV, Schiller PW, Szeto HH (2004) Cell-permeable peptide antioxidants targeted to inner mitochondrial membrane inhibit mitochondrial swelling, oxidative cell death, and reperfusion injury. J Biol Chem 279:34682–34690
Zhao T, Liu X, Singh S, Liu X, Zhang Y, Sawada J, Komatsu M, Belfield KD (2019) Mitochondria penetrating peptide-conjugated TAMRA for live-cell long-term tracking. Bioconjug Chem 30:2312–2316

Plant Phenolics as Dietary Antioxidants: Insights on Their Biosynthesis, Sources, Health-Promoting Effects, Sustainable Production, and Effects on Lipid Oxidation

Pascual García-Pérez and Pedro P. Gallego

1 Phenolic Compounds in Plant Systems

Phenolic compounds (PCs) constitute the largest family of secondary metabolites in plants, with tens of thousands individual compounds described, divided into several subfamilies with heterogenous structures (Kumar et al. 2020). Such vast number of PCs is due to their ubiquitous presence in plant tissues, since PCs contribute to the general adaptative response against plant stress (García-Pérez et al. 2020a, b).

PCs are characterized by the presence of, at least, one phenolic ring in their chemical structure, being considered as polyphenols when presenting more than two phenolic rings (Kumar et al. 2020). As secondary metabolites, PCs are found in low concentrations in plant tissues, and their biosynthesis is induced by environmental stimuli, such as drought, UV-light, extreme temperatures, salinity, heavy metal accumulation and nutritional limitations, collectively known as abiotic stress (García-Pérez et al. 2020a, b), or biological-induced stress, including herbivore attacks and infections, making part of biotic stress (Tak and Kumar 2020). In response to stress, plants must face an acclimatization process to ensure their survival and development under disadvantageous conditions. Such process is driven at a molecular level by reactive oxygen species (ROS), which act as messenger molecules that contribute to the signal transduction associated with stress (García-

P. García-Pérez (✉)
Nutrition and Bromatology Group, Food Science and Technology Faculty, University of Vigo, Ourense, Spain
e-mail: pasgarcia@uvigo.es

P. P. Gallego
Agrobiotech for Health, Plant Biology and Soil Science Department, Biology Faculty, University of Vigo, Vigo, Spain

CITACA, Agri-Food Research and Transfer Cluster, University of Vigo, Ourense, Spain
e-mail: pgallego@uvigo.es

C. Bravo-Diaz (ed.), *Lipid Oxidation in Food and Biological Systems*,
https://doi.org/10.1007/978-3-030-87222-9_18

Pérez et al. 2020b). However, the stress-mediated overproduction of ROS and related molecules may generate the onset of oxidative stress.

In this way, the most relevant bioactivity associated with PCs is their antioxidant activity, since they are synthesized as a response to alleviate the impact of oxidative stress, throughout different antioxidant mechanisms (García-Pérez et al. 2018):

(a) Free-radical scavenging activity, mitigating the oxidative damage attributed to ROS and reactive nitrogen species.
(b) Induction of antioxidant enzymes that metabolize ROS, such as superoxide dismutase and catalase.
(c) Inhibition of lipid peroxidation, preventing the further deleterious injuries caused by oxidative stress at cellular level.
(d) Regulation of cell signaling pathways involved in the onset of oxidative stress.

The antioxidant activity associated with PCs contribute to the health-enhancing properties of these compounds, through diet, especially in the case of chronic diseases, which are characterized by the promotion of oxidative stress at their initial steps, such as obesity, type-2 diabetes, cancer, cardiovascular, neurodegenerative and inflammatory diseases (Leri et al. 2020). Thus, PCs are considered pleiotropic bioactive compounds (BCs), thanks to their antioxidant, anticancer and anti-inflammatory activities.

As stated earlier, PCs are synthesized in response to biotic stress, as well. Consequently, the biosynthesis of these compounds is induced to counter the deleterious effects caused by other organisms, including bacteria, fungi, viruses, herbivores and other plants (Tak and Kumar 2020). In the case of microorganisms, PCs have been extensively reported as antimicrobial agents, exhibiting a wide range of effectiveness against different bacterial and fungal pathogenic species, such as *Escherichia coli*, *Bacillus cereus*, *Staphylococcus aureus*, *Pseudomonas aeruginosa*, *Listeria monocytogenes* and *Aspergillus* and *Penicillium* strains (García-Pérez et al. 2021a). In the case of viruses, and in the context of the current pandemic due to the coronavirus SARS-CoV2, the most recent reports on the antiviral properties of PCs have pointed at their inhibitory mechanisms on specific proteins involved in the COVID-19 disease (Irfan et al. 2021).

As stated above, due to the multiple bioactivities associated to these compounds, PCs constitute one of the major families of plant BCs. Moreover, thanks to their ubiquitous distribution in plants, PCs are found as constituents of dietary vegetables and they can be easily incorporated in the diet. In fact, due to their effectiveness and assessed safety, PCs can be used as additives in a plethora of commercial products, including food, cosmetic, and pharmacological preparations (García-Pérez et al. 2018).

Concerning their biosynthesis, PCs derive from the shikimate-phenylpropanoid pathway that constitutes one of the major biosynthetic pathways on plant secondary metabolism (Quideau et al. 2011). In this pathway, cinnamic acid is produced after the deamination of the amino acid phenylalanine from the shikimate pathway, by the action of the enzyme phenylalanine ammonia lyase (PAL) (Quideau et al. 2011). Afterwards, cinnamic acid is transformed into p-coumaroyl-coenzyme A (CoA),

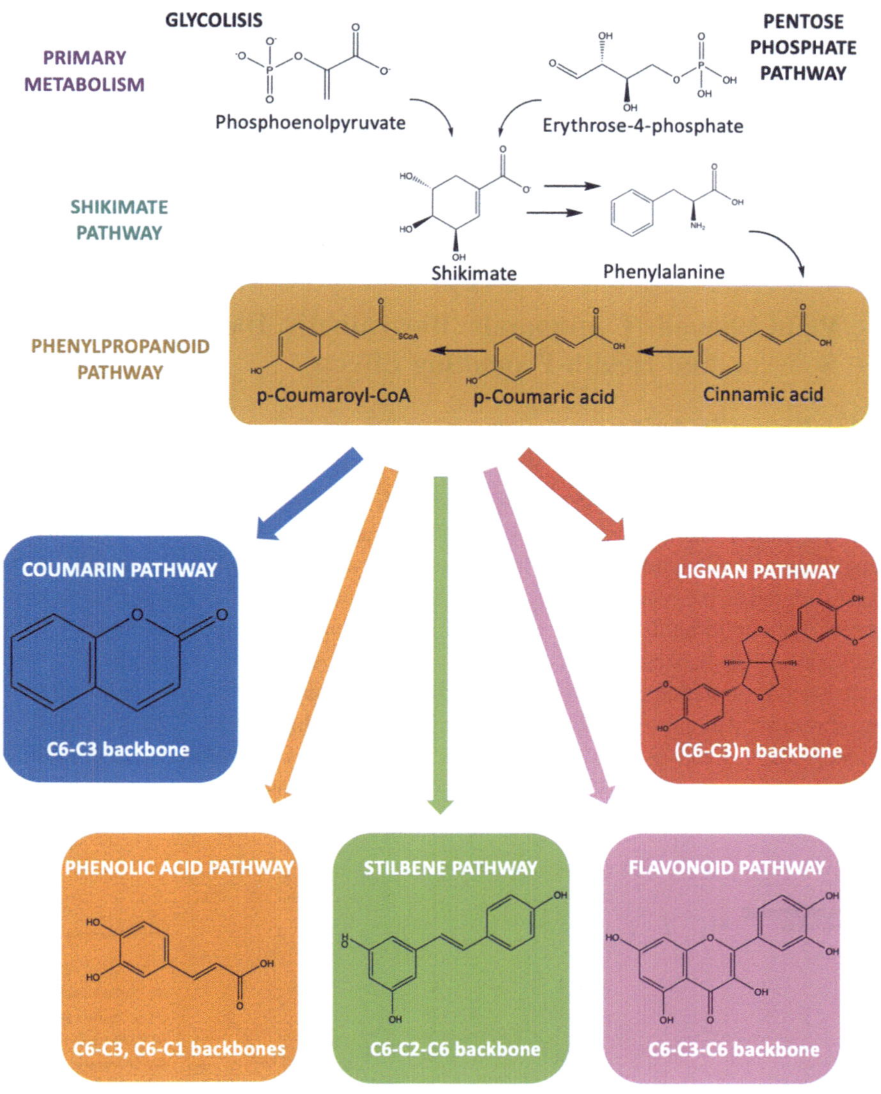

Fig. 1 Overview of the biosynthesis of phenolic compounds

giving rise to the different subfamilies of PCs (Fig. 1). Thus, five subfamilies of PCs are mostly found in higher plants, being considered as dietary PCs, formed along different biosynthetic pathways derived from phenylpropanoids (Fig. 1):

1. Phenolic acids, including compounds with C_6-C_3 (cinnamic acids) and C_6-C_1 (benzoic acids) backbones.
2. Lignans, characterized by a $(C_6\text{-}C_3)_2$ backbone.
3. Coumarins, characterized by a $(C_6\text{-}C_3)$ backbone.

4. Stilbenes, characterized by a (C_6-C_2-C_6) backbone.
5. Flavonoids, characterized by a (C_6-C_3-C_6) backbone.

Other subfamilies, such as naphtoquinones (C_6-C_4) or xanthones (C_6-C_1-C_6), have been also identified in some species in a much lesser extent. Thus, only the subfamilies ubiquitously found in dietary sources and derived products will be considered here.

2 Plant Phenolic Compounds: Biosynthesis, Dietary Sources, and Health-Promoting Effects

2.1 *Phenolic Acids*

Phenolic acids are simple PCs biosynthetically derived from the deamination of phenylalanine, via PAL, to produce cinnamic acid at the transition between the shikimate pathway and the phenylpropanoid pathway (Fig. 1). Thus, cinnamic acid is considered as the common precursor of phenolic acids (Fig. 2). Owing to structural criteria, phenolic acids may be classified into cinnamic acids, possessing a C_6-C_3 backbone, and benzoic acids, possessing a C_6-C_1 backbone. Consequently, the transition of cinnamic acids to benzoic acids is promoted by the side chain shortening through different biosynthetic steps (Fig. 2) (Marchiosi et al., 2020).

Regarding dietary sources, cinnamic acids normally occur alone or in combination with other modifications, such as quinic acid and monosaccharides. In the case of single molecules, p-coumaric, caffeic, ferulic, and sinapic acids are found in high

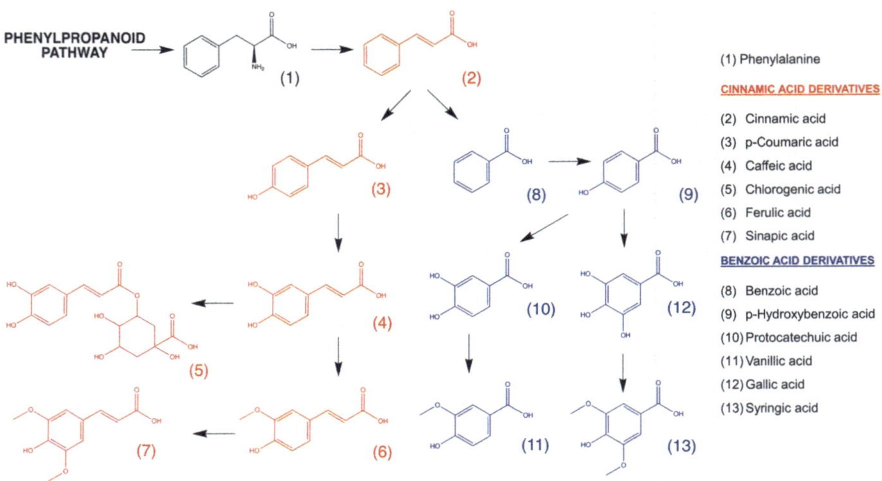

Fig. 2 Biosynthetic pathway of phenolic acids subdivided into cinnamic acid derivatives (red) and benzoic acid derivatives (blue)

concentrations in a plethora of vegetables, such as grains, red lettuce, spinach, radish, cauliflower, and red cabbage; on the other hand, conjugated acids are mostly represented by coumaroyl, feruloyl and caffeoyl derivatives, being chlorogenic acid and rosmarinic acid the most relevant compounds found in dietary sources, as it is the case of broccoli, Brussel sprouts, oregano, salvia, potato and artichoke (Kiokias et al. 2020; Rashmi and Negi 2020 and references therein). In the case of benzoic acids, they are usually found in plants either at single or glycosylated forms, being p-hydroxybenzoic, protocatechuic, vanillic and gallic acid the most abundant compounds from dietary sources. Thus, they are found in high concentrations in parsley, lettuce, mango and pak choi, as well as potato peels (Kiokias et al. 2020; Rashmi and Negi 2020).

As plant secondary metabolites, phenolic acids have been revealed as multifaceted BCs, acting as antimicrobial, antioxidant, cancer preventive, hepatoprotective, cardioprotective, immunomodulators and neuroprotective agents, and metabolic regulators (Kiokias et al. 2020), and they have been selected as excellent candidates to test their effectiveness on human diseases by clinical trials (Salomone et al. 2020).

2.2 *Lignans*

Lignans constitute a wide group of PCs derived from the phenylpropanoid pathway and their structure is based on the condensation of two or more units of phenylpropanoid units or monomers, mostly derived from cinnamic acid, thus forming compounds containing a $(C_6\text{-}C_3)_n$, $(n \geq 2)$ backbone (Fig. 1) (Chen et al. 2020). According to their structure, lignans can be divided into two subfamilies: "classical" lignans are formed upon a β-β link between monomers, while neolignans are formed by non β-β linkages, thus conferring a high structural diversity within this family (Fig. 3) (Cui et al. 2020). Due to the condensation of monomers in different proportions, lignans are found in plants as dimers, trimers and tetramers, and they can be polymerized to give rise to lignin, considered the main constituent of plant cell walls (Del Río et al. 2020).

Both lignans and neolignans are mostly represented by sesamin, matairesinol, and lariciresinol, and they can be predominantly found in flax, sesame and sunflower seeds, as well as black tea, coffee, strawberries, peaches, garlic, carrots and asparagus (Rodríguez-García et al. 2019; Nadeem and Ahmad 2019). With respect to their role as BCs, these compounds have been exploited as scaffolds for the synthesis of

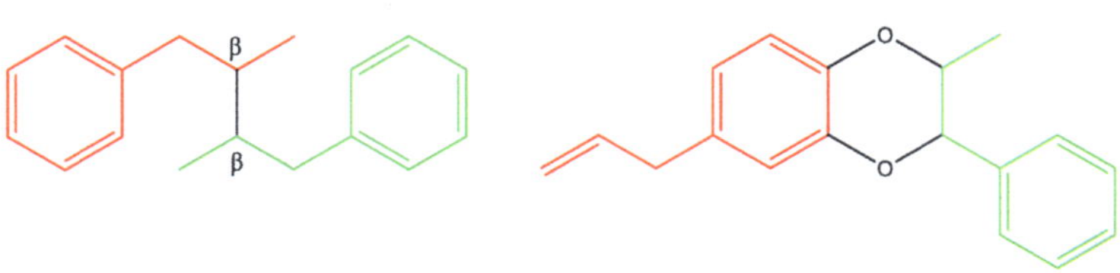

Fig. 3 Basic scaffolds for lignans (left) and neolignans (right). Monomers are displayed in different colors. Adapted from Cui et al. (2020)

novel drugs, in the field of pharmaceutical chemistry (Fang and Hu 2018). Additionally, a plethora of bioactivities were attributed to lignans, ranging from antibacterial, antifungal, antiviral, anticancer, anti-inflammatory and antioxidant properties to other added-value properties such as neuroprotectant, anti-osteoporotic, anti-obesity, antidiabetic, estrogenic, cardiovascular preservatives, and insecticidal compounds (Hano et al. 2021).

2.3 *Coumarins*

Coumarins constitute another subfamily of PCs, characterized by a fused benzopyrone C_6-C_3 backbone scaffold, being 1,2-benzopyrone their parent compound, derived from the phenylpropanoid pathway (Fig. 1) (Robe et al. 2021). Coumarins are structurally classified according to 1,2-benzopyrone substituents, producing simple coumarins, when substituents are single radicals that do not interact between them (Fig. 4). In contrast, when substituents react to form cycled

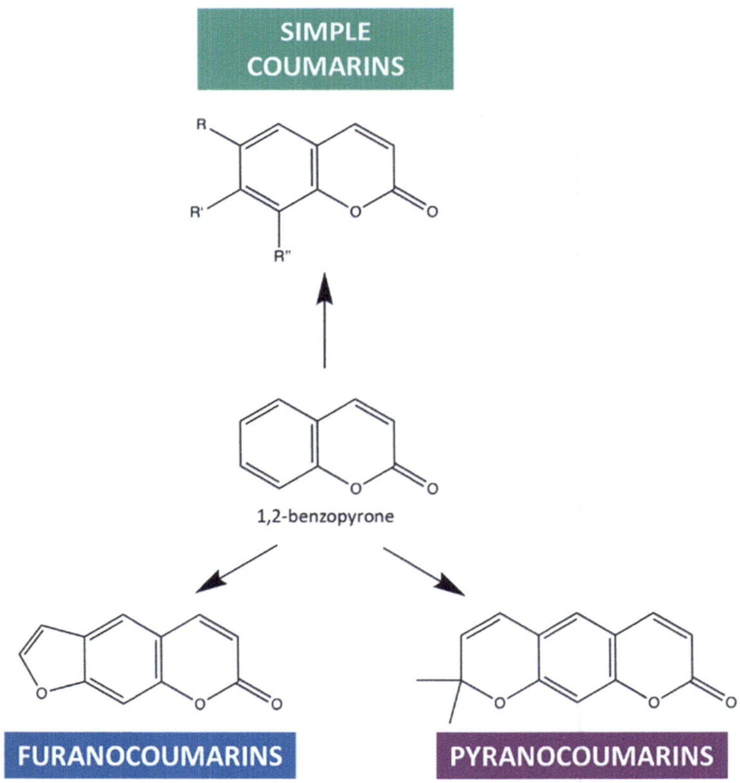

Fig. 4 Major subfamilies of coumarins derived from the 1,2-benzopyrone parental compound

structures, two different subgroups of coumarins are obtained: furanocoumarins, with a basic five-membered furane ring, and pyranocoumarins, presenting a six-membered pyrane ring (Fig. 4) (Zhu and Jiang 2018).

In the case of simple coumarins, such as umbelliferone, esculetin and scopoletin, they are predominantly found in green tea, propolis, lavender honey and plant oils, such as olive, and soy oils (Lončar et al. 2020). Furanocoumarins, in turn, are mostly found in citrus peels, showing that bergamot peel is a potent source of these compounds, together with lemon, grape, and pummelo, and they are mostly represented by psoralen and bergaptol (Lončar et al. 2020). On the other hand, pyranocoumarins are not relevant in dietary sources (Zhu and Jiang 2018).

Coumarins have been widely assessed in terms of their associated bioactive properties, acting as anti-human immunodeficiency virus (HIV), anticancer, antioxidant, antimicrobial, anti-inflammatory, analgesic, and cardiovascular protective agents (Zhu and Jiang 2018). Among them, the cancer preventing properties of coumarins by different cellular mechanisms show a controversial behavior as part of anticancer therapy, since furanocoumarins have been shown to enhance the effectiveness of chemotherapeutics (Sumorek-Wiadro et al. 2020).

2.4 Stilbenes

Stilbenes present a basic C_6-C_2-C_6 structure, biosynthesized from the phenylpropanoid pathway (Fig. 1), considering p-coumaroyl-CoA as the original precursor, which condensates with 3 malonyl-CoA units, by the action of stilbene synthase, enabling the biosynthesis of *trans*-resveratrol, the universal stilbene compound (Tian and Liu 2020).

Stilbenes are especially abundant to grape and its derived products, such as wines, considered one of the major sources of resveratrol (Tian and Liu 2020). Due to the countless health-enhancing properties associated with resveratrol, the most part of the investigations conducted on stilbenes have been focused on this particular molecule. In this sense, resveratrol exerts a wide range of bioactivities including anticancer, anti-inflammatory, antimicrobial, antidiabetic, and antioxidant properties (Fiod Riccio et al. 2020), but it also has been proven to promote additional features as an anti-aging agent, cardiovascular and neurodegenerative protectant and enhancer of liver and renal function (Patra et al. 2020). In addition, the beneficial effects of resveratrol administration on human health were demonstrated by human epidemiological studies (Patra et al. 2020).

2.5 Flavonoids

Flavonoids constitute the widest family of PCs, accounting for more than 10,000 compounds (Arroo et al. 2020), biosynthesized from p-coumaroyl-CoA (Fig. 1),

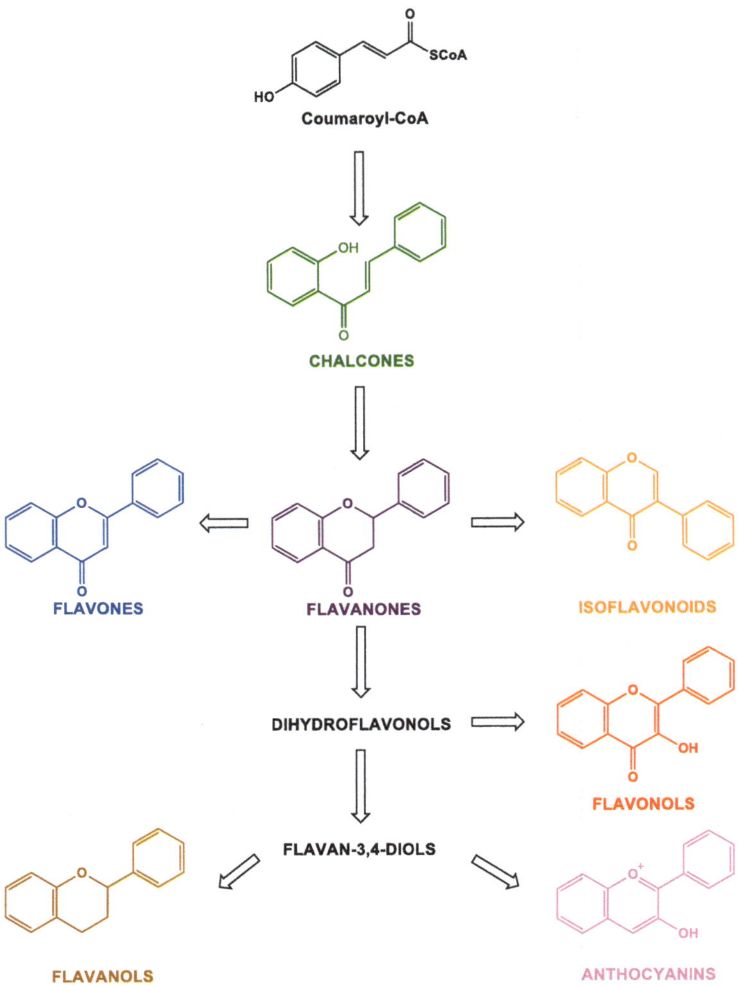

Fig. 5 Schematic representation of flavonoid biosynthesis

where it condensates with 3 molecules of malonyl-CoA by the action of the enzyme chalcone synthase, producing the chalcone scaffold of naringenin, considered the flavonoid precursor (Davies et al. 2020). Afterwards, the cyclation of the C_3 central region gives rise to the basic flavonoid skeleton, characterized by a three-ring structure, being classified by the oxidation degree of the central ring (Fig. 5).

All flavonoid subfamilies exhibit similar associated bioactivities according to the countless studies focused on their functional characterization (Alseekh et al. 2020), in terms of their pleiotropic health-promoting properties, acting as potent antioxidant, antimicrobial, anticancer, anti-inflammatory, cardiovascular and neurodegenerative protectants, immunomodulator and enzymatic modulator agents

(Meshram et al. 2020). In this way, in recent years, the research on flavonoids was mainly focused on the determination of their chemopreventive properties (García-Pérez et al. 2019a).

Due to the wide distribution of flavonoids in the plant kingdom, they constitute the family of PCs with the highest contribution in the diet (Williamson et al. 2018), accounting for seven subfamilies showing a relevant dietary presence, namely: chalcones, flavones, flavanones, isoflavonoids, flavonols, flavanols and anthocyanins.

2.5.1 Chalcones

Chalcones are characterized by the presence of two aromatic rings bound by a linear three-carbon structure possessing a conjugated carbonyl group and a double bond (Fig. 5), which are responsible for their yellow and orange coloration (Mah 2020). Isoliquiritigenin is one of the most relevant compounds belonging to this subfamily, which can be found in high concentration in different dietary sources, including tomato and licorice (Karimi-Sales et al. 2018; Mah 2020).

As members of the flavonoid family, chalcones present a wide range of associated bioactivities, and they have been explored for the development of novel therapeutic strategies. Chalcones present a basic structure that may undergo different chemical modifications for the synthesis of a series of pharmaceutical products involved in the production of novel chalcone-based antidiabetic, hepatoprotective and obesity-preventing drug agents (Karimi-Sales et al. 2018). In addition, the consumption of chalcone-rich products has been assessed in terms of food safety, showing non-toxic effects on the hematological, gastrointestinal and cardiovascular functions (Mah 2020).

2.5.2 Flavones

Flavones present a basic skeleton possessing a double bond in the central ring together with a carbonyl group (Fig. 5), which confers a white-creamy coloration (Hostetler et al. 2017). Among the different flavones found in this subfamily, apigenin, luteolin, and diosmetin are the most relevant compounds, which can be found in a plethora of dietary sources including citrus fruits, celery, pepper, chamomile, mint and different cereals, such as wheat, rice, and sorghum (Hostetler et al. 2017).

In addition to the bioactive properties associated with flavones as part of the flavonoid family, these compounds have been explored for additional features, revealing their efficacy as preservatives of diabetic nephropathy, immunomodulators, and regulators of the metabolic syndrome (Gutiérrez-Grijalva et al. 2020). Additionally, the role of flavones as anticancer compounds has broadened the application of these compounds to cancer therapy, since apigenin can be an excellent candidate to combine with chemotherapy in order to satisfactorily treat resistant cancers (Kashyap et al. 2018).

2.5.3 Flavanones

Flavanones constitute a flavonoid subfamily structurally characterized by the saturation of all carbon atoms of the central ring together with a carbonyl group (Fig. 5), resulting in a dihydropyrone ring, and they are commonly found in citrus fruits, such as lemon, orange, and tangerine (Najmanová et al. 2020).

Among this subfamily, naringenin has been the most in-depth characterized flavanone, by means of their associated bioactivities (Manchope et al. 2017), exhibiting beneficial effects on a plethora of cell signaling pathways, being regarded as an anti-atherogenic, antioxidant, immunomodulatory, anti-inflammatory, hepatoprotective and neuroprotective agent, as proved under in vitro conditions (Salehi et al. 2019). In fact, the immunomodulatory activity attributed to naringenin has arisen its application on the treatment of inflammatory diseases, such as sepsis, hepatitis, fibrosis, cancer (Zeng et al. 2018), and airway diseases like COVID-19 (Tutunchi et al., 2020).

2.5.4 Isoflavonoids

Isoflavonoids constitute a complex subfamily of flavonoids, in which the basic scaffold contains a central ring bound to the third ring at the C-3 position, in contrast to other flavonoid subfamilies in which the third ring is linked at the C-2 position (Fig. 5). The biosynthesis of isoflavanoids is initially developed by the action of 2-hydroxyisoflavanone synthase (2-HIS), that catalyzes the migration of such C-2 bond (Al-Maharik 2019). Hence, this diverse chemical configuration implies that different subfamilies of isoflavonoids are found in the nature, including isoflavones, isoflavanones, and isoflavanols, among others.

Isoflavones have gained much attention because of their associated bioactivities and their high concentrations in dietary sources, mostly soybean and their derived products, and legumes like lentils, chickpeas and beans (Blicharski and Oniszczuk 2017). Daidzein and genistein are considered as the most prevalent dietary isoflavones, as the main compounds of soybean products (Křížová et al. 2019). Besides their bioactive properties as flavonoids, isoflavones stand out by their estrogenic activity, promoting a beneficial effect on related diseases, acting as effective preventive agents of cardiovascular diseases and osteoporosis on menopausal women, breast cancer and obesity, as demonstrated by different clinical trials (Akhlaghi et al. 2017).

2.5.5 Flavonols

Flavonols are characterized by the presence of a hydroxylated central ring which also contains a carbonyl group (Fig. 5), causing their absence of color (Brunetti et al. 2019). Flavonols constitute the largest subfamily of flavonoids and, in consequence, they can be ubiquitously found in countless dietary sources, including fruits- like apple and pomegranate-, vegetables- such as broccoli, onions, and spinach-, cocoa, green and dark tea and red wine (Šeruga and Tomac 2017). Within this subfamily, quercetin, kaempferol and rutin are considered the most prevalent compounds (Dabeek and Marra 2019), being quercetin the major natural flavonoid (El-Saber Batiha et al. 2020). It can be mostly found in broccoli, apple, cauliflower, pepper, and onions (El-Saber Batiha et al. 2020), and it has been largely reported as an antioxidant, hepatoprotective, anti-inflammatory, neuroprotective, cardioprotective, and anticancer compound (Sharma et al. 2018).

Besides the own properties attributed to quercetin as a chemopreventive agent, recent reports are proposing different approaches to include this compound to cancer therapeutical strategies, improving the effectiveness of chemotherapeutic drugs, for instance: enhancing the sensitivity of lung cancer cells to paclitaxel by chitosan nanoparticles (Wang et al. 2021), and the generation of quercetin-loaded micelles to prevent chemotherapeutic-induced toxicity (Casanova et al. 2021).

2.5.6 Flavanols

Flavanols are structurally classified by a hydroxylated saturated central ring without any modification (Fig. 5). Due to this structural feature, they present two asymmetric carbons that confer an enormous plasticity to give rise to different isomers (Das et al. 2019). In addition, flavanols may occur either in a monomeric form or dimeric and polymeric forms, giving rise to proanthocyanidins and tannins, revealing a complex subclassification of this subfamily (Pizzi 2019).

In the case of monomeric flavanols, catechins constitute the most important group within this subfamily, being (+)-catechin, (−)-epicatechin and epigallocatechin gallate (EGCG), three of the highly prevalent compounds from dietary sources (Das et al. 2019), mainly found in green tea, wine, cocoa, grapes, apricots and strawberries (Di Lorenzo et al. 2021). As member of flavonoid family, flavanols were identified as multi-faceted BCs, and they have been selected as candidates for modern combinatorial strategies to treat several diseases, promoting a synergistic effect with chemotherapeutic drugs (Di Lorenzo et al. 2021). In fact, the promising effects of flavanols on human health is already being analyzed by different controlled clinical trials (Akhlaghi et al. 2018).

2.5.7 Anthocyanins

Anthocyanins are structurally formed by a 2-phenylbenzopyrilium skeleton, with a protonated oxygen atom at low pHs (Fig. 5), responsible for the coloration of these compounds, whose color depends on their glycosylated substitutions (Martín et al. 2017).

Cyanidin-3-glucoside is the major anthocyanin found in plants (Khoo et al. 2017), and together with other anthocyanins, such as delphinidin and pelargonidin, they can be predominantly found in berries, grapes, tropical fruits, and red to purplish-colored vegetables (Khoo et al. 2017). In order to take advantage of their associated bioactivities as flavonoid compounds (Eker et al. 2020), and thanks to their colored nature, great efforts are being made to explore novel strategies to incorporate anthocyanins to different food matrices, such as enzymatic biotransformation (Marathe et al. 2021), and the design of anthocyanin-based intelligent packaging (Becerril et al. 2020).

3 Plant Phenolics: Production, Extraction, and Lipid Oxidation Prevention

3.1 Biotechnological Production of PCs

As plant secondary metabolites, PCs are found in little concentrations in plants and, therefore, novel strategies should be developed to increase their biosynthesis and accumulation. As a result, plant tissue culture emerges as a robust technology to achieve such goal, by the controlled induction of plant stress. In this sense, different approaches are undertaken from a biotechnological point of view, as it is the case of the elicitation of plant suspension-cultured cells (PSCCs) (García-Pérez et al. 2021b). In the last decades, the elicitation of PSCCs as a robust strategy to produce PCs has arisen exponentially, by the inclusion of elicitors to these cell-based systems. Elicitors trigger the rate of secondary metabolite production, thanks to their role as signaling molecules that induce plant stress (García-Pérez et al. 2020c). Therefore, jasmonates and salicylates are commonly used alone or in combination for the elicitation of PSCCs from different plant species (Giri and Zaheer 2016). In this sense, elicitation has been largely proposed as an efficient and robust methodology to improve the biosynthesis of PCs, including coumarins, stilbenes, flavonols and anthocyanins at the same time (García-Pérez et al. 2021b).

Nevertheless, thanks to the plasticity of PSCCs, novel elicitors have been recently discovered, as it is the case of cyclodextrins (CDs) (García-Pérez et al. 2019b). CDs act as double-sided elicitors: they promote the induction of secondary metabolism and, at the same time, they act as complexing agents, forming inclusion complexes with PCs, thus facilitating the protection of their antioxidant activity and favoring their solubility in aqueous systems (García-Pérez et al. 2019b). In this way, the

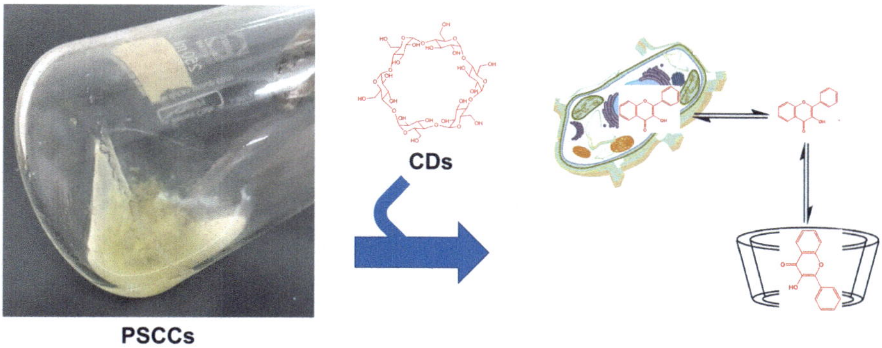

Fig. 6 Complexation of PCs produced by elicited PSCCs into CDs conical structures. PSCCs (left). CD structure (center). Extracellular accumulation of PC-loaded cyclodextrins, forming 1:1 complexes (right)

application of CDs on PSCCs not only leads to the accumulation of PCs, but they also make them more accessible to extraction without interfering with their associated bioactivities, since the conical structure of CDs promote their behavior as host molecules for PCs, which are released by plant cells (Fig. 6) (Almagro et al. 2016; García-Pérez et al. 2019b). As complexing agents, the inclusion of PCs in the CDs cavities follows the formation of inclusion complexes of different stoichiometries, mostly 1:1, 1:2, and 2:1, characterized by an inclusion constant, K_{inc}, given by Eq. (1) for 1:1 complexes. In parallel, as elicitors, CDs promoted a significant enhancement of PCs production by PSCCs at the extracellular culture medium, just after 7 days of culture, achieving increasing rates of ~8-fold for total phenolic content, and ~16-fold for flavonoid content in the case of *Bryophyllum* spp. (García-Pérez et al. 2019b). Moreover, the formation of PC-CD complexes did not affect their antioxidant activity, in terms of radical scavenging activity.

$$\mathrm{PC} + \mathrm{CD} \underset{k_D}{\overset{k_F}{\rightleftharpoons}} \mathrm{PC} - \mathrm{CD} \quad K_{inc} = \frac{k_F}{k_D}\frac{[\mathrm{PC} - \mathrm{CD}]}{[\mathrm{PC}][\mathrm{CD}]} \tag{1}$$

3.2 Sustainable Extraction of PCs

In order to satisfy the consumers' current demands for food naturalness and sustainability, novel advances are leading to the establishment of green and safe production of PCs from plant sources, devoted to the development of environmentally-friendly strategies that overcome the limitations attributed to classical extraction and purification techniques (García-Pérez et al. 2020b). In this sense, the use of sustainable

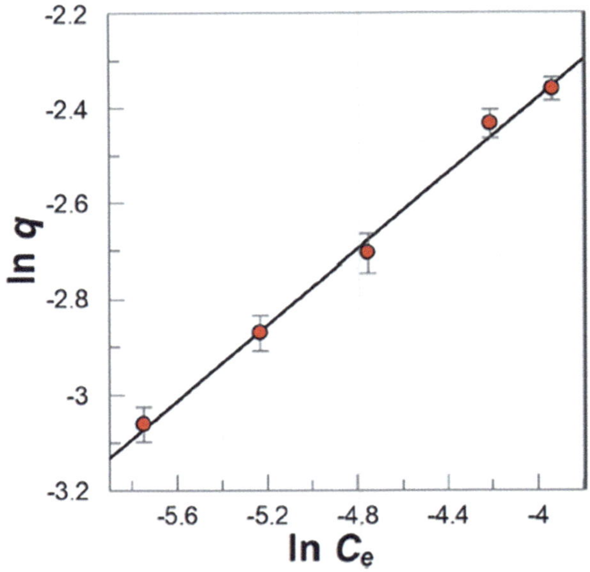

Fig. 7 Freundlich adsorption isotherm for *Bryophyllum* × *houghtonii* aqueous extract at 20 °C. Adapted from García-Pérez et al. (2019c), and reproduced with permission of Scientific Reports, published by Springer Nature

matrices and purification techniques, such as activated carbon, ionic liquids, or supercritical fluids, are emerging in the last decades. Specially, the application of activated carbon (AC) has been recently demonstrated as reliable and sustainable matrix to selectively purify PCs from complex matrices, as it is the case of plant extracts (García-Pérez et al. 2019c). The effectiveness of AC as an adsorptive matrix for PCs is caused by the selective binding of the phenolic rings found in the structures of these compounds, with independence of their chemical substitutions (García-Pérez et al. 2019c).

Moreover, the reliability of AC for the purification of PCs was also assessed in terms of its kinetic and adsorptive behavior, showing that only 7 mg of AC are required to selectively adsorb >99% of PCs from aqueous plant extracts from *Bryophyllum* spp. after nearly 7 hours, following the Freundlich isotherm, given by Eq. (2), where q represents the amount of PCs adsorbed per mass of AC, K_F is a constant relative to the adsorption capacity of AC, Ce represents the PC concentration at the adsorption equilibrium, and n stands for the intensity of adsorption (n values between 0–1 indicate that adsorption is favorable) (García-Pérez et al. 2019c). Moreover, Eq. (2) can be converted into Eq. (3) to elaborate a ln q vs ln Ce plot, in order to represent the Freundlich isotherm behavior followed by plant extracts, as shown by the linear shape of such plot (Fig. 7). These findings, together with those obtained from other authors (Soto et al. 2008; Lehmann et al. 2018) are opening new perspectives in the use of AC in the field of economically-interesting sectors, as it is the case of those related not only with the food industry, but also with the cosmeceutical and pharmacological industries.

$$q = K_F C_e^{\,n} \tag{2}$$

$$\ln q = \ln K_F + n \ln C_e \tag{3}$$

3.3 *PCs as Preservative Antioxidants in Oil-in-Water Emulsions*

Due to the bioactivities associated to PCs, in the last decades there has been an increasing interest on the establishment of different strategies devoted to the inclusion of these compounds on food matrices. In this sense, taking advantage of the antioxidant activity attributed to PCs, they were proposed as preservative agents that may counter the lipid oxidative degradation occurring to lipid-based foods, as it is the case of oil-in-water emulsions (García-Pérez et al. 2018).

More specifically, polyunsaturated fatty acids (PUFAs), which constitute one of the main constituents of both vegetal and fish oils, have gained much attention in the fields of food and nutrition because of their associated health-promoting properties (Liu et al. 2017). However, PUFAs are extremely prone to oxidation, thus causing a deleterious loss of their beneficial properties and, as a consequence, the use of artificial preservatives is a common industrial strategy to avoid their oxidation (García-Pérez et al. 2020d). Nevertheless, in order to overcome the use of artificial additives, plant-based PCs were proposed as natural preservatives of fish oil-in-water emulsions (OWE; Fig. 8) (García-Pérez et al. 2020d).

Fish oil constitutes one of the most relevant sources of omega-3 fatty acids, such as eicosapentaenoic acid (EPA) and docosahexaenoic acid (DHA), and they are commonly found in foods as oil-in-water emulsions (Jacobsen et al. 2013). Thus, the addition of antioxidants to emulsions emerges as a convenient approach to preserve the stability of omega-3 enriched fish oil. Moreover, the establishment of OWE encompasses the generation of a thermodynamically unstable system in which fatty

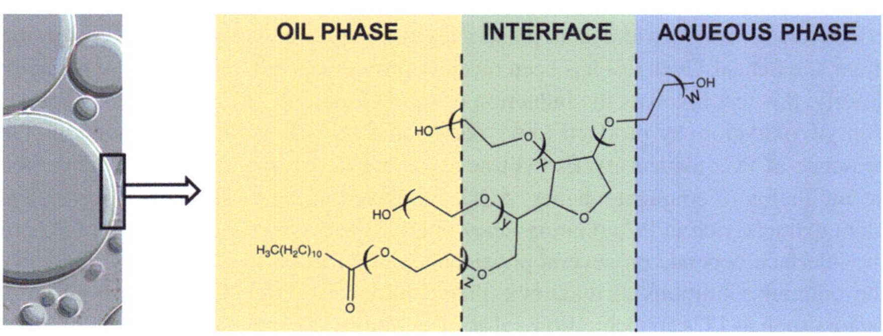

Fig. 8 Optical microscopic picture of fish OWE (left). Triphasic system formed in a fish OWE, stabilized by the inclusion of Tween-80 as surfactant (right)

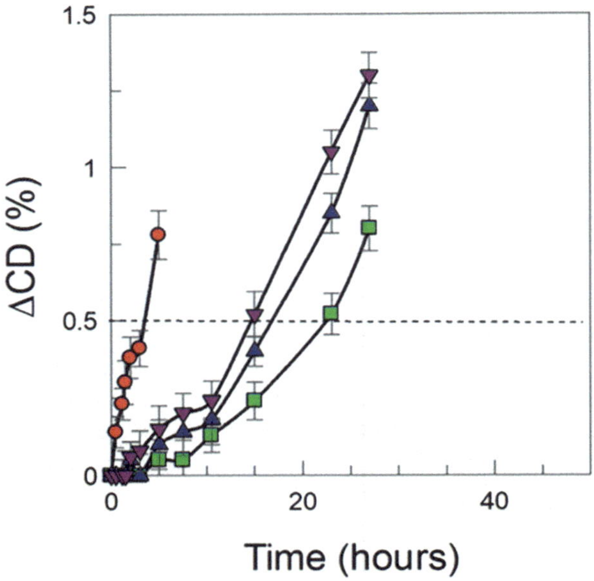

Fig. 9 Accumulation of conjugate dienes, ΔCD, in fish OWE, delayed in the presence of PC-enriched *Bryophyllum* extracts. ΔCD is delayed in the presence of plant extracts (red circles)—untreated OWE, (pink inverted triangles)—OWE treated with *Bryophyllum tubiflorum* extracts, (blue triangles)—OWE treated with *Bryophyllum daigremontianum* extracts, (green squares)—OWE treated with *Bryophyllum* × *houghtonii* extracts. Adapted from García-Pérez et al. (2020d), and reproduced with permission of Plants, published by MDPI

acids form a lipid phase dispersed in an aqueous solution, separated by a thin intermediate layer called interface (McClements et al. 2017). The whole system is then stabilized by the inclusion of surfactants, like Tween-80 or lecithins, which contribute to prevent phase separation, as shown in Fig. 8, thus mitigating the emulsion instability (McClements et al. 2017).

According to the floating peroxyl radical theory, it is now well-known that the peroxyl radicals formed after the oxidation of PUFAs residing in the oil phase migrate towards the interfacial region of the emulsion, suggesting that oxidation takes places in this intermediate region (García-Pérez et al. 2018). Consequently, with the aim of acting as efficient antioxidants, PCs incorporated to emulsions must reach the interfacial region, in which they may quench the radicals formed as a result of lipid oxidation (García-Pérez et al. 2020d). Due to the impossibility of isolating the interface, different indirect methods are developed in order to determine the effectiveness of antioxidants as preservatives of lipid oxidation in OWEs. Among them, the Schaal-Oven test has been recently proved as a reliable method to that aim. Briefly, this test involves the induction of lipid oxidation, promoting the accumulation of oxidation by-products, i.e.: conjugated dienes, which is delayed in the presence of PCs-loaded OWEs (Mitrus et al. 2019), as shown by Fig. 9, where the accumulation of conjugate dienes, ΔCD, is delayed in the presence of *Bryophyllum* plant extracts, rich in PCs (García-Pérez et al. 2020d). Since the diffusion of PCs to the interface depends on several physicochemical factors, it is essential to identify the optimal conditions to maximize their antioxidant efficacy, involving different factors such as PCs concentration, oil-to-water ratio, storage temperature and acidity of the aqueous phase (García-Pérez et al. 2020d).

4 Concluding Remarks

As it has been covered along this chapter, PCs are exceptional BCs that promote a beneficial effect on a plethora of relevant diseases, since they are identified as antioxidants that counter the onset of chronic diseases, characterized by the burst of oxidative stress during their early steps. In this sense, there is a wide evidence pointing at the correlation between the consumption of foods rich in PCs and other bioactive molecules and the prevention of chronic diseases and, furthermore, the delay of degenerative ailments (Martins et al. 2016).

In the same way, it is essential to promote the development of clinical trials and epidemiological studies in order to determine the real effectiveness of the different associated bioactivities attributed to PCs. Such scientific evidence should, then, be aimed at characterizing the gap existing between the effects observed by in vitro and in vivo models and determining the impact of PCs on a dietary basis. Additionally, current studies are underway not only focusing on the direct effect of PCs on human health, but also as complementary agents of therapeutic regimes, contributing to the development of synergistic effects with pharmacological drugs, increasing the sensitive of target organs, or reducing the toxicity associated with severe treatments, such as those of chemotherapy (Casanova et al. 2021).

Consequently, regarding the efficiency of PCs-enriched foods as health-promoting agents, intense efforts should be made in the fields of plant and food science to contribute to the phytochemical and pharmacological valorization of uncharacterized plant-derived matrices, such as medicinal plants, which are employed in the traditional medicine of vast regions worldwide (García-Pérez et al. 2020e). Therefore, novel plant-based products could be identified, thus conferring additional sources of BCs, assisting to the maintenance of a healthy lifestyle and avoiding the excessive consumption of artificial additives, that could be harmful after a long-term consumption.

In this sense, besides plant-derived extracts, individual compounds have also demonstrated their effectiveness as dietary antioxidants, as determined for different subfamilies of PCs, including phenolic acids (Kiokias et al. 2020), lignans (Hano et al. 2017), stilbenes (Matos et al. 2014), and flavonoids (Yang et al. 2015) but also acting as preservatives of OWEs under different storage and manufacturing conditions, providing insight about the great plasticity of these compounds to be used as natural additives, not only of food matrices, but also of cosmetical and pharmacological preparations.

References

Akhlaghi M, Zare M, Nouripour F (2017) Effect of soy and soy isoflavones on obesity-related anthropometric measures: a systematic review and meta-analysis of randomized controlled clinical trials. Adv Nutr 8:705–717. https://doi.org/10.3945/an.117.015370

Akhlaghi M, Ghobadi S, Mohammad Hosseini M, Gholami Z, Mohammadian F (2018) Flavanols are potential anti-obesity agents, a systematic review and meta-analysis of controlled clinical trials. Nutr Metab Cardiovasc Dis 28:675–690. https://doi.org/10.1016/j.numecd.2018.04.001

Almagro L, García-Pérez P, Belchí-Navarro S, Sánchez-Pujante PJ, Pedreño MA (2016) New strategies for the use of *Linum usitatissimum* cell factories for the production of bioactive compounds. Plant Physiol Biochem 99:73–78. https://doi.org/10.1016/j.plaphy.2015.12.009

Al-Maharik N (2019) Isolation of naturally occurring novel isoflavonoids: an update. Nat Prod Rep 36:1156–1195. https://doi.org/10.1039/c8np00069g

Alseekh S, Perez de Souza L, Benina M, Fernie AR (2020) The style and substance of plant flavonoid decoration; towards defining both structure and function. Phytochemistry 174:112347. https://doi.org/10.1016/j.phytochem.2020.112347

Arroo RRJ, Wang MF, Bhambra AS (2020) Chemopreventive potential of flavones, flavonols, and their glycosides. In: Xiao J, Sarker S, Asakawa Y (eds) Handbook of dietary phytochemicals. Springer, Singapore, pp 1–19. https://doi.org/10.1007/978-981-13-1745-3_5-1

Becerril R, Nerin C, Silva F (2020) Bring some colour to your package: freshness indicators based on anthocyanin extracts. Trends Food Sci Technol 111:495–505. https://doi.org/10.1016/j.tifs.2021.02.042

Blicharski T, Oniszczuk A (2017) Extraction methods for the isolation of isoflavonoids from plant material. Open Chem 15:34–45. https://doi.org/10.1515/chem-2017-0005

Brunetti C, Sebastiani F, Tattini M (2019) ABA, flavonols, and the evolvability of land plants. Plant Sci 280:448–454. https://doi.org/10.1016/j.plantsci.2018.12.010

Casanova AG, Prieto M, Colino CI, Gutiérrez-Millán C, Ruszkowska-Ciastek B, de Paz E, Martín Á, Morales AI, López-Hernández FJ (2021) A micellar formulation of quercetin prevents cisplatin nephrotoxicity. Int J Mol Sci 22:1–14. https://doi.org/10.3390/ijms22020729

Chen J, Chen Y, Ye X (2020) Lignans in diets. In: Xiao J, Sarker S, Asakawa Y (eds) Handbook of dietary phytochemicals. Springer, Singapore, pp 1–19. https://doi.org/10.1007/978-981-13-1745-3_38-1

Cui Q, Du R, Liu M, Rong L (2020) Lignans and their derivatives from plants as antivirals. Molecules 25:1–17. https://doi.org/10.3390/molecules25010183

Dabeek WM, Marra MV (2019) Dietary quercetin and kaempferol: Bioavailability and potential cardiovascular-related bioactivity in humans. Nutrients 11:2288. https://doi.org/10.3390/nu11102288

Das AB, Goud VV, Das C (2019) 9 – phenolic compounds as functional ingredients in beverages. In: Grumezescu AM, Holban AM (eds) Value-added ingredients and enrichments of beverages. Academic Press, San Diego, CA, pp 285–323. https://doi.org/10.1016/B978-0-12-816687-1.00009-6

Davies KM, Jibran R, Zhou Y, Albert NW, Brummell DA, Jordan BR, Bowman JL, Schwinn KE (2020) The evolution of flavonoid biosynthesis: a bryophyte perspective. Front Plant Sci 11:1–21. https://doi.org/10.3389/fpls.2020.00007

Del Río JC, Rencoret J, Gutiérrez A, Elder T, Kim H, Ralph J (2020) Lignin monomers from beyond the canonical monolignol biosynthetic pathway: another brick in the wall. ACS Sustain Chem Eng 8:4997–5012. https://doi.org/10.1021/acssuschemeng.0c01109

Di Lorenzo C, Colombo F, Biella S, Stockley C, Restani P (2021) Polyphenols and human health: the role of bioavailability. Nutrients 13:1–30. https://doi.org/10.3390/nu13010273

Eker ME, Aaby K, Budic-Leto I, Brncic SR, El SN, Karakaya S, Simsek S, Manach C, Wiczkowski W, De Pascual-Teresa S (2020) A review of factors affecting anthocyanin bioavailability: possible implications for the inter-individual variability. Foods 9:1–18. https://doi.org/10.3390/foods9010002

El-Saber Batiha G, Beshbishy AM, Ikram M, Mulla ZS, Abd El-Hack ME, Taha AE, Algammal AM, Ali Elewa YH (2020) The pharmacological activity, biochemical properties, and pharmacokinetics of the major natural polyphenolic flavonoid: quercetin. Foods 9:374. https://doi.org/10.3390/foods9030374

Fang X, Hu X (2018) Advances in the synthesis of lignan natural products. Molecules 23:42–45. https://doi.org/10.3390/molecules23123385

Fiod Riccio BV, Fonseca-Santos B, Colerato Ferrari P, Chorilli M (2020) Characteristics, biological properties and analytical methods of trans-resveratrol: a review. Crit Rev Anal Chem 50:339–358. https://doi.org/10.1080/10408347.2019.1637242

García-Pérez P, Lozano-Milo E, Gallego PP, Tojo C, Losada-Barreiro S, Bravo-Díaz C (2018) Plant antioxidants in food emulsions. In: Milani J (ed) Some new aspects of colloidal Systems in Foods. IntechOpen, Rijeka (Croatia), pp 11–29. https://doi.org/10.5772/intechopen.79592

García-Pérez P, Barreal ME, Rojo-De Dios L, Cameselle-Teijeiro JF, Gallego PP (2019a) Bioactive natural products from the genus *Kalanchoe* as cancer chemopreventive agents: a review. In: Rahman A (ed) Studies in natural products chemistry, vol 61. Elsevier, Amsterdam (Netherlands), pp 49–84. https://doi.org/10.1016/B978-0-444-64183-0.00002-6

García-Pérez P, Losada-Barreiro S, Gallego PP, Bravo-Díaz C (2019b) Cyclodextrin-elicited *Bryophyllum* suspension cultured cells: enhancement of the production of bioactive compounds. Int J Mol Sci 20:5180. https://doi.org/10.3390/ijms20205180

García-Pérez P, Losada-Barreiro S, Gallego PP, Bravo-Díaz C (2019c) Adsorption of gallic acid, propyl gallate and polyphenols from *Bryophyllum* extracts on activated carbon. Sci Rep 9:1–9. https://doi.org/10.1038/s41598-019-51322-6

García-Pérez P, Lozano-Milo E, Landin M, Gallego PP (2020a) Machine learning unmasked nutritional imbalances on the medicinal plant *Bryophyllum* sp cultured *in vitro*. Front Plant Sci 11:576177. https://doi.org/10.3389/fpls.2020.576177

García-Pérez P, Lozano-Milo E, Landín M, Gallego PP (2020b) Combining medicinal plant *in vitro* culture with machine learning technologies for maximizing the production of phenolic compounds. Antioxidants 9:210. https://doi.org/10.3390/antiox10030435

García-Pérez P, Lozano-Milo E, Landin M, Gallego PP (2020c) From ethnomedicine to plant biotechnology and machine learning: the valorization of the medicinal plant *Bryophyllum* sp. Pharmaceuticals 13:444. https://doi.org/10.3390/ph13120444

García-Pérez P, Losada-Barreiro S, Bravo-Díaz C, Gallego PP (2020d) Exploring the use of *Bryophyllum* as natural source of bioactive compounds with antioxidant activity to prevent lipid oxidation of fish oil-in-water emulsions. Plan Theory 9:1012. https://doi.org/10.3390/plants9030288

García-Pérez P, Lozano-Milo E, Landin M, Gallego PP (2020e) Machine learning technology reveals the concealed interactions of phytohormones on medicinal plant *in vitro* organogenesis. Biomolecules 10:746. https://doi.org/10.3390/biom10050746

García-Pérez P, Ayuso M, Lozano-Milo E, Pereira C, Dias MI, Ivanov M, Calhelha RC, Soković M, Ferreira ICFR, Barros L, Gallego PP (2021a) Phenolic profiling and *in vitro* bioactivities of three medicinal *Bryophyllum* plants. Ind Crop Prod 162:113141–113149. https://doi.org/10.1016/j.indcrop.2021.113241

García-Pérez P, Miras-Moreno B, Lucini L, Gallego PP (2021b) The metabolomics reveals intraspecies variability of bioactive compounds in elicited suspension cell cultures of three *Bryophyllum* species. Ind Crop Prod 163:113322. https://doi.org/10.1016/j.indcrop.2021.113322

Giri CC, Zaheer M (2016) Chemical elicitors versus secondary metabolite production *in vitro* using plant cell, tissue and organ cultures: recent trends and a sky eye view appraisal. Plant Cell Tissue Organ Cult 126:1–18. https://doi.org/10.1007/s11240-016-0985-6

Gutiérrez-Grijalva EP, Santos-Zea L, Ambriz-Pérez DL, López-Martínez LX, Heredia JB (2020) Flavones and flavonols: bioactivities and responses under light stress in herbs. In: Lone R, Shuab R, Kamili AN (eds) Plant Phenolics in sustainable agriculture. Springer, Singapore, pp 91–115. https://doi.org/10.1007/978-981-15-4890-1_4

Hano C, Corbin C, Drouet S, Quéro A, Rombaut N, Savoire R, Molinié R, Thomasset B, Mesnard F, Lainé E (2017) The lignan (+)-secoisolariciresinol extracted from flax hulls is an effective protectant of linseed oil and its emulsion against oxidative damage. Eur J Lipid Sci Tech 119:1600219. https://doi.org/10.1002/ejlt.201600219

Hano CF, Dinkova-Kostova AT, Davin LB, Cort JR, Lewis NG (2021) Lignans: insights into their biosynthesis, metabolic engineering, analytical methods and health benefits. Front Plant Sci 11:2020–2022. https://doi.org/10.3389/fpls.2020.630327

Hostetler GL, Ralston RA, Schwartz SJ (2017) Flavones: food sources, bioavailability, metabolism, and bioactivity. Adv Nutr 8:423–435. https://doi.org/10.3945/an.116.012948

Irfan M, Siddiqui A, Erum N (2021) Phytomedicine and the COVID-19 pandemic. In: Bhat AB, Hakeem KR, Dervash MA (eds) Phytomedicine: a treasure of pharmacologically active products from plants. Elsevier, Cambridge, MA, pp 693–708. https://doi.org/10.1016/B978-0-12-824109-7.00005-4

Jacobsen C, Sørensen ADM, Nielsen NS (2013) Stabilization of omega-3 oils and enriched foods using antioxidants. In: Woodhead P (ed) Food enrichment with Omega-3 fatty acids. Woodhead Publishing Limited, Amsterdam. https://doi.org/10.1533/9780857098863.2.130

Karimi-Sales E, Mohaddes G, Alipour MR (2018) Chalcones as putative hepatoprotective agents: preclinical evidence and molecular mechanisms. Pharmacol Res 129:177–187. https://doi.org/10.1016/j.phrs.2017.11.022

Kashyap D, Sharma A, Tuli HS, Sak K, Garg VK, Buttar HS, Setzer WN, Sethi G (2018) Apigenin: a natural bioactive flavone-type molecule with promising therapeutic function. J Funct Foods 48:457–471. https://doi.org/10.1016/j.jff.2018.07.037

Khoo HE, Azlan A, Tang ST, Lim SM (2017) Anthocyanidins and anthocyanins: colored pigments as food, pharmaceutical ingredients, and the potential health benefits. Food Nutr Res 61:1361779. https://doi.org/10.1080/16546628.2017.1361779

Kiokias S, Proestos C, Oreopoulou V (2020) Phenolic acids of plant origin-a review on their antioxidant activity *in vitro* (O/W emulsion systems) along with their *in vivo* health biochemical properties. Foods 9(4):534. https://doi.org/10.3390/foods9040534

Křížová L, Dadáková K, Kašparovská J, Kašparovský T (2019) Isoflavones. Molecules 24:1076. https://doi.org/10.3390/molecules24061076

Kumar M, Tak Y, Potkule J, Choyal P, Tomar M, Meena NL, Kaur C (2020) Phenolics as plant protective companion against abiotic stress. In: Shuab R, Kamili AN (eds) Plant Phenolics in sustainable agriculture. Springer, Singapore, pp 277–308. https://doi.org/10.1007/978-981-15-4890-1_12

Lehmann ML, Counce RM, Counce RW, Watson JS, Labbé N, Tao J (2018) Recovery of phenolic compounds from switchgrass extract. ACS Sustain Chem Eng 6:374–379. https://doi.org/10.1021/acssuschemeng.7b02639

Leri M, Scuto M, Ontario ML, Calabrese V, Calabrese EJ, Bucciantini M, Stefani M (2020) Healthy effects of plant polyphenols: molecular mechanisms. Int J Mol Sci 21:1250. https://doi.org/10.3390/ijms21041250

Liu AG, Ford NA, Hu FB, Zelman KM, Mozaffarian D, Kris-Etherton PM (2017) A healthy approach to dietary fats: understanding the science and taking action to reduce consumer confusion. Nutr J 16:1–15. https://doi.org/10.1186/s12937-017-0271-4

Lončar M, Jakovljević M, Šubarić D, Pavlić M, Služek VB, Cindrić I, Molnar M (2020) Coumarins in food and methods of their determination. Foods 9:645. https://doi.org/10.3390/foods9050645

Mah SH (2020) Chalcones in Diets. In: Xiao J, Sarker S, Asakawa Y (eds) Handbook of dietary phytochemicals. Springer, Singapore. https://doi.org/10.1007/978-981-13-1745-3_10-1

Manchope MF, Casagrande R, Verri WA (2017) Naringenin: an analgesic and anti-inflammatory citrus flavanone. Oncotarget 8:3766–3767. https://doi.org/10.18632/oncotarget.14084

Marathe SJ, Shah NN, Bajaj SR, Singhal RS (2021) Esterification of anthocyanins isolated from floral waste: characterization of the esters and their application in various food systems. Food Biosci 40:100852. https://doi.org/10.1016/j.fbio.2020.100852

Marchiosi R, dos Santos WD, Constantin RP, de Lima RB, Soares AR, Finger-Teixeira A, Mota TR, de Oliveira DM, Foletto-Felipe MP, Abrahao J, Ferrarese-Filho O (2020) Biosynthesis and metabolic actions of simple phenolic acids in plants. Phytochem Rev 19:865–906. https://doi.org/10.1007/s11101-020-09689-2

Martín J, Navas MJ, Jiménez-Moreno AM, Asuero AG (2017) Anthocyanin pigments: importance, sample preparation and extraction. In: Soto-Hernández M (ed) Phenolic compounds – natural sources, importance and applications. InTech Open, London. https://doi.org/10.5772/66892

Martins N, Barros L, Ferreira ICFR (2016) In vivo antioxidant activity of phenolic compounds: facts and gaps. Trends Food Sci Tech 48:1–12. https://doi.org/10.1016/j.tifs.2015.11.008

Matos M, Gutiérrez G, Coca J, Pazos C (2014) Preparation of water-in-oil-in-water (W1/O/W2) double emulsions containing trans-resveratrol. Colloid Surface A 442:69–79. https://doi.org/10.1016/j.colsurfa.2013.05.065

McClements DJ, Bai L, Chung C (2017) Recent advances in the utilization of natural emulsifiers to form and stabilize emulsions. Annu Rev Food Sci Technol 8:205–236. https://doi.org/10.1146/annurev-food-030216-030154

Meshram RJ, Bagul KT, Pawnikar SP, Barage SH, Kolte BS, Gacche RN (2020) Known compounds and new lessons: structural and electronic basis of flavonoid-based bioactivities. J Biomol Struct Dyn 38:1168–1184. https://doi.org/10.1080/07391102.2019.1597770

Mitrus O, Zuraw M, Losada-Barreiro S, Bravo-Díaz C, Paiva-Martins F (2019) Targeting antioxidants to interfaces: control of the oxidative stability of lipid-based emulsions. J Agric Food Chem 67:3266–3274. https://doi.org/10.1021/acs.jafc.8b06545

Nadeem H, Ahmad F (2019) Prospects for the use of plant cell culture as alternatives to produce secondary metabolites. In: Akhtar MS, Swamy MK (eds) Natural bio-active compounds, Biotechnology, bioengineering, and molecular approaches, vol 3. Springer, Singapore, pp 153–182. https://doi.org/10.1007/978-981-13-7438-8_6

Najmanová I, Vopršalová M, Saso L, Mladěnka P (2020) The pharmacokinetics of flavanones. Crit Rev Food Sci Nutr 60:3155–3171. https://doi.org/10.1080/10408398.2019.1679085

Patra S, Pradhan B, Nayak R, Behera C, Rout L, Jena M, Efferth T, Bhutia SK (2020) Chemotherapeutic efficacy of curcumin and resveratrol against cancer: chemoprevention, chemoprotection, drug synergism and clinical pharmacokinetics. Semin Cancer Biol. https://doi.org/10.1016/j.semcancer.2020.10.010

Pizzi A (2019) Tannins: Prospectives and actual industrial applications. Biomol Ther 9:344. https://doi.org/10.3390/biom9080344

Quideau S, Deffieux D, Douat-Casassus C, Pouységu L (2011) Plant polyphenols: chemical properties, biological activities, and synthesis. Angew Chemie – Int Ed 50:586–621. https://doi.org/10.1002/anie.201000044

Rashmi HB, Negi PS (2020) Phenolic acids from vegetables: a review on processing stability and health benefits. Food Res Int 136:109298. https://doi.org/10.1016/j.foodres.2020.109298

Robe K, Izquierdo E, Vignols F, Rouached H, Dubos C (2021) The coumarins: secondary metabolites playing a primary role in plant nutrition and health. Trends Plant Sci 26:248–259. https://doi.org/10.1016/j.tplants.2020.10.008

Rodríguez-García C, Sánchez-Quesada C, Toledo E, Delgado-Rodríguez M, Gaforio JJ (2019) Naturally lignan-rich foods: a dietary tool for health promotion? Molecules 24(5):917. https://doi.org/10.3390/molecules24050917

Salehi B, Fokou PVT, Sharifi-Rad M, Zucca P, Pezzani R, Martins N, Sharifi-Rad J (2019) The therapeutic potential of naringenin: a review of clinical trials. Pharmaceuticals 12:1–18. https://doi.org/10.3390/ph12010011

Salomone F, Ivancovsky-Wajcman D, Fliss-Isakov N, Webb M, Grosso G, Godos J, Galvano F, Shibolet O, Kariv R, Zelber-Sagi S (2020) Higher phenolic acid intake independently associates with lower prevalence of insulin resistance and non-alcoholic fatty liver disease. JHEP Rep 2:100069. https://doi.org/10.1016/j.jhepr.2020.100069

Šeruga M, Tomac I (2017) Influence of chemical structure of some flavonols on their electrochemical behaviour. Int J Electrochem Sci 12:7616–7637. https://doi.org/10.20964/2017.08.79

Sharma A, Sharma P, Singh Tuli H, Sharma AK (2018) Phytochemical and pharmacological properties of Flavonols. In: eLS. John Wiley & Sons, Ltd., Chichester, pp 1–12. https://doi.org/10.1002/9780470015902.a0027666

Soto ML, Moure A, Domínguez H, Parajó JC (2008) Charcoal adsorption of phenolic compounds present in distilled grape pomace. J Food Eng 84:156–163. https://doi.org/10.1016/j.jfoodeng.2007.04.030

Sumorek-Wiadro J, Zając A, Maciejczyk A, Jakubowicz-Gil J (2020) Furanocoumarins in anticancer therapy – for and against. Fitoterapia 142:104492. https://doi.org/10.1016/j.fitote.2020.104492

Tak Y, Kumar M (2020) Phenolics: a key defence secondary metabolite to counter biotic stress. In: Shuab R, Kamili AN (eds) Plant Phenolics in sustainable agriculture. Springer, Berlin, pp 309–329. https://doi.org/10.1007/978-981-15-4890-1_13

Tian B, Liu J (2020) Resveratrol: a review of plant sources, synthesis, stability, modification and food application. J Sci Food Agric 100:1392–1404. https://doi.org/10.1002/jsfa.10152

Tutunchi H, Naeini F, Ostadrahimi A, Hosseinzadeh-Attar MJ (2020) Naringenin, a flavanone with antiviral and anti-inflammatory effects: a promising treatment strategy against COVID-19. Phytother Res 34:3137–3147. https://doi.org/10.1002/ptr.6781

Wang Y, Yu H, Wang S, Gai C, Cui X, Xu Z, Li W, Zhang W (2021) Targeted delivery of quercetin by nanoparticles based on chitosan sensitizing paclitaxel-resistant lung cancer cells to paclitaxel. Mater Sci Eng C 119:111442. https://doi.org/10.1016/j.msec.2020.111442

Williamson G, Kay CD, Crozier A (2018) The bioavailability, transport, and bioactivity of dietary flavonoids: a review from a historical perspective. Compr Rev Food Sci Food Saf 17:1054–1112. https://doi.org/10.1111/1541-4337.12351

Yang D, Wang X-Y, Lee JH (2015) Effects of flavonoids on physical and oxidative stability of soybean oil O/W emulsions. Food Sci Biotechnol 24:851–858. https://doi.org/10.1007/s10068-015-0110-3

Zeng W, Jin L, Zhang F, Zhang C, Liang W (2018) Naringenin as a potential immunomodulator in therapeutics. Pharmacol Res 135:122–126. https://doi.org/10.1016/j.phrs.2018.08.002

Zhu JJ, Jiang JG (2018) Pharmacological and nutritional effects of natural coumarins and their structure–activity relationships. Mol Nutr Food Res 62(14):1701073. https://doi.org/10.1002/mnfr.201701073

Olive Oil Phenolic Compounds as Antioxidants in Functional Foods: Description, Sources and Stability

Marlene Costa and Fátima Paiva-Martins

1 Introduction

The mediterranean diet (MD) has been associated with many health benefits and with the lower incidence, in humans that adhere to this diet, of several chronic degenerative diseases, such as, major cardiovascular events (CVD), cancer, type 2 diabetes mellitus and decreased cognitive function, providing protection against overall morbidity and mortality (Mentella et al. 2019; Schwingshackl and Hoffmann 2015). The first indirect evidence indicating that consuming olive oil, within the Mediterranean diet, might increase longevity was originated in the Seven Countries Study (Keys et al. 1986). This was followed by the MONICA study (Tunstall-Pedoe et al. 1999), which showed lower CVD incidence and mortality rates in Mediterranean countries compared with other European regions or the USA. A key component of the Mediterranean diet is virgin olive oil (VOO), and in particularly extra-virgin olive oil, which has been recommended in the prevention and protection against cardiovascular diseases (Estruch et al. 2018; Covas et al. 2015), different types of cancer (Fabiani 2016; Schwingshackl and Hoffmann 2015), type 2 diabetes melittus, obesity and metabolic syndrome (Buckland and Gonzalez 2015; López-Miranda et al. 2010).

VOO presents a unique chemical composition comprising a high monounsaturated fatty acid content and valuable minor components like phytosterols, vitamin E and polyphenols (Kiritsakis and Shahidi 2017). The presence of a relatively high amount of polyphenols gives to this oil an advantage when compared to other common edible oils, including the more recently developed high oleic content oils, because of their possible health promoting properties (Rodríguez-Morató et al. 2015;

M. Costa · F. Paiva-Martins (✉)
REQUIMTE-LAQV, Chemistry and Biochemistry Department, Faculty of Sciences, University of Porto, Porto, Portugal
e-mail: mpmartin@fc.up.pt

C. Bravo-Diaz (ed.), *Lipid Oxidation in Food and Biological Systems*,
https://doi.org/10.1007/978-3-030-87222-9_19

López-Miranda et al. 2015). Emerging evidence have shown that these health promoting properties are not only due to their free radical scavenging capacity, but also to their ability to down-regulate inflammatory mediators by transcriptional or post-transcriptional mechanisms, to modulate the activation of kinases and cell cycle progression (Parkinson and Cicerale 2016; Cicerale et al. 2012) and to affect the expression of genes involved in pathogenesis of many diseases (Piroddi et al. 2017).

Due to the great bioprotective capacity shown by olive oil phenolic compounds, these have become desirable for the development of functional foods. Thus, the search for effective and accessible alternative sources of these compounds as well as the understanding of their activity and stability in food matrices has become a critical issue and the object of several studies.

2 Olive Phenols

2.1 Olive Oil Polar Phenols

Virgin olive oil (VOO) is different from all other edible oils due to its particular composition in phenolic compounds (Garcia et al. 2012; Kiritsakis and Shahidi 2017). Olives contains a high content of phenolic compounds, up to 3% of the fresh pulp weight. The polar phenol classes present in olives are phenolic alcohols, such as hydroxytyrosol (HT) and tyrosol (T), phenolic acids, flavonoids, lignans, hydroxyl-isochromans and secoiridoids. Secoiridoids is a characteristic class of compounds present in Oleacea plants and is the most important class of phenols found in olives, with the glycoside oleuropein being present at levels of up to 14% of the dry weight (Fig. 1) (Amiot et al. 1989)

During the VOO mechanical extraction process, secoiridoid glycosides such as oleuropein and ligstroside present in the fruit are hydrolyzed by β-glucosidases and methylesterases, producing, through complex chemical transformations, a number of more lipophilic derivatives such as oleuropein aglycone, oleacein and oleocanthal that are released into the oil (Fig. 1). In contrast, the more hydrophilic simple phenols and glycosides such as oleuropein, tend to diffuse into the aqueous phase and are lost in great extent in the wastewater phase (Salvador et al. 2003). In fact, only 2% of the phenolic compounds present initially in olives are transferred to olive oil and a considerable amount of phenolic compounds—around 98%—end in olive oil by-products (El-Abbassi et al. 2012). Therefore, the quantitative phenolic composition of olive oil is quite different from the one found in olive fruit (Fig. 2)

Extra virgin olive oil (EVOO) usually contains up to 1200 ppm of phenolic compounds depending on the cultivars and olive maturity stage used for its extraction, pedoclimatic conditions, and extraction conditions (Amiot et al. 1989; Garcia et al. 2012; Piroddi et al. 2017; Salvador et al. 2003) but usual values in commercial EVOO range between 100 and 400 mg/kg. Due to the presence of these compounds, the European Union Commission (Regulation 432/2012, EU, 2012) has approved a health claim for olive oil phenolics—"Olive oil polyphenols contribute to the

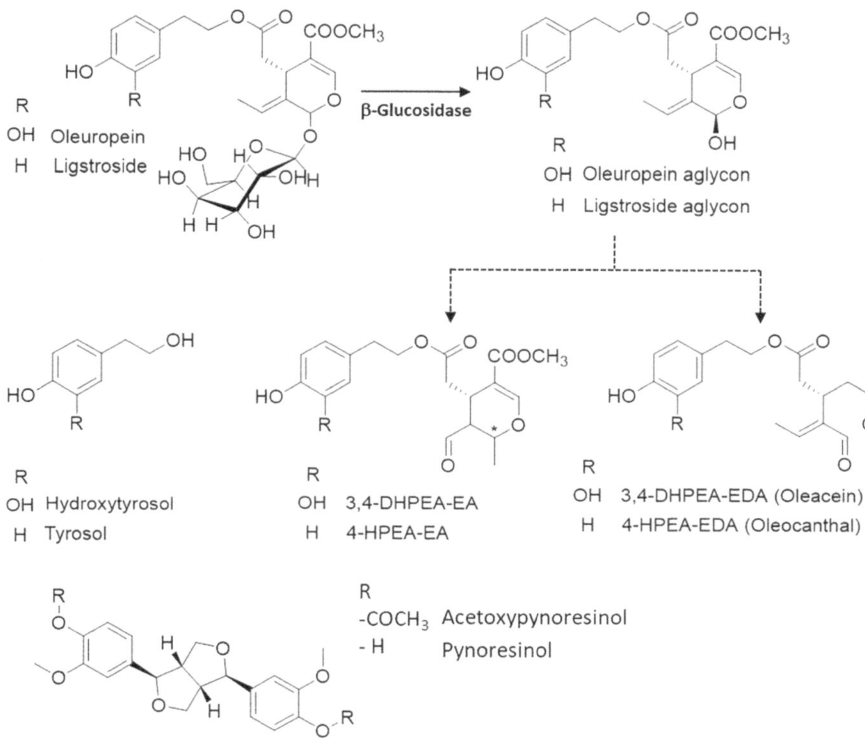

Fig. 1 Most important phenols found in olive oil

Fig. 2 Typical olive oil phenolic extract HPLC chromatogram ($\lambda = 280$ nm). Hydroxytyrosol (A); tyrosol (B); hydroxytyrosol acetate (C); oleacein (D); oleocantal (E); pynoresinol (F); acetoxipynoresinol (G); 3,4-DHPEA-EA (H); p-HPEA-EA (I)

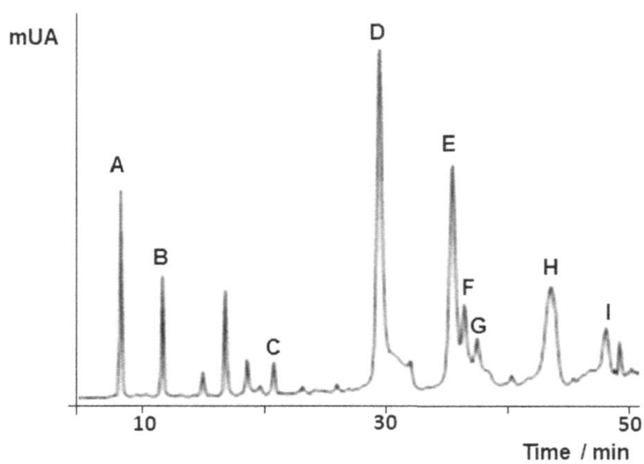

protection of blood lipids from oxidative stress"—that can be used for olive oil that contains at least 5 mg of hydroxytyrosol and its derivatives per 20 g of olive oil. Table 1 presents the main phenols present in virgin olive oil.

Table 1 Main phenols present in virgin olive oil (mg/Kg)

Compound	Range[a]
Hydroxytyrosol	0.0–25
Hydroxytyrosol acetate	21–131
Tyrosol	0.10–123
Vanillic acid	0.09–1
Caffeic acid	0.0–1
Siringic acid	0.0–2.3
p-Cumaric acid	0.04–0.6
Ferulic acid	0.0–2.4
3,4-DHPEA-EDA (Oleacein)	74–840
4-HPEA-EDA (Oleocanthal)	13–86
Tyrosol derivative	0.0–113
Lignans (pynoresinol and acetoxypynoresinol)	112–275
3,4-DHPEA-EA	26–310
Luteolin	0–10

[a] Values obtained from references (Garcia et al. 2012; Kiritsakis and Shahidi 2017; Salvador et al. 2003)

2.2 Tocopherols

Most vegetable edible oils contain important amounts of tocopherols. However, in contrast with other vegetable oils such as soybean or corn oil, α-tocopherol is the most common tocopherol found in olive oil, comprising 90% of the total tocopherol content. α-Tocopherol, commonly known as Vitamin E, has the highest biological activity among the tocopherols and tocotrienols (Woollard and Indyk 2003) and is correlated with preventive action against reactive oxygen species (ROS) in biological systems such as plasma, membranes, and tissues (Woollard and Indyk 2003). A wide range of α-tocopherol content has been reported in olive oils depending on the cultivar, maturity stage and several technological factors involved during transportation, storage and processing of the fruit. However, in VOO, the α-tocopherol content is usually in the range of 50–380 ppm. In olive oils obtained from olives in early maturation stages, this value can go up to 500 ppm (Garcia et al. 2012). It has also been reported low amounts of β-tocopherol (~10 mg/kg), δ-tocopherol (~10 mg/kg) and γ-tocopherol (~20 mg/kg). Although the daily amount of α-tocopherol supplied by the consumption of olive oil is rather low, it may contribute to the antioxidant status in humans.

2.3 Sources of Olive Polyphenols

Olive leaves (OL) are one of the most abundant wastes of the olive oil industry (Şahin and Bilgin 2018; Salta et al. 2007). Nevertheless, as during the extraction of

EVOO a mixture of olive paste and water is obtained, subsequent malaxation of the olive paste allows for the separation of the olive oil from two further by-products, the olive pomace (OP) and the olive wastewater (OMW). These by-products, in particular OP and OMW, are of environmental concern regarding its disposal and management in the Mediterranean countries because large quantities are produced in short periods of time (Roig et al. 2006; Banias et al. 2017; Şahin and Bilgin 2018). However, these by-products, in particular olive leaves, are cheaper and richer alternative sources to olives for the same phenolic compounds found in olive oil (Paiva-Martins and Pinto 2008; Paiva-Martins and Gordon 2001). They can be used as natural antioxidants and preservatives (Şahin and Bilgin 2018; Paiva-Martins and Gordon 2001; Difonzo et al. 2018), to obtain functional foods (Şahin and Bilgin 2018; Rahmanian et al. 2015; Flamminii et al. 2019), or in the development of anti-inflammatory and anti-carcinogenic drugs (Fabiani 2016; Rodríguez-Morató et al. 2015; Cicerale et al. 2012; Segade et al. 2016; Magrone et al. 2018). Therefore, considering the environmental toxicity of these by-products rich in phenols and, simultaneously, the outstanding protective effect of these compounds for human health and as natural antioxidants for food industry, these by products have attracted much attention with alternative and innovative valorization approaches currently being considered.

Olives leaves can be considered the first by-product obtained during the industrial process of olives, representing 10% of the raw material arriving to olive oil mills, (Difonzo et al. 2017). Sometimes olive leaves are also added to olives before milling in order to enrich the resulting oil in polyphenols similar to the ones found in olive oil (Kiritsakis et al. 2017). The chemical composition of olive leaves is widely dependent on several factors such as the cultivar, the soil, the climatic conditions, and degree of contamination with soil (Paiva-Martins and Pinto 2008; Sabry 2014). On the other hand, the phenolic profile of olive leaves is qualitatively and quantitatively influenced by biotic and abiotic factors (Paiva-Martins and Pinto 2008; Talhaoui et al. 2015; Bilgin and Şahin 2013). Whatever the origin of the leaves, the composition of olive leaf phenolic extracts is usually quite different from the one found in olive oil as their major constituent is oleuropein (24%), followed by hydroxytyrosol (1.5%), luteolin-7-glucoside (1.4%), apigenin-7-glucoside (1.37%), verbascoside (1.11%) and tyrosol (0.71%). However, the quantity and the quality of phenolic compounds also depends on both their processing and extraction solvents and techniques used (Paiva-Martins and Pinto 2008; Difonzo et al. 2017; Khemakhem et al. 2017). Several years ago, it was shown that the composition of olive leaves extracts can actually be modulated if leaves are thermally processed prior extraction, with the enrichment of extracts with the most important secoiridoids derivatives found in olive oil, namely oleacein and oleuropein aglycone (Paiva-Martins and Gordon 2001). The storage of leaves at 38 °C reduced the total amount of phenols but changed the phenolic composition of the leaf extract so it was closer to the composition of VOO extract. Ethanol, a less toxic solvent, also improved the yield of phenolic compounds in the extraction. Maceration was found to be the best extraction method, but the yields were improved when trituration was also performed.

The amount of olive mill wastewater produced in a mill, depends on the specific system used for virgin olive oil extraction (Roig et al. 2006). As OMW is a by-product rich in water-soluble phenols, that can be separated and concentrated by industrial membrane technology (Caporaso et al. 2018; Kim et al. 2015) in a low cost-effective purification performed at a low temperature (Cassano et al. 2013). Several studies have proved its content in biologically important phenols and as a source of natural antioxidants. Usually, OMW has no oleuropein but presents high amounts of the more hydrophilic phenols such as tyrosol, protocatechuic acid (3,4-dihydroxybenzoic acid) and hydroxytyrosol. Other phenolic compounds such as verbascoside and oxidation products, as well as higher molecular weight phenols deriving from oxidative polymerization of hydroxytyrosol and elenolic acid, are also present. In fact, the major drawback of this raw material is its low chemical stability caused by the oxidation of phenols. Nevertheless, OMW also contains significant amounts of monosaccharides and polysaccharides, with both prebiotic and complexing properties (Nadour et al. 2015). The iron(II) chelating properties were found to be higher than that observed for EDTA, preventing the production of hydroxyl radicals and the oxidation of catecholic compounds contributing to the antioxidant activity of OMW extracts (El-Abbassi et al. 2012).

Another by-product from olive oil extraction is the semi-solid waste, olive pomace, and represents the highest waste-management problem of the olive oil industry. Depending on the process of extraction used it represents 35–40% of the raw material arriving to olive oil mills (Dermeche et al. 2013). Olive pomace contains high amounts of α-tocopherol (2.63 mg/100 g), and phenols (~200 mg/ 100 g) with HT representing almost 55% and comsegoloside almost 25% (Nunes et al. 2018). Other compounds like verbascoside, caffeoyl-6-secologanoside, tyrosol and pinoresinol also contribute for the phenols content but are present in much lower concentrations.

2.4 Stability of Olive Oil Phenolic Compounds under Storage Conditions

The tocopherol content usually decreases significantly after one year of olive oil storage, about 25–30% of its initial content (Fregapane et al. 2013). In contrast, the decrease in total phenols and o-diphenols throughout the shelf-life of the oil at room temperature is usually small (Fregapane et al. 2013). Nevertheless, a significant increase in simple phenolics (~30%, hydroxytyrosol and tyrosol) is generally observed during one-year storage at this temperature, due to the hydrolytic degradation of oleuropein derivatives (Krichene et al. 2010; Lavelli et al. 2006; Fregapane et al. 2006).

During storage at 5–50 °C, the degradation of secoiridoid phenolics has shown to follow a pseudo-first-order kinetics and to depend on the initial phenolic compounds content (Krichene et al. 2015). In fact, an approximately linear relationship

($r > 0.98$; and at 50 °C, $r = 0.91$; $p < 0.01$) between the initial phenolic concentration and the initial degradation rate was observed when the phenolic concentration is lower than 100 μmol/kg. Moreover, during storage the concentration of the main secoiridoid phenolic compounds, namely oleacein, oleuropein aglycon and oleocanthal, fits well an exponential decay curve ($R^2 > 0.96$). In contrast to complex phenolics, α-tocopherol degradation apparently follows a simpler zero-order kinetic ($r > 0.97$ and $p < 0.01$).

The initial degradation rate has shown to be similar at 5 and 15 C corresponding to less than 30% after a year of storage (Krichene et al. 2015). However, this value increases considerably at 25 C, needing only 3 months for the same decrease, thus, reducing substantially the period of time (up to 40%) an olive oil can be considered with the minimum quantity of phenolics needed to use the health claim. At 50 °C, the degradation of phenolic compounds is even greater and after the same 3 months of storage it was observed less than 30% of the initial phenolic content in the olive oil. In general, tyrosol derivatives show a better stability than hydroxytyrosol derivatives, but this difference decreases as the temperature of storage decreases, being almost inexistent at low temperatures (15 °C) (Lavelli et al. 2006; Daskalaki et al. 2009; Krichene et al. 2015).

3 The Oxidative Stability of Olive Oil

Olive oil has a remarkable stability when compared with other edible oils, not only due to its high monounsaturated-to-polyunsaturated fatty acid ratio, but also due to its phenolic content. Nevertheless, even at room temperature, lipid oxidation occurs slowly, being the main cause of olive oil deterioration. Depending on several factors, such as, fatty acid composition of the oil, phenolic composition and concentration, and storage conditions, stability of VOO may ranges from 9 to more than 18 months (Fregapane et al. 2013; Cinquanta et al. 2001; Gómez-Alonso et al. 2007; Gutiérrez and Fernández 2002; Silva et al. 2015). The European regulation 2568/91 and subsequent amendments (EU, 2013) classify an olive oil in the extra virgin category, when the oxidation indexes peroxide value (PV), K270, and K232 are lower than 20 meq/kg, 0.22 and 2.5, respectively. The K232 index, is an indication of the percentage of hydroperoxides containing a conjugated diene moiety in the oil (primary oxidation products) whereas the K270 index is an indication of the secondary oxidation products formation. At room temperature, these indexes have shown to increase linearly ($R^2 > 0.95$) during ~~a~~ storage at 25 °C in open bottles of almost two years (Gómez-Alonso et al. 2007; Gómez-Alonso et al. 2004). Based on these results, it should therefore be possible to extrapolate and predict the shelf life of OO monitorizing PV or K232 over a relatively short period of storage. In contrast, the kinetic behaviour of oxidation indexes (PV, K232 and K270) and of unsaturated fatty acids (UFA) during storage of different virgin olive oil samples in darkness and at temperatures higher than 25 °C (40–60 °C) showed a stationary phase after a few weeks of storage. However, during the early stages of oxidation and before reaching the plateau, PV and K232 followed a pseudo zero-order kinetics ($R^2 = 0.951$), whereas the behaviour of K270 were better fitted as a pseudo first-order kinetics ($R^2 = 0.926$) (Mancebo-Campos et al. 2014). All the temperature-dependent kinetics of the oxidation indexes and the UFA were well described by the linear Arrhenius equation in this range of temperatures (25–60 °C; $0.960 < R^2 < 0.999$, $p < 0.05$). The best correlation between the degradation of PUFA and the increase of oxidation indexes was obtained with K232 ($0.581 < R^2 < 0.924$). Moreover, K232 was often the first index to exceed the upper limit established by the European regulations for VOO (Gómez-Alonso et al. 2004; Mancebo-Campos et al. 2014; Mancebo-Campos et al. 2008). Since the time required to reach the upper limits for the oxidative indexes correlated well with temperature, there is the possibility to set up an accelerated stability test at temperatures below 60 °C to estimate the potential shelf life of an olive oil under normal storage temperature.

The percentage of unsaturated fatty acids is a decisive factor accounting for the different oxidation rates observed in olive oils. While the oleic (C18:1) acid content remained practically constant during storage at 25–60 °C, linoleic (C18:2) and linolenic (C18:3) acids decreased linearly (Mancebo-Campos et al. 2008). Moreover, the ratio of unoxidized fatty acids C18:1/C18:2, C18:1/C18:3 and C18:2/C18:3 increased slightly over the storage period showing easier oxidation of the more unsaturated fatty acids.

The apparent energy of activation (Ea) calculated from primary oxidation indexes and UFA results was comparable for all samples: ~64.8 kJ/mol for PV, ~69.6 kJ/mol for K232 and ~ 67.6 kJ/mol for PUFA supporting the hypothesis that oxidation of UFA leads to the production of peroxides and conjugated dienes (Mancebo-Campos et al. 2008). On the other hand, a higher activation energy (mean ~ 77.1 kJ/mol) was found for the K270 index.

3.1 Contribution of Olive Oil Phenolic Compounds to the Oxidative Stability of Bulk Olive Oil

The prediction of the susceptibility of olive oil to oxidation using analytical parameters needs to take into account several factors that include, among others, the monounsaturated-to-polyunsaturated fatty acid ratio and their tocopherol and phenolic compounds content (Blekas et al. 2002; Lee et al. 2007; Mateos et al. 2005; Servili and Montedoro 2002). Moreover, the composition, concentration and interaction between phenolic compounds increases the complexity of the oxidation process, making prediction and mathematical modeling difficult. Nevertheless, most differences in the stability of extra virgin olive oils can be explained in terms of their total phenol content and fatty acid composition. Univariate analysis of the data revealed a significant correlation between stability and phenol content ($R = 0.812$) but only a weak correlation with the oleic/linoleic acid content ratio ($R = 0.572$) and initial peroxide value of the samples ($R = -0.348$) (Paiva-

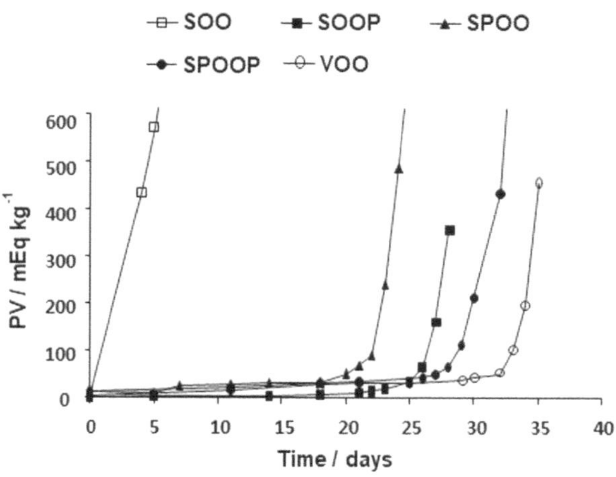

Fig. 3 Effects of the tocopherol and phenolic extracts on the lipid oxidation reaction kinetic of stripped bulk olive oil. *SOO* Stripped olive oil, *SPOO* stripped olive oil containing tocopherols, *SOOP* stripped olive oil containing phenolic compounds, *SPOOP* stripped olive oil containing tocopherols and phenols, *VOO* original virgin olive oil

Martins 2002). However, considering the three factors, and applying a linear regression analysis, a better correlation could be obtained ($R^2 = 0.933$) (Paiva-Martins 2002). Checking the P of F values ($P < 0.0005$), and the plot of residuals against predicted values it was shown that the model was a satisfactory summary of the observations. From the statistical analysis, the total phenolic concentration was the most important factor affecting the stability of olive oil (Sig. = 0.000) followed by the initial PV (Sig. = 0.014) and fatty acid r (Sig. = 0.054). These results are in accordance with several works (Brenes et al. 2001; Lee et al. 2007; Mateos et al. 2005; Romani et al. 2007; Baldioli et al. 1996; Servili and Montedoro 2002; Ninfali et al. 2001; Aparicio et al. 1999) where the concentration of total phenolic compounds has shown to be highly correlated to the VOO shelf-life. Aparicio et al. (Aparicio et al. 1999) also found a good level of correlation between oleic/linoleic acid ratio with stability.

Using olive oil stripped of phenolic compounds and tocopherols, to which was added back the phenolic compounds, it was possible to assess, by the Schaal Oven test performed at 60 °C, the contribution of tocopherols and polyphenols to the stability of olive oil. (Paiva-Martins 2002) Stripped olive oil (SOO) was readily oxidized but stripped olive oil containing only tocopherols (SPOO) still showed high stability compared with the control (SOO) (Fig. 3). Nevertheless, stripped olive oil to which was added back the phenolic compounds (SPOO) (155 ppm, caffeic acid equivalents) showed better stability, proving the more important contribution of phenolic compounds to the olive oil stability. Interesting to note, however, that oil containing only tocopherols to which was added back the phenolic compounds, showed a lower stability than the original virgin olive oil. Theoretically, these samples should have had the same stability, however some phenolic extract may have deteriorated during the extraction procedure.

Most of phenols present in olive oil may act as antioxidants, but the compounds which are mainly responsible for the remarkable stability of olive oil to oxidation, besides tocopherols, are the esters of hydroxytyrosol: oleuropein aglycone, oleacein,

Table 2 DPPH Radical-scavenging capacity at 5 min (EC_{50} values), ABTS radical scavenging capacity (Trolox Equivalent Antioxidant Capacity, TEAC values) and ferric reducing capacity (FRAP values) of olive oil phenolic compounds

Compound	DPPH essay [a]$EC_{50\ \text{at 5 min}}$	ABTS essay TEAC value	FRAP essay (pH 5.5, μM)
Caffeic acid	0.33	Nd	Nd
Hydroxytyrosol	0.34	0.94	1795
Hydroxytyrosol acetate	0.33	0.93	1014
Oleuropein	0.24	0.98	1505
Oleacein	0.29	0.92	1014
3,4-DHPEA-EA	0.13	0.98	1234
Tyrosol	> > 5	0.48	54
α-Tocopherol	0.27	0.99	1776
Trolox	0.26	1.00	Nd

Values obtained from references (Paiva-Martins and Gordon 2001; Paiva-Martins and Gordon 2002; Gordon et al. 2001; Almeida et al. 2016; Costa et al. 2015; Paiva-Martins et al. 2003)
[a] The antiradical activity was defined as the relative concentration of antioxidant required to lower the initial DPPH% concentration by 50% (EC_{50}, M phenolic compound per unit DPPH concentration) obtained after 5 min of reaction. *nd* not determined

decarboxymethyl oleuropein aglycon, and hydroxytyrosol itself (Gómez-Alonso et al. 2003; Baldioli et al. 1996; Gordon et al. 2001; Carrasco-Pancorbo et al. 2006; Romani et al. 2007). This antioxidant activity is mainly attributed to the much better radical scavenging activity observed for these catecholic compounds, when compared to that of monophenols (Table 2) (Paiva-Martins and Gordon 2001; Paiva-Martins and Gordon 2002; Gordon et al. 2001; Almeida et al. 2016; Costa et al. 2015; Paiva-Martins et al. 2003).

The evaluation of the antioxidant capacity of the main olive oil phenolic compounds in striped olive oil (Fig. 4) has shown that HT, 3,4-DHPEA-EA and oleacein may present an antioxidant capacity up to 10 times higher than that of α-tocopherol (Paiva-Martins and Gordon 2002; Paiva-Martins et al. 2003, 2006). Moreover, even in the presence of iron (III), olive oil phenols still showed a remarkable antioxidant activity in bulk olive oil, in particular the oleuropein derivatives oleacein and 3,4-DHPEA-EA. This higher activity showed by secoiridoids in the presence of iron may be related to the lower ferric reducing capacity showed by these compounds (Table 2) (Paiva-Martins and Gordon 2002) when compared with hydroxytyrosol.

The antioxidant activity of phenols is enhanced by the presence of tocopherols (Blekas et al. 1995; Pellegrini et al. 2001). This synergism is more evident when the level of phenols is relatively low. Polar phenols reduce the oxidized forms of alpha-tocopherol and, therefore, tocopherols are more quickly consumed in olive oils with low content in *ortho*-diphenols (Silva et al. 2010a, b). The stability testes performed at temperatures lower than 60 °C are more relevant for the prediction of the antioxidant activity of compounds at room temperature. However, the antioxidant activity of olive oil phenols evaluated at 120 °C in bulk olive oil by the Rancimat

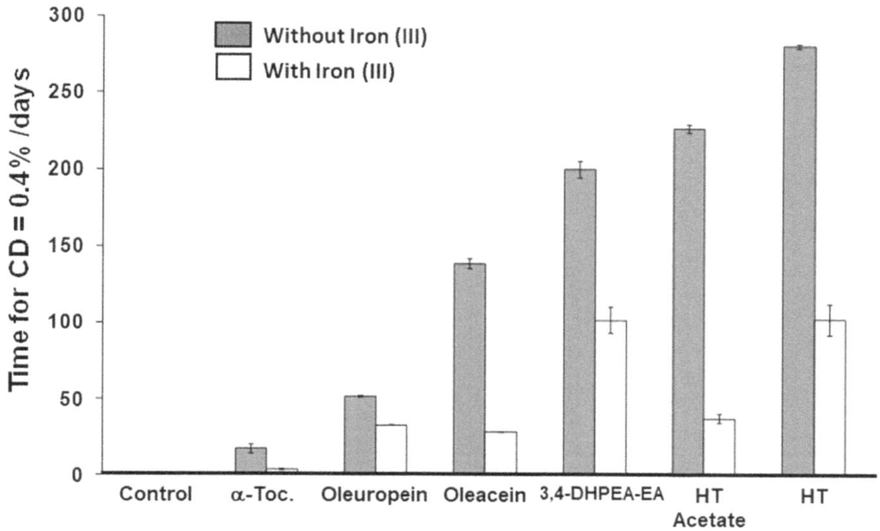

Fig. 4 Time (days) for bulk stripped olive oil stored at 60 °C to reach the CD content of 0.4% in the presence and absence of iron (III) at 1 ppm and selected olive oil polyphenols at 0.6 mmol/kg. Adapted from (Paiva-Martins 2002)

method not only showed an important antioxidant activity of hydroxytyrosol, 3,4-DHPEA-EA and 3,4-DHPEA-EDA (Servili et al. 1996) but also a synergic effect when these phenolic compounds are in association with α-tocopherol (Fig. 5).

In the presence of Cu(II), the antioxidant activity of phenolic compounds in bulk oil was highly affected, with the stability of stripped olive oil samples containing phenolic compounds decreasing by at least 90%, being now the antioxidant activity of hydroxytyrosol, oleacein and 3,4-DHPEA-EA quite similar (Paiva-Martins et al. 2006). Nevertheless, some antioxidant activity was still observed for these compounds in contrast with that of α tocopherol, which did not show any antioxidant activity (Fig. 6).

3.2 Contribution of Olive Oil Phenolic Compounds to the Oxidative Stability of Olive Oil in Multiphase Systems

In 3:7 stripped olive oil-in-water emulsions, olive oil phenolic compounds have a clear effect on the oxidative stability of emulsions both at pH 5.5 and 7.4. In emulsions at pH 5.5, this effect has shown to increase with the increase in concentration of each phenolic compound (Fig. 7).

Moreover, the antioxidant activity of most olive oil phenols was shown to be higher than that of α-tocopherol at higher concentrations, despite the higher

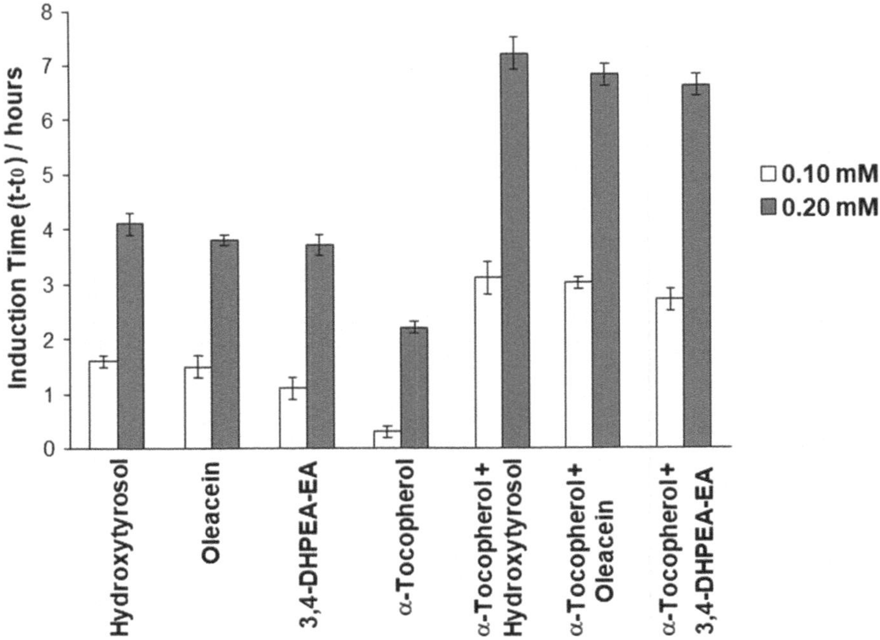

Fig. 5 Effect of individual polyphenols at 0.10 and 0.20 mM either alone or in combination with α-tocopherol on the autoxidation of refined olive oil at 120 °C by the Rancimat method. Adapted from (Baldioli et al. 1996)

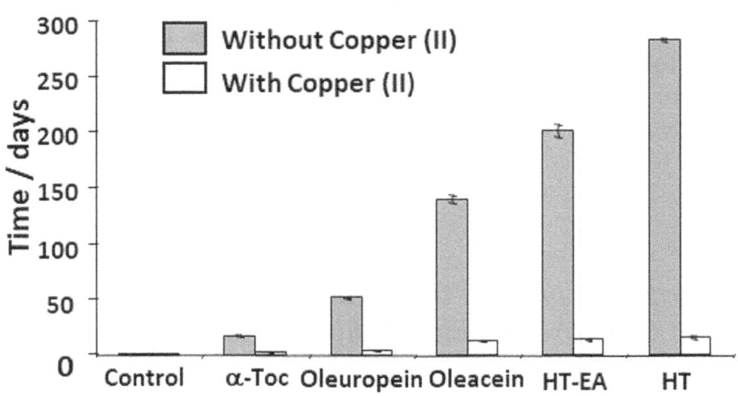

Fig. 6 Time for oil samples containing phenolic compounds at 0.6 mmol/kg, in the absence and presence of Cu(II) at 0.8 mg kg^{-1}, to reach a conjugated diene content of 0.4% at 60 °C. α-Toc., α-tocopherol; HT, hydroxytyrosol; HT-EA, 3,4-dihydroxyphenylethanol- elenolic acid (3,4-DHPEA-EA). Adapted from (Paiva-Martins et al. 2006)

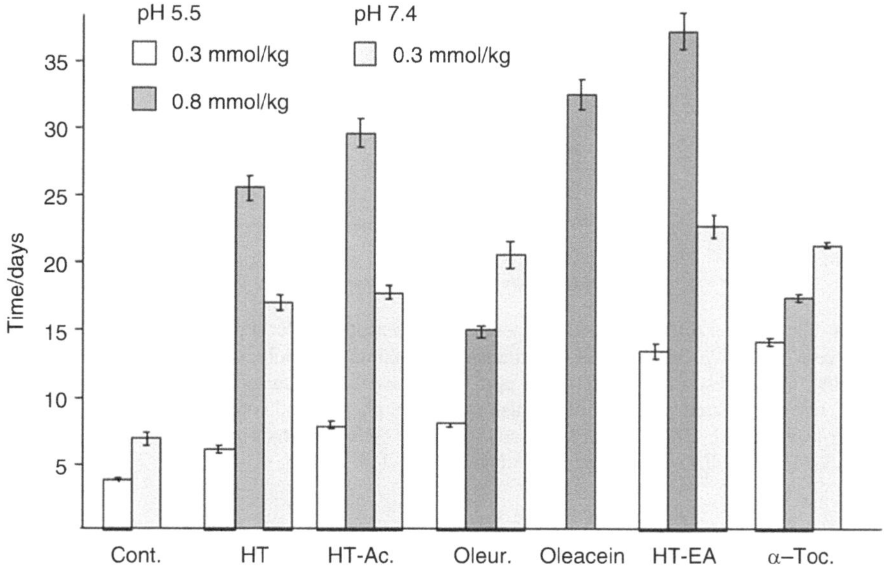

Fig. 7 Time for 3:7 (O/W) striped olive oil/Tween 20/acetate buffer at pH 5.5 or for 3:7 (O/W) striped olive oil/Tween 20/MOPS buffer at pH 7.4 emulsions containing phenolic compounds at 0.3 or 0.8 mmol/kg to reach a conjugated diene content of 0.4% when stored at 60 °C. α-Toc., α-tocopherol; HT, hydroxytyrosol; HT-Ac., hydroxytyrosol acetate; HT-EA, 3,4-dihydroxyphenylethanol- elenolic acid (3,4-DHPEA-EA). Adapted from (Paiva-Martins 2002)

hydrophilicity of the former compounds. This result is not in accordance with the polar paradox and probably results from some pro-oxidant effects already described for α-tocopherol at higher concentrations (Cillard et al. 1980; Bakır et al. 2013). Nevertheless, depending on the composition of the interfacial region, the affinity of phenols for the different regions of the emulsified system is expected to change and so is their antioxidant activity (Almeida et al. 2016; Costa et al. 2020)

In the same emulsified system, but in the presence of iron (III), oleuropein and hydroxytyrosol enhanced the prooxidant effect of ferric chloride at pH 5.5 but not at pH 7.4. In contrast, oleacein and 3,4-DHPEA-EA reduced the prooxidant effect of ferric chloride at both pH 5.5 and pH 7.4 (Paiva-Martins and Gordon 2002). These differences in the antioxidant activity of olive oil phenols may be explained by considering their free radical scavenging activity and ferric reducing capacity (Table 2). The ferric reducing ability at pH 5.5 decreased in the following order: hydroxytyrosol, α-tocopherol > oleuropein >3,4-DHPEA-EA > oleacein. On the other hand, as the iron solubility is very low at alkaline pH, owing to the formation of ferric hydroxide, compounds did not show any ferric-reducing activity at pH 7.4 (Paiva-Martins and Gordon 2002). These results also help to explain the higher antioxidant activity shown by 3,4-DHPEA-EA in the presence of iron, since this compound has the highest radical scavenging activity and a relatively low iron-reducing capacity. In turn, oleacein even having a lower radical scavenging activity,

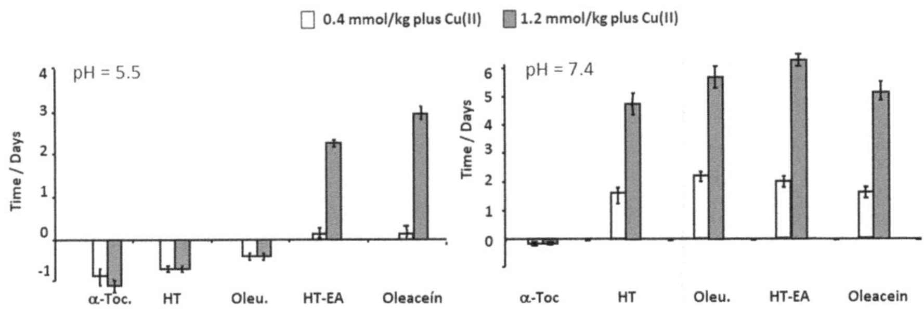

Fig. 8 Time for 3:7 (O/W) stripped olive oil/Tween 20/acetate buffer at pH 5.5 and 3:7 (O/W) olive oil/Tween 20/MOPS buffer at pH 7.4 emulsions containing phenolic compounds at 0.4 and 1.2 mmol/kg in the presence of Cu(II) at 0.8 ppm to increase the conjugated diene content of 0.4 % at 60 °C when compared with control (emulsions without phenols and copper). *α-Toc.* α-tocopherol, *HT* hydroxytyrosol, *HT-Ac.* hydroxytyrosol acetate, *HT-EA* 3,4-dihydroxyphenylethanol- elenolic acid (3,4-DHPEA-EA). Adapted from (Paiva-Martins et al. 2006)

may retain some antioxidant activity in iron-catalyzed oxidation owing to its low iron reducing capacity.

In the presence of copper (II), the lowering in the stability of all emulsified samples was shown to be even more severe than with iron, both at pH 5.5 and 7.4. (Paiva-Martins et al. 2006) (Fig. 8). Moreover, prooxidant effect could be observed in the presence of α-tocopherol both at pH 7.4 and pH 5.5 and in the presence of hydroxytyrosol and oleuropein at pH 5.5 (Fig. 8).

Interestingly, the lower stability of compounds in aqueous solution in the presence of copper and the higher copper reducing capacity shown by all compounds at pH 7.4 could not explain the higher stability of emulsions containing phenolic compounds at this pH value (Paiva-Martins et al. 2006). However, mixtures containing hydroxytyrosol or oleuropein with copper showed higher 1,1-diphenyl-2-picrylhydrazyl radical scavenging activity at pH 7.4 than at pH 5.5. Moreover, oleacein-copper complex showed higher radical scavenging activity then the uncomplexed compound at pH 5.5. It can be concluded that the formation of a copper complex with radical scavenging activity is a key step in the antioxidant action of some olive oil phenolic compounds in an emulsion containing copper ions (Paiva-Martins et al. 2006). The antioxidant activity shown by oleuropein derivatives in the presence of metals is of particular importance in the oxidative stability of food systems containing olive oil since these are the main compounds present in olive oil and traces of iron and copper are naturally present in tap water, equipment and as food constituents.

After roast processing (thermal treatment at 180 °C in an oven for 60 min) using extra virgin olive oil, virgin olive oil and olive oil, the resulting oil phases showed a low level of oxidation, independently of their initial phenolic compound's concentration, even in the presence of meat and potatoes (Silva et al. 2010a). Actually, the presence of food had a protective effect on the olive oils stability, with oil samples processed without food showing a higher level of oxidation than the oil samples processed in the presence of food, probably due to the formation of Maillard reaction products with antioxidant properties. Nevertheless, the concentration of olive phenolic compounds in the oils decreased and this decrease showed to be dramatic in the presence of foods such as meat and potatoes (Silva et al. 2010a, b). The elemental content of foods may have a huge contribution to the loss of phenolic compounds. In fact, many foods are very rich in important elements, such as iron. For example, iron can be found in potatoes in a concentration of higher than 25 ppm (dry matter) and copper can also be present in a concentration of up to 10 ppm (dry matter) (Bethke and Jansky 2008). As seen above, small concentrations, such as 0.4 mmol/kg, of these ions have dramatic effects on the antioxidant activity of phenolic compounds in emulsions and bulk oils because they not only produce more pro-oxidant reduced metallic ions, but they also increase the rate of phenolic compound destruction by oxidation (Paiva-Martins and Gordon 2002, 2005; Paiva-Martins et al. 2006) (Fig. 9a and b). In most studies, hydroxytyrosol has shown the worst stability in the presence of foods probably because hydroxytyrosol forms the least stable complex with metals, has the highest reducing capacity and increases the rate of ferrous ion oxidation to ferric ions by air oxygen (Paiva-Martins and Gordon 2005). This leads to a fast ferrous ion formation in the media and to a higher rate of phenolic compound destruction (Fig. 9b). Besides the higher amount of polyphenolic compounds in EVOO, differences at the polyphenolic content of EVOO and OO samples after processing in the presence of vegetables were not evident.

Changes in the stability and phenolic content of an extra virgin olive oil and an olive oil following boiling operation in the presence of large excess of water and vegetables (potatoes, anions and carrots) were also studied (Silva et al. 2010b). After boil processing for 60 min, none of the isolated olive oil samples were oxidized, independently of the initial concentration of phenolic compounds in the olive oil used. However, in contrast with tocopherols, all other phenolic components decreased in concentration with the thermal treatment and this decrease was much higher in the presence of vegetables. Moreover, the destruction of vegetable antioxidants was even higher in the case of samples prepared with EVOO than those prepared with olive oil. The high reducing capacity of olive oil phenolic compounds not only leads to their own destruction but also to the destruction of other phenols present in foods (Silva et al. 2010b). These deleterious effects on the phenolic concentration, however, could be reduced by processing the olive oil during only 15 min before the end of the boiling process. Nevertheless, a significant loss of phenolic compounds was still observed demonstrating that the use of EVOO instead of OO did not bring benefits to the stability of samples or a higher polyphenolic content in the samples after processing (Silva et al. 2010b).

These results highlight the fact that the antioxidant activity of phenolic compounds in bulk oil or in model food emulsions cannot be extrapolated to real foods as this activity will depend on the presence of other common constituents such as water, metals, or newly formed interfacial regions. Moreover, antioxidants should be used with care in edible oils as their antioxidant activity may change during culinary processing in the presence of other food constituents and tap water. Therefore, the rational use of the several types of olive oil (extra virgin olive oil, olive oil or refined

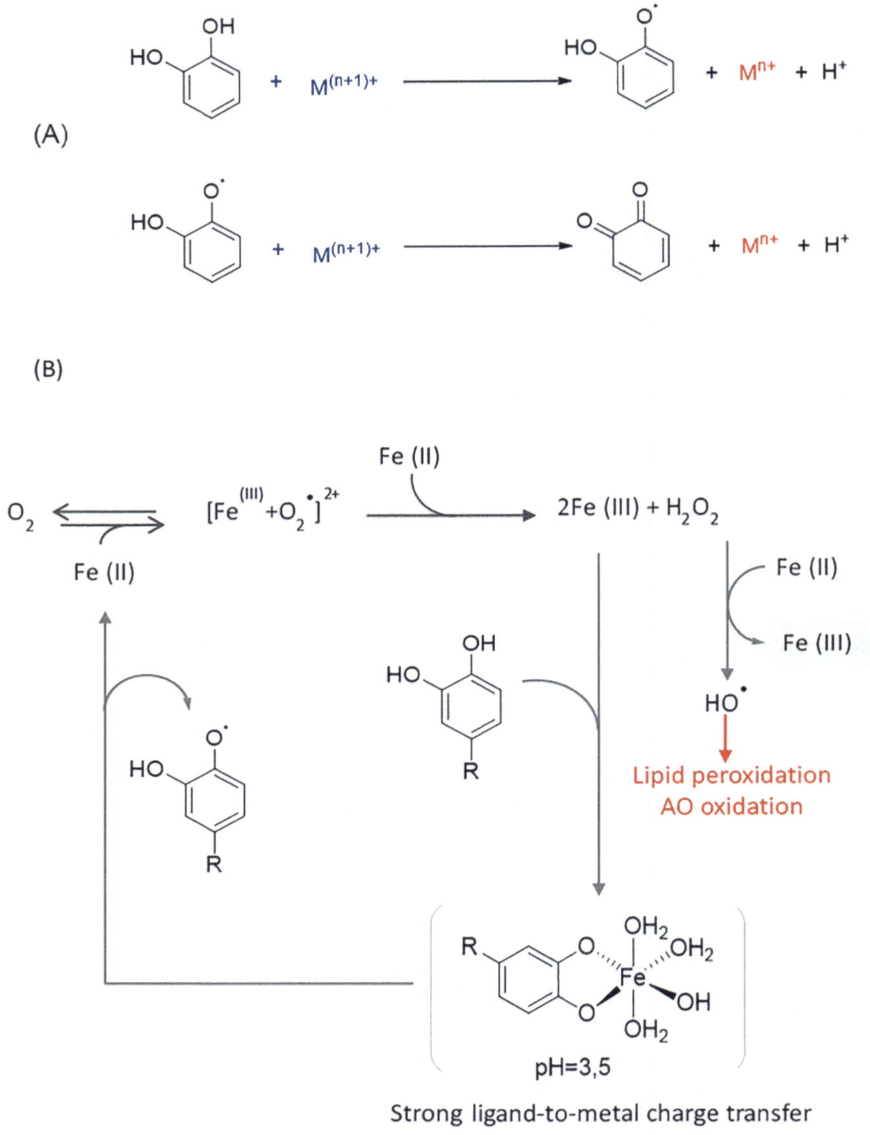

Fig. 9 (**a**) Reduction reactions of metal ions in the presence of catechol antioxidants. (**b**) Illustrative scheme describing the mechanism by which catechols are quickly consumed in the presence of ferric and/or ferrous ions and atmospheric oxygen

olive oil) in our diet to obtain the best benefits in their use is of foremost economic importance as extra virgin olive oil is usually the most expensive edible oil in the market.

4 Application of Olive Phenolic Compounds as Antioxidants in Foods

Several applications have been proposed for pomace and leaf extracts, not only in the food industry, but also in other sectors of commercial interest. Some examples are described in this seccion.

4.1 Vegetable Oils

In order to increase the daily dose of phenolic compounds, without increasing the energy intake or to add back the health benefits of phenolic compounds to refined olive oil, several attempts to enrich olive oils with pomace or leaf extracts have been tried (Difonzo et al. 2017; Paiva-Martins et al. 2003; Rubió et al. 2012; Suárez et al. 2010). In these attempts, both the total phenols and the oxidative stability of olive oils were significantly increased. Paiva-Martins found that a phenolic extract obtained from 1 kg of leaves would be necessary to fortify 50–320 L of ROO. This range results from the different metal concentration of the oil to be fortified, the cultivar, and the time of the year when the leaves were picked, in order to have a ROO with a similar stability to a virgin olive oil without affecting the sensor attributes of the oil (Paiva-Martins et al. 2007).

Since vegetable oils oxidation is a prominent issue, decreasing the sensory, nutritional, and commercial quality of the products, olive pomace and olive leaf phenolic extracts have also been used in the oxidative stabilization of several other refined vegetable oils (maize, soy, high oleic sunflower, sunflower, olive, rapeseed). In all the studies these refined vegetable oils were found to be more stable to oxidation than the control (Rubió et al. 2012; Sánchez de Medina et al. 2011; Abd-ElGhany et al. 2010; Orozco-Solano et al. 2011). Moreover, the oxidative stability of refined canola oil and high oleic sunflower oil containing olive leaf extracts was significantly higher than the control during frying and pan-frying operations (Farag et al. 2007; Orozco-Solano et al. 2011; Zribi et al. 2013). The positive effect against thermal oxidation was correlated to the concentration of leaf phenols. However, bitterness index, chlorophylls and sometimes off-flavours were present in the enriched oils as a consequence of the use of solvents and increased extraction of chlorophylls during phenol extract preparation. Therefore, it should be also taken into consideration the sensorial attributes when considering the feasibility of fortification with these extracts by the food industry.

4.2 Meat Products

Oxidative processes and meat spoilage bacteria are the major factors affecting meat and meat products shelf life. Therefore, olive leaf phenolic extracts could be a good option in the preservation of meat products, such as minced beef and sausages, due to their antioxidant and microbicide activity (Gök and Bor 2012; Djenane et al. 2019; Shalaby et al. 2018; Da Silva et al. 2018). The use of these extracts as effective antioxidant and antimicrobial agents is an interesting strategy for improving the quality and shelf life of meat products without increasing the content in synthetic preservatives. At 200 μg/kg, OLE were able to reduce total viable count during the storage of raw minced beef for 6 days in modified atmosphere packs (MAP) and for 12 days in aerobic storage (Shalaby et al. 2018). OLE added in beef at the same concentration was also able to decrease oxymyoglobin oxidation and lipid oxidation by 78% in aerobic conditions and by 84% under MAP conditions. Moreover, besides antimicrobial activity, OLE showed higher antioxidant activity in meat balls than blueberry and *Z. jujuba* extracts measured by TBARS test (Gök and Bor 2012).

Similar usage was attempted with olive pomace flour in fish burgers. However, although an improvement in antioxidant stability and nutritional quality was obtained, a decrease in the quality of most sensory parameters, such as colour, texture, flavour and taste, was observed (Cedola et al. 2017).

In order to prevent oxidation, several strategies have been developed, either by adding antioxidants directly to food or by protecting the food through coating or through packaging technology (Moudache et al. 2016, 2017; de Moraes et al. 2018). For example, coating chicken nuggets with sodium alginate/calcium chloride-based matrix including OLE was found to be an efficient strategy to reduce lipid oxidation of frozen samples when compared with a control (Ozvural 2019). In another example, immobilizing OLE in the multilayer polyethylene film used in the packaging of fresh minced pork meat was able to extend the shelf-life of the product at 4 °C for about 2 days (Moudache et al. 2017). Similar results were obtained in the preservation of cold-smoked salmon and packed salmon burgers (Albertos et al. 2017; Khemakhem et al. 2018) using edible films containing OLE.

Moraes Crizel et al. (de Moraes et al. 2018) also developed a biodegradable packaging by adding flour and micro-particles of olive pomace flour in chitosan films. The lack of solubility of the flour in the chitosan matrix increased the permeability and water solubility of films; however, it allowed a significant improvement in the antioxidant properties. Edible coatings based on chitosan with polyphenols from olive pomace have also been used as a strategy to extend the shelf-life of other fresh products, such fruit and vegetables (Khalifa et al. 2016).

Another way to modulate the oxidative stability of meat is based on feed supplementation. Chops from pigs that were fed diets containing leaf powder had lower peroxide and conjugated diene contents than chops from pigs fed conventional diets. Moreover, chops from pigs fed with the higher quantity of leaves also showed a lower drip loss. Differences in the oxidative stability of meat was explained by the significantly higher α-tocopherol concentration found in the intramuscular fat and backfat in pigs fed with olive leaf diets (Paiva-Martins et al. 2009; Paiva-Martins et al. 2014). Similar results were obtained with the incorporation of olive leaves in chicken diets and with the incorporation of olive pomace in rabbit's feed. Tyrosol, hydroxytirosol, and verbascoside were found in the chicken meat, with the burgers produced with this meat showing a significant decrease in lipid and protein oxidation during frozen storage (Da Silva et al. 2018). The oxidative stability of rabbit meat was also increased with an effect proportional to the content of antioxidants in the starting olive pomace (Dal Bosco et al. 2012).

Olive pomace has also been used in fish feeding, with the aim of partly replacing fish oil used for animal feed. Cardioprotective effects were obtained by the use of olive pomace due to the higher amount of eicosapentaenoic acid (EPA) and docosahexaenoic acid (DHA) found in the lipid fraction of the fish (Sioriki et al. 2016).

4.3 Dairy Products

Dairy products represent in general the most popular type of functional foods. The application of olive phenols in dairy products is limited due to their bitter taste, some instability in the food matrix, and possible influence in lactic bacteria growth. Pasteurized milk containing olive leaf extract produced a new functional milk with an extended shelf life, (Palmeri et al. 2019) but its bitter taste could not be disguised by other ingredients such as sweeteners or fruit flavours. On the other hand, during the fermentation by lactic bacteria in yogurt, not only a significant destruction of phenolic compounds but also a deep decrease in the sensory evaluation scores was observed (Tavakoli et al. 2018). However, the addition of aqueous phenolic extracts from olive pomace to fermented milk using Streptococcus thermophilus and Lactobacillus acidophilus cultures produced an yogurt like product with higher total phenol content and antioxidant activity when compared with the control (Aliakbarian et al. 2015). Although a deep decrease in the initial phenolic content or in the number of viable functional microorganisms was not observed, a sensory evaluation was not performed in this study.

To overcome the above drawbacks, encapsulation of phenols may be a good strategy to preserve the bioactive substances, as well as masking the unpleasant bitter taste (Katouzian et al. 2017). Microencapsulation and nanoencapsulation seem to be promising techniques to overcome these issues (Ozkan et al. 2019; Paulo and Santos 2021). Furthermore, a controlled release of the bioactive compounds from the delivery system could be also achieved, thus maximizing the biological activity (Ozkan et al. 2019). Nevertheless, only a few studies have been performed with olive phenols encapsulation. Emulsification/internal gelation was proven to be a promising technique to produce olive leave polyphenol extracts-enriched alginate beads for potential food applications (Flamminii et al. 2019, 2020). Enriched nanoliposomes with olive leaf extract showed to have a positive effect on the taste, texture, and overall acceptability of fortified yogurts (Tavakoli et al. 2018). However, more comprehensive studies on the encapsulation of phenolic antioxidants, assessing the effects of process parameters on the antioxidant activities, should be performed. In fact, the effects on antioxidant efficiency, physical and sensory quality of the products needs to be considered during the application of these techniques and the selection of the encapsulation process needs to take in consideration the thermo sensitivity and solubility of the phenolic compounds and the economic feasibility of the process (Ozkan et al. 2019).

The animal diet can also influence the final characteristics of animal food products such as milk, cheese, and eggs. The inclusion of olive pomace in caw feeding resulted in a milk not only with an higher content in oleic acid but also in total tocopherol and hydroxytirosol contents (Terramoccia et al. 2013). Moreover, sensory evaluation did not highlight any difference with the control (Caputo et al. 2015) and did not influence quality, yield, and coagulation parameters of milk (Zilio et al. 2015). In contrast, olive pomace addition in cows' feeding showed some effect on the proteolytic volatile profile with higher levels of the compounds related to fruity, floral and freshness aroma in cheese (Castellani et al. 2018; Castellani et al. 2019).

4.4 Bakery Products

Bakery products are quite prone to physical, chemical and microbiological spoilage. The use of olive phenol extracts and powders in this kind of products is usually aimed at the production of functional bread and biscuits, richer in dietary fibers and polyphenols. A simple way to fortify bread and other bakery products included in the formation of dough, consists in using an aqueous solution of olive antioxidants as water substitute. Bread fortified with 10% of olive pomace showed higher antioxidant activity, due to a higher phenolic acid and flavonoid contents, and an improved glycemic response due to higher fiber content (Cedola et al. 2019; Baiano et al. 2015). Replacement of the water of the bread with olive leaf aqueous extracts with the aim to produce functional bread, allowed the observation that the crumb of the bread showed higher antioxidant activity than the control. However, the addition of olive leaf or olive pomace powders or even OLE may cause significant decrease of bread volume, changes in the crumb colour, which became greyer and darker, and a modification of firmness, smell and taste (earthy, bitterness and astringency) (Cedola et al. 2019, 2020; Cecchi et al. 2019).

Olive pomace powder and OLE has also been added to biscuits and snacks and in general these products have shown a better overall sensory acceptability in texture and appearance, although, depending on the olive pomace powder fraction used, this may lead to a decrease in the viscosity of the fortified dough. This has as consequence a significant reduction in die pressure during extrusion, affecting the feasibility of the obtained extruded snacks (Ying et al. 2017). Nevertheless, baked snacks fortified with OLE showed better antioxidant stability when compared to control and some improvement in the sensory attributes during storage. These products being richer in fiber than controls, showed a significantly lower glycemic index and energy value. Moreover, higher plasmatic concentration of hydroxytyrosol metabolites with beneficial bioactivity was found in consumers (Ying et al. 2017).

5 Conclusions

These chapter describe the antioxidant activity of olive phenols in several lipidic systems and how some factors can affect this activity. This knowledge is of foremost importance when using olive phenols to extend the shelf-life and/or improve the nutritional profile of food products other than olive oil. These compounds can be obtained as a low-cost raw material from phenolic-rich by-products of the olive oil industry, namely, from olive pomace and olive leaves, providing an environmental benefit and an increase in the competitiveness of the olive industry sectors.

Acknowledgements This work has been supported by FCT/MCTES (UIDB/QUI/50006/2020 and PHENOLIVA–PTDC/OCE-ETA/32492/2017–POCI-01-0145-FEDER-032492).

References

Abd-ElGhany ME, Ammar MS, Hegazy AE (2010) Use of olive waste cake extract as a natural antioxidant for improving the stability of heated sunflower oil. World Appl Sci J 11(1):106–113
Albertos I, Avena-Bustillos RJ, Martín-Diana AB, Du W-X, Rico D, McHugh TH (2017) Antimicrobial olive leaf gelatin films for enhancing the quality of cold-smoked Salmon. Food Packag Shelf Life 13:49–55
Aliakbarian B, Casale M, Paini M, Casazza AA, Lanteri S, Perego P (2015) Production of a novel fermented milk fortified with natural antioxidants and its analysis by NIR spectroscopy. LWT – Food Sci Technol 62(1, Part 2):376–383
Almeida J, Losada-Barreiro S, Costa M, Paiva-Martins F, Bravo-Díaz C, Romsted LS (2016) Interfacial concentrations of hydroxytyrosol and its lipophilic esters in intact olive oil-in-water emulsions: effects of antioxidant hydrophobicity, surfactant concentration, and the oil-to-water ratio on the oxidative stability of the emulsions. J Agric Food Chem 64(25):5274–5283
Amiot M-J, Fleuriet A, Macheix J-J (1989) Accumulation of oleuropein derivatives during olive maturation. Phytochemistry 28(1):67–69
Aparicio R, Roda L, Albi MA, Gutiérrez F (1999 Oct 1) Effect of various compounds on virgin olive oil stability measured by Rancimat. J Agric Food Chem 47(10):4150–4155
Baiano A, Viggiani I, Terracone C, Romaniello R, Del Nobile MA (2015) Physical and sensory properties of bread enriched with phenolic aqueous extracts from vegetable wastes. Czech J Food Sci 33(3):247–253
Bakır T, Yıldoğan Beker B, Sönmezoğlu İ, İmer F, Apak R (2013) Antioxidant and prooxidant effects of α-tocopherol in a linoleic acid-copper(II)-ascorbate system. Eur J Lipid Sci Technol 115(3):372–376
Baldioli M, Servili M, Perretti G, Montedoro GF (1996) Antioxidant activity of tocopherols and phenolic compounds of virgin olive oil. J Am Oil Chem Soc 73(11):1589–1593
Banias G, Achillas C, Vlachokostas C, Moussiopoulos N, Stefanou M (2017) Environmental impacts in the life cycle of olive oil: a literature review. J Sci Food Agric 97(6):1686–1697

Bethke PC, Jansky SH (2008) The effects of boiling and leaching on the content of potassium and other minerals in potatoes. J Food Sci 73(5):H80–H85
Bilgin M, Şahin S (2013) Effects of geographical origin and extraction methods on total phenolic yield of olive tree (Olea europaea) leaves. J Taiwan Inst Chem Eng 44(1):8–12
Blekas G, Tsimidou M, Boskou D (1995) Contribution of α-tocopherol to olive oil stability. Food Chem 52(3):289–294
Blekas G, Psomiadou E, Tsimidou M, Boskou D (2002) On the importance of total polar phenols to monitor the stability of Greek virgin olive oil. Eur J Lipid Sci Technol 104(6):340–346
Brenes M, García A, García P, Garrido A (2001) Acid hydrolysis of Secoiridoid Aglycons during storage of virgin olive oil. J Agric Food Chem 49(11):5609–5614
Buckland G, Gonzalez CA (2015) The role of olive oil in disease prevention: a focus on the recent epidemiological evidence from cohort studies and dietary intervention trials. Br J Nutr 113(S2): S94–S101
Caporaso N, Formisano D, Genovese A (2018) Use of phenolic compounds from olive mill wastewater as valuable ingredients for functional foods. Crit Rev Food Sci Nutr 58 (16):2829–2841
Caputo AR, Morone G, Di Napoli MA, Rufrano D, Sabia E, Paladino F et al (2015) Effect of destoned olive cake on the aromatic profile of cows' milk and dairy products: comparison of two techniques for the headspace aroma profile analysis. Ital J Agron 10:15–20
Carrasco-Pancorbo A, Cerretani L, Bendini A, Segura-Carretero A, Gallina-Toschi T, Lercker G, Fernández-Gutiérrez A (2006) Evaluation of individual antioxidant activity of single phenolic compounds on virgin olive oil. Prog Nutr 8(1):28–39
Cassano A, Conidi C, Giorno L, Drioli E (2013) Fractionation of olive mill wastewaters by membrane separation techniques. J Hazard Mater 248–249:185–193
Castellani F, Bernardi N, Vitali A, Marone E, Grotta L, Martino G (2018) Proteolytic volatile compounds in milk and cheese of cows fed dried olive pomace supplementation. J Anim Feed Sci 27(4):361–365
Castellani F, Vitali A, Bernardi N, Marone E, Grotta L, Martino G (2019) Lipolytic volatile compounds in dairy products derived from cows fed with dried olive pomace. Eur Food Res Technol 245(1):159–166
Cecchi L, Schuster N, Flynn D, Bechtel R, Bellumori M, Innocenti M et al (2019) Sensory profiling and consumer acceptance of pasta, bread, and granola Bar fortified with dried olive pomace (Pâté): a byproduct from virgin olive oil production. J Food Sci 84(10):2995–3008
Cedola A, Cardinali A, Del Nobile MA, Conte A (2017) Fish burger enriched by olive oil industrial by-product. Food Sci Nutr 5(4):837–844
Cedola A, Cardinali A, Del Nobile MA, Conte A (2019) Enrichment of bread with olive oil industrial by-product. J Agric Sci Technol 9:119–127
Cedola A, Cardinali A, D'Antuono I, Conte A, Del Nobile MA (2020) Cereal foods fortified with by-products from the olive oil industry. Food Biosci 33:100490
Cicerale S, Lucas LJ, Keast RSJ (2012) Antimicrobial, antioxidant and anti-inflammatory phenolic activities in extra virgin olive oil. Curr Opin Biotechnol 23(2):129–135
Cillard J, Cillard P, Cormier M, Girre L (1980) α-Tocopherol prooxidant effect in aqueous media: increased autoxidation rate of linoleic acid. J Am Oil Chem Soc 57(8):252
Cinquanta L, Esti M, Di Matteo M (2001) Oxidative stability of virgin olive oils. J Am Oil Chem Soc 78(12):1197
Costa M, Losada-Barreiro S, Paiva-Martins F, Bravo-Díaz C, Romsted LS (2015) A direct correlation between the antioxidant efficiencies of caffeic acid and its alkyl esters and their concentrations in the interfacial region of olive oil emulsions. The pseudophase model interpretation of the "cut-off" effect. Food Chem 175:233–242
Costa M, Losada-Barreiro S, Bravo-Díaz C, Vicente AA, Monteiro LS, Paiva-Martins F (2020) Influence of AO chain length, droplet size and oil to water ratio on the distribution and on the activity of gallates in fish oil-in-water emulsified systems: emulsion and nanoemulsion comparison. Food Chem 310:125716

Covas M-I, de la Torre R, Fitó M (2015) Virgin olive oil: a key food for cardiovascular risk protection. Br J Nutr 113(S2):S19–S28
Da Silva SL, Marangoni C, Brum DS, Vendruscolo RG, Silva MS, de Moura HC et al (2018) Effect of dietary olive leaves on the lipid and protein oxidation and bacterial safety of chicken hamburgers during frozen storage. Int Food Res J 25(1):383–391
Dal Bosco A, Mourvaki E, Cardinali R, Servili M, Sebastiani B, Ruggeri S et al (2012) Effect of dietary supplementation with olive pomaces on the performance and meat quality of growing rabbits. Meat Sci 92(4):783–788
Daskalaki D, Kefi G, Kotsiou K, Tasioula-Margari M (2009) Evaluation of phenolic compounds degradation in virgin olive oil during storage and heating. J Food Nutr Res 48(1):31–41
de Moraes CT, de Oliveira RAD, Alves V, Bandarra N, Moldão-Martins M, Hickmann Flôres S (2018) Active food packaging prepared with chitosan and olive pomace. Food Hydrocoll 74:139–150
Dermeche S, Nadour M, Larroche C, Moulti-Mati F, Michaud P (2013) Olive mill wastes: biochemical characterizations and valorization strategies. Process Biochem 48(10):1532–1552
Difonzo G, Russo A, Trani A, Paradiso VM, Ranieri M, Pasqualone A et al (2017) Green extracts from Coratina olive cultivar leaves: antioxidant characterization and biological activity. J Funct Foods 31:63–70
Difonzo G, Pasqualone A, Silletti R, Cosmai L, Summo C, Paradiso VM et al (2018) Use of olive leaf extract to reduce lipid oxidation of baked snacks. Food Res Int 108:48–56
Djenane D, Gómez D, Yangüela J, Roncalés P, Ariño A (2019) Olive leaves extract from algerian oleaster (Olea europaea var. sylvestris) on microbiological safety and shelf-life stability of raw halal minced beef during display. Foods 8(1):10
El-Abbassi A, Kiai H, Hafidi A (2012) Phenolic profile and antioxidant activities of olive mill wastewater. Food Chem 132(1):406–412
Estruch R, Ros E, Salas-Salvadó J, Covas M-I, Corella D, Arós F, et al (2018) Primary prevention of cardiovascular disease with a Mediterranean diet supplemented with extra-virgin olive oil or nuts. N Engl J med [internet]. 378(25):e34. Available from: https://doi.org/10.1056/NEJMoa1800389
Fabiani R (2016) Anti-cancer properties of olive oil secoiridoid phenols: a systematic review of in vivo studies. Food Funct 7(10):4145–4159
Farag RS, Mahmoud EA, Basuny AM (2007) Use crude olive leaf juice as a natural antioxidant for the stability of sunflower oil during heating. Int J Food Sci Technol 42(1):107–115
Flamminii F, Di Mattia CD, Difonzo G, Neri L, Faieta M, Caponio F et al (2019) From by-product to food ingredient: evaluation of compositional and technological properties of olive-leaf phenolic extracts. J Sci Food Agric 99(14):6620–6627
Flamminii F, Di Mattia CD, Nardella M, Chiarini M, Valbonetti L, Neri L et al (2020) Structuring alginate beads with different biopolymers for the development of functional ingredients loaded with olive leaves phenolic extract. Food Hydrocoll 108:105849
Fregapane G, Lavelli V, León S, Kapuralin J, Desamparados SM (2006) Effect of filtration on virgin olive oil stability during storage. Eur J Lipid Sci Technol 108(2):134–142
Fregapane G, Gómez-Rico A, Inarejos AM, Salvador MD (2013) Relevance of minor components stability in commercial olive oil quality during the market period. Eur J Lipid Sci Technol 115 (5):541–548
Garcia B, Magalhães J, Fregapane G, Salvador MD, Paiva-Martins F (2012) Potential of selected Portuguese cultivars for the production of high quality monovarietal virgin olive oil. Eur J Lipid Sci Technol 114(9):1070–1082
Gök V, Bor Y (2012) Effect of olive leaf, blueberry and Zizyphus jujuba extracts on the quality and shelf life of meatball during storage. J Food, Agric Environ 10(2):190–195
Gómez-Alonso S, Fregapane G, Salvador MD, Gordon MH (2003 Jan) Changes in phenolic composition and antioxidant activity of virgin olive oil during frying. J Agric Food Chem 51 (3):667–672

Gómez-Alonso S, Mancebo-Campos V, Desamparados Salvador M, Fregapane G (2004) Oxidation kinetics in olive oil triacylglycerols under accelerated shelf-life testing (25–75 °C). Eur J Lipid Sci Technol 106(6):369–375

Gómez-Alonso S, Mancebo-Campos V, Salvador MD, Fregapane G (2007) Evolution of major and minor components and oxidation indices of virgin olive oil during 21 months storage at room temperature. Food Chem 100(1):36–42

Gordon MH, Paiva-Martins F, Almeida M (2001) Antioxidant activity of Hydroxytyrosol acetate compared with that of other olive oil polyphenols. J Agric Food Chem 49(5):2480–2485

Gutiérrez F, Fernández JL (2002) Determinant parameters and components in the storage of virgin olive oil. Prediction of storage time beyond which the oil is no longer of “extra” quality. J Agric Food Chem 50(3):571–577

Katouzian I, Faridi Esfanjani A, Jafari SM, Akhavan S (2017) Formulation and application of a new generation of lipid nano-carriers for the food bioactive ingredients. Trends Food Sci Technol 68:14–25

Keys A, Mienotti A, Karvonen MJ, Aravanis C, Blackburn H, Buzina R, Djordjevic BS, Dontas AS, Fidanza F, Keys MH, Kromhout D, Nedeljkovic S, Punsar S, Seccareccia F, Toshima H (1986) The diet and 15-year death rate in the seven countries study. Am J Epidemiol 124 (6):903–915. https://doi.org/10.1093/oxfordjournals.aje.a114480

Khalifa I, Barakat H, El-Mansy HA, Soliman SA (2016) Enhancing the keeping quality of fresh strawberry using chitosan-incorporated olive processing wastes. Food Biosci 13:69–75

Khemakhem I, Ahmad-Qasem MH, Catalán EB, Micol V, García-Pérez JV, Ayadi MA et al (2017) Kinetic improvement of olive leaves' bioactive compounds extraction by using power ultrasound in a wide temperature range. Ultrason Sonochem 34:466–473

Khemakhem I, Fuentes A, Lerma-García MJ, Ayadi MA, Bouaziz M, Barat JM (2018) Olive leaf extracts for shelf life extension of salmon burgers. Food Sci Technol Int 25(2):91–100

Kim K, Jung J-Y, Kwon J-H, Yang J-W (2015) Dynamic microfiltration with a perforated disk for effective harvesting of microalgae. J Memb Sci 475:252–258

Kiritsakis A, Shahidi F (2017) Olives and olive oil as functional foods: bioactivity, chemistry and processing. John Wiley & Sons, New York

Kiritsakis K, Rodríguez-Pérez C, Gerasopoulos D, Segura-Carretero A (2017) Olive oil enrichment in phenolic compounds during malaxation in the presence of olive leaves or olive mill wastewater extracts. Eur J Lipid Sci Technol 119(9):1600425

Krichene D, Allalout A, Mancebo-Campos V, Salvador MD, Zarrouk M, Fregapane G (2010) Stability of virgin olive oil and behaviour of its natural antioxidants under medium temperature accelerated storage conditions. Food Chem 121(1):171–177

Krichene D, Salvador MD, Fregapane G (2015) Stability of virgin olive oil phenolic compounds during long-term storage (18 months) at temperatures of 5–50 °C. J Agric Food Chem 63 (30):6779–6786

Lavelli V, Fregapane G, Salvador MD (2006) Effect of storage on Secoiridoid and tocopherol contents and antioxidant activity of Monovarietal extra virgin olive oils. J Agric Food Chem 54 (8):3002–3007

Lee O-H, Kim Y-C, Kim K-J, Kim Y-C, Lee B-Y (2007) The effects of bioactive compounds and fatty acid compositions on the oxidative stability of extra virgin olive oil varieties. Food Sci Biotechnol 16(3):415–420

López-Miranda J, Pérez-Jiménez F, Ros E, De Caterina R, Badimón L, Covas MI et al (2010) Olive oil and health: summary of the II international conference on olive oil and health consensus report, Jaén and Córdoba (Spain) 2008. Nutr Metab Cardiovasc Dis 20(4):284–294

Magrone T, Spagnoletta A, Salvatore R, Magrone M, Dentamaro F, Russo MA et al (2018) Olive leaf extracts act as modulators of the human immune response. Endocr Metab Immune Disord Drug Targets 18:85–93

Mancebo-Campos V, Fregapane G, Desamparados SM (2008) Kinetic study for the development of an accelerated oxidative stability test to estimate virgin olive oil potential shelf life. Eur J Lipid Sci Technol 110(10):969–976

Mancebo-Campos V, Salvador MD, Fregapane G (2014) Antioxidant capacity of individual and combined virgin olive oil minor compounds evaluated at mild temperature (25 and 40°C) as compared to accelerated and antiradical assays. Food Chem 150:374–381
Mateos R, Trujillo M, Pérez-Camino MC, Moreda W, Cert A (2005) Relationships between oxidative stability, triacylglycerol composition, and antioxidant content in olive oil matrices. J Agric Food Chem 53(14):5766–5771
Mentella MC, Scaldaferri F, Ricci C, Gasbarrini A, Miggiano GA (2019) Cancer and mediterranean diet: a review. Nutrients 11(9):2059
Moudache M, Colon M, Nerín C, Zaidi F (2016) Phenolic content and antioxidant activity of olive by-products and antioxidant film containing olive leaf extract. Food Chem 212:521–527
Moudache M, Nerín C, Colon M, Zaidi F (2017) Antioxidant effect of an innovative active plastic film containing olive leaves extract on fresh pork meat and its evaluation by Raman spectroscopy. Food Chem 229:98–103
Nadour M, Laroche C, Pierre G, Delattre C, Moulti-Mati F, Michaud P (2015) Structural characterization and biological activities of polysaccharides from olive mill wastewater. Appl Biochem Biotechnol 177(2):431–445
Ninfali P, Aluigi G, Bacchiocca M, Magnani M (2001) Antioxidant capacity of extra-virgin olive oils. J Am Oil Chem Soc 78(3):243–247
Nunes M, Costa ASG, Bessada S, Santos J, Puga H, Alves RC et al (2018) Olive pomace as a valuable source of bioactive compounds: a study regarding its lipid- and water-soluble components. Sci Total Environ 644:229–236
Orozco-Solano MI, Priego-Capote F, Luque de Castro MD (2011) Influence of simulated deep frying on the antioxidant fraction of vegetable oils after enrichment with extracts from olive oil pomace. J Agric Food Chem 59(18):9806–9814
Ozkan G, Franco P, De Marco I, Xiao J, Capanoglu E (2019) A review of microencapsulation methods for food antioxidants: principles, advantages, drawbacks and applications. Food Chem 272:494–506
Ozvural EB (2019) Fabrication of olive leaf extract and hazelnut skin incorporated films to improve the quality of nuggets during refrigerated and deep freeze storage. Br Poult Sci 60(6):708–715
Paiva-Martins F (2002) Olive oil antioxidant. Thesis PhD
Paiva-Martins F, Gordon MH (2001) Isolation and characterization of the antioxidant component 3,4-Dihydroxyphenylethyl 4-formyl-3-formylmethyl-4-hexenoate from olive (Olea europaea) leaves. J Agric Food Chem 49(9):4214–4219
Paiva-Martins F, Gordon MH (2002) Effects of pH and ferric ions on the antioxidant activity of olive polyphenols in oil-in-water emulsions. J Am Oil Chem Soc 79(6):571–576
Paiva-Martins F, Gordon MH (2005) Interactions of ferric ions with olive oil phenolic compounds. J Agric Food Chem 53(7):2704–2709
Paiva-Martins F, Pinto M (2008) Isolation and characterization of a new Hydroxytyrosol derivative from olive (Olea europaea) leaves. J Agric Food Chem 56(14):5582–5588
Paiva-Martins F, Gordon MH, Gameiro P (2003) Activity and location of olive oil phenolic antioxidants in liposomes. Chem Phys Lipids 124(1):23–36
Paiva-Martins F, Santos V, Mangericão H, Gordon MH (2006) Effects of copper on the antioxidant activity of olive polyphenols in bulk oil and oil-in-water emulsions. J Agric Food Chem 54 (10):3738–3743
Paiva-Martins F, Correia R, Félix S, Ferreira P, Gordon MH (2007) Effects of enrichment of refined olive oil with phenolic compounds from olive leaves. J Agric Food Chem 55(10):4139–4143
Paiva-Martins F, Barbosa S, Pinheiro V, Mourão JL, Outor-Monteiro D (2009) The effect of olive leaves supplementation on the feed digestibility, growth performances of pigs and quality of pork meat. Meat Sci 82(4):438–443
Paiva-Martins F, Ribeirinha T, Silva A, Gonçalves R, Pinheiro V, Mourão JL, et al (2014) Effects of the dietary incorporation of olive leaves on growth performance, digestibility, blood parameters and meat quality of growing pigs. J Sci Food Agric [Internet]. 4(14):3023–9. Available from: https://doi.org/10.1002/jsfa.6650

Palmeri R, Parafati L, Trippa D, Siracusa L, Arena E, Restuccia C et al (2019) Addition of Olive Leaf Extract (OLE) for producing fortified fresh pasteurized milk with an extended shelf life. Antioxidants 8:255

Parkinson L, Cicerale S (2016) The health benefiting mechanisms of virgin olive oil phenolic compounds. Molecules 21(12):1734

Paulo F, Santos L (2021) Deriving valorization of phenolic compounds from olive oil by-products for food applications through microencapsulation approaches: a comprehensive review. Crit Rev Food Sci Nutr 61(6):920–945

Pellegrini N, Visioli F, Buratti S, Brighenti F (2001) Direct analysis of total antioxidant activity of olive oil and studies on the influence of heating. J Agric Food Chem 49(5):2532–2538. https://doi.org/10.1021/jf001418j

Piroddi M, Albini A, Fabiani R, Giovannelli L, Luceri C, Natella F, et al (2017) Nutrigenomics of extra-virgin olive oil: a review. BioFactors [Internet]. 43(1):17–41. Available from: https://doi.org/10.1002/biof.1318

Rahmanian N, Jafari SM, Wani TA (2015) Bioactive profile, dehydration, extraction and application of the bioactive components of olive leaves. Trends Food Sci Technol 42(2):150–172

Rodríguez-Morató J, Xicota L, Fitó M, Farré M, Dierssen M, De la Torre R (2015) Potential role of olive oil phenolic compounds in the prevention of neurodegenerative diseases. Molecules 20:4655–4680

Roig A, Cayuela ML, Sánchez-Monedero MA (2006) An overview on olive mill wastes and their valorisation methods. Waste Manag 26(9):960–969

Romani A, Lapucci C, Cantini C, Ieri F, Mulinacci N, Visioli F (2007) Evolution of minor polar compounds and antioxidant capacity during storage of bottled extra virgin olive oil. J Agric Food Chem 55(4):1315–1320

Rubió L, Motilva M-J, Macià A, Ramo T, Romero M-P (2012) Development of a phenol-enriched olive oil with both its own phenolic compounds and complementary phenols from thyme. J Agric Food Chem 60(12):3105–3112

Sabry OMM (2014) Beneficial health effects of olive leaves extracts. J Nat Sci Res 4(19):1–9

Şahin S, Bilgin M (2018) Olive tree (Olea europaea L.) leaf as a waste by-product of table olive and olive oil industry: a review. J Sci Food Agric 98(4):1271–1279

Salta FN, Mylona A, Chiou A, Boskou G, Andrikopoulos NK (2007) Oxidative stability of edible vegetable oils enriched in polyphenols with olive leaf extract. Food Sci Technol Int 13 (6):413–421

Salvador MD, Aranda F, Gómez-Alonso S, Fregapane G (2003) Influence of extraction system production year and area on Cornicabra virgin olive oil: a study of five crop seasons. Food Chem 80(3):359–366. https://doi.org/10.1016/S0308-8146(02)00273-X

Sánchez de Medina V, Priego-Capote F, Jiménez-Ot C, Luque de Castro MD (2011) Quality and stability of edible oils enriched with hydrophilic antioxidants from the olive tree: the role of enrichment extracts and lipid composition. J Agric Food Chem 59(21):11432–11441

Schwingshackl L, Hoffmann G (2015) Adherence to Mediterranean diet and risk of cancer: an updated systematic review and meta-analysis of observational studies. Cancer Med 4 (12):1933–1947

Segade M, Bermejo R, Silva A, Paiva-Martins F, Gil-Longo J, Campos-Toimil M (2016) Involvement of endothelium in the vasorelaxant effects of 3, 4-DHPEA-EA and 3, 4-DHPEA-EDA, two major functional bioactives in olive oil. J Funct Foods 23:637–646

Servili M, Montedoro G (2002) Contribution of phenolic compounds to virgin olive oil quality. Eur J Lipid Sci Technol 104(9–10):602–613

Servili M, Baldioli M, Miniati E, Montedoro GF (1996) Antioxidant activity of new phenolic compounds extracted from virgin olive oil and their interaction with α-tocopherol and β-carotene. Riv Ital delle Sostanze Grasse 73(2):55–59

Shalaby AR, Anwar MM, Sallam EM (2018) Improving quality and shelf-life of minced beef using irradiated olive leaf extract. J Food Process Preserv 42(11):e13789

Silva L, Pinto J, Carrola J, Paiva-Martins F (2010a) Oxidative stability of olive oil after food processing and comparison with other vegetable oils. Food Chem 121(4):1177–1187

Silva L, Garcia B, Paiva-Martins F (2010b) Oxidative stability of olive oil and its polyphenolic compounds after boiling vegetable process. LWT – Food Sci Technol 43(9):1336–1344

Silva SF, Anjos CAR, Cavalcanti RN, dos Santos CRM (2015) Evaluation of extra virgin olive oil stability by artificial neural network. Food Chem 179:35–43

Sioriki E, Smith TK, Demopoulos CA, Zabetakis I (2016) Structure and cardioprotective activities of polar lipids of olive pomace, olive pomace-enriched fish feed and olive pomace fed gilthead sea bream (Sparus aurata). Food Res Int 83:143–151

Suárez M, Romero M-P, Motilva M-J (2010) Development of a phenol-enriched olive oil with phenolic compounds from olive cake. J Agric Food Chem 58(19):10396–10403

Talhaoui N, Taamalli A, Gómez-Caravaca AM, Fernández-Gutiérrez A, Segura-Carretero A (2015) Phenolic compounds in olive leaves: analytical determination, biotic and abiotic influence, and health benefits. Food Res Int 77:92–108

Tavakoli H, Hosseini O, Jafari SM, Katouzian I (2018 Sep 5) Evaluation of physicochemical and antioxidant properties of yogurt enriched by olive leaf Phenolics within Nanoliposomes. J Agric Food Chem 66(35):9231–9240

Terramoccia S, Bartocci S, Taticchi A, Di Giovanni S, Pauselli M, Mourvaki E et al (2013) Use of dried stoned olive pomace in the feeding of lactating buffaloes: effect on the quantity and quality of the Milk produced. Asian-Australas J Anim Sci 26(7):971–980

Tunstall-Pedoe H, Kuulasmaa K, Mähönen M, Tolonen H, Ruokokoski E (1999) Contribution of trends in survival and coronar y-event rates to changes in coronary heart disease mortality: 10-year results from 37 WHO MONICA project populations. Lancet 353(9164):1547–1557

Woollard DC, Indyk HE (2003) TOCOPHEROLS, properties and determination. In: Caballero BBT-E of FS and N, 2nd edn. Academic Press, Oxford, pp 5789–5796

Ying D, Hlaing MM, Lerisson J, Pitts K, Cheng L, Sanguansri L et al (2017) Physical properties and FTIR analysis of rice-oat flour and maize-oat flour based extruded food products containing olive pomace. Food Res Int 100:665–673

Zilio DM, Bartocci S, Di Giovanni S, Servili M, Chiariotti A, Terramoccia S (2015) Evaluation of dried stoned olive pomace as supplementation for lactating Holstein cattle: effect on milk production and quality. Anim Prod Sci 55(2):185–188

Zribi A, Gargouri B, Jabeur H, Rebaï A, Abdelhedi R, Bouaziz M (2013) Enrichment of pan-frying refined oils with olive leaf phenolic-rich extract to extend the usage life. Eur J Lipid Sci Technol 115(12):1443–1453

Nutritional and Preservative Properties of Polyphenol-Rich Olive Oil: Effect on Seafood Processing and Storage

Lucía Méndez, Raffaele Sacchi, Isabel Medina, and Santiago P. Aubourg

1 Quality Changes of Fish During Processing

Most fish and invertebrate marine species give rise to products of great economic importance in many countries. Additionally, they can lead to food products providing high content of important constituents for the human diet, such as nutritional and digestive proteins, lipidsoluble vitamins (namely, A and D), microelements (*I*, *F*, *Ca*, *Cu*, *Zn*, *Fe* and others) and highly unsaturated fatty acids (Tilami and Sampels 2018). In this context, the marine lipid fraction is now the subject of a great deal of attention due to its high content on *n*-3 polyunsaturated fatty acids (PUFAs), which have shown a positive role in preventing a wide range of human diseases.

However, marine species lead to highly perishable products whose quality rapidly declines post-mortem as a result of processing and storage. Such food deteriorates after death due to the development of different damage pathways that can be summarised as microbiological development, endogenous enzyme activity, non-enzymatic lipid oxidation and browning, and enzymatic browning (Medina et al. 2009a). The relative incidence of each damage mechanism will depend on the kind of technological process applied and on the concrete composition of the marine species involved. In all cases, the different damage pathways will operate simultaneously.

The demand for better quality processed seafood is ever increasing. This has led to the development of a large preservation industry aiming to supply seafood that are safe, nutritious and economical. Therefore, different basic strategies have been employed in order to preserve them from degradation. In a first group of preserving

---

L. Méndez · I. Medina · S. P. Aubourg (✉)
Food Technology Department, Marine Research Institute (CSIC), Vigo, Spain
e-mail: saubourg@iim.csic.es

R. Sacchi
Department of Agricultural Sciences, University of Naples Federico II, Naples, Italy

C. Bravo-Diaz (ed.), *Lipid Oxidation in Food and Biological Systems*,
https://doi.org/10.1007/978-3-030-87222-9_20

strategies, initial characteristics in the raw marine species are pretended to be maintained till the product arrives to the consumer (namely, chilling and refrigerated storage, and freezing and frozen storage). In a second group, a relatively strong process is involved so that initial characteristics of raw marine species are lost and a product offering different properties is attained (namely, heating, salting and drying).

Among the different preserving strategies concerning this second group, thermal treatment occupies a relevant position. Thus, the thermal processing of seafood at moderate temperatures has been reported to improve their hygiene status and shelf life through the destruction of pathogenic microorganisms and the inactivation of undesirable enzymes. Therefore, most attention during thermal treatment has to be given to lipid oxidation and further interaction of oxidised lipids with other constituents, proteins specially, on the basis that proteins are heat denatured and turn into more reactive molecules. Consequently, detrimental effects on nutritional and sensory values due to thermal treatment have been reported, especially if overprocessing is carried out. Among them, heat degradation of nutrients, oxidation of vitamins and lipids, leaching of water-soluble vitamins, minerals and proteins, toughening and drying of fragile marine protein, and Maillard reaction development in shellfish can be mentioned (Aubourg 2001).

It is concluded that, although heat processing of marine foods makes microorganisms and endogenous enzymes inactive, it may also cause the destruction of essential nutrients, which leads to the deterioration of the product quality. Consequently, much attention has to be given to maximising quality retention for a specific reduction in undesirable microorganisms and endogenous enzyme activity. Different strategies have been attempted in order to carry out this optimisation. In this context, a great attention has been accorded to the addition of natural preservative compounds before the heat treatment takes place. Among such natural possibilities, olive oil (OO) represents an important choice on the basis of being a rich source on polyphenol and other preservative compounds.

2 Polyphenol-Rich Olive Oil: Nutritional Aspects Related to Human Health

Polyphenol-rich OO represents a key component of the Mediterranean diet (MD). Thus, extra virgin olive oil (EVOO), being obtained from olive fruits exclusively by physical-mechanical technologies (olive crushing, centrifugation, filtration, etc.), can be considered 'lipid fruit juice'. It has shown a peculiar composition where olive triacylglycerols become a 'solvent' for several natural fruit components (antioxidants, flavour, pigments, etc.) responsible for its sensory and nutritional quality (Sacchi 2008). EVOO is made from pure, cold-pressed olives, contains no more than 0.8% acidity, whereas virgin olive oil (VOO) has an acidity of less than 2% and often uses slighter riper olives.

The polyphenol-rich OO shelf life is higher than other vegetable oils, mainly due to the presence of phenolic molecules having a catechol group, such as hydroxytyrosol (HT) and its secoiridoid derivatives (Servili et al. 2014). Several assays have been used to establish the antioxidant activity of these isolated phenolic compounds. Typical sensory gustative properties of EVOO and VOO, such as bitterness and pungency, have been attributed to secoiridoid molecules. Considering the importance of the phenolic fraction in all kinds of edible OO, especially in VOO and EVOO, high performance analytical methods have been developed in the last years to characterise its complex phenolic pattern (García-Villalba et al. 2010).

Besides being used as food additive, the health benefits of OO in general are well-known so that nutraceutical and nutritional values have been largely reported in literature (Serreli and Deiana 2020). Concretely, VOO and EVOO are important components of the Mediterranean diet (MD) and often pointed out among the main reasons that explain the health benefits attributed to this diet. Particularly, the populations with higher adherence to the MD show a reduction in the incidence of oxidative- and inflammatory-related pathologies, such as cardiovascular diseases, cancer and neurodegenerative disorders, and a longer lifespan (Castro-Barquero et al. 2018). From the viewpoint of composition, all types of OO are primarily divided into two fractions. First, a major saponifiable fraction (about 98–99% of the total weight), mainly in the form of triacylglycerides, including the monounsaturated fatty acid (MUFA) oleic acid (18:1 *n*-9) (55–83%) and other saturated and unsaturated acids (linoleic, palmitic and stearic acids, 3–21%). Secondly, a minor unsaponifiable fraction (about 1–2%), with over two hundred chemical compounds, including a high number of vitamins (tocopherols), polyphenols (mainly HT, tyrosol, and oleuropein) and triterpenes (squalene) (De Pablos et al. 2019). Health properties of OO were attributed to both the high level of oleic acid in triacylglycerols, and to many different minor compounds (squalene, tocopherols, pigments, volatile compounds) and, in particular, to phenolic compounds such as phenyl alcohols, secoiridoids and lignans present in the unsaponifiable fraction of EVOO (Frankel 2011). Thus, several studies correlated the in vitro and in vivo positive actions of EVOO on human health to its chemical composition in the last decade (Frankel 2011).

Some of the main health effects of both saponifiable and unsaponifiable fractions of OO are summarised below.

2.1 Health Effects of the Unsaponifiable Fraction of the Olive Oils

As mentioned before, OO is rich in polyphenolic compounds and, as an essential component of the MD, the total polyphenol intake in the population that regularly follows this diet has been estimated at 90.4 mg daily (11% of total intake) (Tresserra-Rimbau et al. 2013). Polyphenols are maybe the most studied bioactive compounds of the unsaponifiable fraction of OO. For instance, the effect of phenols present in VOO on the formation of carcinogenic heterocyclic amines (HA) was evaluated in a model system by Monti et al. (2001). For it, an aqueous solution of creatinine, glucose and glycine was heated in the presence of two samples of VOO differing only in their composition of phenols. The addition of VOO inhibited the formation of 2-amino-3-methylimidazo[4,5-*f*]quinoxaline, 2-amino-3,8-dimethylimidazo [4,5-*f*]quinoxaline, and 2-amino-3,7,8-trimethylimidazo[4,5-*f*]quinoxaline by 30–50% compared with controls. Freshly-made OO, which contained a higher amount of dihydroxyphenylethanol derivatives, inhibited HA formation in a higher extent than a 1-year-old OO. Inhibition of HA formation was also demonstrated using phenols extracted from VOO. It has been recognised that the polyphenol content in OO plays an important role in the health effects described for this oil, namely antioxidant, anti-inflammatory, anti-tumoural, neuroprotective, cardioprotective and anti-obesity outcomes, among others (Serreli and Deiana 2020). Accordingly, VOO and especially EVOO, which contain high polyphenol amounts, exhibit stronger antioxidant and anti-inflammatory activities than refined OO, with a lower phenolic content. This fact suggests that the phenolic component is the main responsible for those effects, rather than the fat in the oil (Boss et al. 2016).

Approximately, half of the OO phenolic content, which totally ranges between 100–600 mg/kg, corresponds to HT and its derivatives (oleuropein and tyrosol). Therefore, as a major polyphenol in OO (50–200 mg/kg OO), its daily intake in the context of the MD is estimated to be around 0.15 and 30 mg/day (Flori et al. 2019). HT is formed by the hydrolysis of oleuropein during the maturing of olives, storage of OO, and preparation of table olives. It is also produced from oleuropein by microbiota action in the organism after the intake of OO (De las Hazas et al. 2018). In the last few years, HT has received increasing attention due to its multiple pharmacological properties, mainly antioxidant, anti-inflammatory and pro-apoptotic activities (Bertelli et al. 2020). The European Food Safety Authority (EFSA) Panel on Dietetic Products, Nutrition and Allergies (NDA) in 2011 provided a scientific opinion on the scientific substantiation of health claims in relation to polyphenols in olive and protection of LDL particles from oxidative damage, maintenance of normal blood HDL-cholesterol concentrations, maintenance of normal blood pressure: "*anti-inflammatory properties*", "*contributes to the upper respiratory tract health*", "*can help to maintain a normal function of gastrointestinal tract*", and "*contributes to body defences against external agents*". The food constituent, which is the subject of the health claims, is polyphenols in olive (olive fruit, olive mill waste waters or olive oil, *Olea europaea* L. extract and leaf). On the basis of the data presented, the Panel concluded that a cause and effect relationship has been established between the consumption of OO polyphenols (standardised by the content of HT and its derivatives) and protection of LDL particles from oxidative damage. The Panel considers that in order to bear the claim, 5 mg of HT and its derivatives (e.g. oleuropein complex and tyrosol) in OO should be consumed daily (EFSA 2011a).

Besides HT and its derivatives, other components of the unsaponifiable fraction of OO have shown important benefits on human health. Indeed, this fraction is also rich in vitamins, including E, K and A precursor (beta-carotene), pigments as lutein and triterpenes as squalene (Casado-Díaz et al. 2019). Squalene is an aliphatic triterpene and the most abundant hydrocarbon in olive oil (0.7% of the total content; 30–50% of the unsaponifiable fraction). The high squalene content differentiates OO from other oils, which present much lower total amount of this triterpene (between 0.002% and 0.03%) Martínez-Beamonte et al. 2020). Squalene has a high antioxidant capacity, being associated with cardioprotective and antiaging effects (Casado-Díaz et al. 2019).

2.2 *Health Effects of the Saponifiable Fraction of the Olive Oils*

The saponifiable fraction of OO contains both saturated (SFAs) and unsaturated (UFAs) fatty acids, mostly in form of triacylglycerides. OO, particularly EVOO, contains round 85% of UFAs resides, in relation to saturated ones (round 14.5%) (Casado-Díaz et al. 2019). The total intake of oleic acid (the most abundant fatty acid in OO as previously mentioned) in adults ranges between 12 and 18% of energy, reaching up to 29% in Southern European countries (Greece, Italy or Spain), where the adherence to the MD is high and OO is traditionally consumed as part of their regular diet (Flori et al. 2019).

Besides fat quantity intake, it is well-known that fat quality is also a critical aspect for understanding the health effect of this important dietary component. In general, SFAs are associated with cardiovascular morbidity and mortality, and the increase intake of those SFAs is usually implicated in the aetiology and progression of several diet-induced metabolic disorders, which include obesity and obesity-related diseases, but also the severity of hypoxic-damage in the brain and acute myocardial infarct (Flori et al. 2019). On the contrary, the beneficial effect of high consumption of PUFAs, especially *n*-3, has been well-established and their positive effects on lipid profile and systemic inflammation are fully accepted (Calder 2017).

Nevertheless, the evidence supporting the positive effects of MUFAs and particularly that of the most abundant MUFA in OO, namely oleic acid, is scarce and often inconsistent. Several studies have reported similar effects of SFAs in terms of worsening cardiovascular markers. It is noteworthy that, in all those studies, oleic acid was used alone (Casado-Díaz et al. 2019; Flori et al. 2019). In contrast, the benefits of the consumption of large amounts of polyphenol-rich OO, and hence oleic acid, in the context of the MD, were also pointed out. Interestingly, the multicentre study PREDIMED, carried out in Spain, demonstrated a lower cardiovascular risk and a reduced incidence of major cardiovascular events in the group assigned to the MD consuming high oleic acid content as OO (concretely EVOO) or nuts (Tresserra-Rimbau et al. 2013).

In spite of this controversy, the health benefits associated with MUFA content in OO were recognised by the United States Food and Drug Administration (FDA) in

2004, highlighting 'the benefits on the risk of coronary heart disease of eating about two table spoons (23 g) of olive oil daily' (FDA 2004). Health benefits of OO were related with a decrease of total and LDL-cholesterol in serum, diet improvement of endothelial dysfunction, coagulation activity and reduced LDL susceptibility to oxidation (Silva et al. 2015). Because of the cumulative evidence, the EFSA in 2011 (EFSA 2011b), authorised the health claims for OO, suggesting a dose of 20–23 g/day of OO, which should replace to the same amount of SFAs, to decrease the risk of coronary diseases.

3 Polyphenol-Rich Olive Oil: Modifications During Processing and Preservative Behaviour

A wide range of studies have confirmed the active role of EVOO in Mediterranean food processing and their protective role against oxidation and other damage mechanisms of food functional components during processing. An alternative to it would be the use of EVOO polyphenols separated from EVOO during refining (Solinas and Cichelli 1982). Olive mill wastewater, obtained by mechanical compression of olives during oil extraction, is characterised by a high content of polyphenols, which have been studied for their antioxidant properties (Servili et al. 2011). Furthermore, the use of these EVOO phenolics resulted in an inhibition of microbial growth (Fasolato et al. 2015). In particular, oleuropein, HT and aliphatic aldehydes present in olive products inhibited or delayed the growth rate of various bacterial strains (Battinelli et al. 2006).

However, preserving molecules included in OO ought to be resistant to food processing conditions in order to maintain their activity. In this sense, a wide range of studies have been focused on analysing the chemical modifications of OO constituents, especially when strong processing conditions such as thermal treatment are concerned. On the basis of its extended employment, the most studied thermal process has been frying.

3.1 Modifications During Frying

EVOO is frequently employed in traditional Mediterranean cooking in different amounts and time-temperature conditions: i.e., from deep-frying (few minutes at 170–180 °C), to pan-frying (20 min at 120 °C) and sautéing (few minutes at 90–100 °C in the presence of water). In these cooking processes, there are also substantial exchanges of matter between the oil and the cooked food, this leading to complex chemical, physical and chemical-physical interactions (Sacchi et al. 2014). Frying is one of the most popular methods for the preparation of typical Mediterranean dishes based on different kinds of vegetables (potatoes, green peppers, eggplants, zucchini) or meat and fish. The frying improves the sensory quality of foods by formation of aroma compounds, crust and texture, all highly appreciated by consumers (Pedreschi 2012). The most common domestic frying methods are deep-frying; in this case, the food is totally immersed in warm oil, and pan-frying, when the food is cooked in a frying pan with a small amount of oil.

EVOO has demonstrated to be one of the healthiest and most stable frying oil due to the presence of relatively high amounts of monoenoic fatty acids, especially oleic acid, as well as of natural antioxidants such as tocopherols and biophenols (Sacchi et al. 2006; Napolitano et al. 2008). These molecules are especially responsible for inhibiting the chemical reactions induced by the high temperatures used during frying. Phenol compounds acting as primary antioxidants are likely to react with lipid free radicals, formed by the action of oxygen on UFAs, so that relatively stable products that interrupt the propagative stage of oxidative chain reactions are produced (Chiou et al. 2012). Furthermore, biophenolic compounds of EVOO revealed to be quite stable during deep-frying, being detected even after several hours of frying (Ambrosino et al. 2002); such preservative compounds also showed to interact with the food matrix inhibiting the formation of unhealthy compounds such as acrylamide and toxic aldehydes.

The relationship between phenol compounds in VOO and the formation of acrylamide in potato crisps was first investigated by Napolitano et al. (2008). The phenolic composition of twenty VOO samples was screened by LC-MS; thus, four oils having different phenol compound patterns were selected for frying experiments. Slices of potatoes were fried at 180 °C for 5, 10, and 15 min in VOO, and the acrylamide content was determined by LC-MS. It could be observed that VOO phenolic compounds were not degraded during frying, so that the crisps colour was not significantly different when comparing all four VOOs. Acrylamide concentration in crisps increased with frying time, but its formation was faster in the oil having the lowest concentration of phenolic compounds. Moreover, the EVOO having the highest concentration of ortho-diphenolic compounds was able to efficiently inhibit acrylamide formation in crisps from mild to moderate frying conditions. As a result, the use of ortho-diphenolic-rich EVOO was proposed as a reliable strategy in order to reduce acrylamide formation in domestic deep-frying.

Daskalaki et al. (2009) studied the EVOO behaviour when subjected to heat treatment simulating common domestic processing (i.e. boiling, frying and storage). It could be observed that these processes affected the phenolic compounds content of oils to a certain degree, depending on each treatment. Thus, thermal oxidation of oils at 180 °C (frying) caused a significant decrease in HT derivatives (60% reduction after 30 min and 90% reduction after 60 min) and, to a lower degree, in tyrosol derivatives. Contrary, no effect was implied in the content of lignans. On the other hand, thermal oxidation of oils at 100 °C (boiling) for 2 h caused a decrease by less than 20% in all classes of phenolic compounds. The reduction of phenolic compounds during oils storage under environmental conditions was found to correlate with the peroxide value. Provided the peroxide value did not exceed the 20 meq/kg level, as was recorded for low-content linoleic acid oils and low-oxygen availability at the bottles' headspace, a reduction degree of approximately 30% in HT derivatives and 10% in tyrosol derivatives was proved, while lignin levels remained unchanged.

Hydroxy-alkenals can be signalled as other potentially toxic and carcinogenic compounds arising from the frying decomposition of PUFAs (Frankel 1998). Hydroperoxide decomposition products were quantitatively analysed by high-resolution proton nuclear magnetic resonance (^{1}H-NMR) spectroscopy (400–600 MHz) in thermally oxidised oils by Sacchi et al. (2006). Thus, different oils (EVOO, sunflower and soybean) heated in a thermostatic bath fryer (180 °C for 360 min) showed the NMR signals of aldehyde (*n*-alkanals, *trans*-2-alkenals, 4-hydroxy-*trans*-2-alkenals, alka-2,4-dienals) and were monitored in fifteen oil samples (0, 60, 120, 240 and 360 min heating). As a result, 4-hydroxy-2-alkenals were not detected (threshold 0.1 mM/L) in EVOO after 6 h heating but were only in PUFA-rich fried oils. The formation of these compounds, in fact, was related to the decomposition of conjugated hydroperoxydienes arising from the oxidation of PUFAs (Frankel 1998). In the case of OO, such findings could be explained by the low amount of linoleic (5–10%) and linolenic (less than 1%) acids. For the same reason, alka-2,4-dienal compounds were formed in small amounts (1.1 mM/L oil after 6 h heating) compared to polyunsaturated seed oils. In terms of the uptake of phenol compounds in fried-in-EVOO foods, the crust of French fries, when EVOO is used as frying oil in continuous frying, was demonstrated to absorb a significant amount of phenol compounds that can be easily extracted from the fried potatoes and quantified by LC-MS (Savarese et al. 2006). Lozano-Castellón et al. (2020) studied how temperature, time, and their interaction affect the EVOO polyphenolic profile during a domestic pan-frying process; cooking conditions in a home kitchen, without light or oxygen control were applied. Thus, EVOO was processed at two temperatures (120 and 170 °C) for either a short time (30 min) or a long time (60 min) and polyphenol content monitored. It was implied that temperature degraded the polyphenol compounds of EVOO during the sautéing process, whereas time had an important effect on some individual phenols, such as HT, but not on the total phenol content. The authors also concluded that polyphenol content decreased by 40% at 120 °C and 75% at 170 °C compared to raw EVOO.

3.2 Modifications During Other Processing and Model Systems

Brenes et al. (2001) found hydrolysis of the secoiridoid aglycons as the most important change occurring in the phenolic composition of different kinds of VOO varieties (Arbequina, Hojiblanca, and Picual) during storage in darkness at 30 °C. As a result, an increase in the free phenolics HT and tyrosol in the oil was detected. Furthermore, filtration of oil and acidity increase led to a hydrolysis degree increase. Thus, the addition of commercial oleic acid to Hojiblanca and Picual oil varieties increased the hydrolysis rate of secoiridoid aglycons. Contrary, the level of lignans

1-acetoxypinoresinol and pinoresinol remained constant during storage. Additionally, the total molar concentration of the phenolic compounds was slightly reduced (<20%) after a 1-year storage. However, the transformation of the secoiridoid aglycons into free phenolics was expected to modify the oil taste and the antioxidant capacity.

VOO samples with similar oxidative stabilities and fatty acid compositions were stored for 24 months (Psomiadou and Tsimidou 2002); furthermore, changes in the lipid substrate were followed by peroxide value and K232 measurements, as well as by determination of α-tocopherol, pigment, squalene, and total polar phenol levels. It was observed that the loss of α-tocopherol and carotenoid compounds was similar to that of polar phenol content, this suggesting an active participation in autoxidation process. The limited role of squalene in autooxidation was further confirmed using a model system including OO and α-tocopherol. Addition of pheophytin at three different levels to the oil model indicated an antioxidant role that was found concentration-dependent and more pronounced at elevated temperatures; this fact could be partially due to the activity of certain degradation products.

Velasco and Dobarganes (2002) explained the high resistance to oxidative deterioration of EVOO as a result of a low-PUFA triacylglycerol composition and the presence of phenolic antioxidants such as polyphenols and tocopherols. Notably, polyphenol compounds are of greater importance to VOO rancidity stability as compared with other refined oils which are eliminated or drastically reduced during the refining process.

Lavelli et al. (2006) analysed the degradation of secoiridoid and tocopherol, as well as the antioxidant activity in EVOO during a 8-month storage at two temperatures (40 and 25 °C, closed bottles in the dark). The level of secoiridoid aglycones, namely, oleuropein and ligstroside derivatives, and α-tocopherol decreased according to a pseudo-first-order kinetic. In all EVOOs, oleuropein derivatives were less stable than their counterpart ligstroside derivatives and α-tocopherol. In spite of the antioxidant degradation observed, EVOOs with high antioxidant values were still considered as "excellent" after a 240-day storage at 40 °C. It was concluded that beneficial properties of EVOO can be maintained throughout their commercial lives.

Gómez-Alonso et al. (2007) studied the evolution during a 21-month storage at room temperature and darkness of major and minor components and oxidation indices of seven samples of VOO which differed in their initial contents of natural antioxidants.

The evolution of major and minor components and oxidation indices of seven samples of VOO having different initial content of natural antioxidants was studied during a 21-month storage at room temperature and darkness (Gómez-Alonso et al. 2007). Scores of PV, K232 and K270 increased linearly during the storage period, so that the shelf-life of a VOO sample could be predicted by extrapolation from the results obtained. K232 was found to be the first parameter that exceeded the established upper limit for EVOO; therefore, this index would be likely to be the most relevant for monitoring the commercial category of the OO. During the study, the reduction of total phenolic compounds ranged from 43 to 73%, the decrease being higher in samples whose initial phenol contents were greater. HT concentration increased linearly in most cases, whereas the level of its complex forms decreased considerably in most cases.

The preserving effect of oregano (*Origanum vulgare* L.) essential oil on physical and chemical properties in stored EVOO was analysed by Asensio et al. (2011). For it, the essential oil was added into EVOO at a 0.05% concentration, samples being stored (28 days) in three conditions: light exposure (20–23 °C), heating (60 °C; darkness), and darkness (20–23 °C). As a result, OO samples without oregano essential oil stored at 60 °C and exposure to artificial light provided the highest peroxide values. Additionally, higher *p*-anisidine and K232 values after a 7-day storage were detected in temperature, darkness, and light exposure treatments. Light treatment was the main factor that degraded chlorophyll causing loss of colour. Chlorophyll degradation and colour loss showed to be produced especially by light treatment. The highest chlorophyll content (3.9 mg/kg) was observed in samples corresponding to OO with essential oil at the end of storage. In general, OO samples added with oregano essential oil led to a higher rancidity stability (i.e. lower peroxide, conjugated dienes, and *p*-anisidine values) and higher chlorophyll and carotenoid levels.

4 Preserving Effect of Polyphenol-Rich Olive Oil During Fish Processing and Storage

VOO has a series of components that are positively related with the protection of lipid degradation during the different fish technological processes. The natural polyphenols present in OO (VOO and especially EVOO) act as free radical acceptors as well as metal chelators (Visioli et al. 1998). The use of such polyphenol extracts have been proposed successfully for protecting tuna lipids from deterioration during fish thermal processing (Medina et al. 1999). Miraglia et al. (2016) has applied a phenolic extract from olive vegetation water on fresh salmon steaks stored at 4 °C under modified atmosphere. The addition of this phenolic extract resulted in an enhancement of the microbiological quality and a decrement of lipid oxidation of salmon steaks. Additionally, phenolics in an extract derived from olive vegetation water have enhanced the shelf-life of deep-water rose shrimp retarding lipid oxidation, microbial development and the volatile nitrogen compounds formation, in a manner proportionally effective to the dose of extract used (Miraglia et al. 2020). However, the direct effect on melanosis was not evident. The combination of the phenolic extract with sodium metabisulphite was found necessary to significantly delay the formation of black spots with an efficacy equal to that of sodium metabisulphite solution alone at a concentration of 0.5%, which is normally used by manufacturers for anti-melanosis in shrimp.

Ability of antioxidants obtained from natural sources to stabilise foodstuffs containing long *n*-3 fatty acids of marine origin was analysed by Medina et al.

(2003). Emulsified horse mackerel (*Trachurus trachurus*), fish oil-in-water emulsions (4% *n*-3 PUFA) and fish oils (40% *n*-3 PUFA) were studied. Rosemary (*Rosmarinus officinalis*) leaves and EVOO were employed as sources of natural antioxidative phenols. Both vegetable extracts retarded lipid oxidation in the PUFA rich lipid systems. Rosemary extracts with a high content of carnosoic acid showed a marked synergism with fish proteins by reinforcing their antioxidative effectiveness. Fish proteins and EVOO-phenols showed minor cooperative effects for inhibiting oxidation. Antioxidative partition into the different phases in the emulsified systems showed minor amounts of phenols in the aqueous phase, with the exception of HT and tyrosol which migrated more strongly towards this phase, and high adsorption of fish tissue. It was concluded that extracts of vegetable origin may improve the stability of foods such as fish, with high PUFA content.

Isolated components of EVOO, especially HT, have been proposed as single compounds for preserving fish products from oxidation and rancidity during storage or processing. Pazos et al. (2006) have demonstrated that the addition of 100 ppm of HT solutions to fish fillets is efficient for preserving them from oxidation during frozen storage at −10 °C. Aqueous solutions of HT were directly applied by spraying or were incorporated in the glazing layer covering the fish. Authors found that HT was more efficient to prevent oxidation than an extract containing grape procyanidins. A simple spraying of the phenolic compounds onto the fillets showed high antioxidant efficiency. The effectiveness found was much related to the reducing capacity of HT than its chelating capacity and its affinity to be incorporated into the muscle. The distribution of the antioxidants in the food system can also hugely affect their effectiveness. A selective incorporation of antioxidants into the oxidation-sensitive sites should improve its antioxidant efficiency. The membranes, where phospholipids are localised, are suggested as the most susceptible location for the initiation of lipid oxidation in tissues. HT has not a great affinity to membranes according to its weak bonds to food proteins but an efficient distribution of the antioxidant onto the fillet surface can greatly improve its antioxidant efficiency.

Pazos et al. (2008) have also demonstrated that HT could be used as an antioxidant in different in fish lipids foodstuffs as fish oils, fish oil in-water emulsions or products made by fish meal. These products are important in the infant food or elderly food market. HT was evaluated at different concentration levels (10, 50, and 100 ppm), and its antioxidant capacity was compared against that of a synthetic phenolic, propyl gallate. Results proved the efficiency of HT to inhibit the formation of lipid oxidation products in all tested food systems and two different optimal antioxidant concentrations were observed. In bulk oil and oil-in-water emulsions, a higher oxidative stability was achieved by increasing the concentration of HT, whereas an intermediate concentration (50 ppm) showed more efficiency, delaying lipid oxidation in frozen minced fish muscle. The endogenous depletion of α-tocopherol and *n*-3 PUFAs was also inhibited by supplementing HT in minced muscle; however, the consumption of the endogenous total glutathione was not efficiently reduced by either HT or propyl gallate. A concentration of 50 ppm of HT was best to maintain a longer initial level of α-tocopherol (approximately 300 μg/g of fat), whereas both 50 and 100 ppm of HT were able to preserve completely ω-3

PUFAs. HT and propyl gallate showed comparable antioxidant activities in emulsions and frozen fish muscle, and propyl gallate exhibited better antioxidant efficiency in bulk fish oil.

Some recent works have explored the formation of HT conjugates for improving the antioxidant activity of this EVOO compound. Medina et al. (2009a, b) have shown the effect of lipophilisation on the antioxidant efficiency of HT on fish oil enriched systems. HT fatty acid esters with increasing size of the alkyl chain and different lipophilicity showed a significant antioxidant activity in bulk fish oils and fish oil-in-water emulsions. The introduction of a lipophilic chain decreased the antioxidant effectiveness of HT in homogeneous systems as fish oils. However, in emulsion systems, the presence of a short-medium lipophilic chain (acetate, butyrate or octanoate) improved the antioxidant efficiency of HT favouring the physical location of the antioxidant in the interface. A cut up of antioxidant efficiency was found at the chain length of eight carbons for the HT derivative, probably associated with a preferential location of the diorthophenolic moiety in the right geometry. These results are of high importance for the optimum design of effective EVOO antioxidants for *n*-3 enriched foods, which are very susceptible to suffer oxidation and, then, rancidity.

4.1 Preserving Effect During Fish Frying

The frying process has shown to be influenced by a number of variables among which the type of process, i.e., continuous or discontinuous, surface-to-oil volume ratio, temperature, oil saturation degree, composition of the food being fried (namely, lipid composition) and presence of naturally occurring or added minor compounds, are of particular relevance (Weber et al. 2008). If carried out in non-proper conditions, the frying process can lead to substantial oxidative and hydrolytic decomposition of the fat and to the production of smoke and volatile off-odour components and the formation of off-flavours, dark colours, greasy texture and oil foaming. According to the high temperatures concerned and to the highly unsaturated lipid composition of marine species, lipid oxidation in the marine food being fried is likely to be produced, especially if the above-mentioned variables of the process are not carefully applied. Additionally, interactions among components of the seafood and the culinary fat used take place; these exchanges and interactions would imply that the concentrations of some specific fatty acids in the marine species, such as docosahexaenoic acid (DHA) and eicosapentaenoic acid (EPA), could deeply be modified.

In an early study, VOO, OO, soybean oil and sunflower oil were checked for intermittent and repeated open-pan deep-fat frying of fish (22 times) under laboratory and domestic conditions (Varela et al. 1983). Changes in organoleptic, physical, and chemical properties were assessed after 20 and 36, 10 and 22 frying times, respectively. Results obtained indicated no significant changes in the physical and chemical properties of the oils except for rise in viscosity and fall in smoke point;

furthermore, no significant differences were detected by the taste panel in products fried in repeatedly used oils.

Frying process has been considered to be a source of *trans* fatty acids. However, most *trans* fatty acids found in foods would come from the oil used and not from the process itself. To test this, *trans* fatty acid presence in various oils (EVOO, high oleic acid sunflower oil and common sunflower oil) was analysed during deep fat frying of several frozen foods (Romero et al. 2000). A total of twenty sequential fryings were applied whereby frozen fish products were fried in sequence in the oils, with or without replenishment with fresh oil after each frying. Results revealed that non-replenished frying oils had higher contents of *trans* fatty acid than oils that were replenished after each frying; however, in both cases a relatively low concentration of *trans* fatty acids (<5 mg/g oil or fat) was detected. *Trans* linoleic acid isomers were more abundant in sunflower oil, particularly in standard sunflower oil; contrary, elaidic acid was found the main *trans* fatty acid in OO.

Kalogeropoulos et al. (2004) determined the effect of frying in VOO on the calorific value, proximate composition, fatty acid profile and contents of cholesterol and squalene of Mediterranean fish and shell fish. For it, six finfish species (sand smelt, *Atherina boyeri*; anchovy, *Engraulis encrasicholus*; bogue, *Boops boops*; striped mullet, *Mullus barbatus*; picarel, *Spicara smaris*; scad, *Trachurus trachurus*) and two mollusc species (squid, *Loligo vulgaris*; mussel, *Mytilus galloprovincialis*) were pan fried in VOO. Frying resulted in significant water loss as well as increases in total fat and crude protein levels as well as of calorific value. OO absorption during frying was 4.5–14.2%, expressed on fresh weight basis, and was negatively correlated with initial fat content; non-linear inverse correlations were also determined between fat absorption and fish length and weight. Cholesterol contents showed to increase in fried samples from 25–106 to 33–130 mg/100 g, and resulted in a 20–78-fold increase in squalene. Fatty acid profiles were also modified leading to an increased proportion of MUFAs and a lower proportion of SFAs; meantime, significant amounts of *n*-3 PUFA were detected. It was concluded that frying of fish and shell fish in VOO provided a healthy lipid profile, characterised by significant quantities of squalene and a healthy *n*-6/*n*-3 ratio, without providing excess amounts of cholesterol.

The effect of pan-frying in margarine and OO on the fatty acid composition of cod (*Gadus morhua*) and salmon (*Salmo salar*) fillets was analysed by Sioen et al. (2006). The fatty acids of the margarine employed were 55.5% (SFAs), 33.0% (MUFAs) and 11.5% (PUFAs), while those of OO included 15.4% SFAs, 76.1% MUFAs and 8.5% PUFAs. It was observed that using margarine or OO increased the SFA and MUFA percentages, respectively, in both species. For cod fillets (lean species), pan-frying increased the fat content (0.6–4.2 g/100 g and 0.6–2.3 g/100 g before and after pan-frying, with margarine and olive oil, respectively), whereas, for salmon fillets (fatty species), it decreased (13.9–10.6 g/100 g and 15.4–13.0 g/100 g before and after pan-frying with margarine and OO, respectively). As a result, the culinary fat selection showed to affect the fatty acid content and composition of the resulting fish fillet.

Eight species of small Mediterranean finfish (*Engraulis encrasicholus, Mullus barbatus, Merluccius merluccius, Atherina boyeri, Boops boops, Spicara smaris, Trachurus trachurus*, and *Sardina pilchardus*) were prepared and pan-fried in VOO (Kalogeropoulos et al. 2006). Modifications in their proximate composition, energy content, fatty acids profile, and contents on cholesterol, β-sitosterol, squalene, campesterol, and stigmasterol were analysed. Compared with the initial raw samples, pan-fried fish included higher levels of protein, fat, MUFAs, energy, and cholesterol and showed to be enriched in squalene and phytosterols. Results suggested that fish pan-fried in VOO could be classified as highly nutritive according to its protein, major fatty acid classes, *n*-6/*n*-3 ratios, and total fat content; interestingly, such kind of products provided an additional intake of healthy compounds such as squalene and phytosterols for Mediterranean people.

Kalogeropoulos et al. (2007) analysed the polyphenols, hydroxy pentacyclic triterpene acids (HPTA) and alpha-tocopherol contents in the same fresh and fried oils and finfish than in the previously mentioned study (Kalogeropoulos et al. 2006). As a result, raw fish provided no polyphenols and no HPTA, while alpha-tocopherol was present in all raw samples. Besides water loss and oil absorption, pan frying led to partial loss of all the antioxidants analysed in the fried oils, as well as their enrichment in the resulting fried fish. The overall retention of alpha-tocopherol in the fried oil and fish were included in the 30–80% range; meantime, the respective values for polyphenols and HPTA were included in the 51–87% and 46–88% ranges, respectively.

Ansorena et al. (2010) applied a pan-frying process using two types of oil with different lipid profile (EVOO and sunflower oil) to lean (cod, *Gadus morhua*) and fatty (Atlantic salmon, *Salmo salar*) fish species. After frying, the fat content and the total energetic value increased in the lean fish, without relevant changes in the fatty one. Additionally, sunflower oil led to a lower fat absorption rate than EVOO in both fish species. Notably, the frying hardly modified the lipid profile of salmon regardless the oil employed; however, this process led to substantial modifications in fried cod when compared with raw cod. Concerning the *n*-6/*n*-3 ratio, it increased from 0.08 in raw cod to 1.01 and 6.63 in fried cod with OO and sunflower oil, respectively. In the case of salmon, this ratio was 0.38 in raw samples and 0.39–0.58 in fried ones. The amount of EPA + DHA slightly increased with frying processing in cod, but decreased in salmon. Furthermore, the employment of EVOO was found efficient to avoid the intensity of lipid oxidation development during frying in cod but not in salmon. As a general conclusion, a higher influence of the type of oil on the nutritional quality was detected in the lean fish than in the fatty one.

The effects of different pan-frying methods (oil free, EVOO and sunflower oil) on proximate, neutral and polar lipid composition as well as the fatty acid content of picarel (*Spicara smaris*) were determined by Zervou et al. (2012). Pan-frying did not provide substantial modifications in polar lipids and proteins, but did in moisture and ash level; additionally, an increase in the neutral lipids and total sterol scores could be observed after pan frying in EVOO and sunflower oil. Losses of individual phospholipid levels were not significant during pan-frying, but the proportions were affected by the uptake of the fat employed. For EVOO and sunflower oil pan-fried samples, the absorption of oil medium led to a modification of their fatty acid profiles including a substantial increase of C18:*n*-9 and C18:*n*-6 scores and to a decrease of EPA and DHA presence and *n*-3/*n*-6 value. All frying methods influenced the lipid quality indices but their values could be considered satisfactory from a healthiness point of view. Samples corresponding to the free fried treatment showed the lowest triglyceride and cholesterol levels and the highest *n*-3/*n*-6 values, and were concluded to provide the most appropriated frying method for healthy eating.

Pacetti et al. (2015) analysed the effects of deep-fat frying using EVOO, conventional sunflower oil and high-oleic sunflower oil at different frying temperatures (160 and 180 °C) on the composition of phosphatidylethanolamine (PE) and phosphatidylcholine (PC) in sardine (*Sardina pilchardus*). The deep-fat frying process led to substantial modifications on the composition of PE and PC molecular species of the fish fillet. Thus, the deep-fat frying process reflected a marked increase in the relative proportions of the PE and PC species including the combination of DHA and palmitic acid. Meantime, a decrease of the percentage of the PE and PC species including two DHA residues was detected. Both PE and PC compositions were not influenced by the frying temperature, whereas the nature of the culinary fat showed a marked influence on the PC composition. Particularly, frying consisting of conventional sunflower oil led to a marked increase of PC species containing SFAs/PUFAs and a substantial decrease of PC species formed by the combination of two PUFAs.

4.2 Preserving Effect During Fish Canning

Also concerning heat treatments, canning represents one of the most important means of marine species processing and preservation (Aubourg 2001). In it, seafoods are introduced in the presence of other food components (coating oil, minced vegetables, salt, brine, etc.) in sealed hermetic containers. Then, extensive heat treatments consisting of cooking (100–105 °C) and sterilisation (110–130 °C) are applied and lead to a product with different characteristics when compared with the raw material. A wide range of marine species produces excellent canned products, supporting an important role in human nutrition. Precooking should be carried out to prevent further water being released during the sterilisation process, but not to the extent that the fish are overdried or damaged. If water remains in the can it would dilute any added sauce and would make the canned product unattractive, since oil would be released and float to the surface.

The antioxidant carrier is a key point for explaining antioxidant efficiency. Concerning canned fish products, antioxidant activity of natural EVOO phenolic compounds was found to be greater when they were added in an aqueous packing medium than when they were added in an oily packing medium (Frankel 1998). Such higher antioxidant activity of EVOO phenolic compounds in fish heated in brine was related to their greater affinity toward the more polar interface between water and fish lipids than in a bulk olive oil. In addition, the hydrophobic phenolic

EVOO components are more active in oil-in-water emulsions than in bulk oil. Therefore, the use of EVOO polyphenols as components of the packing media of marine products subjected to thermal processing demonstrated to improve their overall nutritional and sensorial quality.

Sacchi et al. (2002) analysed the mechanism of fish fatty acids protection by EVOO during sterilisation. For it, modification and phase behaviour of the major phenol compounds in EVOO during thermal processing was studied in sealed cans filled with oil-brine mixtures (5:1, v/v) in order to simulate canned-in-oil food systems. Levels of HT, tyrosol and a combination of these and other phenols (i.e. dialdehydic form of decarboxymethyl oleuropein and decarboxymethyl ligstroside aglycons, and the oleuropein aglycon) decreased in the oil phase after sterilisation with a marked partitioning towards the brine phase. The increase in total HT and tyrosol content after processing, and the presence of elenolic acid in brine revealed hydrolysis of the ester bond of hydrolysable phenols during thermal processing. It was concluded that both partitioning towards the water phase and hydrolysis of phenols contributed to the loss of phenols presence from EVOO in canned foods, as well as the protection of *n*-3 PUFA in canned-in-EVOO fish products. According to this conclusion, Višnjevec Miklavčič et al. (2020) have recently analysed the level of phenolic compounds in EVOO used as packing medium in marketed canned fish. Results demonstrated that the levels of phenolic in EVOO canned fish were substantially lower and the composition of phenolic compounds was different compared to the usual EVOO.

The effect of different kinds of filling media on lipid oxidation development occurring during tuna (*Thunnus alalunga*) canning was analysed by Medina et al. (1998). For it, EVOO, having a high content of natural polyphenols, and other filling media lacking phenols (refined OO, refined soybean oil and brine) were selected. Different lipid oxidation indices (thiobarbituric acid reactive substances and fluorescent compounds formation) were determined among the four media after thermal processing and storage of canned tuna. As a result, EVOO showed a potential antioxidant activity on fish lipids. The verified antioxidant ability was attributed to the solubilisation of hydrophilic phenols into the water-muscle interface. Thus, phenolic composition from EVOO studied by reversed-phase HPLC showed a marked change after fish processing, this suggesting phenol decompositions and strong interactions between oil phenols and fish muscle components. Remarkably, the aqueous environment built by brine made fish lipids more prone to oxidation, presumably due to accumulation of UFAs at the oil-water interface.

Thirty commercial samples of vegetable and fish canned in OO or EVOO were examined to assess the oxidative and hydrolytic degradation of the oils (Gomes et al. 1998). Results indicated that triglyceride oligopolymers were present in all the oil samples. Their presence in the EVOO could be explained on the basis of the effects of thermal stabilisation procedures and storage times; however, levels were quite low and averaged 0.18%. Overall oxidation of OOs (refined and virgin) used to cover the preserved foodstuffs was determined on the basis of oxidised triglyceride and oligopolymer presence. Preserved fish showed a higher level of lipid oxidation development, with an average value of 2.65%; meantime, the preserved vegetable had an average value of 1.96%. Hydrolytic degradation, measured on the basis of diglyceride content, was found similar for the preserved fish and vegetable (3.57%, and 3.60%, respectively).

The effect of EVOO polyphenol addition as antioxidants in tuna canned in brine and refined OO was analysed by Medina et al. (1999). For it two different processing temperatures (40 and 100 °C) were tested. As a result, a less oxidation degree was detected when they were supplemented with 400 ppm of EVOO polyphenols than in the case of the sample processed with 100 ppm of a 1:1 mixture of the synthetic antioxidants butylated hydroxytoluene and butylated hydroxyanisole. In the case of tuna muscle canned in brine. Additionally, when the EVOO phenolic extract was added to tuna muscle immersed in brine, EVOO antioxidants showed to be active for inhibiting the formation of hydroperoxides and volatiles during both thermal conditions. Notably, the efficiency of EVOO phenolic compounds was higher in tuna brine samples heated at 40 °C than in the same samples heated at 100 °C. During canning processing, complex phenols have been reported to be hydrolysed to simple compounds (tyrosol and HT) (Sacchi et al. 2002). The migration of monomers such as HT and tyrosol into the tuna muscle-water interface has been suggested to contribute to the antioxidant effect observed in EVOO tuna canning (Medina et al. 1998).

With the aim of studying the oil degradation, Caponio et al. (2002) employed high performance size exclusion chromatography to analyse the polar compounds produced in oils (refined and virgin olive, soybean and sunflower) employed as the covering medium in commercial canned tuna fish. In all cases, diffusion of the fatty fraction from the fish into the covering oils was detected. Analysis of the triglyceride oligopolymer and diglyceride fractions showed that soybean and sunflower oils led to a higher *trans*-isomer content and oxidative degradation level but lower hydrolytic oxidation level than both kinds of OOs tested.

Oxidative and hydrolytic stability of vegetable oils used as liquid medium in canned fish was evaluated (Caponio et al. 2003). For it, sixteen canned tuna samples were tested, including two in EVOO, nine in OO (refined OO plus VOO) and five in refined seed oil (soybean oil, corn oil or sunflower oil). As a result, fatty acid profile of dipping oils showed the presence of highly UFAs, typical of fish lipids. Interestingly, *trans* isomers were absent in EVOO, but were present in OO and in refined seed oil. Contents of triglyceride oligopolymers, oxidised triglycerides and diglycerides (polar compounds) ranged within 0.13–1.07, 0.51–2.36 and 0.96–4.02% ranges, respectively. Therefore, a marked quality improvement by employment of EVOO as coating medium was concluded.

The fatty acid composition and the degradation level of the dipping oil (EVOO, OO and refined seed oil) present in canned fish (tuna, sardine, anchovy, and mackerel) was analysed by Caponio et al. (2011). Results showed the lowest extent of both hydrolytic and oxidative degradation in samples containing EVOO. Thus, contents of triacylglycerol oligopolymers, likely to be produced by secondary oxidative degradation, were equal to 0.17%, 0.50% and 0.74% for EVOO, OO and refined seed oil, respectively. OO showed higher hydrolytic degradation than EVOO, with diacylglycerols levels being round a 3.37% score; however, a lower oxidative degradation and *trans* isomers content than for refined seed oil was detected. Finally, the type of fish did not influence the extent of oxidative and hydrolytic degradation, with the only exception of sardine covering oil. This oil, characterised by the highest PUFA content, showed the highest primary oxidation development, i.e. oxidised triacylglycerols (1.32%) and specific absorption at 232 nm.

Thirty-seven different samples of canned sardines and other fish sold in the United Kingdom were analysed for their furan content using a headspace gas chromatography-mass spectrometry procedure (Pye and Crews 2014). As a result, all samples contained detectable furan, with an average level of 26 μg/kg. The maximum furan level was determined in canned fish with tomato sauce and lemon as packaging media, with average values of 49 and 55 μg/kg, respectively. Notably, all fish in brine or in oil contained less than 20 μg/kg furan. Interestingly, furan levels recorded in fish packed in EVOO were low with an average of 2 μg/kg.

Gómez-Limia et al. (2021) compared the effect of different filling media (sunflower oil and OO) in canned eels (*Anguilla anguilla*). The level of vitamin E and the antioxidant capacity was higher in eels packed in sunflower oil and the filled sunflower oil after a 12-month storage. However, canned eels packed in OO presented higher total phenolic contents. These results can be related with the higher vitamin E of sunflower oil and the higher content of polyphenols of OOs. Authors concluded that availability, price, and the kind of market to which the product may be destined should be the most important factors in order select the covering medium.

5 Final Remarks and Future Trends

Research carried out concerning the effect of polyphenol-rich OO (VOO and especially EVOO) and its constituents on seafood has shown marked preserving possibilities. However, their use can be considered as relatively short if we take into account the wide range of seafood processes available nowadays in the market. In order to enlarge the employment of polyphenol-rich OO or its constituents during marine species processing and storage, several aspects ought to be taken into account:

- A wide understanding of the active constituents' distribution between OO phase and marine species would be mandatory. This would require the employment of advanced analytical tools such as MS, NMR, or ESR.
- In order to apply polyphenol-rich OO or its constituents to a wider range of seafood, research studies should be developed to establish the possibilities for taking advantage of all the nutritional and technological properties of these active compounds in different kinds of seafood. Potentially additive and synergistic properties of EVOO and seafood's bioactive effects on human health need also to be deeper explored.

- According to the current needs and lifestyle, minimally processed products are required. Therefore, a special stress ought to be accorded to attractive RTE products, so that the consumers' expectations for taste, flavour and healthiness are encountered. In this sense, active compounds in EVOO can provide new possibilities for quality enhancement and shelf-life time increase.
- Optimisation of the active compounds addition ought to be carried out on the basis of the kind of product concerned (whole fish, fillet, minced, etc.), the kind of species chosen (size, fat content, general composition, skin resistance, wild or cultivated, etc.) and other biological aspects (capture season, maturity, sex, eating state, etc.).
- The search for high-quality products including EVOO can signify an innovative field. With its strong fruity flavour, this oil can lead to attractive food not only including marine species but also other kinds of food matrices. Sustainability of the EVOO processed fish products production lies on an environmentally friendly OO and fish production, combined with the valorisation of olive/fish by-products. These issues are demanded by farmers, fish producers and a society that is conscious of the need for environmental protection.
- Studies of the favourable healthy effects of the combined ingestion of olive oil together with fish products are continuously opening new paths for food markets and nutritional research.
- Innovative and persuasive communication strategies to raise consumers' awareness of quality and healthy effects of polyphenol-rich OO in fish processed foods are required. It is important to increase consumer's willingness to pay a premium price to guarantee a fair income for high quality EVOO and fish producers. In this context, more research is needed to promote the adoption of health claims and to help to present fish processed food including polyphenol-rich OO as a product that is important for human well-being.

References

Ambrosino ML, Della Medaglia D, Paduano A, Sacchi R (2002) Behaviour of phenolic antioxidants of virgin oil during oil heating and frying. In: Bioactive compounds in plant foods: health effects and perspectives for the food industry. European scientific conference of COST action 916, vol EUR 20061. EC Office for Official Publications, Luxembourg, pp 119–120

Ansorena D, Guembe A, Mendizábal T, Astiasarán I (2010) Effect of fish and oil nature on frying process and nutritional product quality. J Food Sci 75:H62–H67

Asensio C, Nepote V, Grosso N (2011) Chemical stability of extra-virgin olive oil added with oregano essential oil. J Food Sci 76:S445–S450

Aubourg S (2001) Review: Loss of quality during the manufacture of canned fish products. Food Sci Technol Int 7:199–215

Battinelli L, Daniele C, Cristiani M, Bisignano G, Saija A, Mazzanti G (2006) In vitro antifungal and anti-elastase activity of some aliphatic aldehydes from *Oleae europaea* L. fruit. Phytomedicine 13:558–563

Bertelli M, Kiani AK, Paolacci S, Manara E, Kurti D, Dhuli K, Bushati V, Miertus J, Pangallo D, Baglivo M, Beccari T, Michelini S (2020) Hydroxytyrosol: a natural compound with promising pharmacological activities. J Biotechnol 309:29–33

Boss A, Bishop KS, Marlow G, Barnett MPG, Ferguson LR (2016) Evidence to support the anti-cancer effect of olive leaf extract and future directions. Nutrients 8:513

Brenes M, García A, García P, Garrido A (2001) Acid hydrolysis of secoiridoid aglycons during storage of virgin olive oil. J Agric Food Chem 49:5609–5614

Calder PC (2017) Omega-3 fatty acids and inflammatory processes: from molecules to man. Biochem Soc Trans 45:1105–1115

Caponio F, Gomes T, Summo C (2002) Use of HPSEC analysis of polar compounds in the ascertainment of the degradation level of oils utilised as liquid medium in canned tuna. Riv Ital Sost Gras 79:285–288

Caponio F, Gomes T, Summo C (2003) Quality assessment of edible vegetable oils used as liquid medium in canned tuna. Eur Food Res Technol 216:104–108

Caponio F, Summo C, Pasqualone A, Gomes T (2011) Fatty acid composition and degradation level of the oils used in canned fish as a function of the different types of fish. J Food Compos Anal 24:1117–1122

Casado-Díaz A, Dorado G, Quesada-Gómez JM (2019) Influence of olive oil and its components on mesenchymal stem cell biology. World J Stem Cells 11:1045–1064

Castro-Barquero S, Lamuela-Raventós RM, Doménech M, Estruch R (2018) Relationship between Mediterranean dietary polyphenol intake and obesity. Nutrients 10:1–13

Chiou A, Kalogeropoulos N, Boskou D, Salta FN (2012) Migration of health promoting microconstituents from frying vegetable oils to French fries. Food Chem 133:1255–1263

Daskalaki D, Kefi G, Kotsiou K, Tasioula-Margari M (2009) Evaluation of phenolic compound degradation in virgin olive oil during storage and heating. J Food Nutr Res 48:31–41

De las Hazas MCL, Rubio L, Macia A, Motilva MJ (2018) Hydroxytyrosol: emerging trends in potential therapeutic applications. Curr Pharm Des. 24:2157–2179

De Pablos RM, Espinosa-Oliva AM, Hornedo-Ortega R, Cano M, Arguelles S (2019) Hydroxytyrosol protects from aging process via AMPK and autophagy; a review of its effects on cancer, metabolic syndrome, osteoporosis, immune-mediated and neurodegenerative diseases. Pharm Res 143:58–72

EFSA (2011a) Panel on Dietetic Products, Nutrition and Allergies. Scientific opinion on the substantiation of health claims related to polyphenols in olive and protection of LDL particles from oxidative damage (ID 1333, *1638*, 1639, 1696, 2865), maintenance of normal blood HDL cholesterol concentrations (ID 1639), maintenance of normal blood pressure (ID 3781), "anti-inflammatory properties" (ID 1882), "contributes to the upper respiratory tract health" (ID 3468), "can help to maintain a normal function of gastrointestinal tract" (3779), and "contributes to body defences against external agents" (ID 3467) pursuant to Article 13 (1) of Regulation (EC) No 1924/2006. EFSA J 9:2033

EFSA (2011b) Panel on Dietetic Products, Nutrition and Allergies. Scientific opinion on the substantiation of health claims related to olive oil and maintenance of normal blood LDL-cholesterol concentrations (ID 1316, 1332), maintenance of normal (fasting) blood concentrations of triglycerides (ID 1316, 1332), maintenance of normal blood HDL cholesterol concentrations (ID 1316, 1332) and maintenance of normal blood glucose concentrations (ID 4244) pursuant to Article 13 (1) of Regulation (EC) No 1924/2006. EFSA J 9:2044

Fasolato L, Cardazzo B, Balzan S, Carrao L, Taticchi A, Montemurro F, Novelli E (2015) Minimum bactericidal concentration of phenols extracted from oil vegetation water on spoilers, starters and food-borne bacteria. Ital J Food Safety 4:4519

FDA (2004) FDA allows qualified health claim to decrease risk of coronary heart disease. http://www.fda.gov/NewsEvents/Newsroom/PressAnnouncements/2004/ucm108368.htm

Flori L, Donnini S, Calderone V, Zinnai A, Taglieri I, Venturi F, Testai L (2019) The nutraceutical value of olive oil and its bioactive constituents on the cardiovascular system. Focusing on main strategies to slow down its quality decay during production and storage. Nutrients 11:1962

Frankel EN (1998) Lipid oxidation. Oily Press, Dundee, Scotland, UK
Frankel EN (2011) Nutritional and biological properties of extra virgin olive oil. J Agric Food Chem 59:785–792
García-Villalba R, Carrasco-Pancorbo A, Oliveras-Ferraros C, Vázquez-Martín A, Menéndez JA, Segura-Carretero A, Fernández-Gutiérrez A (2010) Characterization and quantification of phenolic compounds of extra-virgin olive oils with anticancer properties by a rapid and resolutive LC-ESI-TOF MS method. J Pharm Biomed Anal 51:416–429
Gomes T, Caponio F, Baiano A, de Pilli T (1998) Measurement of oxidative and hydrolytic degradation of olive oils used for covering preserved foodstuffs. Riv Ital Sost Gras 75:77–82
Gómez-Alonso S, Mancebo-Campos V, Salvador MD, Fregapane G (2007) Evolution of major and minor components and oxidation indices of virgin olive oil during 21 months storage at room temperature. Food Chem 100:36–42
Gómez-Limia L, Sanmartín NM, Carballo J, Domínguez R, Lorenzo JM, Martínez S (2021) Oxidative stability and antioxidant activity in canned eels: effect of processing and filling medium. Foods 10:790
Kalogeropoulos N, Andrikopoulos N, Hassapidou M (2004) Dietary evaluation of Mediterranean fish and molluscs pan-fried in virgin olive oil. J Sci Food Agric 84:1750–1758
Kalogeropoulos N, Kotsiopoulou C, Mylona A, Christea M, Andrikopoulos N (2006) Compositional changes and enrichment of Mediterranean finfish during pan frying with virgin olive oil. Ecol Food Nutr 45:171–188
Kalogeropoulos N, Chiou A, Mylona A, Ioannou M, Andrikopoulos N (2007) Recovery and distribution of natural antioxidants (alpha-tocopherol, polyphenols and terpenic acids) after pan-frying of Mediterranean finfish in virgin olive oil. Food Chem 100:509–517
Lavelli V, Fregapane G, Salvador MD (2006) Effect of storage on secoiridoid and tocopherol contents and antioxidant activity of monovarietal extra virgin olive oils. J Agric Food Chem 54:3002–3007
Lozano-Castellón J, Vallverdú-Queralt A, Rinaldi de Alvarenga JF (2020) Domestic sautéing with EVOO: change in the phenolic profile. Antioxidants 9:77
Martínez-Beamonte R, Sanclemente T, Surra JC, Osada J (2020) Could squalene be an added value to use olive by-products? J Sci Food Agric 100:915–925
Medina I, Sacchi R, Biondi L, Aubourg SP, Paolillo L (1998) Effect of packing media on the oxidation of canned tuna lipids. Antioxidant effectiveness of extra virgin olive oil. J Agric Food Chem 46:1150–1157
Medina I, Satué MT, German JB, Frankel E (1999) Comparison of natural polyphenols antioxidant from extra virgin olive with synthetic antioxidants in tuna lipids during thermal oxidation. J Agric Food Chem 47:4873–4879
Medina I, González MJ, Pazos M, Della Medaglia D, Sacchi R, Gallardo JM (2003) Activity of plant extracts for preserving functional food containing n-3-PUFA. Eur Food Res Technol 217:301–307
Medina I, Gallardo JM, Aubourg SP (2009a) Quality preservation in chilled and frozen fish products by employment of slurry ice and natural antioxidants. Int J Food Sci Tehnol 44:1467–1479
Medina I, Lois S, Alcántara D, Lucas R, Morales JC (2009b) Effect of lipophilization of hydroxytyrosol on its antioxidant activity in fish oils and fish oil-in-water emulsions. J Agric Food Chem 57:9773–9779
Miraglia D, Esposto S, Branciari R, Urbani S, Servili M, Perucci S, Ranucci D (2016) Effect of a phenolic extract from olive vegetation water on fresh salmon steak quality during storage. Ital J Food Safety 5(4):6167
Miraglia D, Castrica M, Menchetti L, Esposto S, Branciari R, Ranucci D, Urbani S, Sordini B, Veneziani G, Servili M (2020) Effect of an olive vegetation water phenolic extract on the physico-chemical, microbiological and sensory traits of shrimp (*Parapenaeus longirostris*) during the shelf-life. Foods 9:647

Monti S, Ritieni A, Sacchi R, Skog K, Borgen E, Fogliano V (2001) Characterization of phenolic compounds in virgin olive oil and their effect on the formation of carcinogenic/mutagenic heterocyclic amines in a model system. J Agric Food Chem 49:3969–3975
Napolitano A, Morales F, Sacchi R, Fogliano V (2008) Relationship between virgin olive oil phenolic compounds and acrylamide formation in fried crisps. J Agric Food Chem 56:2034–2040
Pacetti D, Lucci P, Mozzon M, Gagliardi R, Fiorini D, Frega NG (2015) Influence of deep-fat frying process on phospholipid molecular species composition of *Sardina pilchardus* fillet. Food Cont 48:155–162
Pazos M, Alonso A, Bolaños JF, Torres JL, Medina I (2006) Physicochemical properties of natural phenolics obtained from grape and olive oil by-products and their antioxidant activity in frozen fish fillets. J Agric Food Chem 54:366–373
Pazos M, Alonso A, Medina I (2008) Hydroxytyrosol prevents fish lipids-rich foodstuffs from oxidative deterioration. J Agric Food Chem 56:3334–3340
Pedreschi F (2012) Frying of potatoes: physical, chemical, and microstructural changes. Drying Technol 30:707–725
Psomiadou E, Tsimidou M (2002) Stability of virgin olive oil. 1. Autoxidation studies. J Agric Food Chem 50:716–721
Pye C, Crews C (2014) Furan in canned sardines and other fish. Food Addit Contam Part B 7:43–45
Romero A, Cuesta C, Sánchez-Muniz F (2000) *Trans* fatty acid production in deep fat frying of frozen foods with different oils and frying modalities. Nutr Res 20:599–608
Sacchi R (2008) Extravirgin olive oil in kitchen: the secrets of a protagonist of Mediterranean gastronomy, Regione Campania and Università di Napoli Federico II. SBR, Portici, Napoli, Italy, pp 1–64
Sacchi R, Paduano A, Fiore F, della Medaglia D, Ambrosino ML, Medina I (2002) Behaviour of extra virgin olive oil phenolic antioxidants during the thermal processing of a brine-in-oil model system. J Agric Food Chem 50:2830–2835
Sacchi R, Falcigno L, Paduano A, Ambrosino ML, Savarese M, De Giulio B, Addeo F, Paolillo L (2006) Quantitative evaluation of the aldehydes formed in heated vegetable oils using high resolution proton-NMR spectroscopy. Riv Ital Sost Gras 82:257–263
Sacchi R, Paduano A, Savarese M, Vitaglione P, Fogliano V (2014) Extra virgin olive oil: from composition to "molecular gastronomy". In: Zappia V, Panico S, Russo GL, Budillon A, Della Ragione F (eds) Advances in nutrition and cancer, cancer treatment and research 159. Springer, Berlin, pp 325–338
Savarese M, Parisini C, De Marco E, Sacchi R (2006) Detection of biophenols from virgin olive oil in fried French-fries potatoes by high-performance liquid chromatography tandem electrospray ionization mass spectrometry (HPLC-ESI/MS). In: 4th EuroFed lipid congress (oils, fats and lipids for a healthier future), Complutense University of Madrid, Spain
Serreli G, Deiana M (2020) Extra virgin olive oil polyphenols: modulation of cellular pathways related to oxidant species and inflammation in aging. Cells 9:478
Servili M, Esposto S, Veneziani G, Urbani S, Taticchi A, Di Maio I, Selvaggini R, Sordini B, Montedoro GF (2011) Improvement of bioactive phenol content in virgin olive oil with an olive-vegetation-water concentrate produced by membrane treatment. Food Chem 124:1308–1315
Servili M, Sordini B, Esposto S, Urbani S, Veneziani G, Di Maio I, Selvaggini R, Taticchi A (2014) Biological activities of phenolic compounds of extra virgin olive oil. Antioxidants 3:1–23
Silva S, Combet E, Figueira ME, Koeck T, Mullen W, Bronze MR (2015) New perspectives on bioactivity of olive oil: evidence from animal models, human interventions and the use of urinary proteomic biomarkers. Proc Nutr Soc 74:268–281
Sioen I, Haak L, Raes K, Hermans C, de Henauw S, De Smet S, Van Camp J (2006) Effects of pan-frying in margarine and olive oil on the fatty acid composition of cod and salmon. Food Chem 98:609–617

Solinas M, Cichelli A (1982) GLC and HPLC quantitation of phenolic compounds in olive oil: the possible role of tyrosol in assessing virgin olive oil amount in mixture with refined oils. Riv Ital Sci Aliment 11:4–12

Tilami S, Sampels S (2018) Nutritional value of fish: lipids, proteins, vitamins, and minerals. Rev Fish Sci 26:242–253

Tresserra-Rimbau A, Medina-Remón A, Pérez-Jiménez J, Martínez-González MA, Covas MI, Corella D, Salas-Salvadó J, Gómez-Gracia E, Lapetra J, Arós F, Fiol M, Ros E, Serra-Majem L, Pintó X, Muñoz MA, Sáez GT, Ruiz-Gutiérrez V, Warnberg J, Estruch R, Lamuela-Raventós RM (2013) Dietary intake and major food sources of polyphenols in a Spanish population at high cardiovascular risk: the PREDIMED study. Nutr Metab Cardiov Dis NMCD 23:953–959

Varela G, Moreiras-Varela O, Ruiz-Roso B (1983) Utilization of some oils in repeated frying. Changes in fats and sensory analysis of fried foods. Grasas Aceites 34:101–107

Velasco J, Dobarganes MC (2002) Oxidative stability of virgin olive oil. Eur J Lipid Sci Technol 104:661–676

Visioli F, Bellomo G, Galli C (1998) Free radical-scavenging properties of olive oil polyphenols. Biochem Biophys Res Commun 247:60–64

Višnjevec Miklavčič A, Butinar B, Schwarzkopf M, Bučar Miklavčič M (2020) Characterization of olive oils used as medium in canned fish. Sch J Food Nutr Res 3:302–306

Weber J, Bochi VC, Ribeiro CP, Victório AM, Emanuelli T (2008) Effect of different cooking methods on the oxidation, proximate and fatty acid composition of silver catfish (*Rhamdia quelen*) fillets. Food Chem 106:140–146

Zervou A, Sinanoglou V, Papadas G, Thoimadis N, Meimaroglou D, Miniadis-Meimaroglou S (2012) Effects of pan-frying on PUFA, MUFA and individual lipid content in *Spicara smaris*. Acta Alim 41:33–44

The manufacturer's authorised representative in the EU is Springer Nature Customer Service Centre GmbH, Europaplatz 3, 69115 Heidelberg, Germany. If you have any concerns regarding our products, please contact ProductSafety@springernature.com

Printed and bound by CPI Group (UK) Ltd, Croydon, CR0 4YY
23/09/2026
02221532-0012